DELAYED AND NETWORK QUEUES

DELAYED AND NETWORK QUEUES

ALIAKBAR MONTAZER HAGHIGHI
Department of Mathematics
Prairie View A&M University
Member of the Texas A&M University System
Prairie View, Texas, USA

DIMITAR P. MISHEV
Department of Mathematics
Prairie View A&M University
Member of the Texas A&M University System
Prairie View, Texas, USA

WILEY

Published by John Wiley & Sons, Inc., Hoboken, New Jersey

Published simultaneously in Canada

Library of Congress Cataloging-in-Publication Data:

Names: Haghighi, Aliakbar Montazer, author. | Mishev, D. P. (Dimiter P.), author.
Title: Delayed and network queues / Aliakbar Montazer Haghighi, Prairie View A&M University, member of Texas A&M University System, Prairie View, Texas, USA, Dimitar P. Mishev, Prairie View A&M University, member of Texas A&M University System, Prairie View, Texas.
Description: Hoboken, New Jersey : John Wiley & Sons, 2016. | Includes bibliographical references and index.
Identifiers: LCCN 2016015749 (print) | LCCN 2016020734 (ebook) | ISBN 9781119022138 (cloth) | ISBN 9781119022145 (pdf) | ISBN 9781119022152 (epub)
Subjects: LCSH: Routing (Computer network management)–Mathematics. | Computer networks–Mathematical models. | Telecommunication–Traffic. | Queuing networks (Data transmission)
Classification: LCC TK5105.5487 .H34 2016 (print) | LCC TK5105.5487 (ebook) | DDC 004.601/51982–dc23
LC record available at https://lccn.loc.gov/2016015749

Typeset in 10/12pt, TimesLTStd by SPi Global, Chennai, India.

Printed in the United States of America

10 9 8 7 6 5 4 3 2 1

I have always appreciated my lovely wife, Shahin Hamidi, for putting up with me during our about 45 years of marriage. A significant part of what was supposed to be our "quality time" together, I have squandered in research and publication. Fully aware of her involuntary sacrifice, I have always devoted my scholarly work and achievement to her, along with others.

This volume (my 4th in English Language) is also in her honor which she has consented to gracefully share with my late parents, Roghayyeh Jafari and Mohammad Ali Montazer-Haghighi. I am confident, yet continue to pray, that both remain in the peaceful and merciful hands of the Almighty Allah in whom they lived their lives.

"To Allah we belong and to HIM we will return"

Quran Kareem

Aliakbar Montazer Haghighi
Houston, Texas, USA
December 31, 2015

I would like to dedicate the book to my wife, Tatiana, for her patience and encouragement, to my children Elizabeth, Antoan and Ilia, and to my brother Iordan and his family.

Dimitar P. Mishev
Houston, Texas, USA
December 31, 2015

CONTENTS

PREFACE

Queueing theory is considered a part of Operations Research, in which the concept of *stochastic processes* is widely applied. With advancements in manufacturing and production systems, public transportation, logistics, and health care, in this era of technological development made possible by wireless communication and the Internet, performance evaluation of such systems is best accomplished by modeling the system as a network of queues. A network of queues is applied to model a real-world system when a set of resources is shared by some components of the system. Thus, queueing network modeling is a methodology for the analysis of systems such as computer systems. Many examples of networks of queues that appear in our day-to-day lives sometimes go unnoticed. The assembly of an automobile in a car manufacturing plant is a simple example.

In the 1960s, the computer industry was not very aware of the need of queueing network analysis; instead, the time sharing idea was used in the data processing industry. The networks only became fashionable by the end of that decade. New challenges developed in the 1970s and continued when the wireless communication system was studied in the 1980s and has still continued to the present time. By now, the algorithms for evaluating queueing network models are well developed and the mathematical sophistication to analyze systems such as computer systems is not so essential.

However, for long, the networks were based on the assumption of reliable nodes. Breaking down the nodes soon proved the assumption unrealistic in both the Internet and wireless and the industry such as production lines and the automobile industry. This fact is, indeed, a source of delay in the performance analysis of network of queues. This delay is different from the standard waiting time in a service line. Because of the importance of delay that not only affects the time delay but also being an economic factor in a production line, for instance, we have allocated part of Chapter 6 to discuss the unreliability of networks.

This book is an attempt to present some queueing models and networks of queues with delay, in a style that is between a monograph and a textbook. We also hope that it will satisfy the need of a handbook for researchers and practitioners.

The principal audience of the book are senior undergraduates, masters and doctorate students, as well as professionals and researchers in the mathematical sciences, stochastic processes, operations research, and engineering. The materials included in the book should also be of interest to those involved in computer network performance analysis. These materials actually instruct the application of queueing network models as tools to assist in the answering of questions on cost and performance that arise throughout the life of a computer system. Our intention is to provide enough information that readers can fully understand the essential aspects of queues with delay and networks of queues with unreliable nodes that may cause delay. We have omitted numerical examples for some models as the development of the fundamentals of models are sufficiently detailed to understand them.

We have included results of some recent research studies to ensure the contents of the book are current enough for all uses.

Particular advantages of this text are as follows:

1. Treatment of queues with delay and networks of queues with possible breakdown and disruption that may cause delay.
2. Inclusion of recently published materials on the subject so that the book is very Topical.
3. Complementation by modern technology (including current software packages).

We have divided the book into six chapters. In Chapter 1, basics of probability and some preliminary materials are discussed as mind refreshers and background materials for later use. Chapter 2 contains topics in stochastic processes that are the basis for queueing theory. Birth and death processes are discussed, not in detail, in Chapter 3. Some standard queues such as single-server and multiserver queues and queues with feedback with some variations and generalizations are presented in Chapter 4. Chapter 5 is devoted to queues with delay. Different types of models that may cause delay in performance and service are discussed in this chapter. In the final chapter, Chapter 6, networks of queues with delay and models with reliable and unreliable nodes, particularly Jackson's types, are discussed. Additionally, included in this chapter are discussions of assessment of reliability of a network of queues and effect of network service breakdown. The splitting (relatively new and still developing) and blocking features are discussed in both Chapters 5 and 6.

In addition to the calculus and differential equations (to the level taught in most engineering undergraduate schools), some knowledge of difference equations is necessary to better comprehend the contents of this book. Related questions are included in the "Exercises" section at the end of each chapter.

It is our wish that readers of this text have an exciting experience going through this lavish collection of queueing models. We would seize one final opportunity to express our gratitude to John Wiley & Sons, Inc., New Jersey, for offering and awarding us the contract to write this book, as well as those who helped us in this work.

Aliakbar Montazer Haghighi
Dimitar P. Mishev

Houston, Texas, USA
January 25, 2016

1

PRELIMINARIES

1.1 BASICS OF PROBABILITY

1.1.1 Introduction

In this chapter, we introduce some basics of probability that will be needed in the later chapters. We also take the liberty in stating some theorems without presenting proofs and emphasize that the contents of this chapter, by no means, represent all topics of probability that deserve a detailed discussion.

A ***chance*** or ***random experiment*** is an experiment whose outcomes or results of its performance are uncertain. A set of outcomes is called an ***event***. The set of all possible outcomes is referred to as the ***sample space***, denoted by Ω. Thus, an ***event*** is a subset of the sample space and an element of the sample space is a ***sample point***.

Two events A_1 and A_2 are referred to as ***mutually exclusive*** if their intersection is empty. A set $\{A_1, A_2, \dots, A_n\}$ is called a ***partition*** of Ω if the events $A_1, A_2, \dots, A_n$ are mutually exclusive, such that $A_1 \cup A_2 \cup \cdots \cup A_n = \Omega$.

Probability of an event E, denoted by $P(E)$, indicates a number between 0 and 1 (inclusive), describing the likelihood of occurrence of the event E. There are two particular events: an event with probability 1 (referred to as ***almost sure event***) and another with probability 0 (referred to as ***null*** or ***impossible event***).

For a finite sample space with n elements, if all outcomes have the same chance to occur, each member is assigned a probability of $1/n$ and the sample space is called ***equiprobable***. In case of an infinite sample space, elements are with ***uniform measure***.

If a chance experiment is repeated, the chance of occurrence of an outcome is the ratio of the number of occurrences of the outcome to the total number of repetitions. Hence, for a sample space with n ***equiprobable points***, the ***probability of an event***

Delayed and Network Queues, First Edition. Aliakbar Montazer Haghighi and Dimitar P. Mishev.

with k points is k/n, referred to as ***relative frequency***, and it is an approximation of the probability of the event. In other words, for a sample space with equiprobable sample points, if E is an event with k points, the ***probability of the event E*** is given by

$$P(E) = \frac{\text{number of ways the event } E \text{ occurs}}{\text{total number of ways all outcomes could occur}}. \tag{1.1}$$

The number of elements of the event E is referred to as the "***size***" of E. Thus, ***probability of the event E*** may be defined as

$$P(E) = \frac{\text{size of } E}{\text{size of } \Omega}. \tag{1.2}$$

The triplet $(\Omega, \mathfrak{B}, P)$ is called the ***probability space*** associated with the random experiment, where

(a) Ω is the sample space, that is, the set of all outcomes of the random experiment.

(b) $\mathfrak{B}$ is the set function containing all possible events drawn from Ω, which has the structure of the σ field. This means that $\mathfrak{B}$ satisfies the following conditions:

  (i) the empty set is in $\mathfrak{B}$,

  (ii) if $E \in \mathfrak{B}$, then the complement of E is also in $\mathfrak{B}$, and

  (iii) if $E_1, E_2, \ldots \in \mathfrak{B}$, then $\bigcup_{j=1}^{\infty} E_j \in \mathfrak{B}$.

(c) P is the probability (measure) of an event. In fact, P is a function that associates a number $P(E)$ for each element E of $\mathfrak{B}$ with the following properties (called ***axioms of probability***):

  Axiom 1 $0 \leq P(E) \leq 1$, for each event E is $\mathfrak{B}$.

  Axiom 2 $P(\Omega) = 1$.

  Axiom 3 For any sequence $E_1, E_2, \ldots$ of *mutually exclusive* events (disjoint sets, that is, $E_i \cap E_j = \phi$ if $i \neq j$) in $\mathfrak{B}$,

$$P\left\{\bigcup_{i=1}^{\infty} E_i\right\} = \sum_{i=1}^{\infty} P\{E_i\}.$$

1.1.2 Conditional Probability

For the probability space $(\Omega, \mathfrak{B}, P)$, let B be an event (that is, $B \in \mathfrak{B}$) and $P(B) > 0$. Then, given the probability of an event A, the ***conditional probability*** of B, denoted by $P(A \mid B)$, defined on $\mathfrak{B}$, is given by

$$P(A \mid B) = \frac{P(A \cap B)}{P(B)}, \text{ for any event } A \in \mathfrak{B}, \text{ and } P\{B\} \neq 0. \tag{1.3}$$

If $P(B) = 0$, then $P(A \mid B)$ is not defined.

It should be noted that conditional probability exhibits properties similar to ordinary probability, but restricted to a smaller space.

One of the concepts often needed is the "independence" of events. We offer the definition for two events that can be easily expanded. Hence, we have the following:

Two events A and B are ***independent*** if and only if

$$P(A \cap B) = P(A)P(B), \tag{1.4}$$

In other words, occurrence of one does not affect the chance of occurrence of the other. Relation (1.4) can be expanded for an arbitrary number of events. Hence, the arbitrary family of events $\{E_i, i \in \mathbb{N}\}$, where $\mathbb{N}$ is the set of natural numbers, is independent if

$$P(E_{i_1} \cap \cdots \cap E_{i_n}) = P(E_{i_1}) \cdots P(E_{i_n}), \tag{1.5}$$

for every finite subset of indices $\{i_1, \ldots, i_n\} \subset \mathbb{N}$.

As a consequence of Equation (1.4), it can be easily proved that if two events A and B are independent and $P(B) > 0$ then

$$P(A \mid B) = P(A); \tag{1.6}$$

and conversely, if $P(B) > 0$ and Equation (1.6) holds to be true, then A and B are independent.

As another application of independence, the following is called the ***multiplicative law***. Although we offer the definition for only two events, it can be expanded for any finite number of events. Thus, for any two events A and B with conditional probability $P(B \mid A)$ or $P(A \mid B)P(A)$ and as long as $P(A)$ or $P(B)$ is nonnegative, we have

$$P(A \cap B) = P(B \mid A)P(A) = P(A \mid B)P(B). \tag{1.7}$$

Another property of conditional probability that can be easily verified called ***the law of total probability*** or ***total probability theorem*** is that if $E_1, E_2, \ldots, E_n$ is a partition of the sample space Ω, that is, n mutually exclusive events whose sum is unity, then for any given arbitrary event E, we have

$$P(E) = \sum_{i=1}^{n} P(E_i)P(E \mid E_i). \tag{1.8}$$

Using Equation (1.8) and the conditional probability, a very important theorem called the ***Bayes' theorem*** can be proved. It can be stated as follows: If E is an event and $E_1, E_2, \ldots, E_n$ a partition of the sample space Ω, then

$$P(E_i \mid E) = \frac{P(E_i)P(E \mid E_i)}{\sum_{j=1}^{n} P(E_j)P(E \mid E_j)}. \tag{1.9}$$

Relation (1.9) is referred to as ***Bayes' formula***. The conditional probability $P(E_i \mid E)$ is called the ***posterior probability***. The original probability of E_i is called the ***prior probability*** of E_i.

1.2 DISCRETE RANDOM VARIABLES AND DISTRIBUTIONS

Although sample points are numerical in many practical problems, there are nonnumerical in many others. For practical purposes, a numerical sample space is more desirable. The tool to quantify a sample space to a numerical one is the ***random variable***.

Before defining a random variable, we define a countable set. A set is called ***countable*** if it has the same number of elements (cardinality) as some subset of the set of natural numbers $\mathbb{N} = \{1, 2, \ldots\}$. In case of a finite subset of $\mathbb{N}$, the set is sometimes referred to as ***finitely countable*** or ***at most countable***. For the case of the same cardinality as $\mathbb{N}$, it is sometimes referred to as ***countably infinite***. A set that is not countable is called ***uncountable*** or ***denumerable***. Throughout the book, we may use any of the terms as may be appropriate. Thus for us, the set of natural numbers $\mathbb{N} = \{1, 2, \ldots\}$, the set of natural numbers including zero $\mathbb{N}_0 = \{0, 1, 2, \ldots\}$, the set of integers $\mathbb{Z} = \{\ldots, -2, -1, 0, 1, 2, 3, \ldots\}$, and the set of rational numbers (reactions) $\mathbb{Q}$ are all countable or infinitely countable sets. However, $\{1, 2, 3\}$ is at most countable or finitely countable. We note that all infinitely countable sets are of the same size, infinite. However, the set of real numbers $\mathbb{R}$ and an interval [0, 1] are not countable. The latter was first proved by George Cantor using the ***diagonal argument*** method.

Thus, a random variable is defined as a function (mapping) that assigns a numerical value (or a set of values) to a sample point. Hence, If X is a random variable, it assigns a value $X(\omega)$ to each outcome ω in Ω. If the function X is on an at most countable sample space into the set of real numbers, $\mathbb{R}$, the random variable X is called a ***discrete random variable.*** Thus, a random variable is discrete if it takes at most countably many values. In other words, if there is a finitely countable set of real numbers, say A, such that $P\{X \in A\} = 1$.

For the sake of convenience, we may, allow a discrete random variable to assume positive infinity, as one of its values such as waiting time for an event to occur for the first time. This is because the event may never occur.

We leave it as an exercise for the reader to prove the following properties of discrete random variables. If X and Y are two discrete random variables, then $X \pm Y$, XY, and X / Y are also random variables, for the last one, provided Y is nonzero.

Probability distribution of a discrete random variable indicates the assignment of probabilities over the entire values of a random variable.

It is important to note that probabilities are nonnegative real numbers and total assignment of probabilities must be 1. Hence, these two simple properties establish two conditions for a function to be a probability distribution function.

Let the discrete random variable X be defined on a sample space Ω with a typical element x. Then, the ***probability mass function (pmf) of X***, denoted by p_x, is defined as $p_x \equiv P(X = x)$. If no specific value is given, we denote it by p_X.The pmf p_X is sometimes described by a table or matrix. For instance, if X is a random variable with the general vale x and specific values $x_1, x_2, x_3, \ldots$, that are assigned probabilities $p_1, p_2, p_3, \ldots$, then the pmf can be written as in Table 1.1.

It is important to note that according to the second axiom of probability, $\sum_x p_x = 1$, where x varies over all possible values of X.

When two random variables X and Y have the same distribution, say p_X and p_Y, we say X and Y as ***equally distributed***.

TABLE 1.1 pmf

X	x_1	x_2	x_3	$\cdots$
$P_x = P(X = x)$	p_1	p_2	p_3	$\cdots$

Example 1.2.1 Bernoulli Distribution The simplest chance experiment (that is, an experiment whose outcomes are determined by chance, that is also called a ***trial***) is an experiment with only two outcomes. Such an experiment is called a ***Bernoulli Trial***. Independent repetitions of such trials are referred to as ***Bernoulli Trials***. The random variable representing a Bernoulli trial is called a ***Bernoulli random variable***. Thus, the sample space for a Bernoulli random variable has two sample points, referred to as ***success*** and ***failure***. Denoting the probability of a success by p, $0 \leq p \leq 1$, the probability of failure will be $q = 1 - p$. Thus, if X is a Bernoulli random variable with values 1 and 0, corresponding to success and failure, respectively, then distribution of X is given by

$$P(X = 1) = p; \quad P(X = 0) = 1 - p, \tag{1.10}$$

with $p + (1 - p) = 1$.

Relation (1.10) can also be expressed as

$$P(X = k) = p^k(1 - p)^{1-k}, \;\; k = 0, 1. \tag{1.11}$$

In order to determine whether Equation (1.11) defines a distribution function, in this case, the ***Bernoulli probability distribution function***, we note that

$$p^k(1 - p)^{1-k} > 0 \text{ and } \sum_{k=0}^{1} p^k(1 - p)^{1-k} = p + (1 - p) = 1.$$

Example 1.2.2 The Indicator Function For an event A in the probability space, the indicator function of A is defined as the random variable

$$I_A(\omega) = \begin{cases} 1, & \text{if } \;\omega \notin A, \\ 0, & \text{if } \;\omega \notin A. \end{cases} \tag{1.12}$$

Hence, for every $\omega \in \Omega$, $I_\Omega(\omega) = 1$ and $I_\phi(\omega) = 0$. In other words, distribution function of an indicator function is a Bernoulli distribution with parameter $p = P(A)$. Because of this fact, sometimes, a Bernoulli random variable is called the ***indicator random variable***. The random variable X, in this case, is an indicator of the event $\{X = 1\}$. This is because, when $X = 1$, ω is in the event A, the first part of Equation (1.12). For example, suppose two dice are rolled. Let X_1 and X_2 be random variables representing the numbers shown on the first and second rolling, respectively. Thus, the sample space Ω for each of these discrete random variables is $\Omega = \{1, 2, 3, 4, 5, 6\}$, and for both will be the cross-product $\Omega \times \Omega$, containing 36 elements, which are ordered pairs. In other words, $(k, l) \in \Omega \times \Omega$, $k, l = 1, 2, 3, 4, 5, 6$.

Suppose that we are interested in the sum of X_1 and X_2 be at least 7.Thus, we have $Y = X_1 + X_2$. Let the sum of interest be denoted by Z. That is, $Z = I_{Y \geq 7}$, or

$$Z = \begin{cases} 1, & \text{if } Y \geq 7, \\ 0, & \text{if } Y < 7. \end{cases} \tag{1.13}$$

Thus, we have to compute $P(Z = 1)$. Now, for the sum to be at least 7, we have to achieve the following ordered pairs: (1,6), (2,5), (3,4), (4,3), (5,2), and (6,1). That is, the sample space of Z, which is a subset of $\Omega \times \Omega$ contains (1,6), (2,5), (3,4), (4,3), (5,2) and (6,1). Probability of achieving each of these pairs is 1/36. Thus, $P(Z = 1) = 1/6$.

Example 1.2.3 Binomial Distribution Consider Example 1.2.1. Let us assume that a Bernoulli trial is independently repeated n times and also that the random variable $\boldsymbol{X}$ represents the number of successes occurred in the n trials. Obviously, the sample space, in this case, will have 2^n sample points. The number of ways k success can occur in n trials is $\binom{n}{k}$. Requiring k success in n trials leaves $n - k$ failures. With the trials being independent, the probability of a succession of successes and failures will be the product $p^k(1-p)^{n-k}$. Thus, the pmf of X, denoted by $\mathcal{B}_k \equiv b(k; n, p)$, and called the ***binomial random variable*** is given by

$$\mathcal{B}_k \equiv b(k; n, p) = P(X = k) = \binom{n}{k} p^k (1-p)^{n-k}, \quad k = 0, 1, 2, \ldots, n. \tag{1.14}$$

Relation (1.14) is, indeed, a probability distribution function, called the ***binomial distribution function*** with ***parameters n*** and ***p***. We leave the proof as an exercise to the reader.

Example 1.2.4 Poisson Distribution The random variable X with probability distribution function (or pmf)

$$p_x = P(X = k) = \frac{e^{-\lambda}\lambda^k}{k!}, \quad k = 0, 1, 2, \ldots, \tag{1.15}$$

where λ is a constant, is called a ***Poisson random variable*** and Equation (1.15) is called ***Poisson distribution function (or Poisson pmf)*** *with* ***parameter*** λ. We leave it as an exercise to show that Equation (1.15) is, actually, a distribution function.

The ***discrete joint pmf*** of the random variables $X_1, \ldots, X_n$, with real values $x_1, x_2, \ldots, x_n$, denoted by $p_{x_1 x_2 \cdots x_n}$, is defined by

$$p_{x_1 x_2 \cdots x_n} \equiv P(X_1 = x_1, X_2 = x_2, \ldots, X_n = x_n), \tag{1.16}$$

where $p_{x_1 x_2 \cdots x_n}$ is nonnegative and $\sum_{X_1} \sum_{X_2} \cdots \sum_{X_n} p_{x_1 x_2 \cdots x_n} = 1$.

From Equation (1.16), each individual probability, $p_{x_i} \equiv P\{X_i = x_i, \}, i = 1, 2, \ldots, n$, is called a ***marginal mass function***, that is,

$$p_{x_i} = \sum_{X_1} \sum_{X_2} \cdots \sum_{X_{i-1}} \sum_{X_{i+1}} \cdots \sum_{X_n} p_{x_1 x_2 \cdots x_n}, \quad i = 1, 2, \ldots, n, \tag{1.17}$$

where the summation is over all possible n-tuples with the ith component held fixed at a specific value, say x_i.

We list two properties of a random variable.

Property 1.2.1 We consider two random variables X and Y, with their respective pmfs p_X and p_Y and their joint mass function is denoted by $p_{X,Y}$. Then, the ***conditional mass function of X given Y*** is defined by

$$p_{X\,|\,Y} = P(X = x\,|\,Y = y) = \frac{P(X = x \text{ and } Y = y)}{P(Y = y)} = \frac{p_{X,Y}}{p_Y}, \tag{1.18}$$

provided that $p_Y > 0$.

As observed in defining the conditional probability function (1.18), random variables can be substituted for events. Thus, Equation (1.18) produces a different probability distribution for each value of y.

The pmf of the ***conditional random variable*** $X\,|\,Y = y$ (read as "X given $Y = y$") is given by

$$P\{(X\,|\,Y = y) = x\} = P(X = x\,|\,Y = y), \tag{1.19}$$

for values x of X.

Example 1.2.5 Conditional pmf Let X and Y be jointly distributed as shown in Table 1.2.

TABLE 1.2 $p_{X,Y} = P(X = x \text{ and } Y = y)$

Y	X			
	2	4	6	$p_y = P(Y = y)$ Marginal pmf of Y
0	0.10	0.03	0.04	$p_0 = 0.17$
1	0.01	0.16	0.07	$p_1 = 0.24$
2	0.15	0.06	0.00	$p_2 = 0.21$
3	0.05	0.09	0.24	$p_3 = 0.38$
$p_x = P(X = x)$ Marginal pmf of X	$p_2 = 0.31$	$p_4 = 0.34$	$p_6 = 0.35$	1

From Table 1.1, $p_1 = P(Y = 1) = 0.24$. Then, the conditional mass function of X given $Y = 1$ is

$$p_{X\,|\,Y=1}(x) = P(X = x\,|\,Y = 1) = \begin{cases} \dfrac{0.01}{0.24}, & x = 2, \\ \dfrac{0.16}{0.24}, & x = 4, \\ \dfrac{0.07}{0.24}, & x = 6. \end{cases} \tag{1.20}$$

Keeping Equation (1.5) for events in mind, we leave it as an exercise to prove that two random variables X and Y, with their respective pmfs p_X and p_Y, and their joint mass function $p_{X,Y}$, are independent if and only if

$$p_{XY} = p_X p_Y. \tag{1.21}$$

Let X, Y, and Z be discrete random variables. Then, X and Y are said to be ***conditionally independent***, given Z, if

$$P(X \le x, Y \le y\,|\,Z = z) = P(X \le x\,|\,Z = z)P(Y \le y\,|\,Z = z), \tag{1.22}$$

for all x, y, and z such that $P(Z = z) > 0$. As a consequence of Equation (1.22), if X and Y are conditionally independent, given Z, then

$$P(X = x\,|\,Y = y, Z = z) = P(X = x\,|\,Z = z), \tag{1.23}$$

for all x, y, and z such that $P(Z = z) > 0$.

Property 1.2.2 Functions of a Random Variable Let X be a discrete random variable with pmf, denoted by $F_X(x)$, where x is a real number representing the values of X. Let y be a real-valued function of the real variable x, denoted by $y = g(x)$. Hence, $Y = g(X)$ transforms the random variable X into the random variable Y, whose values are denoted by y. Thus, as X maps a sample point s to a real number, so does $Y = g(X(s))$. This indicates that cumulative distribution function (cdf) of Y depends on $g(x)$ and cdf of X.

We note that the domain of g should contain the range of X. Hence, pmf of Y can be expressed as

$$F_Y(y) = P(Y \le y) = P[g(X) \le y] = P[X \le I_y], \tag{1.24}$$

where I_y is the ***indicator function***.

1.3 DISCRETE MOMENTS

For k integers $n_1, n_2, \ldots n_k$, the ***arithmetic average***, denoted by $\overline{n}$, is defined as the sum of integers divided by k. In other words, the arithmetic average of the k integers is obtained by multiplying each number by $1/k$ and add, that is,

$$\overline{n} = n_1 \cdot \frac{1}{k} + n_2 \cdot \frac{1}{k} + \cdots + n_k \cdot \frac{1}{k}. \tag{1.25}$$

Similarly, let X be a random variable with values $x_1, x_2, \ldots,$ and x_k with probabilities $p_1, p_2, \ldots,$ and p_k, respectively. Then, the ***mean*** (or ***weighted average*** or ***expected value*** or ***expectation*** or ***mathematical expectation***) of X, denoted by $E(X)$, is defined as

$$E(X) = \sum_{i=1}^{k} x_i p_i. \tag{1.26}$$

Note that each p_i, $i = 1, 2, \ldots, k$, is the weight for each value $x_1, x_2, \ldots,$ and x_k, respectively.

Example 1.3.1 The Indicator Function We leave it as an exercise to prove that

$$E(I_A) = P(A). \tag{1.27}$$

Relation (1.26) may be expanded for a random variable with infinite values. In other words, let X be a random variable with values $x_0, x_1, x_2, \ldots,$ and respective probabilities $p_0, p_1, p_2, \ldots,$ that is $P(X = x_i) = p_i$, $i \in \mathbb{N}_0$. Then, the ***mean*** (or ***weighted average*** or ***expected value*** or ***expectation*** or ***mathematical expectation***) of X is defined as

$$E(X) = \sum_{i=0}^{\infty} x_i\ p_i, \tag{1.28}$$

provided that the infinite series in Equation (1.28) converges absolutely; otherwise, the series does not exist, and hence, the expected value of X does not exist.

The following are some of the properties of Equation (1.28):

1. The only way Equation (1.28) would not exist is when

$$\lim_{n\to\infty} \sum_{i=0}^{n} i p_i = +\infty. \tag{1.29}$$

2. Let X and Y be random variables and a, b, and c constants, then

$$E(aX + bY + c) = aE(X) + bE(Y) + c. \tag{1.30}$$

3. Let X and Y be two independent random variables with marginal pmf p_X and p_Y, respectively. Assume $E(X)$ and $E(Y)$ exist. Then,

$$E(XY) = E(X)E(Y). \tag{1.31}$$

Relation (1.31) can be expanded for a finite number of independent random variables.

4. For random variables X and Y,

$$E(X \mid Y = y) = \sum_x xP(X = x \mid Y = y), \tag{1.32}$$

where x and y are values of X and Y, respectively.

It is important to note that Equation (1.32) is a function of y, which implies that $E(X \mid Y)$ is a random variable in its own right.

Example 1.3.2 Conditional Expected Value Consider Example 1.2.5 and Table 1.2. In Example 1.2.5, we found the conditional pmf of X given $Y = 1$. Now we want to find the expected value of this conditional variable. From Table 1.2 and Equation (1.20), we have

$$E(X \mid Y = 1) = 2 \cdot \frac{0.01}{0.24} + 4 \cdot \frac{0.16}{0.24} + 6 \cdot \frac{0.07}{0.24} = \frac{1.08}{0.24} = 4.50.$$

5. For two random variables X and Y, we have

$$E[E(X \mid Y)] = E(X). \tag{1.33}$$

6. For a random variable X, with x representing its values, the ***nth moment***, denoted by M^n, is defined as

$$E_n = E(X^n) = \sum_x x^n p_X, \tag{1.34}$$

where n is a nonnegative integer.

(a) If $n = 1$, then Equation (1.34) reduces to the expected value or the first moment of X. With this note in mind, the variance of the random variable X, denoted by Var(X), can be defined as

$$\text{Var}(X) = \sum_x [x - E(X)]^2 p_X, \tag{1.35}$$

where x represents values of X. The following are some properties of variance (relation (1.35)):

(b) Let X and Y be random variables and a, b, and c constants, then

$$\text{Var}(aX \pm bY + c) = a^2\text{Var}(X) + b^2\text{Var}(Y). \tag{1.36}$$

(c) For $n = 2$, then Equation (1.34) reduces to the second moment and that in turn leads to the variable of X, that is

$$\text{Var}(X) = E_2 - E_1^2. \tag{1.37}$$

7. ***Central Moment.*** The expected value of the random variable of X is denoted by μ, that is, $\mu = E(X)$. The nth moment of the random variable $X - \mu$, that is,

$E[(X-\mu)^n]$, is called the ***central moment*** of X. The random variable $X-\mu$ measures the ***deviation of X from its expectation***. This deviation can be positive or negative depending on the values of $X-\mu$. Hence, its absolute value gives an ***absolute measure of deviation*** of the random variable X from its mean μ. Yet a better measure, the ***mean square deviation***, that is, $E[(X-\mu)^2]$, called the ***second central moment of X***, is, indeed, the variance of X. Thus, the variance measures the ***average deviation*** or ***dispersion*** of the random variable from its mean. As the variance measures the deviation from the mean by squares, in order to adjust for this squaring process, the square root is used. The positive square root of the variance of a random variable X is called the ***standard deviation***.

8. **Generating Function.** Let X be a discrete random variable with nonnegative integer values with p_X as the pmf of X. The ***probability generating function (pgf)*** of X, denoted by $G(z)$,where z is a complex number, that was defined in Chapter 1, can be redefined as

$$G(z) \equiv E(z^X) = \sum_{x=0}^{\infty} p_X z^x, \tag{1.38}$$

where the power series in Equation (1.38) converges absolutely at least for all z, such that $|z| \leq 1$. The idea of applying generating function to probability is intended to encapsulate all the information about the random variable.

(a) When X takes k as its value, z^X takes z^k as its value.

(b) The power series in Equation (1.38) follows all the rules of convergence of power series with nonnegative coefficients.

(c) The product of two generating functions $G(z)$ and $H(z)$ is given by

$$G(z)H(z) = \left(\sum_{i=0}^{\infty} a_i z^i\right)\left(\sum_{j=0}^{\infty} b_j z^j\right) = \sum_{k=0}^{\infty} c_k z^k, \tag{1.39}$$

where

$$c_k = \sum_{i=0}^{k} a_i b_{k-i}. \tag{1.40}$$

The sequence $\{c_k\}$ defined in Equation (1.40) is called the ***convolution*** of the two sequences $\{a_i\}$ and $\{b_j\}$. In general, the convolution initiates two independent random variables X and Y with pmfs p_X and p_Y, respectively. Here it is defined as follows. Let X and Y be two independent discrete random variables with pmfs of p_X and p_Y, respectively. Then, the convolution of p_X and p_Y is a pmf $p_Z = p_X * p_Y$ such that

$$p_l = \sum_k p_k \cdot p_{i-k}, \quad l = \ldots, -2, -1, 0, 1, 2, \ldots. \tag{1.41}$$

The pmf defined in Equation (1.41) is distribution function of the random variable $Z = X + Y$.

(d) For a random variable X that takes $k, k = 0, 1, \ldots$, as its values, pmf of X is recovered by the derivatives of the generating function as follows:

$$p_k = P(X = k) = \frac{G^{(k)}(0)}{k!}, \; k = 0, 1, 2, \ldots . \tag{1.42}$$

(e) If p_X is the pmf of a random variable X, then

$$E(1) = G(1^-) = \lim_{z\to 1} G(z) = \sum_{x=0}^{\infty} p_x = 1. \tag{1.43}$$

(f) If $z = 1$, then

$$G(1) = \sum_{n=0}^{\infty} p_n = 1. \tag{1.44}$$

(g) For a random variable X,

$$E(X) = G'(1^-) \tag{1.45}$$

$$\text{Var}(X) = G''(1^-) + G'(1^- - [G'(1^-)]^2. \tag{1.46}$$

(h) The ***nth factorial moment***, denoted by $E[(X)_n]$, is defined as

$$E[(X)_n] = E[X(X-1)\cdots(X-n+1)], \tag{1.47}$$

and for a random variable X, it is given by

$$E[(X)_n] = E\left(\frac{X!}{(X-n)!}\right) = G^{(n)}(1^-), \tag{1.48}$$

where

$$(X)_n = X(X-1)\cdots(X-n+1), \tag{1.49}$$

All the moments of X can be obtained from Equation (1.48). For instance, if $n = 1$, then we obtain Equation (1.43). If $n = 2$, we obtain $E(X(X-1)) = E(X^2) - E(X)$, which leads to the variance given in Equation (1.44).

9. ***The Moment generating function*** of a random variable X, denoted by $M_X(t)$, with the pmf p_X is defined as

$$M_X(t) = E(e^{tX}) = G_X(e^t) = \sum_x e^{tx} p_x, \tag{1.50}$$

where $G_X(t)$ is the generating function of X and t is a nonnegative real number.

It is important to note the following points:

(a) It is possible that t is a complex number with nonnegative real part.

(b) The moment generating function generates all the moments of X, as did the generating function.

(c) If $t = 0$, then

$$M_X(0) = \sum_{x=0}^{\infty} p_x = 1. \tag{1.51}$$

Recall that for generating function, we had relation (1.42), that is, if $z = 1$, then $G(1) = \sum_{x=0}^{\infty} p_x = 1$.

(d) The nth moment of the random variable X can be obtained from the moment generating function as

$$E(X^n) = \sum_x x^x p_x = \lim_{t \to 0} \frac{d^n M_X(t)}{dt^n}. \tag{1.52}$$

Recall that for generating function, we had a similar relation (1.49), that is, to obtain the nth factorial moment of a random variable X, we need to evaluate the nth derivative of the generating function at 1. Hence, for higher moments than the second, the moment generation yields direct results.

(e) For computational purposes, it is more convenient to use the following relations:

$$E(X) = \lim_{t \to 0} \frac{d[\ln M_X(t)]}{dt}. \tag{1.53}$$

and

$$\text{Var}(X) = \lim_{t \to 0} \frac{d^2[\ln M_X(t)]}{dt^2}. \tag{1.54}$$

1.4 CONTINUOUS RANDOM VARIABLES, DENSITY, AND CUMULATIVE DISTRIBUTION FUNCTIONS

A sample space is called ***continuous*** if the outcomes of a chance experiment can assume real values, that is, Ω is the entire real number line $\mathbb{R}$ or a part of it. However, in practice, a smaller set of subsets can be taken to contain all events of our interest. The smallest such set, denoted by $\mathcal{B}$, is referred to as the ***Borel set***. For instance, the set of all rational numbers in the interval [0, 1] is a Borel set for [0, 1]. The Borel set leads to a new ***probability space*** $(\Omega, \mathcal{B}, P)$, where P is the probability of evens in the Borel set.

We define the function $f_X(x)$ on the set of real numbers, $\mathbb{R}$, such that $f_X(x) \geq 0$, for all real x, and $\int_{-\infty}^{\infty} f_X(x)dx = 1$. Then, $f_X(x)$ is called a ***continuous probability density function (pdf) on*** $\mathbb{R}$. Let Ω be a continuous sample space. A random variable X, which takes its values for such Ω, is called a ***continuous random variable***. In other words, a continuous random variable maps Ω to a subset of the real line, that is, X is a real-valued function defined on Ω such that $X : \Omega \to R \subset \mathbb{R}$. In terms of pdf, $f_X(x)$, we have the following properties:

$$P(a \leq X \leq b) = \int_a^b f_X(x)dx, \quad \text{for any interval } [a, b], \tag{1.55}$$

$$P(X = a) = \int_a^a f_X(x)dx = 0, \qquad \text{for any fixed real}\, a, \tag{1.56}$$

$$P(X < b) = \int_{-\infty}^{b} f_X(x)\,dx. \tag{1.57}$$

Suppose X is a continuous random variable defined on a sample space Ω with a probability density function $f_X(x)$. Then, the ***cdf*** of X, denoted by $F_X(x)$, is defined as

$$F_X(x) = P(X \leq x) = \int_{-\infty}^{x} f_X(t)dt. \tag{1.58}$$

As it can be seen, if the density function $f_X(x)$ is continuous, then $f_X(x) = \frac{dF_X(x)}{dx}$. We note that the distribution $F_X(x)$ is nondecreasing, right continuous, and exhibits the following properties:

$$\lim_{x\to-\infty} F_X(x) = 0\, \text{and} \lim_{x\to\infty} F_X(x) = 1. \tag{1.59}$$

If there is no danger of confusion, we will suppress the subscript X from $f_X(x)$ and $F_X(x)$. Equation (1.55) can be rewritten using Equation (1.55) as

$$P(a \leq X \leq b) = F(b) - F(a). \tag{1.60}$$

It is important to note that $0 \leq F(x) \leq 1$. Thus, for an event, the longer the elapse time, the greater would be the probability of occurrence of the event. If $1 - F(x)$ is denoted by $S(x)$, that is, $S(x) = P(X > x)$, then $S(x)$ is called the ***survival probability***. In other words, if x represents the time, then $S(x)$ is the probability that the event will not happen until the time x. The ratio of density function and survival probability is referred to as ***hazard or failure rate***. Denoting the hazard rate by $h(x)$ the hazard rate will be defined as

$$h(x) = \frac{f(x)}{S(x)}. \tag{1.61}$$

Similar to the discrete random variable, we define the convolution of two density functions. Let X and Y be two continuous random variables with density functions $f(x)$ and $g(y)$, respectively, for all real numbers x and y. Then, the ***convolution*** of functions f and g, denoted by $h \equiv f * g$, is defined by

$$\begin{aligned} h(z) = (f * g)(z) &= \int_{-\infty}^{\infty} f(z-y)g(y)dy \\ &= \int_{-\infty}^{\infty} g(z-x)f(x)dx. \end{aligned} \tag{1.62}$$

Furthermore, we note that Equation (1.62) is the pdf of sum of two independent random variables, $Z = X + Y$.

Example 1.4.1 Uniform Random Variable Let X be a continuous random variable with pdf as follows:

$$f_X(x) = \begin{cases} \frac{1}{b-a}, & a \leq x \leq b, \\ 0, & \text{otherwise.} \end{cases} \tag{1.63}$$

Relation (1.63) is the pdf of the ***uniform distribution*** over the interval $[a, b]$. The random variable with this pdf is called a ***uniform random variable***.

Example 1.4.2 Exponential Random Variable A continuous random variable X with pdf

$$f_X(x) = \begin{cases} \mu e^{-\mu x}, & x \geq 0, \\ 0, & \text{otherwise,} \end{cases} \tag{1.64}$$

is referred to as ***negative exponential*** (or ***exponential***) ***random variable***, where μ is a parameter of the *pdf*.

The following is an important property of exponential distribution:

Theorem 1.4.1 Memoryless Property A random variable $X : \Omega \to (0, \infty)$ has an exponential distribution if and only if it exhibits the following property, referred to as the ***memoryless*** (or ***forgetful***) ***property***:

$$P\{X > s + t | X > s\} = P(X > t), \quad \text{for all } s,\ t \geq 0. \tag{1.65}$$

Proof: Let X have an exponential distribution with parameter $\mu > 0$. Then, we have

$$P\{X > s + t\,|\,X > s\} = \frac{P\{X > s + t \text{ and } T > t\}}{P\{X > s\}} = \frac{P\{X > s + t\}}{P\{X > s\}}$$

$$= \frac{e^{-\mu(s+t)}}{e^{-\mu s}} = e^{-\mu t} = P\{X > t\}.$$

The converse follows because the function $g(t) = P\{X > t\}$ satisfies

$$g(s + t) = g(s)g(t), \quad \text{for all } s, t \geq 0 \text{ and } g(0) = 1.$$

Example 1.4.3 Hyperexponential Random Variable Let $Y_i, i = 1, 2, \ldots, n$, be n independent exponentially distributed random variables with pdf f_{Y_i}, $i = 1, 2, \ldots, n$, and parameters $\mu_i, i = 1, 2, \ldots, n$, with $\mu_i \neq \mu_j$, for $i \neq j$. Furthermore, let p_i, $i = 1, 2, \ldots, n$, be positive constants such that $\sum_{i=1}^{n} p_i = 1$. Then, a continuous random variable X is called a ***hyperexponential random variable*** if its mixed pdf has the following form:

$$f_X(x) = \sum_{i=1}^{n} p_i f_{Y_i}(x), \tag{1.66}$$

with $2n$ parameters μ_i and p_i, $i = 1, 2, \ldots, n$.

Example 1.4.4 Gamma Random Variable As a generalization of Equation (1.63), we consider a continuous random variable with two-parameter pdf as follows:

$$f_X(x;\mu,t)=\begin{cases}\frac{\mu^t x^{t-1}}{\Gamma(t)}e^{-\mu x}, & x>0,\mu,t>0\\ 0, & \text{otherwise,}\end{cases} \tag{1.67}$$

where

$$\Gamma(t)=\int_0^\infty x^{t-1}e^{-x}dx \tag{1.68}$$

is the ***gamma function*** and t is a positive real number. In Equation (1.67) if $t=1$, we will obtain the exponential density function with parameter μ defined by Equation (1.64).

Example 1.4.5 Erlang Random Variable In Equation (1.67) if t is a nonnegative integer, say n, it reduces to pdf of another random variable, called the ***Erlang pdf of order r***. In other words, a random variable X has ***Erlang*** pdf with shape parameter r and rate parameter μ, denoted by $e(x;\, r,\, \mu)$, which is defined as

$$e(x;\, r,\, \mu)=\frac{\mu^r x^{r-1}e^{-\mu x}}{(r-1)!},\quad \text{for } x,\, \mu\geq 0. \tag{1.69}$$

In Equation (1.69), if $r=1$, then we will have the exponential pdf. In addition to the shape parameter r, if rather than the rate parameter μ, a scale parameter, $\beta=1/\mu$, is of interest, then an alternative, but equivalent, pdf is defined as follows:

$$e(x;\, r,\, \beta)=\frac{x^{r-1}e^{-x/\beta}}{\beta^r(r-1)!},\ \text{for } x,\, \beta\geq 0. \tag{1.70}$$

We note that when $\beta=2$, the Erlang distribution simplifies to the chi-squared distribution with $2r$ degrees of freedom. Thus, Equation (1.70) can be regarded as pdf of a generalized chi-squared distribution for even numbers of degrees of freedom. We also note that because of the factorial function in the denominator of Eqs. (1.69) and (1.70), the Erlang distribution is only defined when the shape parameter r is a positive integer.

Example 1.4.6 Normal Distribution Function A continuous random variable X with mean μ, variance σ^2, and pdf

$$f_X(x)=\frac{1}{\sigma\sqrt{2\pi}}e^{\frac{-(x-\mu)^2}{2\sigma^2}},\quad -\infty<x<\infty, \tag{1.71}$$

is called a ***normal*** random variable, denoted by $X\sim N(\mu,\, \sigma^2)$ or $X\sim\Phi(\mu,\, \sigma^2)$. If $\mu=0$ and $\sigma^2=1$, then X is called a ***standard normal*** random variable.

1.5 CONTINUOUS RANDOM VECTOR

The idea of random vector is a multidimensional generalization of the random variable. Thus, let Ω be the sample space. A **random vector** X is an n-dimensional vector, whose components are random variables, that is $X = (X_1, X_2, \dots, X_n)$. Continuous random vectors are described by the ***joint pdf***. For instance, consider two continuous random variables X and Y with pdf $f_X(x)$ and $f_Y(x)$, respectively. The ***joint cumulative bivariate cdf of X and Y***, denoted by $F_{X,Y}(x, y)$ or simply $F(x, y)$, is defined as

$$F(x, y) = P(X \leq x, Y \leq y). \tag{1.72}$$

The joint pdf for X and Y, denoted by $f_{X,Y}(x, y)$, is obtained as

$$f_{X,Y}(x, y) = \frac{\partial^2 F_{X,Y}(x, y)}{\partial x \partial y}. \tag{1.73}$$

Relations (1.72) and (1.73) can be finitely extended for many random variables as in the discrete case. We note that $f_{X,Y}(x, y) \geq 0$, for every $(x, y) \in \mathbb{R}^2$. We also note that

$$\int_{-\infty}^{\infty} \int_{-\infty}^{\infty} f_{X,Y}(x, y)dx\,dy = 1.$$

Finally, we note that for any region D in the xy plane, the following condition holds:

$$P\{(X, Y) \in D\} = \int\int_D f_{X,Y}(x, y)dxdy.$$

From the joint pdf, individual pdfs, called ***marginal pdf***,

$$f_X(x) = \int_{-\infty}^{\infty} f_{X,Y}(x, y)dy \text{ and } f_Y(y) = \int_{-\infty}^{\infty} f_{X,Y}(x, y)dx. \tag{1.74}$$

The joint bivariate cdf for X and Y is

$$F_{X,Y}(x, y) = P(X \leq x, Y \leq y). \tag{1.75}$$

Similar to the discrete case, from the joint pdf, we can define the conditional pdf. Let us consider two continuous random variables X and Y, with their respective pdfs f_X and f_Y and joint mass function $f_{X,Y}$. Then, for any y such that $f_Y(y) > 0$, the ***conditional pdf of X given Y=y***, denoted by $f_{X\,|\,Y=y}(x)$, is defined by

$$f_{X\,|\,Y=y}(x) = \frac{f_{X,Y}(x, y)}{f_Y(y)}, \tag{1.76}$$

for all $x \in \mathbb{R}$. The quantity $f_{X\,|\,Y=y}(x)$ is not defined when $f_Y(y) = 0$.

Example 1.5.1 Marginal pdf An industry produces two types of productions. There are defective items in each production. Let us denote the proportion of defective items of Type 1 production by X and Type 2 production by Y. The joint pdf of X and Y, denoted by $f_{X,Y}(x, y)$ is given by

$$f_{X,Y}(x, y) = \begin{cases} 1 - x^2, & 0 \le x \le 1, 0 \le y \le 1, \\ 0, & \text{otherwise.} \end{cases}$$

We need to determine

1. marginal probability $P(0 \le X \le 0.3)$,
2. marginal probability $P(0.1 \le Y \le 0.7)$,
3. joint probability $P(0 \le X \le 0.5, 0.4 \le Y \le 0.9)$,
4. marginal pdf f_X,
5. marginal pdf f_Y, and
6. conditional probability $P(X \le 0.3 | Y = 0.6)$.

The answers to the aforementioned questions are as follows:

1. $P(0 \le X \le 0.3) = \int_0^{0.3} (1 - x^2)dx = 0.291.$
2. $P(0.1 \le Y \le 0.7) = \int_{0.1}^{0.7} (1 - x^2)dy = 0.6(1 - x^2).$
3. $P(0 \le X \le 0.5, 0.5 \le Y \le 0.9) = \int_0^{0.5} \int_{0.4}^{0.9} (1 - x^2)dy\,dx = 0.2292.$
4.
$$f_X(x) = \int_{-\infty}^{\infty} f_{X,Y}(x, y)dy = \int_{-\infty}^{\infty} (1 - x^2)I_{\{0 \le x \le 1, 0 \le y \le 1\}}dy$$
$$= \int_0^1 (1 - x^2)I_{\{0 \le x \le 1\}}dy = (1 - x^2)I_{\{0 \le x \le 1\}}$$
$$= \begin{cases} (1 - x^2), & \text{if } 0 \le x \le 1, \\ 0, & \text{otherwise.} \end{cases}$$

5.
$$f_Y(y) = \int_{-\infty}^{\infty} f_{X,Y}(x, y)dx = \int_{-\infty}^{\infty} (1 - x^2)I_{\{0 \le x \le 1, 0 \le y \le 1\}}dx$$
$$= \int_0^1 (1 - x^2)I_{\{0 \le y \le 1\}}dx = \left[x - \frac{x^3}{3}\right]_0^1 I_{\{0 \le y \le 1\}}$$
$$= \begin{cases} \frac{2}{3}, & \text{if } 0 \le y \le 1, \\ 0, & \text{otherwise.} \end{cases}$$

We note that the indicator functions are used in answers (4) and (5) to cover various cases at once and, thus, make the calculations easier.

6.

$$P(X \leq 0.3|Y = .6) = \int_{-\infty}^{0.3} f_{X|Y=0.6}(x)dx$$

$$= \int_{0}^{0.3} \frac{1-x^2}{\frac{2}{3}}dx = \frac{3}{2}\left[x - \frac{x^3}{3}\right]_0^{0.3} = 0.4485.$$

Referring to Equation (1.61), a distribution function $F(x)$ is called ***heavy-tailed***, if $1 - F(x) = S(x) > 0$, $x \geq 0$ and

$$\lim_{x\to\infty} P\{X > x + y\,|\,X > x\} = \lim_{x\to\infty} \frac{S(x+y)}{S(x)} = 1, \quad \forall y \geq 0. \tag{1.77}$$

In other words, if X ever exceeds a given larger value, then it is likely to exceed any larger value than that value. The Pareto distribution function with $S(x) = X^{-\alpha}$, $x \geq 1$, where $\alpha > 0$ is a parameter, is a standard example. On the contrary, an exponential distribution function with parameter μ is not a heavy-tailed, because

$$\lim_{x\to\infty} \frac{S(x+y)}{S(x)} = \frac{e^{-\mu(x+y)}}{e^{-\mu x}} = e^{-\mu y}, \quad \forall x, y \geq 0, \tag{1.78}$$

which becomes 1 only for $y = 0$ and not for all values of y.

1.6 FUNCTIONS OF RANDOM VARIABLES

Let us consider a sample space Ω with a general element ω. Let X be a random variable, that is, a function relating the real number $X(\omega)$ to the outcome ω, with pdf $f_X(x)$ and cdf $F_X(x)$, if X is continuous, and p_X if X is discrete. Let $y = g(x)$ be a real-valued function of the real variable x, and $Y = g(X)$ be a transformation of the random variable X into the random variable Y. Thus, as $X(\omega)$ is a random variable, $g(X(\omega))$ is called ***function of the random variable X***. Then, the question is how to find pdf $f_Y(y)$ and cdf $f_Y(y)$ of $g(X(\omega))$. We note that the domain of function g should contain the range of X.

The idea of a function of a random variable may be expanded for a finite number of variables. Thus, if $\mathbf{X} = (X_1, X_2, \ldots, X_n)$ is a random vector of n random variable that associates the sample space Ω to the space $\mathbb{R}^n$ of real n-tuples, then the joint distribution function of X, for the continuous case, is defined as

$$F_{X_1,X_2,\ldots,X_n}(x_1, x_2, \ldots, x_n) = P\{X_1 \leq x_1, X_2 \leq x_2, \ldots, X_n \leq x_n\}, \tag{1.79}$$

and pmf, for the discrete case, is defined as

$$p_{X_1,X_2,\ldots,X_n}(x_1, x_2, \ldots, x_n) = P\{X_1 = x_1, X_2 = x_2, \ldots, X_n = x_n\}. \tag{1.80}$$

For the continuous case, the pdf is defined as

$$f_{X_1,X_2,\ldots,X_n}(\mathrm{x}) = \frac{\partial^n}{\partial x_1 \cdots \partial x_n} F_{X_1,\ldots,X_n}(x_1, \ldots, x_n). \tag{1.81}$$

As in the case of one variable, the marginal pdf for each random variable is obtained as follows:

$$f_{X_1}(x_1) = \int \cdots \int f_{X_1}, \ldots, x_n(x_1, \ldots, x_n) dx_2 \cdots dx_n. \tag{1.82}$$

For the discrete case, summation will replace the integrals.

Example 1.6.1 Sum of Random Variables Consider the random variable Z as the sum of two random variables X and Y. Hence,

$$P\{Z \le z\} = P\{X + Y \le z\} \tag{1.83}$$

and

$$F_Z(z) = \int_{-\infty}^{\infty} \int_{-\infty}^{z-y} f_{X,Y}(x, y)\, dx\, dy. \tag{1.84}$$

Thus,

$$f_Z(z) = \frac{dF_Z(z)}{dz} = \int_{-\infty}^{\infty} f_{X,Y}(x, z - x)\, dx. \tag{1.85}$$

If X and Y are independent random variables, then the pdf of Z is the convolution of the two pdfs. That is,

$$f_Z(z) = \int_{-\infty}^{\infty} f_X(x) f_Y(z - x) dx. \tag{1.86}$$

If g is a function of a discrete random variable X, $g = g(X)$, then expected value of $g(X)$ can be computed as

$$E[g(X)] = \sum_k g(x_k) P(X = x_k). \tag{1.87}$$

We will discuss the continuous case in the next section.

We now define, without proof, three types of convergence in the theory of probability. The first has two versions. It is important to note that in a classical sense, when we say a sequence $\{X_n\}$ converges to X as n approaches infinity, we mean that the difference of X_n and X, in absolute value, gets closer to 0 as n becomes larger and larger.

Theorem 1.6.1 Convergence in Probability **Version a:** The sequence of random variables $\{X_n, n \in \mathbb{N}\}$, not necessarily all defined on the same sample space Ω, ***converges in probability*** to a real number c, is denoted by $X_n \xrightarrow{\text{prob}} X$, if and only if

$$\lim_{n\to\infty} P(|X_n - c|) \geq \epsilon) = 0, \tag{1.88}$$

for any arbitrary positive number ε.

Version b: The sequence of random variables $\{X_n, n \in \mathbb{N}\}$ defined on a sample space Ω ***converges in probability*** to a random variable X, defined on a sample space Ω, is denoted by $X_n \xrightarrow{\text{prob}} X$, only if

$$\lim_{n\to\infty} P(|X_n - X|) \geq \epsilon) = 0, \tag{1.89}$$

for any arbitrary positive number ε In other words, as n increases, $P(|X_n - X|) > \varepsilon)$ decreases, and X is called the ***probability limit*** of the sequence.

It is important to note that if X is a constant c rather than a random variable, then both versions are the same. Furthermore, if each of the two sequences $\{X_n, n \in \mathbb{N}\}$ and $\{Y_n, n \in \mathbb{N}\}$ converges in probability to X and Y, respectively, with all random variables defined on the same sample space Ω, that is, $X_n \xrightarrow{\text{prob}} X$, and $Y_n \xrightarrow{\text{prob}} Y$, then, we leave it as an exercise to the readers to prove that $(X_n + Y_n) \xrightarrow{\text{prob}} (X + Y)$.

We note that the idea of convergence in probability can be extended to a sequence of random vectors.

Theorem 1.6.2 Almost Surely Convergence The sequence of random variables $\{X_n, n \in \mathbb{N}\}$, defined on the sample space Ω, ***converges almost surely*** (or ***with probability one***) to a random variable X, denoted by $X_n \xrightarrow{\text{a.s.}} X$ (or $X_n \to X$ with probability one) if there is a set $A \subset \Omega$ such that

$$\lim_{n\to\infty} X_n(\omega) = X(\omega), \text{ for all } \omega \in A, \tag{1.90}$$

and $P(A) = 1$, or

$$P(\{\lim_{n\to+\infty} X_n(\omega) = X(\omega), \forall\omega \in \Omega\}) = 1. \tag{1.91}$$

We leave it as an exercise that the ***almost sure convergence*** implies ***convergence in probability***, but the converse is not true. We also note that the almost sure convergence reduces the idea of convergence of deterministic sequences. We further note that convergence almost surely is the probabilistic version of pointwise convergence in real analysis. Finally, we note that the random variables X_n, $n \geq 1$, are generally highly dependent.

Example 1.6.2 Convergence in Probability A random experiment (per day income) runs over time n. Let the income on day n be denoted by R_n. Let the income on the first n days be denoted by S_n, that is,

$$S_n = \sum_{k=1}^{n} R_k. \tag{1.92}$$

Finally, we denote the lifetime income by S, that is,

$$S = \sum_{k=1}^{\infty} R_k. \tag{1.93}$$

Assuming S_k is positive for all k, we have $S_n \xrightarrow{\text{a.s.}} S$.

Theorem 1.6.3 Convergence in Distribution The sequence of random variables $\{X_n, n \geq 1\}$ ***converges in distribution*** to a random variable X, denoted by $X_n \xrightarrow{\text{dist}} X$, if

$$\lim_{n\to\infty} F_{X_n}(x) = F_X(x), \tag{1.94}$$

for every $x \in \mathbb{R}$ at which $F_X(x)$ is continuous.

We also state the following two main limit theorems in the theory of probability, without proof.

Theorem 1.6.4 Law of Large Numbers Let $\{X_n, n \geq 1\}$ be a sequence of independent, identically distributed random variables (iid), sometimes referred to as a ***random sample***, such that $\mu \equiv E[|(X_1)|] = E[|(X_2)|] = \cdots$ is finite. Then,

$$\overline{X}_n \equiv \frac{X_1 + \cdots + X_n}{n} \xrightarrow{\text{a.s.}} \mu. \tag{1.95}$$

Theorem 1.6.4 states that the mean of the infinite sequence of iid random variables $\{X_n, n \geq 1\}$ converges to the expected value as n approaches to infinity. If the convergence is almost possible, the law is called the ***strong law of large numbers***; if it is in distribution, it is called the ***weak law of large numbers***. The strong and weak laws of large numbers, respectively, state that

$$P\{\lim_{n\to\infty} |\overline{X}_n = \mu|\} = 1 \tag{1.96}$$

and

$$\lim_{n\to\infty} P\{|\overline{X}_n = \mu| > \epsilon\} = 0, \text{ for any } \epsilon > 0. \tag{1.97}$$

For the proof of the strong law of large number, one may refer to Grimmet and Stirzaker (2004).

We note that Equation (1.95) can be generalized replacing random variables by functions of random variables. Thus, let g be a function from reals to reals, that is, $g : \mathbb{R} \to \mathbb{R}$ such that $\mu = E[g(X_1)] = E[g(X_2)] = \cdots$ exits and

$$\frac{g(X_1) + \cdots + g(X_n)}{n} \to \mu = \int_{-\infty}^{\infty} g(x) f_{X_1}(x)\,dx, \; as\, n \to \infty. \tag{1.98}$$

Theorem 1.6.5 Central Limit Theorem Let $\{X_n, n \geq 1\}$ be a sequence of iid random variables, such that $E[|X_1^2|](E[|X_1^2|] = E[|X_2^2|] = \cdots)$ is finite. Let $\mu \equiv E(X_1) = E(X_2)\cdots$, and $\sigma^2 \equiv Var(X_1) = Var(X_2) = \cdots$. Then,

$$\frac{X_1 + \cdots + X_n - n\mu}{\sigma\sqrt{n}} \xrightarrow{\text{dist}} \Phi(0, 1). \tag{1.99}$$

In other words, the left-hand side of Equation (1.99) approaches standard normal distribution when n approaches infinity.

1.7 CONTINUOUS MOMENTS

We offer definition of moments for a continuous random variable similarly to discrete random variables (Eqs. (1.26) and (1.27)). Thus, let X be a continuous random variable with pdf $f_X(x)$ and a real-valued function $g(x)$. Then, the expected value of $g(X)$ is defined as

$$E[g(X)] = \int_{-\infty}^{\infty} g(x) f_X(x)dx, \tag{1.100}$$

provided the integral exists, that is, $E[g|X|]$ is finite. In particular, if $g(X) = X$, the mean μ and variance σ^2 of X are obtained as follows:

$$\mu = E(X) = \int_{-\infty}^{\infty} x f_X(x)\,dx, \tag{1.101}$$

and

$$\sigma^2 = \text{Var}(X) = \int_{-\infty}^{\infty} (x - \mu)^2 f_X(x)dx. \tag{1.102}$$

Example 1.7.1 Expected Value of a Function of a Variable Let X be a normal random variable with $\mu = 0$, variance σ^2, and $g(x) = e^{\tau x}$, where τ is a constant. Then, $E[g(x)]$ can be computed as follows:

$$\begin{aligned} E[g(x)] &= \frac{1}{\sqrt{2\pi\sigma^2}} \int_{-\infty}^{\infty} e^{\tau x} e^{-\frac{x^2}{2\sigma^2}}\,dx \\ &= \frac{1}{\sqrt{2\pi\sigma^2}} e^{\frac{\sigma^2\tau^2}{2}} \int_{-\infty}^{\infty} e^{-\frac{(x-\sigma^2\tau)^2}{2\sigma^2}}\,dx \\ &= e^{\frac{\sigma^2\tau^2}{2}}. \end{aligned}$$

Example 1.7.2 Characteristic Function The characteristic function of a real-valued random variable X, denoted by $\phi_X(t)$, is defined by

$$\phi_X(t) = E[e^{itx}], \tag{1.103}$$

where i is the imaginary unit and t is a real number. If $F_X(x)$ is the cumulative distribution of X, then

$$\phi_X(t) = \int_{\mathbb{R}} e^{itx} dF_X(x), \tag{1.104}$$

where $\mathbb{R}$ is the set of real numbers, and the integral in Equation (1.104) is of the Riemann–Stieltjes kind. All moments of X can be found using derivatives of Equation (1.103) with respect to t, evaluated at $t = 0$. For instance, we leave it to the reader as an exercise to show that for the Poisson random variable with parameter λ, its characteristic function is

$$\phi(t) = e^{\lambda(e^{it} - 1)}, \tag{1.105}$$

where λ is the mean.

Now, let $f_X(x)$ be a real-valued continuous function of a real variable about a number c. Then, the nth moment of $f_X(x)$, denoted by μ_n, is defined as

$$\mu_n = \int_{-\infty}^{\infty} (x - c)^n f_X(x)\, dx. \tag{1.106}$$

If c is the mean, then Equation (1.106) is referred to as the ***central moments***. The nth moment of the pdf $f_X(x)$ when $c = 0$ in Equation (1.106) is the expected value of X^n and is called a ***raw*** or ***crude*** moment. In this case, the nth ***moment*** of the cdf of X is defined by the Riemann–Stieltjes integral as

$$\mu_n = E(X^n) = \int_{-\infty}^{\infty} x^n dF_X(x). \tag{1.107}$$

If the value of the integral in Equation (1.107) for the absolute value of X^n is infinite, then the moment does not exist. We note that if $n = 0$, then the moment of the pdf is 1. When $n = 1$ and $n = 2$, we obtain the mean μ and variance σ^2 of X using Eqs. (1.101) and (1.102), respectively.

Example 1.7.3 Hyperexponential We leave it as an exercise to the reader to show that the first two moments and variance of a hyperexponential distribution function are respectively

$$E(X) = \int_{-\infty}^{\infty} x f_X(x) dx = \sum_{i=1}^{n} p_i \int_{0}^{\infty} x \mu_i e^{-\mu_i x} dx = \sum_{i=1}^{n} \frac{p_i}{\mu_i}, \tag{1.108}$$

$$E(X^2) = \int_{-\infty}^{\infty} x^2 f_X(x) dx = \sum_{i=1}^{n} p_i \int_{0}^{\infty} x^2 \mu_i e^{-\mu_i x} dx = \sum_{i=1}^{n} \frac{2p_i}{\mu_i}, \tag{1.109}$$

and

$$\operatorname{Var}(X) = E(X^2) - [E(X)]^2 = \left[\sum_{i=1}^{n} \frac{p_i}{\mu_i}\right]^2 + \sum_{i=1}^{n}\sum_{j=1}^{n} p_i p_j \left(\frac{1}{\mu_i} - \frac{1}{\mu_j}\right). \tag{1.110}$$

1.8 DIFFERENCE EQUATIONS

1.8.1 Introduction

Recently, there has been an increasing interest in the calculus of difference and differential-difference equations, particularly for solving some Markovian queueing models. Some other particular interests in the theory of difference equations are: because of the emerge of development and advancement of high-speed computers; numerous applications of difference equations to engineering, sciences (such as physics, chemistry, biology, probability, and statistics), economics, and psychology; and the mathematical theory in view of the analogy of the theory to differential equations. One main interest in difference equations versus differential equations is that the former involve discrete changes of an unknown function while the latter involve instantaneous rates of changes of an unknown function. In other words, difference equations are the discrete analogs of differential equations.

The theory and solutions of difference equations in many ways are parallel to the theory and solutions of differential equations. In fact, derivative of a function of one variable defined as a limit of a difference quotient is the main reason for many analogies between the differential calculus and the calculus of finite differences. In other words, as we may recall from differential calculus, the derivative of a given function $f(x)$ is defined as

$$f'(x) = \lim_{h\to 0} \frac{f(x+h) - f(x)}{h} = \lim_{h\to 0} \frac{\Delta[f(x)]}{h},$$

if the limit exists, where Δ denotes the ***difference operator***, that is, $\Delta f(x) = f(x+h) - f(x)$. For higher order differences, we may continue the Δ operator.

1.8.2 Basic Definitions and Properties

Definition 1.8.1 This definition consists of the following six parts:

1. Given a function F and a set S, for k in S and some positive integer n, a ***difference equation*** over a given set S is an equation of the form

$$F(k; y_k, y_{k+1}, \ldots, y_{k+n}) = 0. \tag{1.111}$$

A rewording of this statement is as follows: A difference equation is the one that involves differences. We write y_k for $y(x)$ to indicate the value of y at k. The range of values of the integer k must be stated. It should be noted that shifting indices has no bearing in the order of the difference equation. Hence,

$$F(k + r; y_{k+r}, y_{k+r+1}, \dots, y_{k+n+r}) = 0 \tag{1.112}$$

still is of order n, which is equivalent to Equation (1.111).

2. The ***order*** of a difference equation is the difference between the highest and lowest indices that appear in the equation.
3. A ***solution*** of the difference equation (Equation (1.111)) indicates a sequence $\{y_k\}$, which satisfies the difference equation Equation (1.111) for all values of k in S. The solution is referred to a ***general solution*** of Equation (1.111) if it involves exactly n arbitrary constants. A ***particular solution*** of Equation (1.111) is obtained from the general solution by assigning values to the arbitrary constants.
4. A difference equation of the form

$$f_0(k)y_{k+n} + f_1(k)y_{k+n-1} + \cdots + f_{n-1}(k)y_{k+1} + f_n(k)y_k = f(k), \tag{1.113}$$

where each of f and f_i, $i = 0, 1, 2, \dots, n$, is a function of k defined for all values of $k \in S$, is called ***linear*** over the set S. If a difference equation is not linear, it is called ***nonlinear***. The linear difference equation (1.113) is called of ***order n***, if both $f_0(k)$ and $f_n(k)$ are different from zero at each point of S. In other words, the order of the difference equation (1.113) is the difference between the highest and lowest indices that appear in the equation, if and only if

$$f_0(k)f_n(k) \neq 0, \forall k \in S. \tag{1.114}$$

If $f(k)$ on the right-hand side of Equation (1.113) is zero, then the equation is called a ***homogeneous linear difference equation of order n***; otherwise, it is called ***nonhomogeneous*** or ***inhomogeneous***. In general, an nth ***order linear homogeneous difference equations with constant coefficients*** is an equation of the form

$$y_{k+n} + a_1 y_{k+n-1} + \cdots + a_{n-1}y_{k+1} + a_n y_k = 0, \tag{1.115}$$

where $a_1, a_2, \dots, a_n$ are n given constants with $a_n \neq 0$.

5. In addition to differences in an equation, there may be derivatives or integrals. In such cases, we refer to the equation as ***differential-difference equation*** or ***integral-difference equation***, respectively.
6. Consider $z = f(x, y)$ as a function with two variables x and y. We write differences of z with respect to x and y. That is, we consider y as a constant and denote the difference as

$$\Delta_x f(x, y) = f(x + h, y) - f(x, y). \tag{1.116}$$

Similarly, we consider x as a constant and denote the difference as

$$\Delta_y f(x, y) = f(x, y + k) - f(x, y). \tag{1.117}$$

Relations (1.116) and (1.117) are called ***partial differences*** of the function $z = f(x, y)$. An equation that contains partial differences is referred to as a ***partial difference equation***.

Theorem 1.8.1 Existence and Uniqueness Theorem If there are n initial conditions $y_0, y_1, \ldots, y_{n-1}$ for Equation (1.113), then corresponding to each of these conditions, Equation (1.113) has a unique solution.

Proof: The proof for this is left as an exercise for the reader.

1.9 METHODS OF SOLVING LINEAR DIFFERENCE EQUATIONS WITH CONSTANT COEFFICIENTS

In general, not every method mentioned below could be applied to solve every equation. Perhaps hence different methods have been developed.

1.9.1 Characteristic Equation Method

Homogeneous Case For the nth order linear homogeneous difference equation with constant coefficient given by Equation (1.115), the nth degree polynomial equation

$$r^n + a_1 r^{n-1} + \cdots + a_{n-1} r + a_n = 0 \tag{1.118}$$

is called ***characteristic equation*** (or ***auxiliary equation***).

Theorem 1.9.1 If r_i, $i = 1, 2, \ldots, n$, is a root of the characteristic equation (Equation 1.118), then

(a)
$$y_k^{(i)} = r_i^k \tag{1.119}$$

is a solution of Equation (1.118).

(b) If all n roots of the characteristic equation (1.118) are distinct real numbers, then the general solution of the equation is

$$y_k = c_1 y_k^{(1)} + c_2 y_k^{(2)} + \cdots + c_n y_k^{(n)}, \tag{1.120}$$

where c_i, $i = 1, 2, \ldots, n$, are n arbitrary constants.

(c) If roots r_i of the characteristic equation (1.118) are with multiplicity m_i, $i = 1, 2, \ldots, l$, such that

$$\sum_{i=1}^{l} m_i = n, \tag{1.121}$$

then the general solution of the equation is

$$y_k = r_1^k(c_1^{(1)} + c_2^{(1)}k + \cdots + c_{m_1}^{(1)}k^{m_1-1}) + r_2^k(c_1^{(2)} + c_2^{(2)}k + \cdots + c_{m_2}^{(2)}k^{m_2-1}) + \cdots + r_{m_l}^k(c_1^{(1)} + c_2^{(1)}k + \cdots + c_{m_l}^{(1)}k^{m_l-1}), \tag{1.122}$$

where $c_i^{(j)}$, $i = 1, \ldots, l; j = 1, 2, \ldots, m_i$, are arbitrary constants.

Proof: See Mickens (1990), Second Edition, p. 124.

Theorem 1.9.2 For arbitrary constants $c_{i,j+1}$ and c_j, let

$$y(x) = \sum_{i=1}^{m}\left(\sum_{j=0}^{n_i-1} c_{i,j+1}x^j\right)e^{r_i x} + \sum_{j=(n_1+\cdots+n_m)+1}^{n} c_j e^{r_j x}, \tag{1.123}$$

where $n_i \geq 1$, $i = 1, 2, \ldots, m$, with $n_1 + n_2 + \cdots + n_m \leq n$, the general solution of the nth order linear differential equation

$$\frac{d^n y(x)}{dx^n} + a_1\frac{d^{n-1}y(x)}{dx^{n-1}} + \cdots + a_{n-1}\frac{dy(x)}{dx} + a_n y(x). \tag{1.124}$$

In the above equation, a_i, $i = 1, 2, \ldots, n$, $a_n \neq 0$, are constants. In addition, let the characteristic equation (1.118) has r_j simple roots and r_i roots with multiplicity n_i, $i = 1, 2, \ldots, m$. Furthermore, let y_k be the general solution of the difference equation (1.115). Then,

$$y_k = \left.\frac{d^k y(x)}{dx^k}\right|_{x=0}, \tag{1.125}$$

and

$$y_k = \sum_{i=1}^{m}\left(c_{i1} + \sum_{l=1}^{n_i-1}\gamma_{i,l}k^l\right)r_i^k + \sum_{j=(n_1+\cdots+n_m)+1}^{n} c_j r_j^k, \tag{1.126}$$

Proof: See Mickens (1990, p. 139).

Nonhomogeneous Case Consider the first-order linear nonhomogeneous difference equation

$$y_{k+1} - ay_k = b, k = 0, 1, 2, \dots, \tag{1.127}$$

with initial condition $y_0 = c$, where a, b, and c are constants and $a \neq 0$. The characteristic equation of the homogeneous part, when $b = 0$,

$$y_{k+1} - ay_k = 0, \quad k = 0, 1, 2, \dots, \tag{1.128}$$

of Equation (1.127) is $r - a = 0$, with only one root $r = a$. Hence, the general solution of Equation (1.128) is

$$Y_k = Ca^k, k = 0, 1, 2, \dots, \tag{1.129}$$

where C is an arbitrary constant. In order to find the particular solution of Equation (1.127), different cases for values of a and b may be considered.

1.9.2 Recursive Method

Considering the nonhomogeneous case, we rewrite Equation (1.127) as

$$y_{k+1} = ay_k + b, \quad k = 0, 1, 2, \dots. \tag{1.130}$$

Now we consider different values of k, starting with 0.

$k = 0$: This value of k gives the initial value $y_0 = c$, substituting into Equation (1.130) gives

$$y_1 = ay_0 + b = ac + b. \tag{1.131}$$

$k = 1$: Substituting this value of k into Eqs. (1.130) and (1.131), we obtain

$$y_2 = ay_1 + b = a(ac + b) + b = a^2c + b(1 + a). \tag{1.132}$$

$k = 2$: Substituting $k=2$ into Eqs. (1.130–132) results in

$$y_3 = ay_2 + b = a[a^2c + b(1 + a) + b] + b = a^3c + b(1 + a + a^2). \tag{1.133}$$

From Equation (1.131) to Equation (1.133), we can obtain a pattern that would conjecture the general case and prove it by mathematical induction. Thus, we choose the solution for the initial problem (Equation 1.127) as

$$y_k = a^kc + b(1 + a + a^2 + \cdots + a^{k-1}). \tag{1.134}$$

We leave it as an exercise for the reader to prove Equation (1.134) that would be the proof of the following theorem:

Theorem 1.9.3 Existence and Uniqueness Theorem The linear first-order initial-valued differential equation given by Equation (1.127) has a unique solution given by Equation (1.134).

The following theorem will provide a special case of Equation (1.127), that is, linear first-order nonhomogeneous case. We leave its proof as an exercise to the reader.

Theorem 1.9.4 If the sequence $\{y_k, k = 0, 1, \ldots\}$ is the solution of Equation (1.127) when $a = 1$, that is, the linear first-order nonhomogeneous equation

$$y_{k+1} = y_k + b, k = 0, 1, 2, \ldots, y_0 = c, \tag{1.135}$$

then,

(a) if $b < 0$, it will diverge to $-\infty$,
(b) if $b = 0$, $\{y_k\}$ is a constant sequence, and
(c) if $b > 0$, the solution sequence will diverge to $+\infty$.

We will provide an example later in this chapter after the Laplace transform method to show how the recursive method works.

1.9.3 Generating Function Method

Let $\{a_n\}$ be a sequence of real numbers such that $a_n = 0$, for $n < 0$. If the infinite series

$$G(z) \equiv \sum_{n=0}^{\infty} a_n z^n, \tag{1.136}$$

where z is a dummy variable, converges for $|z| < R$, where R is the radius of convergence, then $G(z)$ is called the ***generating function*** of the sequence $\{a_n\}$. If the sequence $\{a_n\}$ is bounded, then the series (Equation 1.136) converges for at least $|z| < 1$.

If $\{a_n\}$ is the probability function $P_n(t)$, $n = 0, 1, 2, \ldots$, of a nonnegative discrete random variable $X(t)$, Equation (1.136) is called the ***pgf of X(t)***, denoted by $G(z, t)$, and is defined as

$$G(z, t) \equiv E(z^n) = \sum_{n=0}^{\infty} P_n(t) z^n, \ |z| < 1. \tag{1.137}$$

The essence of the method of pgf is to encapsulate all the information about the random variable. The following are some properties of pgf:

(a)

$$G(1, t) = \sum_{k=0}^{\infty} P_n(t) = 1. \tag{1.138}$$

(b) If X is a random variable, all its moments can be found via Equation (1.137) because

$$\frac{\partial^k G(z, t)}{\partial z^k} = \sum_{k=n}^{\infty} k! P_k(t) \frac{z^{k-n}}{(n-k)!} \tag{1.139}$$

and, hence,

$$\begin{aligned} \left.\frac{\partial^k G(z, t)}{\partial z^k}\right|_{z=1} &= \sum_{k=n}^{\infty} k! P_k(t) \frac{z^{k-n}}{(n-k)!} \\ &= E[X(X-1)\cdots(X-n+1)]. \end{aligned} \tag{1.140}$$

For instance, if $n = 1$, from Equation (1.140) we have $E(X)$ and if $k = 2$, we have $E[X(X-1)] = E(X^2) - E(X)$. Thus, variance can be obtained as

$$\text{Var}(X) = E(X^2) - [E(X)]^2 = E[X(X-1)] + E(X) - [E(X)]^2. \tag{1.141}$$

(c) Applying Taylor expansion in Equation (1.136), we obtain

$$\left.\frac{\partial^k G(z, t)}{\partial z^k}\right|_{z=0} = n! P_k(t). \tag{1.142}$$

Theorem 1.9.5 The generating function of sum of n independent random variables is the product of the generating function of each variable.

Proof: The proof of this is left as an exercise for the reader.

Example 1.9.1 An Application of pgf This example is to show how the pgf method works, particularly on a system of differential difference equation. This example is extracted from Example 4.5.2 of Haghighi and Mishev (2013).

Let us solve the following system of differential difference equations using the pgf method:

$$\begin{cases} P_0'(t) = -\lambda P_0(t), \\ P_n'(t) = -\lambda P_n(t) + \lambda P_{n-1}(t),\ n = 1, 2, \ldots, \end{cases} \tag{1.143}$$

with initial condition $P_0(0) = \delta_{i0}$, where δ_{i0} is ***Kronecker's delta function*** defined as

$$\delta_{i0} = \begin{cases} 1, & \text{if } i = 0, \\ 0, & \text{if } i \neq 0, \end{cases} \tag{1.144}$$

and normalizing equation $\sum_{n=0}^{\infty} P_n(t) = 1$. We need to find $P_n(t)$, $n = 0, 1, 2, \ldots$.

Using Equation (1.137), we first find the pgf of Equation (1.144) that gives

$$G(z, 0) = z^i. \tag{1.145}$$

Differentiating Equation (1.137) term by term, within the region of convergence, with respect to t, gives

$$\frac{\partial}{\partial t}G(z,t) = \sum_{n=0}^{\infty} \frac{\partial}{\partial t}[P_n(t)]z^n = P_0'(t) + \sum_{n=1}^{\infty} P_n'(t)z^n. \tag{1.146}$$

Now, multiplying the first equation of system (Equation 1.137) by z^n and summing over $n = 1, 2, \ldots$, we have

$$\frac{\partial}{\partial t}G(z,t) - P_0'(t) = -\lambda[G(z,t) - P_0(t) - zG(z,t)]. \tag{1.147}$$

Using the second equation of system (Equation 1.137), we have

$$\frac{\partial}{\partial t}G(z,t) = G(z,t)[\lambda(z-1]. \tag{1.148}$$

Solving the system of two equations (Eqs. 1.147 and 1.148) together with Equation (1.145), we have

$$G(z,t) = z^i e^{\lambda(z-1)t}. \tag{1.149}$$

Thus, $P_n(t)$, $n = 0, 1, 2, \ldots$, are the coefficients of the Taylor expansion of Equation (1.149), which are obtained as

$$P_n(t) = \begin{cases} 0, & n = 0, 1, 2, \ldots, i-1, \\ \dfrac{(\lambda t)^{n-1}}{(n-1)!}e^{-\lambda t} & n = i, i+1, i+2, \ldots. \end{cases} \tag{1.150}$$

1.9.4 Laplace Transform Method

Portrait of Pierre-Simon Laplace

Before we start the discussion of Laplace transform, it would be instructive to give a brief history of this great mathematician. **Pierre-Simon, marquis de Laplace** (23 March 1749–5 March 1827) was a French mathematician and astronomer, who was born in Beaumont-en-Auge, Normandy. He formulated Laplace's equation and pioneered the Laplace transform, which appears in many branches of mathematical physics, a field that he took a leading role in forming. The Laplacian differential operator, widely used in applied mathematics, is also named after him. He is remembered as one of the greatest scientists of all time, sometimes referred to as a French Newton or Newton of France, with a phenomenal natural mathematical faculty superior to any of his contemporaries (http://en.wikipedia.org/wiki/Pierre-Simon_Laplace).

Laplace transform has been a powerful method for solving linear ordinary differential equations, their corresponding initial-valued problems, and systems of linear differential equations. In order to obtain functions from their Laplace transforms, the inverse Laplace transform was introduced at the same time. Therefore, we start with the definition of Laplace transform.

The ***Laplace transform*** (or ***unilateral Laplace transform***) of a function $f(t)$, $t \geq 0$, is an integral operator to this function on $[0, \infty)$, denoted by $F(s)$, $\mathcal{L}(f)(s)$, $\mathcal{L}\{f(t)\}$ or $F*(s)$ (and many other notations used by different authors). The parameter s is arbitrary and is taken as in the definition (Equation 1.151) below:

$$F(s) = \mathcal{L}(f)(s) = \mathcal{L}\{f(t)\} = \int_0^{\infty} e^{-st} f(t)\, dt. \tag{1.151}$$

The ***bilateral Laplace transform*** of a function $f(t)$, $t \geq 0$, (although not often used) is defined as

$$F^{(2)}(s) = \mathcal{L}^{(2)}(f)(s) = \mathcal{L}^{(2)}\{f(t)\}(s) = \int_{\infty}^{\infty} e^{-st} f(t) dt. \tag{1.152}$$

The Laplace–Stieltjes transform of a distribution function $F(x)$ of a random variable X, denoted by $F^*(s)$, is defined as

$$F^*(s) = E[e^{-sX}] = \int_0^{\infty} e^{-sx} dF(x), \tag{1.153}$$

where s is a complex number.

The ***inverse Laplace transform*** of $F(s)$, denoted by $f(t)$, $\mathcal{L}^{-1}(F)(t)$ or $\mathcal{L}^{-1}\{F(s)\}$, is defined as

$$f(t) = \mathcal{L}^{-1}(F)(t) = \mathcal{L}^{-1}\{F(s)\}, \tag{1.154}$$

if the function $F(s)$ is determined from the function $f(t)$ in (1.151). See Lerch (1903).

One of the basic properties of Laplace transform is that it is a ***linear operator***, that is

$$\mathcal{L}\{af(t) + bg(t)\} = a\mathcal{L}\{f(t)\} + b\mathcal{L}\{g(t)\}, \tag{1.155}$$

for any constant a and b, provided that both $\mathcal{L}\{f(t)\}$ and $\mathcal{L}\{g(t)\}$ exist.

There are various tables for Laplace transforms and their inverses; Table 1.3 is one such table.

TABLE 1.3 Laplace Transforms of Some Elementary Functions

Formula #	$f(t)$	$F(s) \equiv \mathcal{L}\{f(t)\}$
1	$f(at)$	$\frac{1}{s}F\left(\frac{s}{a}\right), \quad s > a,$
2	$f'(t)$	$s\mathcal{L}\{f(t)\} - f(0)$
3	$f''(t)$	$s^2\mathcal{L}\{f(t)\} - sf(0) - f'(0)$
4	1	$\frac{1}{s}$
5	$\frac{t^n}{n!}, \quad n = 0, 1, 2, \ldots$	$\frac{1}{s^{n+1}}, \quad n = 0, 1, 2, \ldots$
6	$\frac{t^{a-1}}{\Gamma(a)}, \quad a \geq 0$	$\frac{1}{s^a}, \quad a > 0$
7	$u(t-a)$ Heaviside function	$\frac{1}{s}e^{-as}$
8	$u(t-a)f(t-a)$	$e^{-as}F(s)$
9	$\delta(t)$, unit impulse at $t = 0$.	1
10	$\delta(t-a)$ Dirac delta function	e^{-as}
11	$e^{at}, \quad a \geq 0$	$\frac{1}{s-a}, \quad s > a$
12	$e^{at}f(t), \quad a \geq 0$	$F(s-a), \quad s > a,$
13	$\frac{(ac+d)e^{at} - (bc+d)e^{bt}}{a-b}$	$\frac{cs+d}{(s-a)(s-b)}, \quad a \neq b$
14	$\frac{1}{t}(e^{at} - e^{bt})$	$\ln\left(\frac{s-b}{s-a}\right)$
15	$\sin(at)$	$\frac{a}{s^2+a^2}$
16	$\cos(at)$	$\frac{s}{s^2+a^2}$

Theorem 1.9.6 Existence of Laplace Transform Let $f(t)$ be a function that (1) it is defined on $[0, \infty)$; (2) it is piecewise continuous on $[0, \infty)$; and (3) it satisfies

$$|f(t)| \leq Me^{kt}, \quad t \in [0, \infty], \tag{1.156}$$

for some constants M and k, with $M > 0$. Then, $f(t)$'s Laplace transform $\mathcal{L}(f)$ exists for $s > k$.

Proof: Proof of this is left as an exercise for the reader.

The following example is extracted from Example 4.5.2 of Haghighi and Mishev (2013):

Example 1.9.2 Let us consider the system (Equation 1.143) discussed in Example 1.9.1. We now solve the same system using recursive and Laplace methods combined and find $P_n(t)$, $n = 0, 1, 2, \ldots$.

Let us denote by $P_0^*(s)$ and $P_n^*(s)$, $n = 1, 2, \ldots$, the Laplace transforms of $P_0(t)$ and $P_n(t)$, $n = 1, 2, \ldots$, respectively. Then, using formula # 2 in Table 1.3 we will have:

$$\mathcal{L}[P_0'(t)] = sP_0^*(s) - P_0(0) = sP_0^*(s) - 1 \tag{1.157}$$

and

$$\mathcal{L}[P_n'(t)] = sP_n^*(s) - P_n(0) = sP_n^*(s). \tag{1.158}$$

Substituting Eqs. (1.157) and (1.158) into Equation (1.137), we have

$$\begin{cases} sP_0^*(s) - 1 = -\lambda P_0^*(s), \\ sP_n^*(s) = -\lambda P_n^*(s) + \lambda P_{n-1}^*(s), \quad n = 1, 2, \ldots, \end{cases} \tag{1.159}$$

or

$$\begin{cases} (s+\lambda)P_0^*(s) = 1 \\ (s+\lambda)P_n^*(s) = \lambda P_{n-1}^*(s), \quad n = 1, 2, \ldots. \end{cases} \tag{1.160}$$

Recursively, we will have the following:

$$\left.\begin{aligned} P_0^*(s) &= \frac{1}{(s+\lambda)}, \\ P_1^*(s) = \frac{\lambda}{(s+\lambda)}P_0^*(s) &= \frac{\lambda}{(s+\lambda)}\frac{\lambda}{(s+\lambda)} \\ P_0^*(s) &= \frac{\lambda}{(s+\lambda)^2}, \\ &\vdots \\ P_n^*(s) &= \frac{\lambda^n}{(s+\lambda)^{n+1}}. \end{aligned}\right\} \tag{1.161}$$

Then, inverting Equation (1.161), we have

$$P_n(t) = \mathcal{L}^{-1}[P_n^*(s)] = \mathcal{L}^{-1}\left[\frac{\lambda^n}{(s+\lambda)^{n+1}}\right] = \lambda^n \frac{t^n e^{-\lambda t}}{n!}, \quad n = 1, 2, \ldots, \tag{1.162}$$

and

$$P_0(t) = \mathcal{L}^{-1}[P_0^*(s)] = \mathcal{L}^{-1}\left[\frac{1}{(s+\lambda)}\right] = e^{-\lambda t}. \tag{1.163}$$

Combining Eqs. (1.162) and (1.163), we will have

$$P_n(t) = \frac{(\lambda t)^n}{n!} e^{-\lambda t}, \quad n = 0, 1, 2, \ldots, \tag{1.164}$$

which is the Poisson probability distribution function.

EXERCISES

1.1 If X and Y are two discrete random variables, then prove that $X \pm Y$, XY, and X / Y are also random variables, for the last one, provided Y is nonzero.

1.2 Prove that relation (1.14) is, indeed, a probability distribution function.

1.3 Prove that relation (1.15) is, indeed, a probability distribution function.

1.4 Prove that two random variables X and Y, with their respective pmfs p_X and p_Y and joint mass function, denoted by $p_{X,Y}$, are independent if and only if $p_{XY} = p_X p_Y$.

1.5 Prove that: $E(I_A)P(A)$, where I_A is the indicator function of A.

1.6 If each of the two sequences $\{X_n, n \in \mathbb{N}\}$ and $\{Y_n, n \notin \mathbb{N}\}$ converges in probability to X and Y, respectively, with all random variables defined on the same sample space Ω, that is, $X_n \xrightarrow{\text{prob}} X$, and $Y_n \xrightarrow{\text{prob}} Y$, then prove that $(X_n + Y_n) \xrightarrow{\text{prob}} (X + Y)$.

1.7 Show that almost sure convergence implies convergence in probability, but the converse is not true.

1.8 Give the detail that the first two moments and variance of a hyperexponential distribution function are, respectively, Eqs. (1.79), (1.109), and (1.110).

1.9 Prove Theorem 1.8.1.

1.10 Prove relation (1.134).

1.11 Prove Theorem 1.9.4.

1.12 Prove Theorem 1.9.5.

1.13 Prove Theorem 1.9.6.

1.14 Let the discrete time random process $\{X_n\}$ be defined by $X_n = s^n, \quad n = 0, 1, \ldots,$ where s is selected randomly from (0,1).

a. Find the cumulative distribution of X_n.

b. Find the joint cumulative distribution of X_n and X_{n+1}.

1.15 Let $X(t) = At + B$, where A and B are independent random variables.

a. Find pdf of $X(t)$, $t > 0$.

b. Find the mean of $X(t)$.

1.16 Let X_n consist of an iid sequence of random variables with Poisson distribution with parameter λ, $\lambda > 0$, and $S_n = X_1 + X_2 + \cdots + X_n$.

a. Find the pmf of S_n.

b. Find the joint pmf of S_n and S_{n+k}.

c. Find the covariance of S_n and S_{n+k}.

2

STOCHASTIC PROCESSES

2.1 INTRODUCTION AND BASIC DEFINITIONS

Recall from Chapter 1 that a random variable was defined as a function on the sample space Ω, as part of the probability space $(\Omega, \mathfrak{B}, P)$. In this chapter, let Θ be a set (countable, denumerable, or an interval), in particular, a subset of the real line $\mathbb{R}$, that is, the interval $(-\infty, \infty)$ or the nonnegative part of the real line, that is, $[0, \infty)$. Let $X(t)$ be a random variable, where $t \in \Theta$. t is referred to as a ***parameter***, ***index***, ***time***, or an ***epoch***. The set Θ is called the ***parameter space*** or the ***index set***.

A sequence of real numbers $\{t_n, n \in \mathbb{N}, t_1 \geq 0\}$ is called a ***point process*** if $t_1 < t_2, \ldots$ and $\lim_{n \to \infty} t_n = +\infty$, and the numbers t_n, $n = 1, 2, \ldots$, are called ***event times*** or ***epochs***. We note that a point process is a strictly increasing sequence of real numbers, which does not have a finite limit point. The following are some examples of point processes: arrival time points of tasks at service stations such as supermarkets, failure time points of machines, and time points of traffic accidents. In these examples, t_n are times at which the events occur, that is, the arrival times.

It is important to note that epochs t_n, $n = 1, 2, \ldots$, do not necessarily have to be points of time.

Let the number of events occurred, in an interval $(0, t]$, $t > 0$, at event times t_n be $X(t)$, $t > 0$. Then, $X(t) = \max\{n, t_n \leq t\}$. The sequence $\{X(t), t > 0\}$ is called the ***counting process*** belonging to the point process $\{t_n, n \in \mathbb{N}, t_1 \geq 0\}$. If it is assumed that no more than one event can occur at a time, the process is called a ***simple counting process***. When the time points t_n are random variables, then $\{t_n, n \in \mathbb{N}, t_1 \geq 0\}$ is called ***random point process*** with $P\{\lim_{n \to \infty} t_n = +\infty\} = 1$. If the time points are of different types, such as arrival of different types of jobs to a service station, the process is called a ***marked point process***.

Delayed and Network Queues, First Edition. Aliakbar Montazer Haghighi and Dimitar P. Mishev.

A family or sequence of random variables $X(t)$ depending on an index set, say Θ, denoted by $\{X(t), t \in \Theta\}$, is called a ***stochastic*** or ***random process***. Hence, a stochastic process models evolution of a random system in time, discretely or continuously. As the process depends on both Ω and Θ, it is sometimes denoted by $\{X(t, \omega), t \in \Theta, \omega \in \Omega\}$. For each fixed $\omega \in \Omega$, the random variable $X(t, \omega)$ is a function of t and it is called a ***sample function***. However, for each $t \in \Theta$, $X(t, \omega)$ is just an ordinary random variable, and hence we ignore ω and simply use the notation $\{X(t), t \in \Theta\}$.

Depending on whether Θ is countable (usually $\mathbb{N}$ or $\mathbb{Z}$) or an interval of $\mathbb{R}$ (often $(0, \infty)$ or $\mathbb{R}$), the process is called ***discrete-time*** (or ***discrete-parameter***) or ***continuous-time*** (or ***continuous-parameter***) process, respectively. For a discrete-time process, if the index set is the set of nonnegative integers, the process is written as $\{X_n, n = 0, 1, 2, \dots\}$, while for the continuous-time one, it is written as $\{X(t), t \geq 0\}$.

For a stochastic process, the set of all possible values of the random variable $X(t)$, denoted by S, is called the ***state space of the process*** (sometimes called the ***sample space of the process***). The state space may be countably finite or infinitely countable or uncountable. Thus, in terms of discrete-/continuous-time dichotomy, we have all four cases: discrete time, discrete space; discrete time, continuous space; continuous time, discrete space; and continuous time, continuous space.

The starting value of the process is called the ***initial state***. If the process starts at $t = 0$, then the initial state is $X(0)$.

Any set of instances of $\{X_n, n = 0, 1, 2, \dots\}$ or $\{X(t), t \geq 0\}$ can be regarded as a ***path*** of a certain random move in the state space, S, its position at the epoch n being X_n or time t being $X(t)$. These paths are called ***sample paths or realization of the stochastic process***.

Example 2.1.1 Discrete-Time Process Consider a game of moving among three numbers 1, 2, and 3 in such a way that probability of moving from 1 to 2 is 1; from 3 to either 1 or 2 is 1/2; from 2 to 3 is 1/3, and to stay put at 2 is 2/3. Thus, we have a discrete-time process $\{X_n, n = 0, 1, 2, \dots\}$ with a finite state space $S = \{1, 2, 3\}$. We can display the probability distribution in a matrix form, say $\mathbf{Q}$, as follows:

$$\mathbf{Q} = \begin{array}{c} \\ 1 \\ 2 \\ 3 \end{array} \begin{array}{c} \begin{array}{ccc} 1 & 2 & 3 \end{array} \\ \begin{pmatrix} 0 & 1 & 0 \\ 0 & 2/3 & 1/3 \\ 1/2 & 1/2 & 0 \end{pmatrix} \end{array} . \tag{2.1}$$

We are interested in finding the probability that, after the third move, the process will be at state 2. The answer can be obtained by raising $\mathbf{Q}$ to the third power and choosing the second Column. That is, (0.6111, 0.6852, 0.6389) from matrix $\mathbf{Q}^3$, where

$$\mathbf{Q}^3 = \begin{array}{c} \\ 1 \\ 2 \\ 3 \end{array} \begin{array}{c} \begin{array}{ccc} 1 & 2 & 3 \end{array} \\ \begin{pmatrix} 0.1667 & 0.6111 & 0.2222 \\ 0.1111 & 0.6852 & 0.2037 \\ 0.0833 & 0.6389 & 0.2778 \end{pmatrix} \end{array} . \tag{2.2}$$

The matrix in 2.2 can be used to find similar probabilities. For instance, the probability of starting the game at any state and be at state 1 after three moves will be found by the first column. The probabilities of being at state 3 are given by the third column.

Example 2.1.2 Continuous-Time Process Consider passengers arriving, one at a time, at an airline pavilion in an airport. We are interested in measuring the interarrival times of the passengers. Let us denote the interarrival times of the passengers by positive random variables $\tau_1, \tau_2, \ldots$. That is, assuming the process starts at time $t = 0$, τ_1 is the time of the first arrival (it is the time between the start and the arrival of the first passenger), τ_2 is the time between the first and second arrivals, and so on. Hence, the time it takes for k passengers to arrive, denoted by T_k, will be the sum of the k random variables $\tau_1, \tau_2, \ldots, \tau_k$, that is,

$$T_k \equiv \tau_1 + \tau_2 + \cdots + \tau_k. \tag{2.3}$$

Let us denote the number of arrivals in the time interval $[0, t]$ by $X(t)$. Hence, we represent k (a nonnegative integer) arrivals by time t as $X(t) = k$ if and only if $T_k < t < T_{k+1}$, for each t in $[0, \infty)$, and $X(t) = 0$, if $t < \tau_1$. Hence, for each $t \geq 0$, $X(t)$ is a random variable whose value is a nonnegative integer. Therefore, $\{X(t), t \geq 0\}$ is a continuous-time process with nonnegative integral values state space. The graph of this process is nondecreasing and right continuous, where its increase is by unit jumps at epochs $\tau_1, \tau_2, \ldots, \tau_k$.

Definition 2.1.1 We say a process $\{X(t), t \geq 0\}$ with index set Θ has ***stationary increments*** if for $s, t \in \Theta$ with $s \leq t$, the increment $X(t) - X(s)$ has the same distribution as $X(t - s)$. We also say that the process $\{X(t), t \geq 0\}$ has ***stationary independent increments*** if for $t_1, t_2, \ldots, t_n \in \Theta$ with $t_1 < t_2 < \cdots < t_n$, the increments $X(t_1), X(t_2) - X(t_1), \ldots, X(t_n) - X(t_{n-1})$ are independent. In discrete-time case, we can choose $\Theta = \mathbb{N}$ and make slight changes accordingly. For the process $\{X(t), t \geq 0\}$, if the sequence of increments is a sequence of *iid* random variables, then the random point process $\{t_1, t_2, \ldots\}$ is called a ***recurrent point process***.

Let $\tau_i = t_i - t_{i-1}$, $i = 1, 2, \ldots, t_0 = 0$. When random variables τ_i, $i = 1, 2, \ldots$, are nonnegative, the recurrent point process $\{\tau_1, \tau_2, \ldots\}$ is called an *ordinary* ***renewal process***. The intervals τ_i, $i = 1, 2, \ldots$, are called ***renewal periods*** or ***renewal cycle length***. If we further assume that $F_1(t) \equiv P\{\tau_1 \leq t\}$ and τ_i, $i = 2, 3, \ldots$, are identically distributed as $P\{\tau_i \leq t\} \equiv F(t)$, $i = 2, 3, \ldots, F(t) \neq F_1(t)$. Then, the sequence $\{\tau_1, \tau_2, \ldots\}$ is called a ***delayed renewal process***. The random time point at which the nth renewal occur is given by

$$T_n = \sum_{i=1}^{n} \tau_i, \quad n = 1, 2, \ldots. \tag{2.4}$$

The random point process $\{T_1, T_2, \ldots\}$ is called the ***time points of the renewal process***.

Consider a sequence of real variable $\{t_n, n \in \mathbb{N}, t_1 \geq 0\}$ and let $X(t) = \max\{n, t_n \leq t\}$ be the random number of events occurring in the interval $(0, t]$. A continuous-time stochastic process $\{X(t), t \geq 0\}$ with state space $S = \{0, 1, 2, \dots\}$ is called the ***random counting process*** belonging to the random point process $\{t_1, t_2, \dots, t_n, \dots\}$ if it satisfies the following three properties:

(i) $X(0) = 0$;
(ii) $X(t_1) \leq X(t_2)$, for $t_1 \leq t_2$; and
(iii) for any t_1 and t_2 such that $0 \leq t_1 \leq t_2$, the number of events occurred in $(t_1, t_2]$, denoted by $X(t_1, t_2)$, is equal to the increment $X(t_2) - X(t_1)$.

Note that $X(t) = X(0, t)$ and also that every stochastic process $\{X(t), t \geq 0\}$ in continuous-time exhibiting properties i–iii is the counting process of a certain point process $\{t_1, t_2, \dots, t_n, \dots\}$. Thus, statistically speaking, the stochastic processes $\{t_1, t_2, \dots, t_n, \dots\}$ and $\{X(t), t \geq 0\}$ are equivalent.

Example 2.1.3 Continuous-Time Poisson Process A continuous-time stochastic process $\{N(t), t \geq 0\}$ defined on a probability space $(\Omega, \mathfrak{B}, P)$ is called a ***continuous-time Poisson process*** with parameter λ, $\lambda > 0$, if it satisfies the following properties:

1. $N(0) = 0$,
2. for any n, $n \geq 1$, and any $0 \leq \tau_1 < \tau_2 < \cdots < \tau_n$, the increments $N(\tau_n) - N(\tau_{n-1}), \dots, N(\tau_2) - N(\tau_1)$, are independent random variables, and
3. for any $0 \leq s < t$, the increment $N(t) - N(s)$ has a Poisson distribution with parameter $\lambda(t - s)$, that is, for a fixed constant λ, $\lambda > 0$, we have

$$P\{N(t) - N(s) = k\} = \frac{[\lambda(t-s)]^k}{k!} e^{-\lambda(t-s)}, \quad k = 0, 1, 2, \dots . \tag{2.5}$$

In perspective of the terminology of Example 2.1.2, a Poisson process may be constructed as below. Let $\{\tau_n, n \geq 1\}$ be a sequence of random variables representing interarrival times with exponential pdf of parameter λ. For $n \geq 1$, let us also define T_n as in Equation (2.3). That is, we set $T_0 = 0$ and $T_n = \tau_1 + \tau_2 + \cdots + \tau_n$, for $n \geq 1$. In fact, $\lim_{t\to\infty} T_n = \infty$ and because of the strong law of large numbers, we have

$$\lim_{n\to\infty} \frac{T_n}{n} = \frac{1}{\lambda}. \tag{2.6}$$

As defined earlier, $N(t)$ is the random variable representing the sum of the number of arrivals by time t, that is

$$N(t) = \sum_{n=1}^{\infty} n I_{T_n < t < T_{n+1}}, \tag{2.7}$$

where I_A is the indicator function of A. Then, the stochastic process $\{N(t), t \geq 1\}$, the arrival process associated with the interarrival times τ_n, is a Poisson process with parameter λ, $\lambda > 0$. We leave the proof of the last statement as an exercise for the reader. We will return to this example later in this chapter for a pure birth case.

We end this section by introducing some other very important processes.

Definition 2.1.2 Suppose that events occur at epochs $\tau_1, \tau_2, \ldots, \tau_n, \ldots$ in $[0, \infty)$. Assume that the time differences $\tau_n - \tau_{n-1}$, $n = 1, 2, \ldots$, $\tau_0 = 0$, are independent positive random variables with distribution functions

$$\begin{cases} \widehat{F}(x) = P\{\tau_1 \leq x\}, \\ F(x) = P\{\tau_n - \tau_{n-1} \leq x\}, n = 2, 3, \ldots. \end{cases} \tag{2.8}$$

Let us denote the number of events that occur in (0,t] by $\upsilon(t)$. Thus, $\upsilon(0) = 0$. Then, the process $\{\upsilon(t), 0 \leq t \leq \infty\}$ is called a ***recurrent process***. If $F(x) = \widehat{F}(x)$, then the recurrent process $\{\upsilon(t), 0 \leq t < \infty\}$ is called the ***Palm process***. In addition, if for a positive integer m and a positive parameter λ,

$$F(x) = \widehat{F}(x) = \begin{cases} 1 - \sum_{n=0}^{m} e^{-\lambda x \frac{(\lambda x)^n}{n!}}, & x \geq 0, \\ 0, & x < 0, \end{cases} \tag{2.9}$$

then, $\{\upsilon(t), 0 \leq t < \infty\}$ is called the ***Erlang process***. In particular, if $m = 1$, then the Erlang process reduces to the ***exponential process***.

If the mean of $F(x)$, denoted by μ, is finite, then let us define

$$F^*(x) = \begin{cases} \dfrac{1}{\mu} \displaystyle\int_0^{\infty} [1 - F(y)]dy, & x \geq 0, \\ 0, & x < 0, \end{cases} \tag{2.10}$$

If $F^*(x) = \widehat{F}(x)$, then the recurrent process $\{\upsilon(t), 0 \leq t < \infty\}$ is called ***homogeneous recurrent process***.

2.2 MARKOV CHAIN

So far, we have dealt with independent events and independent random variables. In other words, knowledge of the outcome of an experiment at a time does not affect the prediction of the outcome of the next experiment. However, not all random experiments (trials) and random variables are independent. We now know that when a sequence of chance experiments is conducted independently, the possible outcomes for each experiment are the same, and they occur with the same probabilities.

Andrey Markov (1856–1922)

However, if the assumption of independence is relaxed, the probabilities vary from one experiment to the next. In other words, the knowledge of previous outcomes influences prediction for future experiments' outcomes.

This idea was first studied in 1907 by Andrey (Andrei) A. Markov (Markoff), a Russian probabilist, who contributed much to the theory of stochastic processes, and hence a process of this type carries his name as Markov chain and Markov process, for discrete and continuous cases, respectively.

Markov processes are used in important random systems such as transportation networks, queueing systems, communication systems, image segmentation and analysis, biological systems, and DNA science analysis.

Consider a discrete-time stochastic process or a ***chain*** $\{X_n, n = 0, 1, 2, \ldots\}$ with state space $S = \{x_1, \ldots, x_n\}$, $n = 0, 1, 2, \ldots$. In other words, the process starts from a state in S and moves successively from one state to another, including itself. Each move is called a ***step***. The probability of moving from state x_i to state x_j in one step is denoted by $p_{i,j}$. According to this notation, the probability of the chain stay put in one step will be $p_{i,i}$. As a convention, we may refer to states as i and j, where $i,j = 0, 1, 2, \ldots$. The probabilities $p_{i,j}$ are called ***transition probabilities*** of the chain.

The $p_{i,j}$ denotes the probability of the chain being in its current state x_j, which does not depend on the previous state of the chain. For instance, economic mobility in a society may be modeled in this way. This is because, an individual's chance of being in a certain economic class only depends on his/her parents' economic class. Therefore, the chain exhibits the following property:

$$P\{X_n = x_n | X_{n-1} = x_{n-1}, \ldots, X_1 = x_1, X_0 = x_0\} = P\{X_n = x_n | X_{n-1} = x_{n-1}\}. \quad (2.11)$$

The right-hand side of Equation (2.11) is called ***one-step transition probability*** of moving from state x_{n-1} to state x_n, that is, $p_{n-1,n}$, where

$$p_{n-1,n} = P\{X_n = x_n | X_{n-1} = x_{n-1}\}, \text{ if } P\{x_{n-1}\} > 0. \quad (2.12)$$

Property (2.11) is called ***forgetful property***, ***forgetfulness property***, or ***Markov property***. This is because, essentially, a chain with Markov property "forgets" the history except its previous state. That is, if the left-hand side of Equation (2.11) exists, then it shows the dependence between the random variables X_n, $n = 0, 1, \ldots$ and implies that given the present state of the chain, the future is independent of the past. Hence, we have the following definition:

A discrete-time stochastic process $\{X_n, n = 0, 1, 2, \ldots\}$ with finite or infinite state space S is called a ***Markov chain*** with stationary transition probabilities if it satisfies the forgetful property (Equation 2.11). If the state space is finite, the chain is called a ***finite state Markov chain*** or a ***finite Markov chain***. A Markov chain is called ***homogenous*** if

$$P\{X_n = j|X_{n-1} = i\} = P\{X_1 = j|X_0 = i\}, \text{ for all } n, i, j, \tag{2.13}$$

That is, transitions do not depend on time. In this case, Equation (2.13) defines the transition probabilities of a homogeneous Markov chain and are denoted by $p_{i,j}$.

In general, an n-step transition from state x_i to state x_j is denoted by $p_{i,j}^{(n)}$, $n = 1, 2, \ldots, i, j = 0, 1, 2, \ldots . p_{i,i}^{(n)}$, and indicates probability of return to state i after n transitions, not necessarily the first return. It is assumed that $p_{i,i}^{(0)} = 1$.

Transition probabilities for a Markov chain may be demonstrated by a matrix, called ***transition matrix***. A one-step time-independent (stationary) transition matrix is denoted by $\mathbf{P}$ and n-step one is denoted by $\mathbf{P}^{(n)}$, for $n = 1, 2, \ldots$, with $\mathbf{P}^{(1)} \equiv \mathbf{P}$. Thus, elements of $\mathbf{P}$ and $\mathbf{P}^{(n)}$ are one-step and n-step transition probabilities, respectively, as shown below:

$$\mathbf{P} = \begin{pmatrix} p_{0,0} & p_{0,1} & p_{0,2} & \cdots & p_{0,j} & \cdots \\ p_{1,0} & p_{1,1} & p_{1,2} & \cdots & p_{1,j} & \cdots \\ \vdots & \vdots & \vdots & \cdots & \vdots & \cdots \\ p_{i,0} & p_{i,1} & p_{i,2} & \cdots & p_{i,j} & \cdots \\ \vdots & \vdots & \vdots & \cdots & \vdots & \ddots \end{pmatrix} \tag{2.14}$$

and

$$\mathbf{P}^{(n)} = \begin{pmatrix} p_{0,0}^{(n)} & p_{0,1}^{(n)} & p_{0,2}^{(n)} & \cdots & p_{0,j}^{(n)} & \cdots \\ p_{1,0}^{(n)} & p_{1,1}^{(n)} & p_{1,2}^{(n)} & \cdots & p_{1,j}^{(n)} & \cdots \\ \vdots & \vdots & \vdots & \cdots & \vdots & \cdots \\ p_{i,0}^{(n)} & p_{i,1}^{(n)} & p_{i,2}^{(n)} & \cdots & p_{i,j}^{(n)} & \cdots \\ \vdots & \vdots & \vdots & \cdots & \vdots & \ddots \end{pmatrix} \tag{2.15}$$

When a transition matrix, say $\mathbf{P}$, with state space S exhibits two properties

1. $0 \leq p_{i,j} \leq 1$ and
2. $\sum_{j \in S} p_{i,j} = 1, \forall i \in S$, (i.e., sum of each row is 1),

it is called a ***stochastic matrix***. If the sum of each column is also 1, then it is called a ***doubly stochastic matrix***.

Elements of **P** given in Equation (2.14) are denoted by $(p_{i,j})$, that is, $\mathbf{P} = (p_{i,j})$, and of $\mathbf{P}^{(n)}$ given in Equation (2.15) are denoted by $(p_{i,j}^{(n)})$. Elements of $\mathbf{P}^{(n)}$ may be interpreted as

$$p_{i,j}^{(n)} = P\{X_{n+m} = j | X_m = i\}, \quad n = 1, 2, \ldots. \tag{2.16}$$

It should be noted that

P(transition from i to j in n steps) = sum of probabilities of all paths from i to j in n steps,

or

$$p_{i,j}^{(n+m)} = \sum_k p_{i,k}^{(m)} p_{k,j}^{(n)}. \tag{2.17}$$

In terms of transition matrix, Relation (2.17) can be rewritten as:

$$\mathbf{P}^{(m+n)} = \mathbf{P}^{(m)} \cdot \mathbf{P}^{(n)}. \tag{2.18}$$

In particular, if $m = 0$, then we will have

$$\mathbf{P}^{(n)} = \mathbf{I} \cdot \mathbf{P}^{(n)} = \mathbf{P}^n. \tag{2.19}$$

Thus, we have proved the following theorem.

Theorem 2.2.1 For a Markov chain with state space $S = \{x_0, x_1, \ldots\}$ and transition matrix **P** as defined in Equation (2.14), the (i, j)th entry of matrix $\mathbf{P}^n$, $n = 0, 1, 2, \ldots$, will be the probability that the Markov chain starting in state x_i will be in state x_j after n steps, with $\mathbf{P}^0 = \mathbf{I}$, where **I** is the identity matrix. That is,

$$\mathbf{P}^{(n)} = \mathbf{P}^n, \quad m = 0, 1, 2, \ldots. \tag{2.20}$$

The limiting value of a power matrix is denoted by π and it indicates that the chain has achieved its ***probabilistic or stochastic equilibrium*** or it has achieved ***stationary***. In other words, the probability that the chain will be in state j after running for a long time is π.

The stochastic equilibrium of a Markov chain, π, ***may also be interpreted as the proportion of time the chain spends in the state j after running for a long time***. Hence, let I_n be the indicator random variable defined as

$$I_n = \begin{cases} 1, & \text{if } X_n = j, \\ 0, & \text{if } X_n \neq j. \end{cases} \tag{2.21}$$

Thus, the number of visits to state j in the first k transitions is $\sum_{n=1}^{k} I_n$. Keeping in mind the properties of the indicator random variable, starting with state i, we will have the following:

$$\begin{aligned} E\left(\sum_{n=1}^{k} I_n|X_0 = i\right) &= \sum_{n=1}^{k} E(I_n|X_0 = i) \\ &= \sum_{n=1}^{k} [1 \cdot P(I_n = 1|X_0 = i) + 0 \cdot P(I_n = 0|X_0 = i) \\ &= \sum_{n=1}^{k} P(X_n = j|X_0 = i) = \sum_{n=1}^{k} p_{i,j}^{(n)}, \end{aligned}$$

which leads to the expected proportion of time in state j, $\frac{\sum_{i=1}^{k} p_{i,j}^{(n)}}{k}$. In other words,

$$\lim_{k\to\infty} \frac{1}{k} \sum_{i=1}^{k} p_{i,j}^{(n)} = \pi_j. \tag{2.22}$$

Thus, when a Markov chain is in stationary case, then

$$\lim_{n\to\infty} p_{i,j}^{(n)} = \pi_j, \tag{2.23}$$

where π_j is the liming probability of state j. Hence,

$$p_{i,j}^{(n)} = \sum_{k} p_{i,k}^{(n-1)} p_{k,j}, \quad \forall n. \tag{2.24}$$

As n approaches infinity, we obtain

$$\pi_j = \sum_{k} \pi_k p_{k,j}, \quad \pi_j \geq 0 \forall j \in S \text{ and } \sum_{j} \pi_j = 1, \tag{2.25}$$

or, in matrix form, we have

$$\pi = \pi \mathrm{P}. \tag{2.26}$$

Thus, ***the stationary distribution for a Markov chain uniquely exists only if all the three following conditions are satisfied***:

$$\text{(a)}\, \pi = \pi \mathrm{P}, \quad \text{(b)}\, \pi_j \geq 0, \quad \text{and} \quad \text{(c)} \sum_{j} \pi_j = 1. \tag{2.27}$$

It should be noted that satisfying conditions stated in Equation (2.27) are not sufficient for the Markov chain to possess a limiting distribution. We also note that the chain is in the stationary case, because if the initial distribution is $\boldsymbol{\delta}^{(0)} = \boldsymbol{\pi}$, then $\boldsymbol{\delta}^{(n)} = \boldsymbol{\pi}$, as well. We further note that $\lim_n \sum_j p_{i,j}^{(n)}$ may not be equal to $\sum_j \lim_n p_{i,j}^{(n)}$, and hence

$$\sum_j p_{i,j}^{(n)} = 1 \text{ does not necessary apply that } \sum_j \pi_{i,j} = 1. \tag{2.28}$$

Example 2.2.1 Stationary Distribution Consider a transition probability matrix

$$\mathbf{P} = \begin{matrix} & \begin{matrix} 1 & 2 & 3 \end{matrix} \\ \begin{matrix} 1 \\ 2 \\ 3 \end{matrix} & \begin{pmatrix} 1/2 & 1/4 & 1/4 \\ 1/2 & 0 & 1/2 \\ 1/4 & 1/4 & 1/2 \end{pmatrix} \end{matrix}. \tag{2.29}$$

Let us find the stationary distribution $\langle \pi_1, \pi_2, \pi_3 \rangle$, with $\pi_1, \pi_2, \pi_3 \geq 0$, and $\pi_1 + \pi_2 + \pi_3 = 1$, that is, solve for elements of the vector $\langle \pi_1, \pi_2, \pi_3 \rangle$.

From Equation (2.27), we can write

$$\langle \pi_1, \pi_2, \pi_3 \rangle = \langle \pi_1, \pi_2, \pi_3 \rangle \begin{pmatrix} 1/2 & 1/4 & 1/4 \\ 1/2 & 0 & 1/2 \\ 1/4 & 1/4 & 1/2 \end{pmatrix}. \tag{2.30}$$

From Equation (2.30) and conditions mentioned in relation 2.29, we have the following system of equations:

$$\begin{cases} \pi_1 = \frac{1}{2}\pi_1 + \frac{1}{2}\pi_2 + \frac{1}{4}\pi_3 \\ \pi_2 = \frac{1}{4}\pi_1 + 0 + \frac{1}{4}\pi_3 \\ \pi_3 = \frac{1}{4}\pi_1 + \frac{1}{2}\pi_2 + \frac{1}{2}\pi_3 \\ \pi_1 + \pi_2 + \pi_3 = 1. \end{cases} \tag{2.31}$$

Solving the system of Eqs. (2.31), we will have

$$\pi_1 = \frac{2}{5}, \quad \pi_2 = \frac{2}{5}, \quad \pi_3 = \frac{1}{5}. \tag{2.32}$$

It is obvious that the third condition mentioned in Equation (2.29) is also satisfied. Hence, $\pi = \left\langle \frac{2}{5}, \frac{2}{5}, \frac{1}{5} \right\rangle$ is the stationary distribution for the chain.

It should be noted that, as we will see later in this book, there are cases of Markov chain where it would be difficult to find explicit form for the stationary probabilities. There are also cases that transition probabilities would be difficult to be defined and/or be used. In most of these cases, the generating function method is used. A well-known process is provided as an example for such a use of probability generating function (pgf).

Example 2.2.2 Discrete-Time Galton–Watson Process (GWP)

Francis Galton (1822–1911)

Henry W. Watson (1827–1903)

The discrete-time Galton–Watson process (GWP) is the earliest known example of the theory of stochastic processes, the branching process. Branching processes have been applied to different models such as population growth, survival of mutant genes, electron multipliers, and neutron chain reactions. The discrete-time GWP is a population size model that is a Markov chain with its state space as a countably infinite set. However, study of the branching process, although it is a part of Markov chain, uses the generating function rather than the transition matrix method.

Historically, in the 1870s, two scientists, Reverend Henry William Watson (a clergyman and mathematician) and Francis Galton (a biometrician), were interested in studying the survival of family names (Mode 1971). In 1873, Galton submitted two questions to the *Educational Times* (Mode 1971). The contents of the problem with assumptions and two questions are as follows (Allen 2003):

Assumptions

1. Adult males in a population each have different last name.
2. In each generation, for adult males, the percent of children who survive to adulthood is c_0 with none, c_1 with one, c_2 with two, c_3 with three, c_4 with four, and c_5 with five.

Questions

1. What is the proportion of last names that become extinct after n generations?
2. How many instances are there of the same last name being held by k persons?

Although complete answers to the questions were provided in the 1930s by mathematicians such as Fisher, Haldane, Erlang, and Steffenson, see Mode (1971), the process carries the Galton–Watson name with it. However, some believe that Bienaymé has contributed to the solution and, hence, refer to the process as Bienaymé–Galton–Watson process. See Schinazi (1999).

We now cast the essence of the problem in a rigorous mathematical formalism. Let X_0 be a random variable representing the total population size of the initial generation. We also denote by $X_1, X_2, \ldots, X_n, \ldots$ the random variables representing the number of male descendants of an individual in the first, second, $\ldots$, and nth generation after him. In other words, X_n is the total size of population of males at the nth generation. Hence, we consider a sequence $\{X_n, n = 0, 1, 2, \ldots\}$ with state space $\{0, 1, \ldots\}$. We assume that each individual in the nth generation will leave Y (a general random variable) sons, that is, in the $(n+1)$st generation. Thus,

$$X_{n+1} = Y_1 + Y_2 + \cdots + Y_{X_n}. \tag{2.33}$$

That is, X_{n+1} is the sum of X_n *iid* random variables Y_k, $k = 1, 2, \ldots, X_n$. We also assume that the number of sons of all individuals in all generations are independent random variables with pmf $\{p_k\}$. In other words, p_k, $k = 0, 1, 2, \ldots$, is the probability that any given individual male in his lineage has k sons, that is,

$$P\{Y = k\} = p_k, \quad k = 0, 1, 2, \ldots. \tag{2.34}$$

The process $\{X_n, n = 0, 1, 2, \ldots\}$ is referred to as a ***branching process***. The probability that a lineage has died in n or fewer generations is the same as the probability that there are no nth generation descendants. In other words, if the total population size of the generation n is zero, that is, $X_n = 0$, then $X_{n+r} = 0$, for $r = 1, 2, \ldots$. In other words, we have a sure one-step transition probability from population 0 to 0, $p_{0,0} = 1$. This is the only transition probability that was easy to write. It is not easy to write one-step transition probabilities for states $2, 3, \ldots$.

Now, let $X_0 = 1$, that is, the total size of population of males at the initial generation is 1. Let us denote the pgf of X_n, $n = 0, 1, 2, \ldots$, by $G_n(z)$, $n = 0, 1, 2, \ldots$. Then, for X_0, we have

$$G_0(z) = z. \tag{2.35}$$

That is, in the next generation, each individual male will have k sons with probability p_k. In addition, pgf of $X_1(= Y_1)$ is

$$G_1(z) = \sum_{k=0}^{\infty} p_k z^k. \tag{2.36}$$

From Equation (2.33), we have

$$X_2 = Y_1 + Y_2 + \cdots + Y_{X_1}, \tag{2.37}$$

that is, each of the X_1 individuals will have Y sons and the sum of these sons is X_2. Hence (see Allen 2003), transition probability of the chain be in the current generation X_1 with population size i and the next generation with population size j. This results in the following relation:

$$P\{X_2 = j \,|\, X_1 = i\} = P\left\{\sum_{k=1}^{i} Y_k = j\right\}. \tag{2.38}$$

From Equation (2.37), X_2 is the sum of X_1 *iid* random variables. Thus, from properties of pgf (Chapter 1), we have

$$G_2(z) = G_1(G_1(z)).$$

Similarly, for Equation (2.33), we have

$$G_{n+1}(z) = G_n(G_1(z)).$$

Hence, as $G_n(z) = G_{n-1}(G_1(z))$, we have

$$G_{n+1}(z) = G_1(G_1(\cdots(G_1(z))\cdots)), \tag{2.39}$$

where $G_1(z)$ is given in Equation (2.36). Thus, the probability that a lineage has died in n or fewer generations is $G_n(0)$, that is, the probability that there are no nth generation descendants.

For the sake of convenience, if we denote $G_1(z)$ by $g(z)$, then $G_n(z)$ is the n-fold composition of $g(z)$, that is,

$$g^n(z) = g(g(\cdots(g(z))\cdots)). \tag{2.40}$$

Derivation of $G_n(z)$, as in Equation (2.40), was based on the assumption of $X_0 = 1$. However, in general, if $X_0 = \nu$, where ν is a natural number, then $G_0(z) = z^\nu$ and the chain begins with ν independent branches. Hence,

$$G^m(z) = |f^n(z)|^\nu, \quad X_0 = \nu \in \mathbb{N}. \tag{2.41}$$

Thus, for $X_0 = 1$, we choose $\{p_k\}$ as (Allen 2003)

$$p_k = \begin{cases} 1/3, & k = 0, \\ 2/3, & k = 1, \\ 0, & k = 2, 3, \ldots. \end{cases} \tag{2.42}$$

Hence, there is either no son or one son. From Eqs. (2.36), (2.42), and (2.40), pgf of X_1 is

$$G_1(z) = g(z) = \frac{1}{3} + \frac{2}{3}z.$$

For X_2, the pgf is

$$G_2(z) = g^2(z) = g(g(z)) = \frac{1}{3}\left(1 + \frac{2}{3}\right) + \left(\frac{2}{3}\right)^2 z.$$

For X_n, the pgf is

$$\begin{aligned} G_n(z) = g^n(z) &= \frac{1}{3}\left(1 + \frac{2}{3} + \cdots + \left(\frac{2}{3}\right)^{n-1}\right) + \left(\frac{2}{3}\right)^n z \\ &= 1 - \left(\frac{2}{3}\right)^n + \left(\frac{2}{3}\right)^n z. \end{aligned}$$

Thus, if we denote the probability that the population size is k in generation n by $q_k(n)$, then we have

$$q_0(n) = 1 - \left(\frac{2}{3}\right)^n \text{ and } q_1(n) = \left(\frac{2}{3}\right)^n. \tag{2.43}$$

It should be noted that when $X_0 = 1$, then $p_k = q_k(1)$, $k = 0, 1, \ldots$.
In general, when $X_0 = \nu \in \mathbb{N}$, we have

$$[g^n(z)]^\nu = \binom{\nu}{0}[q_0(n)]^\nu + \binom{\nu}{1}[q_0(n)]^{\nu-1}q_1(n)z + \cdots + \binom{\nu}{\nu}[q_1(n)]^\nu z^\nu. \tag{2.44}$$

From Equation (2.44), it should be noted that the following items address Galton's question 1 after r generations and just a point about the second question. For more details about GWP, see Allen (2003).

(i) The probability that all last names have disappeared $= [q_0(r)]^\nu$.
(ii) The probability that $\nu - 1$ last names have disappeared $= \nu[q_0(r)]^{\nu-1}q_1(r)$.
(iii) $\vdots$ …
(iv) The probability that no last names have disappeared $= [q_1(r)]^\nu$.
(v) From the above results, it can be seen that the expected proportion of last names that have disappeared by generation r is $q_0(r)$. With distribution given in Equation (2.42), this expected proportion is

$$q_0(r) = 1 - \left(\frac{2}{3}\right)^r. \tag{2.45}$$

Of course, as r approaches infinity, we will have

$$\lim_{r\to\infty} q_0(r) = 1. \tag{2.46}$$

(vi) When $X_0 = 1$, the probability that there are exactly n same last names in generation $r = q_n(r)$.

With distribution given in Equation (2.42), the probability that there are two or more same last names in any generation is zero.

We now define some terminologies that are of use later.

2.2.1 Classification of States

1. A subset C of the space state S is called ***closed*** if $p_{i,j} = 0$, for all $i \in C, j \notin C$. In other words, the chain cannot escape from C. That is, $\sum_{i\in C} p_{i,j} = 1, \forall j \in C$. This indicates that in a Markov chain, a set of states is closed if it is impossible to access any state outside of that class from inside by only one transition.
2. State j is said to be ***accessible*** from state i, denoted by $i \to j$, if $p_{i,j}^{(n)} > 0$, for some $n \geq 1$. In other words, there is a chance of reaching state j from state i in some number of steps. If state j is not accessible from i, then $p_{i,j}^{(n)} = 0$, for all n, that is, the chain starts from state i and never visits state j. Reason for $p_{i,j}^{(n)} = 0$, is as follows:

$$\begin{aligned} P\{\text{ever visits state } j|X_0 = i\} &= P\left\{\bigcup_{n=0}^{\infty}\{X_n = j\}\,|X_0 = i\right\} \\ &\leq \sum_{n=0}^{\infty} P\{X_n = j|X_0 = i\} = 0. \end{aligned}$$

For a finite Markov chain with k states, state j is accessible from state i if and only if

$$p_{i,j}^{(1)} + p_{i,j}^{(2)} + \cdots + p_{i,j}^{(k)} > 0. \tag{2.47}$$

If $p_{i,j}^{(0)} = 1$, then every state is accessible from itself.

3. If state i is accessible from j and state j is accessible from i, then we say that i and j ***communicate***, and it is denoted by $i \leftrightarrow j$. The communication relation is an equivalence relation. In other words, it satisfies the following relations:
   (i) $i \leftrightarrow i$;
   (ii) $i \leftrightarrow j$ implies $j \leftrightarrow i$; and
   (iii) $i \leftrightarrow k$ and $k \leftrightarrow j$ imply $i \to j$.

   In order to prove relation (iii), we assume $i \leftrightarrow k$ and $k \leftrightarrow j$. Hence, there exist nonnegative integers n and m such that $p_{i,k}^{(n)} > 0$ and $p_{k,j}^{(m)} > 0$. In other words, the chain moves from state i to state k in n steps and from state k to state j in m steps or from state i to state j in $n+m$ steps. Hence,

$$p_{i,j}^{n+m} \geq p_{i,k}^{n} p_{k,j}^{m} > 0. \tag{2.48}$$

4. The accessibility property of states divides them into ***classes***. All states communicate among each other within each class. However, no two states from different classes communicate with each other.

   It is important to note that if every state of a Markov chain is accessible from every other state, then there is only one class for the chain. This relation ***partitions*** the state space S into disjoint, but not necessarily closed classes in such a way that two states i and j belong to the same class if and only if they communicate.

5. A Markov chain with only one class is called ***irreducible***. For a finite Markov chain with n states, all transition probabilities are positive.

6. The state space S of a Markov chain can be ***decomposed*** uniquely as

$$S = U \cup C_1 \cup C_2 \cup \cdots, \tag{2.49}$$

   where U is the set of transient states and each C_i, $i = 1, 2, \ldots$, is a closed and ever pare of i and j in C communicate (that is, C is an irreducible subset of S).

7. The probability of a Markov chain starting from state i and returning to state j after n transitions for the ***first time*** is denoted by $f_{i,j}^{(n)}$.

   We note that the difference between $p_{i,j}^{(n)}$ and $f_{i,j}^{(n)}$ is that $p_{i,j}^{(n)}$ is the probability that the chain starting from state i will reach the state j after n steps, but it may have visited state j along the path, while $f_{i,j}^{(n)}$ is the probability that the chain visits j for the first time. By the law of total probability, it can be shown that a relation between $p_{i,j}^{(n)}$ and $f_{i,j}^{(n)}$ is

$$p_{i,j}^{(n)} = \sum_{k=1}^{n} f_{i,j}^{(k)} p_{i,j}^{(n-k)}, \text{ with } p_{j,j}^{(0)} = 1, \quad \forall j \in S. \tag{2.50}$$

   Thus, ***the first passage time*** may be calculated, recursively, from the following:

$$f_{i,j}^{(n)} = p_{i,j}^{(n)} - \sum_{k=1}^{n-1} f_{i,j}^{(k)} p_{i,j}^{(n-k)}, \quad n = 2, 3, \ldots. \tag{2.51}$$

   We will discuss the first passage time in Section 2.4 with random walk.

8. Probability of a Markov chain starting from state i returning to state i after n transitions for the ***first time*** is denoted by $f_i^{(n)}$ with $f_i^{(0)} = 0$, by convention.

9. As n approaches infinity, the n from $f_i^{(n)}$ is dropped and f_i is to indicate the ***probability of eventual return*** to the state i. Hence,

$$f_i = f_i^{(1)} + f_i^{(2)} + \cdots + f_i^{(n)} + \cdots. \tag{2.52}$$

In other words, f_i is the probability that state i ever returns to itself, that is

$$f_i = P\{\text{ever return to state } i \,|\, X_0 = i\}. \tag{2.53}$$

10. If $f_i = 1$, the state i is called ***recurrent*** and if $f_i < 1$, it is called ***transient***. In any class, all states are either recurrent or transient. In addition, if a chain is irreducible, then all states are either recurrent or transient. It can be shown that state i is recurrent if and only if

$$\sum_{n=1}^{\infty} p_{i,i}^{(n)} = \infty. \tag{2.54}$$

It can also be proved that state i is transient if and only if

$$\sum_{n=1}^{\infty} p_{i,i}^{(n)} < \infty. \tag{2.55}$$

We note that starting from a state, a Markov chain visits a recurrent state infinitely many times or not at all. It can also be proved that two states in the same communicating class are either both recurrent or both transient. It can be further shown that the expected number of visits to state i if the chain started with that state is

$$E(\text{number of visits to state } i \,|\, X_0 = i) = \frac{1}{1 - f_i}. \tag{2.56}$$

11. Let the greatest common divisor (gcd) of the times n for which $p_{i,i}^{(n)}$ is positive be denoted by d_i. That is

$$d_i = gcd\{n \in \mathbb{N} \,:\, p_{i,i}^{(n)} > 0\}. \tag{2.57}$$

A state i is called ***periodic*** with period d_i, if d_i, as defined in Equation (2.57), is greater than one; that is, if it can be revisited only at multiples of an integer greater than one. If $d_i = 1$, the state i is called ***aperiodic***.

It is important to note that for a Markov chain with countable state space S and transition matrix $\mathbf{P}$, the ***period*** of a state i, $i \in S$, is d_i., and that an irreducible Markov chain is either periodic or aperiodic. Furthermore, it should be noted that every finite irreducible Markov chain has a unique stationary distribution, and if the chain is aperiodic, then

$$\lim_{n\to\infty} p_{i,i}^{(n)} = \pi_i, \text{ for each } i. \tag{2.58}$$

12. State i for which $p_{i,i} = 1$ (so that $p_{i,j} = 0$, for $j \neq i$, in other words, P (leaving an absorbing state) = 0) is called an ***absorbing state*** (or an ***absorbing barrier***).

    It is important to note that if a state is a nonabsorbing one, it is a transient state.

13. Suppose θ_i denote the time of the first return to state i. Then, $\{f_i^{(n)}; n \geq 1\}$ is the distribution of θ_i. The mean of the first return time to state i or the ***mean recurrence time*** of state i is denoted by μ_i. A recurrent state which has an infinite mean recurrence time is called a ***null recurrent*** state; if mean recurrence time is finite, the state is called ***positive recurrent*** state.

    It can be shown that every finite closed class is positive recurrent. It can also be shown that every finite Markov chain is positive recurrent. We also note that if i and j communicate, then i and j are either both transient, null recurrent, and positive recurrent or neither. Furthermore, if i and j communicate, then both have the same period.

Example 2.2.3 Irreducible Markov Chain Let us consider a Markov chain with three states 0, 1, and 2, with the transition matrix **P** given by

$$\mathbf{P} = \begin{matrix} & \begin{matrix} 0 & 1 & 2 \end{matrix} \\ \begin{matrix} 0 \\ 1 \\ 2 \end{matrix} & \begin{pmatrix} \frac{1}{2} & \frac{1}{2} & 0 \\ \frac{1}{2} & \frac{1}{4} & \frac{1}{4} \\ 0 & \frac{1}{3} & \frac{2}{3} \end{pmatrix} \end{matrix}. \tag{2.59}$$

The chain is irreducible because $0 \leftrightarrow 1$ and $1 \leftrightarrow 2$ in one or more steps.

Example 2.2.4 Irreducible Markov Chain Let us consider a Markov chain with four states 0, 1, 2, and 3, with the transition matrix **P** given by

$$\mathbf{P} = \begin{matrix} & \begin{matrix} 0 & 1 & 2 & 3 \end{matrix} \\ \begin{matrix} 0 \\ 1 \\ 2 \\ 3 \end{matrix} & \begin{pmatrix} 1/2 & 1/2 & 0 & 0 \\ 1/2 & 1/2 & 0 & 0 \\ 0 & 0 & 1/4 & 3/4 \\ 0 & 0 & 0 & 1 \end{pmatrix} \end{matrix}. \tag{2.60}$$

This chain is not irreducible, because the transition matrix in (2.60) divides into three classes $\{0, 1\}$,$\{2\}$, and $\{3\}$.

Example 2.2.5 Recurrent States Consider Example 2.2.4. Then, state 3 is a recurrent one since $f_3 = 0$. On the other hand, state 2 is transient since $f_2 = 1/4$.

Example 2.2.6 Recurrent State Suppose for a Markov chain with two states 0 and 1, the transition matrix **P** is given by

$$\mathbf{P} = \begin{matrix} & \begin{matrix} 0 & 1 \end{matrix} \\ \begin{matrix} 0 \\ 1 \end{matrix} & \begin{pmatrix} 1-\alpha & \alpha \\ \beta & 1-\beta \end{pmatrix} \end{matrix}. \tag{2.61}$$

Let us find the ways visiting state 0 for the first time after $1, 2, \ldots,$ or n steps. From Equation (2.20), we can see that to visit state 0 starting from state 0 is the first entry of Equation (2.52), that is, $f_0^{(1)} = 1 - \alpha$. After two steps, the only possible way is the path 0 to 1 in the first step and from 1 to 0 in the second step, that is, $f_0^{(2)} = \alpha\beta$. For $n \geq 3$, the first return to state 0 happens after n transitions if (1) the first transition is from state 0 to 1, (2) the next $n-2$ transitions are from state 1 to 1, and (3) the nth transition is from state 1 to 0. The path in this case is $0 \xrightarrow{\text{step } 1} 1 \xrightarrow{\text{step } 2,3,\ldots,n-1} 1 \xrightarrow{\text{step } n} 0$. Hence, we have

$$f_0^{(n)} = \alpha(1-\beta)^{n-2}\beta. \tag{2.62}$$

As n approaches infinity, we have

$$\begin{aligned} f_0 = \sum_{n=0}^{\infty} f_i^{(n)} &= 1 - \alpha + \sum_{n=2}^{\infty} \alpha(1-\beta)^{n-2}\beta \\ &= 1 - \alpha + \alpha\beta \sum_{m=0}^{\infty} (1-\beta)^m \\ &= 1 - \alpha + \frac{\alpha\beta}{1-(1-\beta)} = 1. \end{aligned} \tag{2.63}$$

Relation (2.63) shows that state 0 is a recurrent state.

Theorem 2.2.2 If a finite Markov chain with state space $S = \{0, 1, 2, \ldots, k\}$ is aperiodic, then for all states i and j, we have

$$\lim_{n\to\infty} p_{i,j}^{(n)} = \pi_j, \tag{2.64}$$

where $\boldsymbol{\pi} = \langle \pi_0, \pi_1, \ldots \pi_k \rangle$ is the unique stationary distribution of the chain.

The basic limit theorem for Markov chain can be stated as follows:

Theorem 2.2.3 For each state i of a recurrent, irreducible, and aperiodic Markov chain, we have

$$\lim_{n\to\infty} p_{i,i}^{(n)} = \frac{1}{\sum_{n=0}^{\infty} n f_i^{(n)}} = \frac{1}{\mu_i}. \tag{2.65}$$

From Equation (2.58), we can see that for a finite irreducible aperiodic Markov chain, we have

$$\pi_i = \frac{1}{\mu_i}. \tag{2.66}$$

Hence, the distribution is stationary and we can find the mean return times from Equation (2.66).

Example 2.2.7 Stationary Distribution Let us go back to Example 2.2.6 with $\alpha, \beta \neq 0$ or 1. Then, the stationary distribution is

$$\boldsymbol{\pi} = \langle \pi_0, \pi_1 \rangle = \left\langle \frac{\beta}{\alpha + \beta}, \frac{\alpha}{\alpha + \beta}, \right\rangle. \tag{2.67}$$

Hence,

$$\mu_0 = \frac{\alpha + \beta}{\beta} \text{ and } \mu_1 = \frac{\alpha + \beta}{\alpha}. \tag{2.68}$$

2.3 MARKOV PROCESS

In order to discuss the Markov process, we simply extend the notion of Markov chain to that of continuous time with discrete- and continuous-state spaces.

2.3.1 Markov Process with Discrete Space State

Thus, we start with the memoryless or Markov property of Markov chain (Equation 2.11) for continuous-time process that is as follows:

$$P\{X(s+t) = j | X(s) = i, X(s_n) = i_n, \ldots, X(s_1) = i_1\} = P\{X(s+t) = j | X(s) = i\}$$
$$\text{for all } t > 0,\ s > s_n > \cdots > s_1 \geq 0 \text{ and } i, j, j_k \in S. \tag{2.69}$$

The memoryless property for the process indicates that once the process arrives at a certain state, the future behavior is always the same, regardless of how the process arrived in the state.

Definition 2.3.1 A continuous-time stochastic process $\{X(t), t \geq 0\}$ with finite or infinite index set, say $T = [0, \infty)$ or a subset of it, with the set of values of $X(t)$ in a state space S (still discrete) is called a ***Markov Process*** if it satisfies the Markov property (Equation 2.69).

In summary, a Markov process is characterized by a random variable $X(t)$, indexed over time t and a state space S such that $X(t) \in S$, for all t. There are three ways to verify a stochastic process is a Marko process. They are as follows: (1) show that the Markov property holds; (2) show that the process possesses independent increments; and (3) show that the process is a function of a Markov process.

Transition probability of moving from state i to state j, in this case, in the time interval t, denoted by $p_{i,j}(t)$, is defined by

$$p_{i,j}(t) = P\{X(t) = j \,|\, X(0) = i\}. \tag{2.70}$$

The parameter $p_{i,j}(t)$ is referred to as the ***transition probabilities of moving from state i to state j at time t***. In general, this transition probability within the interval of length s is defined as follows:

$$p_{i,j}(s, t+s) = P\{X(t+s) = j|X(s) = i\}. \tag{2.71}$$

The parameter $p_{i,j}(t)$ defined in Equation (2.70) constitutes the general element for the ***transition probability matrix***, denoted by $\mathbf{P}(t)$, for the Markov process. In other words,

$$\mathbf{P}(t) = (p_{i,j}(t)), \quad t \geq 0, \tag{2.72}$$

where

$$p_{i,j}(0) = \begin{cases} 1, & i = j, \\ 0, & i \neq j, \end{cases} \tag{2.73}$$

that is,

$$\mathbf{P}(0) = \mathbf{I}, \tag{2.74}$$

where $\mathbf{I}$ is the identity matrix.

We note that the distinction between discrete- and continuous-time Markov process is that in the former, there is a "jump" to a new state at times $1, 2, \ldots$, while in the latter, a new state may be entered at any time $t \geq 0$. Hence, most terminologies and properties of a Markov chain are carried for a Markov process, considering this distinction.

A Markov process is called ***time-homogeneous*** (or if there is no danger of confusion, just ***homogeneous***) if for $0 \leq s \leq t$,

$$P\{X(t) = j\,|\,X(s) = i\} = P\{X(t-s) = j\,|\,X(0) = i\}. \tag{2.75}$$

That is, the behavior of the process does not depend on when it is observed. In particular, the transitions between states are independent of the time at which the transitions occur.

It can be proved that if the process $\{X(t), t \geq 0\}$ has stationary and independent increments, then it is a homogeneous Markov process.

Thus, if the process visits a state i in the space state at time t and the next state will be visited at time $t + \theta$, then θ is referred to as the ***holding time in state i***. However, by the Markov property, at any time epoch τ, the holding time (the time until the next state change) is independent of the previous holding time. That is, holding times are memoryless. However, we already know that the only distribution that has the memoryless property is the exponential distribution. Hence, the holding time in any state of a Markov process is an exponentially distributed random variable.

Thus, at any time epoch τ, the probability that there is a transition in the interval $(\tau, \tau + \Delta t)$, that is of length Δt, is $\lambda_i \Delta t + o(\Delta t)$, where λ_i is the distribution parameter in state i. From Equation (2.70), if the transition from state i to state j occurs within a time interval Δt with probability $p_{i,j}$, then, by the Markov property, this transition should only depend on i and j. Hence, for $i, j \in S$, $i \neq j$, we have

$$P\{X(t + \Delta t) = j\,|\,X(t) = i\} = \lambda_i p_{i,j} \Delta t + o(\Delta t). \tag{2.76}$$

The terms $\lambda_i p_{i,j}$ are referred to as ***instantaneous transition rates***. If the mean holding time between a transition from i to j is exponentially distributed with parameter μ, then $\lambda_i p_{i,j} = \mu$. In this case, Equation (2.76) can be rewritten as

$$P\{X(t + \Delta t) = j \,|\, X(t) = i\} = \mu \Delta t + o(\Delta t). \tag{2.77}$$

We should note that μ is the parameter of the distribution that governs the holing times.

The probability that a process is in state i at time t, denoted by $p_i(t)$, that is, $p_i(t) = P\{X(t) = i\}$, is called the ***absolute state probability*** at time t. $\{p_i(t), i \in S, t \geq 0\}$ is referred as the (one-dimensional) ***absolute distribution*** of the process at time t, and $\{p_i(0), i \in S\}$ is called the ***initial probability distribution*** of the process. According to the law of total probability, we have

$$p_j(t) = \sum_{i \in S} p_i(0) p_{i,j}(t), \quad j \in S. \tag{2.78}$$

For a Markov process $\{X(t), t \geq 0\}$ with state space S, the equation

$$p_{i,j}(t + s) = \sum_j p_{i,k}(t) p_{k,j}(s), \quad i, k, j \in S, \, t, \, s \geq 0 \tag{2.79}$$

or in matrix form

$$\mathbf{P}(t + s) = \mathbf{P}(t)\mathbf{P}(s), \quad t, s \in [0, \infty), \tag{2.80}$$

are called the ***Chapman–Kolmogorov Equations.***

From Equation (2.79) we have

$$\frac{p_{i,j}(t + s) - p_{i,j}(t)}{s} = \sum_{k \neq i} \frac{p_{i,k}(s)}{s} p_{i,j}(t) - \frac{1 - p_{i,i}(s) p_{i,j}(t)}{s}. \tag{2.81}$$

Taking the limit of Equation (2.81) as $s \to 0$ and letting

$$q_i = \lim_{s \to 0} \frac{1 - p_{i,i}(s)}{s} \text{ and } q_{i,j} = \lim_{s \to 0} \frac{p_{i,j}(s)}{s}, \, i \neq j, \tag{2.82}$$

if $p_{i,i}(0) = 1$ (and, thus, $p_{i,j}(0) = 0, \, i \neq j$), we have

$$p'_{i,j}(t) = \sum_{k \neq i} q_{i,k} p_{k,j}(t) - q_i p_{i,j}(t), \quad t \geq 0. \tag{2.83}$$

Thus,

$$p'_{i,i}(0) = \lim_{t \to 0} \frac{dp_{i,i}(t)}{t} = -q_i \text{ and } p'_{i,j}(0) = \lim_{t \to 0} \frac{dp_{i,j}(t)}{t} = -q_{i,j}, \quad i \neq j. \tag{2.84}$$

The quantities q_i and $q_{i,j}$ defined in Equation (2.82) are called the ***transition rates*** of the Markov chain. In fact, q_i is the ***unconditional rate*** of leaving the state i to make a transition to any other state, and $q_{i,j}$ is the ***conditional rate*** of making a transition from state i to state j. Equation (2.83) is called ***Kolmogorov's backward equations for the transition probabilities.***

Switching places of t and s in Equation (2.79) leads to

$$p'_{i,j}(t) = \sum_{k \neq j} p_{i,j}(t) q_{k,j} - q_j p_{i,j}(t), \quad t \geq 0. \tag{2.85}$$

Equation (2.85) is referred to as the ***Kolmogorov's forward equations for the transition probabilities.***

It is trivial that we have to accept the fact $p_{i,j}(t) \geq 0$. However, if we also assume that

$$\sum_{i \neq j} p_{i,j}(t) = 1, t \geq 0, \quad i = 0, \pm 1, \pm 2, \ldots,$$

then

$$\sum_{(i, i \neq j)} q_{i,j}(t) = q_i, \quad x = 0, \pm 1, \pm 2, \ldots. \tag{2.86}$$

We note that a similar discussion of the above equations can be done for discrete-time Markov chains. In addition, a Markov process whose transition probabilities $p_{i,j}(t)$ are time independent is called ***stationary*** or ***time-homogeneous*** or simply **homogeneous.**

2.4 RANDOM WALK

One of the simplest models for a random process (in particular, of a Markov chain or discrete-time Markov process) is a ***random walk*** (termed by Karl Pearson in 1906). Here is a very simple description of random walk. Consider the integers $\{\ldots, -2, -1, 0, 1, 2, \ldots\}$. Suppose one is standing at 0 when a coin is flipped. If the coin comes up heads, one moves to the right by one step. If it comes up tails, he or she moves to the left by one step. The procedure is repeated several times.

George Pólya (1887–1985)

Must one come back to zero? This is the question Pólya laid on the table and then answered. He essentially expressed the idea of random walk in a very simple way.

Perhaps because of his simplicity of expression, George Pólya is considered the Father of Problem Solving in Mathematics Education, although he is a mathematician who made other significant contributions to probability theory, number theory, the theory of functions, and the calculus of variations.

The answer to the aforementioned question is that *with probability one, the random walker will return to zero in a finite number of steps.*

Another way of random walk was illustrated as follows: A drunken man walks on a road that runs east and west, that is, a fifty percent probability in either of the two directions. Being intoxicated, he is likely to take a step forward (east) as backward (west). He decides to start walking. From each new position he is again as likely to go forward as backward. Each of his steps is of the same length, but of random direction – forward or backward. The position of the walker or the ***state of the random walk after n steps*** can be found. This example shows that a random walk is a process consisting of a sequence of discrete steps of fixed length. For example, the random thermal perturbations in a liquid form a random walk process, known as Brownian motion. The collisions of molecules in a gas also form a random walk that is responsible for diffusion.

Another application of a random walk is in transport molecules that play a crucial role for cell viability: among others, linear motors transport cargos along rope-like structures from one location of the cell to another in a probabilistic manner. Therefore, each step of the motion, either forward or backward, bridges a fixed distance and requires several biochemical transformations, which are modeled as internal states of the motor. While moving along the rope, the motor may detach and the walk can be interrupted. Many complex processes take place in living cells. Transport of cargos across the cytoskeleton is fundamental to cell viability and activity. In order to move cargos between the different cell parts, cells use **molecular motors**, which are responsible for a huge variety of tasks, ranging from cell division to DNA replication and chemical transport.

A special class of such motors operates by transporting cargos along the so-called ***cellular microtubules***, namely rope-like structures that connect, for instance, the cell nucleus and outer membrane. One particular example of such motors is Kinesin V, common in eukaryotic cells. Because of the periodic molecular structure of the microtubules, all the steps of Kinesin have the same length equal to 18 nm. Under normal conditions present in living cells, this motor performs a random walk in one dimension with a drift on the microtubule possibly stopped by detachment from the tubule. A mathematical formalization of such dynamics would be a random walk process research project. This project will show that the process is an extension of random walks to which we would need to add an absorbing state to model the detachment of the motor from the rope. Investigating particular properties of such processes would be a subject of interest. *A result to describe would be the maximal distance reached from the starting point and the position from which detachment takes place.*

A class of Markov chains is called *killed Quasi-Random Walk*. It is a generalization of the usual random walk on the integers allowing at each position internal states on which further movement depends. This could be seen as the so-called *Markov-modulated random walk* or a random walk in a random environment. This is a relatively new field in research that can develop into a variety of properties. A birth and death process is another application of a random walk that could be quite interesting and is discussed in Chapter 3.

A formal description of a (simple, one-dimensional) random walk process is as follows:

1. The process consists of a subset Ω of the nonnegative integers, called the ***state space***, where each integer is called a ***state***.
2. Each state $x \in \Omega$ is assigned an ***initial probability***, $p_x(0)$, where $\sum_{x\in\Omega} p_x(0) = 1$.
3. Each state x has associated with it, a random experiment, which results in an outcome of $(x, x-1)$, (x, x), or $(x, x+1)$, with respective probabilities $p_{x,x-1}$, $p_{k,k}$, and $p_{k,k+1}$, with $p_{k,k-1} + p_{k,k} + p_{k,k+1} = 1$. Some, but not all, of $p_{k,k-1}$, $p_{k,k}$, and $p_{k,k+1}$ may be zero.
4. The probabilities in (3) are the ***transition probabilities*** defined as below:
   (i) $p_{k,k-1} = P$ (transition from state k to state $k-1$),
   (ii) $p_{k,k} = P$ (transition from state k to itself), and
   (iii) $p_{k,k+1} = P$ (transition from state k to state $k+1$).

Question: What is the position of the walk at a given time, that is, a finite number of discrete time? We may define the random walk as follows:

Definition 2.4.1 Let $\{X_k\}$, $k = 1, 2, \ldots$, be a sequence of iid discrete random variables representing outcomes of independent random experiments. We also let S_n, $n = 1, 2, \ldots$, denote the sum $X_1 + X_2 + \cdots + X_n$. Then, the sequence $\{S_n\}$, $n = 1, 2, \ldots$, is called a ***random walk***. If the common values of X_k, $k = 1, 2, \ldots$, are in $\mathbb{R}^m$, then the sequence $\{S_n\}$, $n = 1, 2, \ldots$, is a random walk in $\mathbb{R}^m$.

On the basis of this definition, the probability of any particular finite sequence of X_k, $k = 1, 2, \ldots, K$ (from the common distribution of X_k) is the product of probabilities that each individual X_k takes on the specified value in the sequence. That is how the aforementioned question can be answered. Hence, when $m = 1$, we have the random walk described earlier. Thus, a ***one-dimensional random walk*** may be described as a particle walking on the integral points of the real line.

We now provide some examples.

Example 2.4.1 Random Walk on $\mathbb{R}$ Suppose that there is a game machine playing a game of repeatedly tossing a coin (fair or possibly biased), with a counter set at the initial fortune of S_0 points. Each time a head appears, a positive point is added to the counter and each time a tail appears, a negative point is added. Hence, we define a random variable X_n taking two values, +1 for a positive point with probability p and

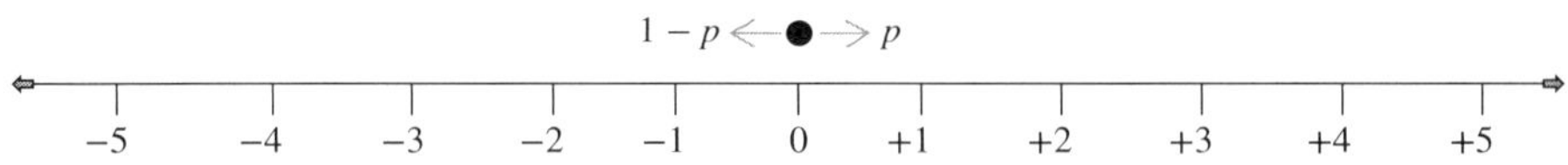

Figure 2.1 One-dimensional random walk.

-1 for a negative point with probability of $1-p, p+(1-p)=1$. In other words, the pmf of X is

$$p_X(x) = \begin{cases} p, & x = +1, \\ 1-p, & x = -1. \end{cases} \tag{2.87}$$

This is an example of a one-dimensional random walk (Fig. 2.1).

It is assumed that outcome of each toss is independent of all outcomes of the previous tosses. Let us denote the fortune of the machine after n tosses by S_n, where S_n could be a negative, zero, or positive integer. Thus, $S_{n+1} = S_n + X_{n+1}$ and

$$S_n = S_0 + \sum_{i=1}^{n} X_i. \tag{2.88}$$

That is, $X_1, X_2, \ldots$, is a sequence of independent Bernoulli random variables taking values of $+1$ and -1, with probability p and $1-p$, respectively. The sum, S_n, defined in Equation (2.88), in this case, identifies the ***position*** of the particle after n moves. If $p = 1/2$, the random walk is called ***symmetric***. In other words, the pmf for this case is

$$p_X(x) = \begin{cases} \dfrac{1}{2}, & x = \pm 1, \\ 0, & \text{otherwise}. \end{cases} \tag{2.89}$$

The position of the random walk, in this case, is the number of heads minus the number of tails. As mentioned earlier, the probability of each path leading to the position after n tosses is $1/2^n$.

In the discussion of Markov chain, we mentioned the first passage time. We now state the following theorem for a symmetric random walk.

Theorem 2.4.1 For a symmetric random walk, the probability of a return to the origin at discrete time n, denoted by u_n, occurs when n is an even integer, say $n = 2k$, $k \in \mathbb{Z}$, given by

$$u_{2k} = \binom{2k}{k} \frac{1}{2^{2k}}. \tag{2.90}$$

Proof: For a symmetric random walk, it is clear that a return to the origin, assuming that the walk starts from the origin, occurs when $S_n = 0$. However, $S_n = 0$ only if n is an even integer. Hence, if n is an integer, say $n = 2k$, $k \in \mathbb{Z}$, then, there will be $\binom{2k}{k}$

path to make the walk return to the origin. Each of these paths has a probability of $1/2^{2k}$, as discussed before and this completes the proof.

On the basis of this theorem, we can observe that the ***first return at time*** $n = 2k$, $k \in \mathbb{Z}$, denoted by f_{2k}, $k \in \mathbb{Z}$, with $f_0 = 0$, occurs when $k > 0$ and $S_{2l} \neq 0$ for all $l < k$. The relation between a return and the first return to the origin is given by the following theorem.

Theorem 2.4.2 The relationship between the probabilities u_{2k} and f_{2k}, $k \in \mathbb{N}$, for a symmetric random walk is as follows:

$$u_{2k} = f_0 u_{2k} + f_2 u_{2k-2} + \cdots + f_{2k} u_0, \ k \in \mathbb{N}. \tag{2.91}$$

In addition,

$$f_{2k} = \frac{u_{2k}}{2k-1} = \frac{\binom{2k}{k}}{(2k-1)2^{2k}}. \tag{2.92}$$

Proof: Let the origin and a return point after $2k$ discrete time units to it be denoted by (0, 0) and $(2k, 0)$, respectively. There are $u_{2k}2^{2k}$ paths between these two points, which are the endpoints of the paths. The sum of paths, use of generating function (or the Binomial Theorem), and the fact that

$$\sum_{k=0}^{\infty} \binom{2k}{k} z^k = \frac{1}{\sqrt{1-z}}, \tag{2.93}$$

lead to the completion of the proof of the theorem. We leave the complete proof as an exercise for the reader.

Example 2.4.2 First Passage Time in a Random Walk on the Real Line, $\mathbb{R}$ We have already discussed the first return to the origin. We now want to discuss the ***first passage time***. An application of this example is presented in a later chapter. Let us consider a random walk on the integer number points $\ldots, -2, -1, 0, +1, +2, \ldots$ on the real line, $\mathbb{R}$. At each integer point, the walker takes either a unit step to the left with probability q or a unit step to the right with probability p, $p + q = 1$. The objective is to find the expected time for the first passage of the walker from the site $+1$.

The problem may be stated in terms of gambling process as follows. A gambler wins a dollar with probability p and loses a dollar with probability q. We are to find the expected time that the gambler's net earning reaches $+1$ for the first time, if he starts with 0 dollar amount.

We answer this question as follows:

Let the random variable X_j be defined for $j = 1, 2, \ldots$, as

$$X_j = \begin{cases} +1, & \text{if } j\text{th step is to the right,} \\ -1, & \text{if } j\text{th step is to the left.} \end{cases} \tag{2.94}$$

TABLE 2.1 First Passage Time Probabilities

n	$\mathcal{F}_n$	ψ_n
1	$\{H\}$	p
3	$\{THH\}$	p^2q
5	$\{THTHH, TTHHH\}$	$2p^3q^2$
$\vdots$	$\vdots$	$\vdots$

Assuming that the walker starts from the origin, we also define the position of the walker at time n as

$$S_n = X_1 + X_2 + \cdots + X_n, \quad S_0 = 0. \tag{2.95}$$

Again, in terms of gambling, S_n is the net earnings after n bets. Now, let us denote the ***first passage*** of the walker to +1 at time n by $\mathcal{F}_n$. Thus, $\mathcal{F}_n$ is the following event:

$$S_1 \leq 0, S_2 \leq 0, \ldots, S_{n-1} \leq 0, \quad S_n = +1. \tag{2.96}$$

We denote the probability of the event defined in Equation (2.96) by ψ_n. We illustrate the walk as left and right, by the case when we flip a fair coin with heads, H, denoting by "+1" and a tails, T, denoting by "−1". It is clear that no even number for n is possible since the walker must walk on odd integer points to reach +1 for the first time, that is, $\mathcal{F}_{2n} = \emptyset$. Thus, for odd values of n, we summarize the result of the walk, in terms of H and T, in Table 2.1.

Let us now define the generating function of ψ_n, denoted by $\Psi(z)$, as

$$\Psi(z) = \sum_{n=0}^{\infty} \psi_n z^n, \quad |z| \leq 1. \tag{2.97}$$

From Table 2.1 and relation (2.97), we observe that

$$\Psi(z) = pz + p^2qz^3 + 2p^3q^2z^5 + \cdots. \tag{2.98}$$

Therefore, we need to find $\Psi(z)$. For this, let A_1 and A_2 be two disjoint events partitioning the space Ω, that is, $\Omega = A_1 \cup A_2$. We define the random variable $X : \Omega \to \mathbb{R}$. Then, the expected value of X can be written in terms of the conditional expected values as follows:

$$E(X) = E(X\,|\,A_1)P(A_1) + E(X\,|\,A_2)P(A_2). \tag{2.99}$$

Let N be a random variable representing the first subscript n for which $S_n = +1$. In other words, for a given $\omega \in \Omega$, let $N(\omega)$ be equal to the first integer n for which $S_n(\omega) = +1$. If no such n exists, then $N(\omega) = \infty$. It is important to note that $N(\omega) = 1, \forall\omega = (\omega_1, \omega_2, \ldots), \omega_1 = 1$. Then,

$$E(z^N) = \sum_{n=0}^{\infty} P(N = n)z^n = \sum_{n=0}^{\infty} \psi_n z^n = \Psi(z). \tag{2.100}$$

From Equation (2.99) and $A_1 = \{X = +1\}$ and $A_2 = \{X = -1\}$, we have

$$E(z^N) = E\{z^N \mid X = +1\}P(X = +1) + E\{z^N \mid X = -1\}P(X = -1). \tag{2.101}$$

It is clear that given that $X = +1$, $N(\omega) = 1$ so that

$$E\{z^N \mid X = +1\} = z. \tag{2.102}$$

On the other hand, since $p = P(X = +1)$ and $q = P(X = -1)$, from Eqs. (2.101) and (2.102), we have

$$E\{z^N\} = pz + qE\{z^N \mid X = -1\}. \tag{2.103}$$

When $X = -1$, that is, stepping to the left, let us denote the number of steps required to return to the origin, that is, increase the partial sum by one step from -1 to 0 by N_1. Similarly, we denote the number of subsequent steps required to move from the origin to 1, that is, to increase the partial sum by one step from 0 to 1 by N_2. Obviously, N_1 and N_2 are independent since they depend upon different sets of X_j, which are independent. N_1 and N_2 are also independent of X since they depend only on the subsequent steps.

It is important to note that the distributions of N_1 and N_2 are similar to that of N. Furthermore, we note that the first passage time from 0, starting at -1, has the same distribution as the first passage time from $+1$, starting at 0, and so on.

Thus, starting moving left, the random variable N is

$$N = 1 + N_1 + N_2. \tag{2.104}$$

Using Equation (2.104), the independences mentioned earlier, and the same distribution for N_1 and N_2, we can write

$$\begin{aligned}
E\{z^N|X = -1\} &= E\{z^{1+N_1+N_2}|X = -1\}\\
&= E\{zz^{N_1}z^{N_2}|X = -1\}\\
&= zE\{z^{N_1}z^{N_2}|X = -1\}\\
&= zE\{z^{N_1}|X = -1\}E\{z^{N_2}|X = -1\}\\
&= zE\{z^{N_1}\}E\{z^{N_2}\}.
\end{aligned}$$

Therefore,

$$E\{z^N \mid X = -1\} = z\Psi^2(z). \tag{2.105}$$

Substituting Equation (2.105) into Equation (2.103), we have

$$\Psi(z) = pz + qz\Psi^2(z). \tag{2.106}$$

It is important to note that from Equation (2.106), we have $\Psi(1) = 1$ and $\Psi(0) = 0$.Solving the quadratic equation (Equation 2.106) for $\Psi(z)$, we have

$$\Psi(z) = \frac{1 - \sqrt{1 - 4pqz^2}}{2qz}. \tag{2.107}$$

We did not choose $\frac{1+\sqrt{1-4pqz^2}}{2qz}$, because solution $\frac{1-\sqrt{1-4pqz^2}}{2qz}$ allows $\Psi(0) = 0$, as noted earlier.

Using Taylor expansion of the radical in Equation (2.107) and using the conventions

$$(-1)!! = +1, \tag{2.108}$$

and

$$(2n+1)!! = (2n+1)(2n-1)\cdots 5 \cdot 3 \cdot 1, \tag{2.109}$$

we have

$$\psi_{2n-1} = \frac{(2n-3)!!}{n!} 2^{n-1} p^n q^{n-1}. \tag{2.110}$$

For instance, let p_{2n-1} denote the probability that the first passage time occurs in time less than or equal to $2n - 1$. Then,

$$p_{2n-1} = \psi_1 + \psi_3 + \cdots + \psi_{2n-1}. \tag{2.111}$$

Simulating Equation (2.111), we obtain the following table of values:

$2n - 1$	p_{2n-1}
1	0.5000
3	0.6250
5	0.6875
7	0.7266
9	0.7539
11	0.7744
101	0.9212
1,001	0.9748
100,001	0.9920

From Equation (2.106), after some manipulations and using the fact that if $p + q = 1$, then $1 - 4pq = (p - q)^2$, we have

$$\Psi(1) = \frac{1 - |p - q|}{2q} = \begin{cases} \frac{p}{q}, & p < q, \\ 1, & p \geq q. \end{cases} \tag{2.112}$$

In other words, if $p < q$, then the probability that the walker stays to the left of "+1" is $(1 - p/q)$. However, if $p \geq 1$, then the walker will visit the site +1 for sure, that is, with probability one. Thus,

$$E(N) = \sum_{n\geq 1} n\psi_n = \Psi'(1)$$
$$= \frac{1 - |p-q|}{2q|p-q|}, p < q. \tag{2.113}$$

For $p \geq q$, however, we have

$$E(N) = \begin{cases} \frac{1}{p-q}, & p > q, \\ \infty, & p = q = 1/2. \end{cases} \tag{2.114}$$

Thus, a walker with equal chance of going to the left or the right will, with probability one, visit the site "+1." However, the expected waiting time for this first passage will be infinite.

It is important to note the following two facts: (1) the expected value being infinite indicates that the series $\sum_{n\geq 1} n\psi_n$ diverges and (2) "with the probability one" statement indicates that $\sum_{\geq 1} n\psi_n$ converges to "+1."

2.5 UP-AND-DOWN BIASED COIN DESIGN AS A RANDOM WALK

As an application of a random walk process, we end this section by proposing a relatively new idea that has a vast impact on patient treatment. Consider a clinical system consisting of one bed and an infinite source of patients, hereafter, we call them subjects. Samples of sizes S subjects will be chosen to undergo treatment. A subject occupies the bed while undergoing a treatment. The sequence of experiments consists of treatment of subjects with the proper dose levels of a medicine. A ***treatment***, ***stimulus***, or ***stress*** is applied at a finite number of levels (or dosages), and the cumulative number of responses at each level is observed after the end of each trial. Treatment levels with a high probability of response are undesirable. When the total number of subjects is fixed and a set of possible treatment levels is given, an experimental design is completely specified by the rule that allocates treatment to a subject.

Nancy Flournoy

Aliakbar Montazer Haghighi

Up-and-down designs are rules that allocate a treatment to a subject. That is, an up-and-down design specifies the treatment for the next trial to be one level higher, one level lower, or stay the same as the treatment selected for the current trial. (See Durham et al. 1993).

Each treatment is a Bernoulli trial with outcomes of a ***response*** (**toxic**) or a ***no-response*** (***nontoxic***). The system starts treatment of a sample with an initial fixed dose level. Therefore, the treatment selection is in fact random, and hence a random walk process.

Thus, we denote these Bernoulli random variables by $Y(n)$, $n = 0, 1, 2, \ldots$, where an arbitrary outcome of a treatment without explicit mention of its position in the sequence of treatments is denoted by Y that has two values +1 and –1, standing for observing a toxic and nontoxic observation. That is, outcomes of the treatment on the nth subject will be $Y(n) = +1$, if the outcome of the treatment on the nth subject is toxicity and $Y(n) = -1$, if the outcome of the treatment on the nth subject is nontoxicity. An arbitrary dose level without explicit mention of its position in the sequence of treatments is denoted by X. Observation of a subject stops at the first sign of a response or at some fixed time, whichever comes first. Then, the next subject is admitted for treatment. The probability of toxicity and nontoxicity at level x are denoted by

$$\psi(x) = P\{Y = +1 \,|\, X = x\} \text{ and } \phi(x) \equiv 1 - \psi(x), \tag{2.115}$$

respectively, and $\psi(x)$, referred to as ***response function***, is assumed to be strictly increasing in x. It is also assumed that given the level x, $\psi(x)$ is constant over all trials $n = 0, 1, 2, \ldots$.

Before we explain the model further, we define the term quantile to define our model. Let V be a continuous random variable with pdf $f(v)$ and cdf $F(v)$. Hence, $F(v) = P\{V \le v\}$, for all v. Let r be a number in (0, 1). Then, **r*th quantile*** of V or of F, is a number x such that $F(v) = r$, or generally $F(v) \ge r$ and $P\{V \ge v \ge 1 - r$. If F is strictly increasing in a neighborhood of v, then rth quantile is uniquely determined, and is denoted by v_r. If it is continuous and increasing on a subset of $\mathbb{R}$, say J, for which $0 < F(v) < 1$, then the inverse of F, denoted by F^{-1}, from (0, 1) onto J exists such that $F^{-1}(r) = x$, for $0 < r < 1$.

For the case of discrete random variable V, rth quantile is the smallest value v_r such that $F(v) = P\{V = v_r\} = r$, if there is at least one such value.

We now return to our example. We suppose that there are K possible dose levels and that the nth subject has been given dose level x_k, $k = 1, 2, \ldots, K$, a value of the random variable representing the dose level used at the nth trial, denoted by $X(n)$, $n = 0, 1, 2, \ldots, x_0$ is the initial dose level used. We denote the probability of using dose level x_k, $k = 1, 2, \ldots, K$, at trial n, by $P_k(n)$.

The transition probabilities associated with changing the treatment levels depend on the outcome of the experiment that follows the treatment. From dose level x_j, the one-trial and n-trial transition probabilities are denoted by p_{jk} and $p_{jk}^{(n)}$, respectively. In other words,

$$p_{j,k} \equiv P\{X(n) = x_k \,|\, X(n-1) = x_j\} \tag{2.116}$$

and

$$p_{j,k}^{(n)} \equiv P\{X_n = x_k | X_0 = x_j\}, \; n = 0, 1, \ldots; j, k = 1, 2, \ldots, K, \tag{2.117}$$

with $p_{j,k}^{(0)} = \delta_{j,k}\delta_{j,k}$ is the Kronecker's delta function. Hence, $p_{j,k} = 0$, for $|j - k| > 1$. Therefore, the process is determined by the probabilities $p_{k,k+1}, p_{k,k-1}$ and $p_{k,k}$ that the dose levels will move up from level x_k to $x_k + 1$, down from level x_k to $x_k - 1$, and stay at level x_k, respectively, for $k = 1, 2, \ldots, K$, with $p_{k,k-1} + p_{k,k} + p_{k,k+1} = 1$, for $k = 1, 2, \ldots, K$. At the boundaries, $p_{1,0} = 0$ and $p_{K,K} = 0$, that is, the process will stay away from both states 0 and K. The matrix of transition probabilities, denoted by $\mathbf{P} \equiv (p_{j,k})$, in this case, is a $K \times K$ dimensional matrix. The transition probabilities to higher levels are monotone decreasing, that is, $\{p_{1,2} > \cdots > p_{K-1,K} > p_{K,K+1} = 0\}$ and the transition probabilities to lower levels are monotone increasing, that is, $\{p_{1,0} = 0 < p_{2,1} < \cdots < p_{K-1,K-2} < p_{K,K-1}\}$ with $p_{1,2} > 0$ and $p_{K,K-1} > 0$. Thus, as the sequence of dose levels moves higher, it becomes increasingly less likely that this trend will continue, and similarly, as the sequence of dose levels moves lower, the chance of continuing this trend diminishes. It appears that these opposing forces would cause the dose levels to concentrate in some particular region in the middle.

Let μ be the target quantile and Γ be the target probability of a response of the target quantile μ. In other words, $\psi(\mu) = \Gamma$. We fix the target probability response, Γ, between 0 and 1 and assume that the spacing between dose levels is one unit. Given a proportion Γ, **the purpose of the problem** is to consider designs that will generate distributions about the dose level μ, for which

$$\psi(\mu) = \Gamma, \; \Gamma \equiv P\{Y = +1 \,|\, X = \mu\}. \tag{2.118}$$

From Equation (2.118), it can be seen that $\psi^{-1}(\Gamma) = \mu$. Without loss of generality, we let $0 \le \Gamma \le 1/2$. By symmetry, analogous results can be obtained for $1/2 \le \Gamma \le 1$. We note that the idea of the proportion of responses at the end of the last trial was introduced by Johnson and explored by Narayana in 1957. Durham et al. (1993) used stationary distribution to estimate the unknown dose, target quantile, μ.

The rule for allocating dose levels to subjects produces a sequence $X(n)$, $n = 0, 1, 2, \ldots$, that forms a random walk on a finite sample space $\Omega_x = \{x_1, x_2, \ldots, x_K\}$. Different rules exist for allocating treatment levels to subjects. We discuss a recent method called the ***Biased Coin Design*** (***BCD***), which is a simple sequential design that requires small, possibly sample sizes without loss of accuracy in the estimation; they have several advantages such as targeting any arbitrary percentile (not only the median); and it converges quickly and has minimum variance among a large class of up-and-down designs (Sheng and Sheng 2009).

There are variations of this rule that was originally developed by Stephen Durham and Nancy Flournoy in 1993 and 1994. We choose the following version and refer to it as the ***Biased Coin Design I*** (***BCD I***) (Fig. 2.2). This rule is designed to center the dose-level frequency distribution around arbitrary unknown target quantile μ, independently of the response function, using a biased coin. It will asymptotically center the allocation proportions unimodally around μ, whenever μ is properly contained

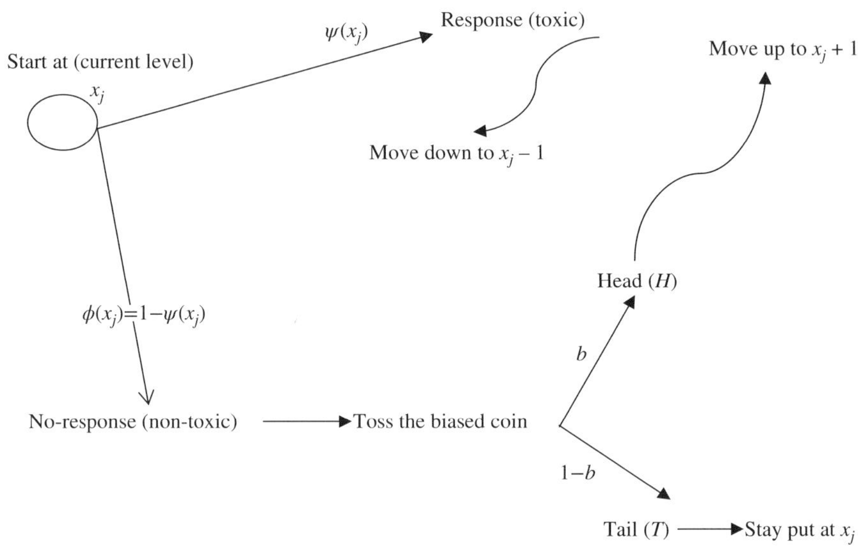

Figure 2.2 Treatment rule, biased coin design I (BCD I).

in the dose space (Durham and Flournoy 1994). It assigns a dose level to the next subject according to the instructional steps below:

Step 1. Start the experiment with dose level x_j, $j = 1, 2, \ldots, K$.

Step 2. If the outcome of the initial treatment is toxicity, then move down to the level $x_j - 1$.

$$X(1) = [x_{j-1} \mid Y(0) = -1]. \tag{2.119}$$

Step 3. If the outcome of the initial treatment is nontoxicity, toss a biased coin, with the probability of getting a head (H) and tail (T) equal to b and $1 - b$, respectively, that is,

$$P\{Y = +1\} = b \text{ and } P\{Y = -1\} = 1 - b, \tag{2.120}$$

where

$$b = \frac{\Gamma}{1+\Gamma} \text{ and } 1 - b = \frac{1}{1+\Gamma},\ 0 \le b \le \frac{1}{2}. \tag{2.121}$$

Step 3.a. If the coin turns head (H), then move up one level to $x_j + 1$.

$$X(1) = [x_j + 1] \mid [X(0) = x_j \text{ and } H]. \tag{2.122}$$

Step 3.b. If the coin turns tail (T), then do not move, stay put at the current level x_j.

$$X(1) = x_j \mid [X(0) = x_j \text{ and } T]. \tag{2.123}$$

Thus, on the basis of these steps, the transition probabilities in this case are

$$\begin{aligned} p_{1,0} &= 0, p_{1,1} = 1 - bP_i, p_{k,k} = (1-b)P_k, \quad k = 2, 3, \ldots, K-1, \\ p_{K,K} &= P_K, p_{k,k+1} = bP_k, \quad k = 1, 2, \ldots, K-1, \\ & p_{k,k-1} = \psi(x_k), k = 2, \ldots, K, \text{ and } p_{K,K+1} = 0. \end{aligned} \tag{2.124}$$

We note that in Dixon and Mood's (1948) up-and-down design, *BCD* is used with $b=1$, that is, it is targeting the 50th percentile (median), $\Gamma = \psi(\mu) = 0.50$. For a detailed discussion of this and other rules, see Durham and Flournoy (1993, 1994) and Durham et al. (1993).

Let the probability distribution of X_n be represented by the K-dimensional vector

$$\mathbf{P}(n) = \langle P_1(n), P_2(n), \ldots, P_k(n) \rangle, n = 0, 1, 2, \ldots, \tag{2.125}$$

where $P_k(n) \equiv P\{X(n) = x_k\}$, that is, the probability that the trial n uses dose level x_k. If the initial treatment is fixed at x_j, then

$$P_j(0) = 1. \tag{2.126}$$

Let

$$\pi_k = \lim_{n \to \infty} P\{X(n) = x_k\}, \quad k = 1, 2, \ldots, K. \tag{2.127}$$

Stationary distribution of the dose-level frequencies for our design with the aforementioned allocation rules has been studied in Durham et al. (1993). Durham and Flournoy (1994) provided the limiting treatment probability distributions with *BCD* in terms of response function as follows:

$$\pi_1 = \lambda_1, \tag{2.128}$$

$$\pi_k = \lambda_1 \left(\frac{\Gamma}{1-\Gamma} \right)^k \prod_{i=2}^{k} \frac{P(x_{i-1})}{\psi(x_i)}, \quad k = 2, 3, \ldots, K, \tag{2.129}$$

where

$$\lambda_1 = \frac{1}{1 + \sum_{i=2}^{K} \left(\frac{\Gamma}{1-\Gamma} \right)^i \prod_{l=2}^{i} \frac{P(x_{l-1})}{\psi(x_l)}}. \tag{2.130}$$

Thus, the expected value and variance of the dose levels when the sample size is large are

$$E_\pi(X) = \sum_{k=1}^{K} x_k \pi_k \text{ and } \mathrm{Var}_\pi(X) \sum_{k=1}^{K} x_k^2 \pi_k - \left(\sum_{k=1}^{K} x_k \pi_k \right)^2, \tag{2.131}$$

respectively.

Now let us consider a numerical case for this example. Suppose that the cdf of the unknown responses is a ***logistic*** distribution function

$$\psi(x_k) = \frac{1}{1+e^{-\frac{x_k-\alpha}{\beta}}}, \quad k = 1, 2, \ldots, K, \tag{2.132}$$

where K is the maximum dose level and α and β are the location and scale parameters, respectively, of the logistic distribution. For the target quantile, we have

$$\Gamma = \frac{1}{1+e^{-\frac{\mu-\alpha}{\beta}}}, \quad k = 1, 2, \ldots, K, \tag{2.133}$$

Let $K = 10$, that is, $\Omega = \{1, 2, \ldots, 10\}$, $\alpha = 5.5$ and $\beta = 1.5$. Taking parameters as such, from Equation (2.132) we have $\psi(5.5) = 0.50$ and $\psi(8.8) = 0.90$. In other words, we set the median of the dose levels at 5.5 and the 90th percentile at 8.80. It is important to note that this choice of parameters centers the dose-level assignments unimodally around any target quantile of interest (Durham et al. 1997). For all dose levels, we have the values in Table 2.2.

From Equation (2.127), we have

$$\mu = \psi^{-1}(\Gamma) = \alpha + \beta \ln\left(\frac{\Gamma}{1-\Gamma}\right). \tag{2.134}$$

Thus, for different values of our target response function, Γ, using logistic distribution, we have Table 2.3.

Let us now choose ***Weibull*** distribution for the unknown responses function, that is

$$\psi(x_k) = 1 - e^{-\left(\frac{x_k}{\beta}\right)^{\alpha}}, \tag{2.135}$$

where α and β are location and scale parameters, respectively. Thus, we have

$$\Gamma = \psi(\mu) = 1 - e^{-\left(\frac{\mu}{\beta}\right)^{\alpha}}, \; \mu \geq 0. \tag{2.136}$$

Therefore,

$$\mu = \beta\left[\ln\left(\frac{1}{1-\Gamma}\right)\right]^{\frac{1}{\alpha}}. \tag{2.137}$$

Once again, by choosing values for α and β, we have Tables 2.4 and 2.5.

TABLE 2.2 Response Function Values Under Logistic CDF

k	1	2	3	4	5	6	7	8	9	10
$\psi(x_k)$	0.0474	0.0884	0.1589	0.2689	0.4174	0.5826	0.7311	0.8411	0.9116	0.9526

TABLE 2.3 Target Quantile Values, μ, Under Logistic CDF

Γ	0.10	0.20	0.25	0.33	0.50	0.90
μ	2.2042	3.4206	3.8521	4.4377	5.5000	8.7958

TABLE 2.4 Target Quantile Values, μ, Under Weibull CDF, $\alpha = 5.5$, and $\beta = 1.5$

Γ	0.10	0.20	0.25	0.33	0.50	0.90
μ	0.9963	1.1420	1.1959	1.2701	1.4033	1.7456

TABLE 2.5 Target Quantile Values, μ, Under Weibull CDF, $\alpha = 7.0$, and $\beta = 3$

Γ	0.10	0.20	0.25	0.33	0.50	0.90
μ	2.5378	2.8249	2.9293	3.0711	3.3215	3.9429

EXERCISES

2.1 In Example 2.1.3, prove the following statement: The stochastic process $\{N(t), t \geq 1\}$, the arrival process associated with the interarrival times τ_n, is a Poisson process with parameter λ, $\lambda > 0$.

2.2 Give the complete proof of Theorem 2.4.2.

2.3 Let the discrete-time random process $\{X_n\}$ be defined by $X_n = s^n$, $n = 0, 1, \ldots$, where s is selected randomly from $(0, 1)$. Find the mean and autocovariance of X_n.

Definition 1: An event occurring at time $t + k$, $k > 0$, is said to ***lag behind*** the event occurring at time t, the extent of the lag being $\boldsymbol{k}$. An event occurring k time units before another may be regarded as having a ***negative lag***.

Definition 2: A stationary processes is referred to as ***weakly stationary***, if it only requires that the first moment and autocovariance do not vary with respect to time.

Definition 3: Let $\{X_n; n \in \mathbb{Z}\}$ be a stochastic process such that $\text{Var}(X_n) < \infty, \forall n \in \mathbb{Z}$. Then, $\gamma_X(n_1, n_2) = \text{cov}(X_{n1}, X_{n2})$ is called the ***autocovariance function*** of the stochastic process $\{X_n; n \in \mathbb{Z}\}$. In other words, the autocovariance is a function that gives the covariance of the process with itself at pairs of time points.

Definition 4: If $\{X_n\}$ is weakly stationary, the covariance of X_n and X_{n+k} depends only on the ***lagk***.

From Definitions 3 and 4, the relation $\gamma(k) = cov(X_{n+k}, X_n)$ is called ***autocovariance***.

Note that $\gamma(0) \geq |\gamma(k)|, \forall k$. Note also that correlation between X_n and X_{n+k} is referred to autocorrelation, denoted by $\rho(k)$, and is defined by $\rho(k) = \frac{\gamma(k)}{\gamma(0)}$.

3

BIRTH AND DEATH PROCESSES

3.1 OVERVIEWS OF THE BIRTH AND DEATH PROCESSES

In the previous chapter, we discussed both discrete- and continuous-time Markov processes with some examples. Before we discuss standard queues, we need to discuss another example of Markov chain and Markov process, which is the ***birth and death* (*B–D*) *processes***. These processes may be discrete- or continuous-time. The discrete case is sometimes referred to as ***birth and death chain***. Our discussion, however, is mainly on a continuous-time case, with a brief mention of discrete-time as well. Although the B–D process is important in itself, it has also been proven to be a useful tool for modeling queueing, circuit switches with a limited number of outgoing channels, reliability, inventory, performance evaluation of servers, demography, epidemiology, biology, and others. B–D processes are also good models for the flow of radioactive, cosmic, and other particles. In the economical sciences, B–D processes are used for describing the development of a number of enterprises in a particular area and manpower fluctuations.

B–D processes and chains with a finite or countable number of states also play a central role in stochastic modeling for applied probabilities. A number of books are available on various aspects of B–D models, for example, Gross and Harris (1998), Karlin and Taylor (1975), Kijima (1997), and Medhi (2003). Some classical probability texts such as Feller (1968) and Hoel et al. (1972) also contain discussions of B–D chains in terms of gambler's ruin probabilities.

The concept of random walk defined and discussed in the previous chapter is the most general concept of the B–D process. Random walk, however, is a ***semi-Markov process*** since the holding time in each state has an arbitrary probability distribution function that does not exhibit the forgetfulness property. But, if we think of time in

Delayed and Network Queues, First Edition. Aliakbar Montazer Haghighi and Dimitar P. Mishev.

distinct increments of ones, $\Delta t = 1, t = 1, 2, \ldots$, then a random walk also exhibits two properties: (1) the Markov property and (2) stationary transition probabilities. This will allow us to study random walk as a B–D process with the position of the walk as the state of the process, that is, the population size. In the context of queueing theory, which is discussed later, we observe the random walk position as the number of tasks or jobs in the system.

Note also that a B–D process may be called a ***continuous-time random walk***. Therefore, if the walker takes only forward steps such as $0 \to 1 \to 2 \to 3 \cdots$, the chain is called a ***pure birth process***. In this case, we denote the transition rate from state x to its upper (or right) neighboring state $x+1$ by λ_x. If the walker is allowed to stay put at state x, the transition rate for the event is denoted by r_x and the process is called a ***birth process***. We associate $\lambda_x > 0$ and $r_x > 0$ with this case. On the other hand, if only backward transitions, such as $0 \leftarrow 1 \leftarrow 2 \leftarrow 3 \leftarrow \cdots$, denoted by μ_x, are allowed, the process is called a ***pure death process***. If the walker is allowed to stay put, then the process is called a ***death process***. In this case, $\mu_x > 0, r_x > 0$ and $\lambda_x = 0$. If only forward and backward steps are allowed, the process is called a ***pure birth and death process***. In that case, $r_x = 0$. If staying put is allowed, the process is called a ***birth–death process***.

It is assumed that the birth and death events are independent of each other.

If we had assumed the birth process to be a Poisson process and the times between deaths be exponentially distributed, based on the properties of these two processes, we could make the following probability statements for an occurrence of events during the time interval $[t, t + \Delta t]$, while the system is not empty. Thus, for a birth during a time interval Δt, we have

$$\begin{cases} P(\text{no birth}) = 1 - \lambda_x \Delta t + o(\Delta t), \\ P(\text{one birth}) = \lambda_x \Delta t + o(\Delta t), \\ P(\text{more than one birth}) = o(\Delta t), \end{cases} \tag{3.1}$$

where $o(\Delta t)$ is called the ***little o*** of Δt, that is

$$\lim_{\Delta t \to 0} \frac{o(\Delta t)}{\Delta t} = 0. \tag{3.2}$$

Similarly, for a death during a time interval Δt, we have

$$\begin{cases} P(\text{no death}) = 1 - \mu_x \Delta t + o(\Delta t), \\ P(\text{one death}) = \mu_x \Delta t + o(\Delta t), \\ P(\text{more than one death}) = o(\Delta t), \end{cases} \tag{3.3}$$

where $o(\Delta t)$ is defined in Equation (3.2).

We note that with the above assumptions, one of the important properties that a B–D process exhibits is that the times between births and the time between the deaths are each distributed exponentially, when the system is not empty. This is because they are Markovian. We also note that under the above assumptions, a B–D process does not allow transitions of more than one step. Transitions such as $x \to x+2, x-3 \leftarrow x-1$, are not allowed. In other words, the probability of a transition of more than one step is considered to be negligible.

The finite pure birth process case of Equation (3.1) may be considered backward through the following example.

Suppose the probability exists that a finite population, such as an organism or species, produces a new member in a very small time interval Δt is $\lambda \Delta t$, where λ is a constant, independent of the size of the population, say n, where $n = 1, 2, \ldots, N$.

Note that the population cannot exist with the original size of zero; otherwise, no birth or death can occur. Thus, we assume the initial population size to be $n_0, n_0 \geq 1$. In other words,

$$P_n(0) = n_0, \quad n_0 = 1, 2, \ldots, N, \text{ and } P\{n_0 \geq 1\} = 1. \tag{3.4}$$

Thus, if the maximum size of the population is given by N, then the population size could be $n_0, n_0 + 1, n_0 + 2, \ldots, N$.

We purposely choose Δt sufficiently small so that no more than one birth may occur in it, that is, the probability of more than one birth is negligible. Then, starting at time t, in order to have the size of the population at time $t + \Delta t$, that is, Δt time unit later, we have the following two possibilities:

(i) The size at time t is already n. Thus, no birth should occur in the time interval $(t, t + \Delta t)$.
(ii) The size at time t is $n - 1$. Thus, exactly one birth should occur during the time interval $(t, t + \Delta t)$ to make up the size to be n.

Since the probability of a birth from each member within a time interval $(t, t + \Delta t)$, as mentioned, is $\lambda \Delta t$, then from the entire population, it will be $n\lambda \Delta t$. Hence, the probability of transition from size n to size $n + 1, n_0 \leq n + 1 \leq N$ is $n\lambda \Delta t$. Therefore, the probability of no birth within the same interval, that is, the size remaining the same, is $1 - n\lambda \Delta t$. Similarly, we may argue about the transition from size $n - 1$ to n, that is, $(n - 1)\lambda \Delta t$. Thus, if we denote by $P_n(t)$ the probability that the population size is n at time t, then we can write the following:

$$\begin{aligned} P_n(t + \Delta t) &= P_n(t) \cdot P\{\text{no birth occuring in } (t, t + \Delta t)\} \\ &\quad + P_{n-1}(t) \cdot P\{\text{exactly one birth occuring in } (t, t + \Delta t)\}. \\ &= P_n(t) \cdot (1 - n\lambda \Delta t) + P_{n-1}(t) \cdot (n - 1)\lambda \Delta t. \end{aligned} \tag{3.5}$$

After some algebra, Equation (3.5) can be written as

$$\frac{dP_n(t)}{dt} = (n - 1)\lambda P_{n-1}(t) - n\lambda P_n(t), \quad 1 \leq n \leq N, \tag{3.6}$$

with $P_n(0)$ defined in Equation (3.4).

To solve Equation (3.6), we use the probability generating function method and obtain:

$$P_n(t) = \binom{n-1}{n_0} e^{n_0\lambda t}(1 - e^{-\lambda t})^{n-n_0}, \tag{3.7}$$

where $P_n(0) = n_0, n = n_0, n_0 + 1, n_0 + 2, \ldots, N, n_0 = 1, 2, \ldots, N$.

If we denote $p = e^{\lambda t}, n_0 = k, n - n_0 = r$, then Equation (3.7) can be rewritten as

$$P_n = \binom{k+r-1}{k} \cdot (1-p)^r p^k, \tag{3.8}$$

which is a pmf of negative binomial distribution.

Thus, the mean and variance of $P_n(t)$ given in Equation (3.7), denoted by $L(t)$ and $\sigma^2(t)$, can be found, respectively, as follows:

$$L(t) = \frac{(n-n_0)e^{\lambda t}}{1 - e^{\lambda t}} \tag{3.9}$$

and

$$\sigma^2(t) = \frac{(n-n_0)e^{\lambda t}}{(1-e^{\lambda t})^2}, n = n_0, n_0 + 1, \ldots, N, \quad n_0 = 1, 2, \ldots, N. \tag{3.10}$$

It should be noted that if $n_0 = 1$, Equation (3.7) is the pmf of a geometric distribution.

Similarly to a finite pure birth process, we leave it as an exercise for a finite pure death process and consequently for a B–D process. We discuss a general form of a finite B–D process in a later section.

Figure 3.1 is a graphical presentation illustrating the idea of a pure B–D process with birth rate $\lambda_x, x = 0, 1, 2, \ldots$, and death rate $\mu_x, x = 1, 2, \ldots$, which is a standard transition diagram of a pure B–D process with infinite population.

Let us denote by $p_{x,y}$ the probability of transition between neighboring states x, which is defined as

$$p_{x,y} = \begin{cases} \lambda_x, & \text{if only one birth occurs, that is } y = x + 1, \\ \mu_x, & \text{if only one death occurs, that is } y = x - 1, \\ 0, & \text{otherwise.} \end{cases} \tag{3.11}$$

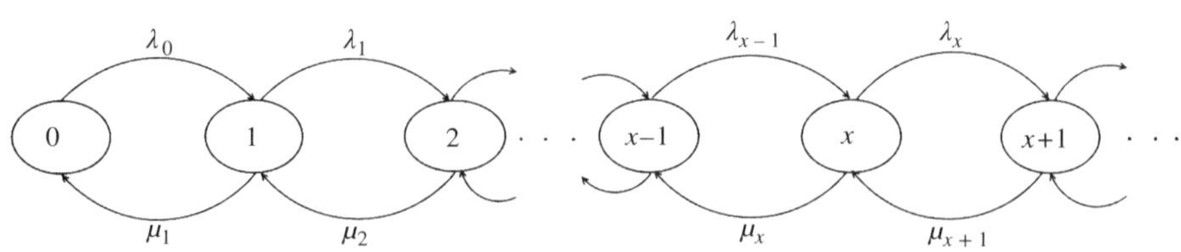

Figure 3.1 Transition diagram of a birth and death processes. Arrows represent the possible transitions between states and the labels on the arrows represent the state transition rates between states.

Note:

1. From Equation (3.11), it is clear that revising each state is only possible from a neighboring state and that may occur infinitely many times until the process is more than one step away from its neighbors. We also note that from Equation (3.11), transitions from state x to itself would not be a choice. This is the case when birth and death occur simultaneously.
2. If we assume a positive probability for a case of no birth and no death for a random length of time, such as in a random walk process, then after the process enters state x, it may stay (sojourns) there for some random length of time. Let us assume that this length of time is exponentially distributed with parameter $(\lambda_x + \mu_x)$. Then, when the process leaves the state x, it enters either state $x + 1$ with probability $\lambda_x/(\lambda_x + \mu_x)$ or state $x - 1$ with probability $\mu_x/(\lambda_x + \mu_x)$. Let us assume that the process chooses its next state as $x + 1$. Then, again, it is assumed that, the length of stay of the process in this new state will have an exponential distribution with parameter $(\lambda_{x+1} + \mu_{x+1})$, where λ_{x+1} and μ_{x+1}, are the birth and death rates in state $x + 1$, respectively, and the process will choose the next state and so on.

Now, let $X(t)$ describe the states of the B–D process at time t that are nonnegative integers. In other words, a state of a B–D process is the population size at time t that cannot be negative. Let us also denote by $p_x(t)$, the probability that the process is in state x at time t, that is,

$$p_x(t) = P\{X(t) = x\}.$$

We further let $p_{x,y}(t)$ be the transition probability of the process moving from state x to state y within the time interval $[0, t]$. In general, this interval could be chosen as $[t, t + \Delta t]$ with the interval length of Δt. Thus, by transition probability $p_{x,y}(\Delta t)$, we indicate the conditional probability

$$p_{x,y}(\Delta t) = P\{X(t + \Delta t) = y|X(t) = x\}$$

or

$$p_{x,y}(\Delta t) = P\{X(t + \Delta t) = y|X(t) = x\} = q_{x,y}\Delta t + o(\Delta t), \tag{3.12}$$

where $o(\Delta t)$ is defined by Equation (3.2). What Relation (3.12) indicates that when Δt is small, the transition probability from state x to state y in the time interval $[t, t + \Delta t]$ is proportional to the length, Δt, of the time interval, relative to absolute time t, with proportionality constant $q_{x,y}$. For a small Δt, $o(\Delta t)$ is near zero. Hence, dropping the $o(\Delta t)$ part, Equation (3.12) may be rewritten as

$$p_{x,y}(\Delta t) = P\{X(t + \Delta t) = y|X(t) = x\} \approx q_{x,y}\Delta t. \tag{3.13}$$

Because of Equation (3.13), $p_{x,y}$ is called the ***probability transition rate***.

The transition probabilities, $p_{x,y}(t), x, y = 0, 1, 2, \ldots$, may be found by solving the Kolmogorov forward and backward system of differential difference equations

(DDEs, e.g., see Gross and Harris 1985). Letting $\mathbf{P}(t)$ denote the matrix of transitions with elements $p_{xy}(t)$, that is,

$$\mathbf{P}(t) = [p_{xy}(t)], \tag{3.14}$$

and $\mathbf{Q}$, representing the transition rate matrix with birth rate $\lambda_x, x = 0, 1, 2, \ldots$, and death rate $\mu_x, x = 1, 2, \ldots$, the uniformly bounded, for the B–D process, we can write

$$\mathbf{P}'(t) = \mathbf{Q} \cdot \mathbf{P}(t) = \mathbf{P}(t) \cdot \mathbf{Q}, \tag{3.15}$$

where

$$\mathbf{P(t)} = \begin{bmatrix} p_{0,0}(t) & p_{0,1}(t) & \cdots & p_{0,n}(t) & \cdots \\ p_{1,0}(t) & p_{1,1}(t) & \cdots & p_{1,n}(t) & \cdots \\ \vdots & \vdots & \ddots & \vdots & \cdots \\ p_{n,0}(t) & p_{n,1}(t) & \cdots & p_{n,n}(t) & \cdots \\ \vdots & \vdots & \vdots & \vdots & \ddots \end{bmatrix}, \tag{3.16}$$

and

$$Q = \begin{bmatrix} -\lambda_0 & \lambda_0 & 0 & 0 & 0 & \cdots \\ \mu_1 & -(\lambda_1 + \mu_1) & \lambda_1 & 0 & 0 & \cdots \\ 0 & \mu_2 & -(\lambda_2 + \mu_2) & \lambda_2 & 0 & \cdots \\ 0 & 0 & \mu_3 & -(\lambda_3 + \mu_3) & \lambda_3 & \cdots \\ \vdots & \vdots & \vdots & \vdots & \vdots & \cdots \end{bmatrix}. \tag{3.17}$$

Then, the solution of the Kolmogorov backward equation is well known as

$$\mathbf{P}(t) = e^{\mathbf{Q}t}. \tag{3.18}$$

For example, see Bhattacharya and Waymire (1990).

It should be noted that if we let

$$M = \sup_i |\lambda_i + \mu_i| < \infty \tag{3.19}$$

and

$$\mathbf{S} \equiv \frac{1}{M}\mathbf{Q} + \mathbf{I}, \tag{3.20}$$

where $\mathbf{S}$ is a stochastic matrix. Then

$$\mathbf{Q} = M(\mathbf{S} - \mathbf{I}). \tag{3.21}$$

Now, from Equations (3.18) and (3.21), we have

$$\begin{aligned} \mathbf{P}(t) = e^{\mathbf{Q}t} &= \sum_{n=0}^{\infty} \frac{(\mathbf{Q}t)^n}{n!} \\ &= e^{M(\mathbf{S}-\mathbf{I})t} = e^{M\mathbf{S}t}e^{-Mt} \end{aligned}$$

$$=e^{-Mt}\sum_{n=0}^{\infty}\frac{(Mt)^n}{n!}\mathbf{S}^n$$

$$=e^{-Mt}\sum_{n=0}^{\infty}\frac{(Mt)^n}{n!}[s_{x,y}]^n, \tag{3.22}$$

where $s_{x,y}, x, y = 0, 1, 2, \ldots$, are elements of matrix **S**, defined in Equation (3.20).

Krinik et al. (2005), dedicated to M. M. Rao, proposed a ***dual process*** (sometimes referred to as ***inverse process*** or ***reverse process***) to the B–D, as seen in Figure 3.2:

The purpose of the duality is to facilitate finding the transition probability distributions discretely. Hence, the state "−1" in Figure 3.2 has no physical meaning and/or interpretation. Denoting the transition probability functions of the dual process presented by Figure 3.2 by $p^*_{x,y}(t)$, through Theorem 3.1.1, they showed that the relationship (3.23) holds between a general B–D process and its duality. Thus, if the transient probability functions in either the original B–D process or dual B–D system are known, then the transient probability functions in the other system are known as well.

Theorem 3.1.1 If $p_{x,y}(t)$ and $p^*_{x,y}(t)$ are the transient probability functions of the B–D processes corresponding to Figures 3.1 and 3.2, respectively, then, assuming

$$P^*_{-1,k}(t) = 0, \quad \text{for } k > -1,$$

we have

$$P_{x,y}(t) = \sum_{k=x}^{\infty}[P^*_{y,k}(t) - P^*_{y-1,k}(t)] \text{ and } P_{x,y}(t) = \sum_{k=0}^{x}[P_{y,k}(t) - P_{y+1,k}(t)], \tag{3.23}$$

for all states $x, y = 0, 1, 2, \ldots$.

Proof: The proof of this theorem appears as Proposition 2.3 on page 269 of Anderson (1991). It is essentially based on the forward and backward Kolmogorov equations. See also Krinik et al. (2005) for example. The outline of proof of Theorem 3.1.1 is as follows:

Consider the finite recurrent B–D chain having transition probabilities depicted in Figure 3.3.

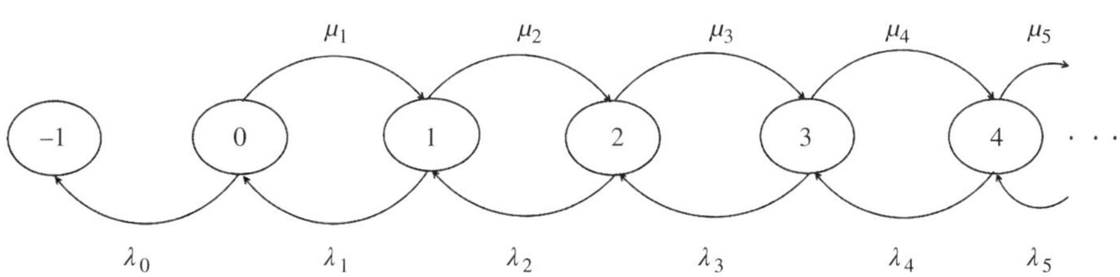

Figure 3.2 Transition diagram of the dual process of the birth–death process of Figure 3.1. Arrows represent the possible transitions between states and the labels on the arrows represent the state transition rates between states.

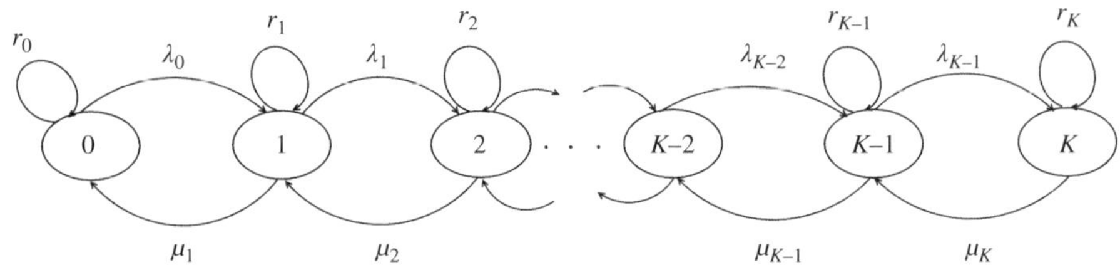

Figure 3.3 Transition diagram of a birth and death process with staying put allowed. Arrows represent the possible transitions between states and the labels on the arrows represent the state transition rates between states.

For the B–D process represented by its transition diagram (Fig. 3.3), assuming that all characters are fractions between 0 and 1, inclusively, we should note the following:

$$
\begin{aligned}
r_0 + \lambda_0 &= 1, \\
r_1 + \lambda_1 + \mu_1 &= 1, \\
r_2 + \lambda_2 + \mu_2 &= 1, \\
&\vdots \\
r_{K-1} + \lambda_{K-1} + \mu_{K-1} &= 1, \\
r_K + \mu_K &= 1.
\end{aligned}
$$

In order to ensure that the traffic intensities in the dual processes are less than 1, it is also assumed that

$$
\begin{aligned}
\lambda_0 + \mu_1 &\le 1, \\
\lambda_1 + \mu_2 &\le 1, \\
\lambda_2 + \mu_3 &\le 1, \\
&\vdots \\
\lambda_{K-1} + \mu_K &\le 1.
\end{aligned}
$$

In case that in the general B–D process, that is, $r_x \ge 0$, which we described as the random walk in which staying put is allowed, Krinik et al. (2005) showed that a theorem similar to Theorem 3.1.1 holds, that is, Theorem 3.1.2 for B–D chains. Although statement of the theorem is on a finite state space, it also holds for infinite B–D chains.

Figures 3.3 and 3.4 are similar transition probability diagrams as Figures 3.1 and 3.2. Figure 3.4 is the absorbing B–D chain dual to the B–D process represented by Figure 3.3.

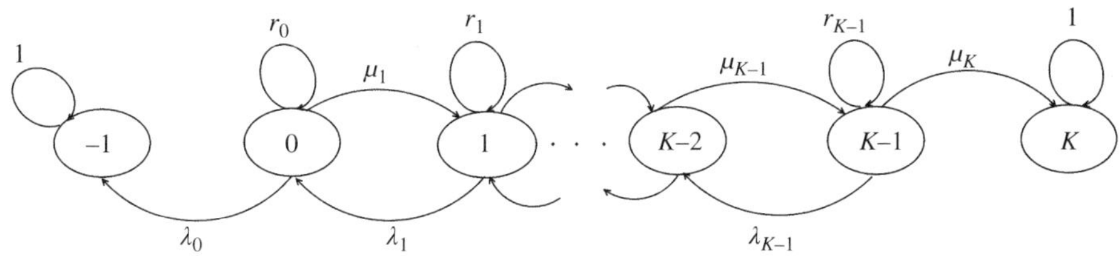

Figure 3.4 Transition diagram of the dual birth–death process depicted in Figure 3.3. It is an absorbing B–D at states −1 and K. Arrows represent the possible transitions between states and the labels on the arrows represent the state transition rates between states.

In Figure 3.4, it is assumed that all characters are fractions between 0 and 1, inclusively, and that

$$\begin{aligned}\lambda_0 + r_0 + \mu_1 &= 1,\\ \lambda_1 + r_1 + \mu_2 &= 1,\\ &\vdots\\ \lambda_{K-1} + r_{K-1} + \mu_K &= 1.\end{aligned}$$

Theorem 3.1.2 If $p_{x,y}^{(n)}(t)$ and $p_{x,y}^{*(n)}(t)$ are the n-step transition probabilities of the B–D processes corresponding to Figures 3.3 and 3.4, respectively, then, assuming

$$P_{-1,k}^{*(n)}(t) = 0, \quad \text{for } k > -1,$$

we have

$$P_{i,j}^{(n)}(t) = \sum_{k=j}^{K} [P_{j,k}^{*(n)}(t) - P_{j-1,k}^{*(n)}(t)] \text{ and } P_{i,j}^{*(n)}(t) = \sum_{k=0}^{i} [P_{j,k}^{(n)}(t) - P_{j+1,k}^{(n)}(t)], \tag{3.24}$$

For $n \geq 0$ and all states, $i, j = 0, 1, 2, \ldots, K$.

Proof: See Krinik et al. (2004).

It is important to note that Reisman and Trivedi (1988) discussed numerical transient analysis of Markov models. They pointed out difficulties facing numerical solution of such systems. However, the transient probability distribution of the population of a finite-state B–D process has been solved using a variety of methods over decades. Obtaining analytic explicit solution is almost impossible at that time. Lederman and Reuter (1954) and Takács (1962) offered the corresponding solutions. However, the methods were very cumbersome. Sharma and Dass (1988) presented a transient solution to a special category of Markovian model using eigenvalues of matrices. In this chapter, we use the method offered by Mohanty et al. (1993), using DDEs in an elementary way. This method uses the technique used by Sharma and Dass (1988), but offers a much simpler way to compute the solutions. The method engages Laplace transformation to obtain a system of algebraic equations, that is, solved the system by Cramer's rule. The results are then inverted using the method of partial fraction.

3.2 FINITE B–D PROCESS

We present this example to emphasize the fact that the various techniques used in this chapter are quite standard and well known. In spite of the remarkable simplicity of the suggested procedure, it has not been commonly discussed in textbooks by the time the paper (see Mohanty et al. 1993),, including this method, was published. This is perhaps due to the unavailability of high-speed computer facilities in the past. Thus, our intention is to popularize the procedure by using it in applications.

Let us consider a B–D process with $N+1$ states, namely $0, 1, 2, \ldots, N$, with B–D rates λ_n and μ_n, respectively, when the process is in state n, and initially start the analysis when the population size is i. We denote the probability that the population size is n at time t by $P_n(t)$. We further denote by $\Psi_n(s)$ the Laplace transform of $P_n(t)$. We leave it as exercise for the reader to show that $\Psi_n(s), 0, 1, 2, \ldots, N$, satisfies the following set of DDEs:

$$\begin{cases} (\lambda_0 + s)\Psi_0(s) - \mu_1\Psi_1(s) = \delta_{i,0}, \\ (-\lambda_{n-1}\Psi_{n-1}(s) + (\lambda_n + \mu_n + s)\Psi_n(s) - \mu_{n+1}\Psi_{n+1}(s) = \delta_{i,\ n}, \quad 1 \le n \le N-1, \\ -\lambda_{N-1}\Psi_{N-1}(s) + (\mu_N + s)\Psi_N(s) = \delta_{i,N}, \end{cases} \tag{3.25}$$

where $\delta_{i,j}$ is Kronecker's delta function. In order to solve the system (3.25), we note that it may be written in a matrix form. Let us denote the determinant of the coefficient matrix of the system (Equation 3.25) by $A_N(s)$. After some algebra, $A_N(s)$ may be written as follows:

$$A_n(s) = C_N(s)$$
$$= \begin{bmatrix} \lambda_0 + s & \sqrt{\lambda_0\mu_1} & & & & & \\ \sqrt{\lambda_0\mu_1} & \lambda_1 + \mu_1 + s & \sqrt{\lambda_1\mu_2} & & & & \\ & \sqrt{\lambda_1\mu_2} & \lambda_2 + \mu_2 + s & \sqrt{\lambda_2\mu_3} & & & \\ & & \sqrt{\lambda_2\mu_3} & \lambda_3 + \mu_3 + s & & & \\ & & & \ddots & & & \\ & & & & & \lambda_{N-1} + \mu_{N-1} + s & \sqrt{\lambda_{N-1}\mu_N} \\ & & & & & \sqrt{\lambda_{N-1}\mu_N} & \mu_N + s \end{bmatrix}. \tag{3.26}$$

Let $T_k(s)$ and $B_k(s)$ be the determinants of the $k \times k$ matrices formed at the top-left and bottom-right corners of the coefficient matrices, respectively. Set $T_0(s) = B_0(s) = 1$. Thus, using Cramer's rule, the solution of Equation (3.25) becomes

$$\Psi_n(s) = \left(\prod_{j=n+1}^{i} \mu_j\right) \frac{T_n(s)B_{N-i}(s)}{A_N(s)}, \quad 0 \le n \le i,$$

$$= \left(\prod_{j=i}^{n-1} \lambda_j\right) \frac{T_i(s)B_{N-n}(s)}{A_N(s)}, \quad i+1 \le n \le N, \tag{3.27}$$

where the first product may be interpreted as

$$\prod_{j=k}^{n} u_j = 1, \tag{3.28}$$

whenever $n < k$.

In order to invert Equation (3.27), we need to express it as a partial fraction. For this, we note that the right-hand side of Equation (3.27) is a ratio of two polynomials with the degree of numerator less than that of the denominator. It is important to note that s is a zero of the matrix $C_N(s)$ defined by Equation (3.26), if and only if $-s$ is an eigenvalue of the matrix $E_N(s)$ defined as

$$E_N(s) = \begin{bmatrix} \lambda_0 & \sqrt{\lambda_0\mu_1} & & & & & \\ \sqrt{\lambda_0\mu_1} & \lambda_1+\mu_1 & \sqrt{\lambda_1\mu_2} & & & & \\ & \sqrt{\lambda_1\mu_2} & \lambda_2+\mu_2 & \sqrt{\lambda_2\mu_3} & & & \\ & & \sqrt{\lambda_2\mu_3} & \lambda_3+\mu_3 & & & \\ & & & \ddots & & & \\ & & & & & \lambda_{N-1}+\mu_{N-1} & \sqrt{\lambda_{N-1}\mu_N} \\ & & & & & \sqrt{\lambda_{N-1}\mu_N} & \mu_N \end{bmatrix}. \tag{3.29}$$

It should be noted that $s = 0$ is an eigenvalue of $E_N(s)$. But each off-diagonal element of $E_N(s)$ is nonzero. Hence, all the eigenvalues are distinct (see Parlett (1980), p. 124). Hence, s is the only zero of $C_N(s)$. In addition, since all the minor elements of $E_N(s)$ are positive, by the Sturn sequence property, all other eigenvalues are positive (see Wilkinson 1965, p. 300). Therefore, all eigenvalues of the positive semi-definite matrix $E_N(s)$ are real, distinct, and nonnegative. Hence, $A_N(s)$ has exactly $N+1$ distinct zeros, one of which is zero and the rest are negatives of the eigenvalues of $E_N(s)$.

We now denote the zeros of $A_N(s)$ by $z_k, k = 0, 1, 2, \ldots, N$, with $z_0 = 0$. Letting the numerator of $\Psi_n(s)$, defined in Equation (3.27), be $G_n(s)$, Equation (3.11) may be rewritten as

$$\Psi_n(s) = \frac{G_n(s)}{A_N(s)}, \quad n = 0, 1, 2, \ldots, N, \tag{3.30}$$

and, in turn, it can be expressed in partial fraction as follows:

$$\Psi_n(s) = \frac{\beta_{n,k}}{s - z_k}, \quad n = 0, 1, 2, \ldots, N, \tag{3.31}$$

where

$$\beta_{n,k} = \frac{G_n(s)}{\prod_{j=0, j\neq k}^{N} (z_k - z_j)}, \quad n = 0, 1, 2, \ldots, N, \tag{3.32}$$

or

$$\beta_{n,k} = \frac{\prod_{j=0}^{N-n}(z_k - s_j)}{\prod_{j=0, j\neq k}^{N}(z_k - z_j)}, \quad n = 0, 1, 2, \ldots, N, \tag{3.33}$$

where s_j are the roots of $G_n(s)$. Thus, inverting Equation (3.33), we have the distribution of the population as

$$P_n(t) = \beta_{n,0} + \sum_{k=1}^{N} \beta_{n,k} e^{z_k t}, \quad n = 0, 1, 2, \ldots, N, \tag{3.34}$$

where $\beta_{n,k}$ are given in Equation (3.32).

For the stationary distribution, denoted by P_n, letting $t \to \infty$, we obtain

$$P_n = \lim_{t\to\infty} P_n(t) = \beta_{n,0} = \frac{G_n(0)}{\prod_{j=1}^{N}(-z_j)}. \tag{3.35}$$

We leave it as exercise for the reader to show that

$$\beta_{n,0} = \beta_{0,0} \prod_{k=1}^{n} \frac{\lambda_{k-1}}{\mu_k}, \quad 1 \le n \le N. \tag{3.36}$$

Using the normalization equation, we have

$$\beta_{0,0} = \left[1 + \sum_{n=1}^{N} \prod_{k=1}^{n} \frac{\lambda_{k-1}}{\mu_k}\right]^{-1}. \tag{3.37}$$

Example 3.2.1 B–D Process with Finite Number of States In order to obtain a numerical distribution of the number of tasks in the B–D process with $N + 1$ states, as described earlier, we will use the numerical example in Mohanty et al. (1993), in which programming language FORTRAN was used.

It is important to note that the old FORTRAN computer program is still being used and many researchers suggest it for scientific programming, although MatLab is also a readily available software. For instance, the code ARTEMIS for the simulation of nuclear reactors, currently under development in Aldo Dall'Osso's company in Paris, France (Aldo Dall is a nuclear engineer), is written in FORTRAN 95. Other codes under development employ C++ and FORTRAN 95, with appropriate interfaces. FORTRAN 2003 is an object-oriented program. Some scientists prefer C++

over FORTRAN, because one can structure the source code better without losing performance.

For this example, the program follows the following procedure:

1. The eigenvalues of E_k are computed by the IMSL subroutine DEVLSB.
2. The zeros of $A_K(s)$, denoted by z_k, are the negatives of the eigenvalues of E_k, except that z_0 is set to zero.
3. From Equation (3.31), having values of z_k, we can compute $G(z_k)$ and having $\prod_{j=0,j\neq k}(z_k - z_j)$, we can compute the $\beta_{n,k}$.
4. Alternatively, having z_k and s_j, we can compute $\prod_{j=0,j\neq k}^{K}(z_k - z_j)$ and $\prod_{j=0}^{K-n}(z_k - z_j)$, and thus, $\beta_{n,k}$. The probabilities are obtained once the $\beta_{n,k}$ are known.
5. We follow the same algorithm for the distribution of a busy period.

The FORTRAN program is run on a VAX 8300 (VMS 4.7). We checked our method for various values of N. Some of the probabilities computed by the program together with the negative of the eigenvalues are presented in Tables 3.1–3.4. In Tables 3.1, 3.2, and 3.4, we have chosen $N=7$. Table 3.1 is given as a sample of the numerical values for larger K, that is, $K=39$. It is important to note that for the reason given in remark 3.29, we have chosen $p = 1$ and $\alpha = 0$ in Table 3.4. We also checked our method for various values of the parameters involved, but we have reported examples with the numerical values of the parameters so that we would have $\frac{\lambda}{c\mu} < 1$, $\frac{\lambda}{c\mu} < 1$, and $\frac{\lambda}{c\mu} < 1$, with various probabilities of balking, as well as various reneging rates. The time units in Tables 3.1–3.4 are cut off at an instance when the steady-state solutions are reached; that is, when convergences have been observed; in these tables, some rows appear to be identical due to round-off. Table 3.4 gives the execution times to obtain the steady-state solution for our method for different values of K.

TABLE 3.1 Transient Probability Distribution of the Number of Tasks in the System with $K=39$ and $c=3$

	k or n						
	0	1	2	12	25	38	39
	$\lambda = 4, \mu = 1, p = 1, \alpha = 0$						
z_k	0.00000	−0.09446	−0.16227	−2.95394	−9.40526	−13.83590	−13.90509
$p_n(0)$	1.00000	0.00000	0.00000	0.00000	0.00000	0.00000	0.00000
$p_n(1)$	0.07951	0.19954	0.24663	0.00011	0.00000	0.00000	0.00000
$p_n(2)$	0.02901	0.09606	0.15497	0.00534	0.00000	0.00000	0.00000
$p_n(66)$	0.00001	0.00002	0.00004	0.00037	0.00546	0.18341	0.24438
$p_n(200)$	0.00000	0.00000	0.00001	0.00011	0.00445	0.18750	0.25000

TABLE 3.2 Transient Probability Distribution of the Number of Tasks in the System with $K = 7$ and $c = 3$

	k or n							
	0	1	2	3	4	5	6	7
				$\lambda = 1, \mu = 4, p = 0.5, \alpha = 5$				
z_k	0.00000	−4.01223	−8.25403	−13.26090	−17.25717	−22.05825	−27.48892	−34.66849
$p_n(0)$	1.00000	0.00000	0.00000	0.00000	0.00000	0.00000	0.00000	0.00000
$p_n(1)$	0.78241	0.19203	0.02357	0.00193	0.00006	0.00000	0.00000	0.00000
$p_n(2)$	0.77892	0.19467	0.02433	0.00203	0.00006	0.00000	0.00000	0.00000
$p_n(3)$	0.77886	0.19471	0.02434	0.00203	0.00006	0.00000	0.00000	0.00000
$p_n(4)$	0.77886	0.19471	0.02434	0.00203	0.00006	0.00000	0.00000	0.00000
				$\lambda = 1, \mu = 4, p = 0.9, \alpha = 5$				
z_k	0.00000	−4.00410	−8.10124	−12.59525	−17.13745	−22.19106	−28.16837	−36.40253
$p_n(0)$	1.00000	0.00000	0.00000	0.00000	0.00000	0.00000	0.00000	0.00000
$p_n(1)$	0.78239	0.19202	0.02356	0.00193	0.00010	0.00000	0.00000	0.00000
$p_n(2)$	0.77888	0.19466	0.02432	0.00203	0.00010	0.00000	0.00000	0.00000
$p_n(3)$	0.77882	0.19470	0.02434	0.00203	0.00010	0.00000	0.00000	0.00000
$p_n(4)$	0.77882	0.19470	0.02434	0.00203	0.00010	0.00000	0.00000	0.00000
				$\lambda = 10, \mu = 9, p = 0.9, \alpha = 8$				
z_k	0.00000	−9.11231	−18.09908	−27.54689	−39.12215	−53.10727	−69.48002	−91.53230
$p_n(0)$	1.00000	0.00000	0.00000	0.00000	0.00000	0.00000	0.00000	0.00000
$p_n(1)$	0.33001	0.36663	0.20366	0.07542	0.01939	0.00406	0.00072	0.00011
$p_n(2)$	0.32997	0.36663	0.20368	0.07544	0.01940	0.00406	0.00072	0.00011
$p_n(3)$	0.32997	0.36663	0.20368	0.07544	0.01940	0.00406	0.00072	0.00011
				$\lambda = 10, \mu = 9, p = 0.5, \alpha = 8$				
z_k	0.00000	−9.85973	−20.81287	−28.86287	−39.74234	−51.04649	−61.74538	−79.93031
$p_n(0)$	1.00000	0.00000	0.00000	0.00000	0.00000	0.00000	0.00000	0.00000
$p_n(1)$	0.33404	0.37113	0.20617	0.07636	0.01091	0.00127	0.00012	0.00001
$p_n(2)$	0.33401	0.37113	0.20618	0.07636	0.01091	0.00127	0.00012	0.00001
$p_n(3)$	0.33401	0.37113	0.20618	0.07636	0.01091	0.00127	0.00012	0.00001

				$\lambda = 2, \mu = 2, p = 0.8, \alpha = 7$				
z_k	0.00000	−2.18470	−5.08370	−9.00690	−13.34179	−20.21162	−28.49880	−40.07249
$p_n(0)$	1.00000	0.00000	0.00000	0.00000	0.00000	0.00000	0.00000	0.00000
$p_n(1)$	0.42176	0.36568	0.15945	0.04724	0.00544	0.00041	0.00002	0.00000
$p_n(2)$	0.37739	0.37138	0.18298	0.06027	0.00737	0.00059	0.00003	0.00000
$p_n(3)$	0.37251	0.37184	0.18561	0.06179	0.00760	0.00061	0.00004	0.00000
$p_n(4)$	0.37197	0.37189	0.18591	0.06196	0.00763	0.00061	0.00004	0.00000
$p_n(5)$	0.37190	0.37190	0.18594	0.06198	0.00763	0.00061	0.00004	0.00000
$p_n(6)$	0.37190	0.37190	0.18595	0.06198	0.00763	0.00061	0.00004	0.00000
				$\lambda = 3, \mu = 1, p = 0.9, \alpha = 1$				
z_k	0.00000	−0.93329	−3.55767	−7.05954	−14.57042	−26.84474	−40.29274	−51.74159
$p_n(0)$	1.00000	0.00000	0.00000	0.00000	0.00000	0.00000	0.00000	0.00000
$p_n(1)$	0.14561	0.25681	0.18890	0.02268	0.02575	0.03599	0.07856	0.24570
$p_n(2)$	0.05129	0.10499	0.08478	0.01097	0.01650	0.04157	0.14770	0.54220
$p_n(17)$	0.00026	0.00079	0.00119	0.00119	0.00801	0.04325	0.19462	0.75069
$p_n(18)$	0.00026	0.00079	0.00119	0.00119	0.00801	0.04325	0.19462	0.75069
$p_n(19)$	0.00026	0.00079	0.00119	0.00119	0.00801	0.04325	0.19462	0.75069
				$\lambda = 4, \mu = 1, p = 0.9, \alpha = 1$				
z_k	0.00000	−1.48510	−4.55386	−8.53413	−20.56462	−34.76467	−50.43036	−63.66725
$p_n(0)$	1.00000	0.00000	0.00000	0.00000	0.00000	0.00000	0.00000	0.00000
$p_n(1)$	0.07458	0.16791	0.15176	0.01826	0.02106	0.03305	0.09875	0.43463
$p_n(2)$	0.01554	0.03889	0.03751	0.00484	0.00805	0.02823	0.14417	0.72273
$p_n(11)$	0.00004	0.00015	0.00030	0.00041	0.00365	0.02629	0.15777	0.81138
$p_n(12)$	0.00004	0.00015	0.00030	0.00041	0.00365	0.02629	0.15777	0.81138

TABLE 3.3 Transient Probability Distribution of the Number of Tasks in the System by Randomization Method with $K = 7$, $c = 3$, and m being Preassigned According to $m = ft + 4\sqrt{ft} + 5$

	k	0	1	2	3	4	5	6	7	$p_k(t)$
					$\lambda = 1, \mu = 4, p = 1, \alpha = 0$					
$m = 5$	$p_k(0)$	1.000000	0.000000	0.000000	0.000000	0.000000	0.000000	…	0.00000	1.000000
$m = 32$	$p_k(1)$	0.782350	0.191995	0.023555	0.001924	0.000156	0.000012	…	0.00000	0.999994
$m = 51$	$p_k(2)$	0.778820	0.194639	0.024321	0.002026	0.000169	0.000014	…	0.00000	0.999991
$m = 68$	$p_k(3)$	0.778750	0.194686	0.024336	0.002028	0.000169	0.000014	…	0.00000	0.999984
$m = 85$	$p_k(4)$	0.778748	0.194687	0.024336	0.002028	0.000169	0.000014	…	0.00000	0.999913
$m = 102$	$p_k(5)$	0.778750	0.194688	0.024336	0.002028	0.000169	0.000014	…	0.00000	0.999987

Note: For this table, we have fixed $p_0(0) = 1$.

TABLE 3.4 Execution Time to Obtain the Steady-State Solution for the Present Method for Various Values of N with $c = 3$

$\lambda = 4, \mu = 1, p = 1, \alpha = 0$	
N	Execution time
7	099 s
10	2.075 s
20	11.24 s
50	3 min and 52 s
75	21 min and 53 s
100	53 min and 58 s
125	2 h and 37 min

Note

1. Computing eigenvalues involves errors to some degree.
2. For analysis of the errors and numerical examples, the reader is referred to Parlett (1980), Wilkinson (1965), and Murphy and O'Donohoe (1975).
3. In our case, to reduce the error, since we know one of the eigenvalues is zero, we have set z_0 and u_0 to zero in the program.
4. The steady-state probabilities obtained from the transient case, using Equations (3.27) and (3.28) and the direct formula, match exactly.
5. The IMSL performance index in all cases did not exceed 0.06, which, according to IMSL (1987), is considered to be "excellent."

Remarks

1. The randomization method, together with an algorithm provided in Grassmann (1977a,b), is available to obtain the transient solution, and it works well for nonstiff systems (that is, systems in which the diagonal elements do not vary considerably in size), only it does not provide exact values of the probabilities. Given the error tolerance, the randomization algorithm determines where the computation should be truncated, that is, the value of m has to be preassigned according to the error tolerance. m is the truncation level in the relation

$$p_k(t) \approx \sum_{n=0}^{m} \pi_k^n (f\,t)^n \frac{e^{-f\,t}}{n!}, \tag{3.38}$$

where $f = \max|a_{ii}|$ is the coefficient matrix of the system of DDEs

$$\pi_k^n = \sum_{i=0}^{K} \pi_i^{n-1} p_{i,j} \text{ and } p_{i,j} = \frac{a_{i,j}}{f}, i \neq j. \tag{3.39}$$

For example, for $K = 7, c = 3, \lambda = 1, \mu = 4, p = 1$, and $\alpha = 0$, using the randomization method, Table 3.4 provides values of probabilities for different truncation levels. In Table 3.4, m is calculated according to Grassmann (1977a) as $m = ft + 4\sqrt{ft} + 5$. By pushing K for larger values, an approximation for the infinite case can be obtained as well.

For specific values of parameters given in Table 3.4, the steady-state solution appears to be reached at $m = 102$. As expected, this shows that, in order to obtain the steady-state solution using randomization, one cannot truncate arbitrarily and m has to be determined for the specific values of λ, μ, p, and α, while for our method, the steady-state distribution is exactly and easily obtained.

For $K = 7, \lambda = 1, \mu = 4, p = 1$ and $\alpha = 0$, randomization took 39.21 s to be executed to reach the steady state, while for our method it took only 0.99 s.

2. Our approach in this section yields the steady-state solution analytically with ease, while the randomization method does so less easily.
3. Computationally, the method presented in this section offers a reasonably good approximation for the steady state for infinite case, although it may take a longer execution time for larger K (see Table 3.4).
4. For $p = 1$ and $\alpha = 0$, the method presented in this section works for the finite case perfectly, even for a relatively large K (at least $K = 125$). For $K = 50$, it took only 232 s ≪4 min) to obtain the steady-state solution. This is in contrast to the claim by Grassmann (1990) that "Cramer's rule is impractical for all but very small problems." This, perhaps, is one of the small problems, since we are dealing with a tridiagonal matrix. Grassmann (1990) also stated that, for $K = 50$, "a modern supercomputer requires an execution time of 10^{44} years and this time exceeds the age of the universe by many orders of magnitude" – this, however, seems not to be our experience.
5. For the so-called ***stiff system***, Whitlock mentioned that no finite f can be found, and thus, the randomization algorithm is not applicable, while our method works regardless. It is for this reason that we have chosen $p = 1$ and $\alpha = 0$, in Table 3.4. However, for some values of the parameters, the diagonal elements became large, thereby making the eigenvalues large, and thus causing overflow problem while running the program.

3.3 PURE BIRTH PROCESS (POISSON PROCESS)

As mentioned earlier, a pure birth process, at any time t, either remains in its current state, $x \to x$, or transits to the next higher state, $x \to x + 1$. Any other transition is not allowed, that is, occur with probability 0. Thus, a birth process is a Markov chain for which the probability of transiting from state x to state $x + 1$ in a time interval $[t, t + \Delta t]$ is $\lambda\Delta t + o(\Delta t)$, where $o(\Delta t)$ is the *little o* of Δt defined by Equation (3.2).

In Example 2.1.3, we considered a continuous-time stochastic process defined on a probability space $(\Omega, \mathfrak{B}, P)$ called a continuous-time Poisson process. We now elaborate that process. The process counts the number of events occurred (births) up to and including time t. For this reason, it is called a pure birth process. Hence, a continuous-time Markov process $\{X(t), t \geq 0\}$ on the state space $\Omega = \{0, 1, 2, \dots\}$ is called a ***Poisson (birth) process with transition rate*** (or ***mean***) λ, $\lambda > 0$, provided that the following properties hold:

(i)

$$\begin{cases} P\{X(0) = 0\} = 1, \\ P\{X(t + \Delta t) = x + 1 | X(t) = x\} = p_{x,x+1}(\Delta t) = \lambda \Delta t + o(\Delta t). \end{cases} \tag{3.40}$$

This indicates that the probability that the number of births increases by a unit in a small time interval $[t, t + \Delta t]$ is approximately proportional to the length, Δt, of the time interval. The transition rates remain constant throughout a Poisson process, that is,

$$\lambda \equiv q_{0,1} = q_{1,2} = q_{2,3} = \cdots. \tag{3.41}$$

(ii)

$$P\{X(t + \Delta t) > x + 1 | X(t) = x\} = o(\Delta t). \tag{3.42}$$

If Δt is small, the probability of 2 or more births in the time interval $[t, t + \Delta t]$ is negligible. Equivalently, this may be stated as $q_{x,y} = 0$ for $y > x + 1$.

(iii)

$$P\{X(t + \Delta t) < x | X(t) = x\} = 0. \tag{3.43}$$

In a Poisson process, the number of births remains the same or increases over time; it will never decrease. In other words, ***Poisson process is a pure birth process***.

(iv)

$$\begin{aligned} P\{X(t + \Delta t) = x | X(t) = x\} &= 1 - [\lambda \Delta t + o(\Delta t)] - o(\Delta t) \\ &= 1 - \lambda \Delta t - 2o(\Delta t) = 1 - \lambda \Delta t - o(\Delta t). \end{aligned} \tag{3.44}$$

We note that a Poisson process will have a ***stationary independent increment*** based on Example 2.1.3. That is, for arbitrary time epochs $t_0 = 0 < t_1 < t_2 < \cdots < t_{n-1} < t_n$, the number of events, X, occurring in disjoint intervals

$$X(t_1) - X(t_0), X(t_2) - X(t_1), X(t_3) - X(t_2), \dots, X(t_n) - X(t_{n-1}) \tag{3.45}$$

are independent random variables. In other words, the number of events that occur in one time interval is independent of the number of events occurring in a time interval that is disjoint from the first time interval.

Now, applying the conditions of the Poisson process, we state the following theorem:

Theorem 3.3.1 For a Poisson process with transition rate λ and the initial condition $p_0(0) = 1$, let

$$p_x(t) = P\{\text{the process is in state } x \text{ at time } t\}, \quad x = 0, 1, 2, \dots. \tag{3.46}$$

Then, the following system of DDEs holds:

$$\begin{cases} \frac{dp_0(t)}{dt} = -\lambda p_0(t), & p_0(0) = 1, \\ \frac{dp_x(t)}{dt} = \lambda p_{x-1}(t) - \lambda p_x(t), & p_x(0) = 0, \quad x = 1, 2, \ldots. \end{cases} \tag{3.47}$$

Proof: See Haghighi and Mishev (2014).

Solving the system (3.47) yields the distribution function of the number of births up to the time instant t for a Poisson process. To solve the system, we note that the first equation of the system (3.47) is a linear first-order differential equation, whose solution, under the given initial condition, is

$$p_0(t) = p_0(0)e^{-\lambda t} = e^{-\lambda t}. \tag{3.48}$$

For the second equation of the system, we solve one equation at a time for each state. Then, using mathematical induction, we have the solution of the system (3.47) as

$$p_x(t) = \frac{(\lambda t)^x e^{-\lambda t}}{x!}, \quad x = 0, 1, 2, \ldots, \tag{3.49}$$

which is a Poisson distribution with parameter λt. The result is summarized in the following theorem:

Theorem 3.3.2 In a Poisson process with transition rate λ, the probability distribution, *pmf*, of the number of events occurred up to time t, is Poisson, which can be written as follows:

$$p_x(t) = \{P\{X(t) = x\} = \frac{(\lambda t)^x e^{-\lambda t}}{x!}, \quad x = 0, 1, 2, \ldots. \tag{3.50}$$

This theorem states that for $t > 0$ and $\Delta t \geq 0$, the random variable $X(t + \Delta t) - X(\Delta t)$ representing the number of events occurring (independently) between the time interval Δt and $t + \Delta t$ follows the Poisson distribution, that is,

$$P\{X(t + \Delta t) - X(\Delta t) = x\} = \frac{(\lambda t)^x e^{-\lambda t}}{x!}, \quad x = 0, 1, 2, \ldots. \tag{3.51}$$

3.4 PURE DEATH PROCESS (POISSON DEATH PROCESS)

We now consider a finite population of size N that has no capability of generating additional members (that is no birth). Hence, occurrences of events diminish the population one at a time. This process could be considered as a Poisson process with a transition state μ, where the constant μ is the mean of the number of occurrences of events (deaths). In this case, the state 0 is an absorbing state, that is, the state 0 represents extinction of the population.

Therefore, we let

$$p_x(t) = P\{\text{the population size at time } t \text{ is } x\}, \quad x = 0, 1, 2, \dots, N. \tag{3.52}$$

Thus, for the state 0, $p_0(t)$ is the probability that the population has extinct at time t, that is,

$$p_N(0) = 1. \tag{3.53}$$

Therefore, as in the Poisson birth process, the system of DDEs is as follows:

$$\begin{cases} \frac{dp_0(t)}{dt} = \mu p_1(t), & p_0(0) = 1, \\ \frac{dp_x(t)}{dt} = \mu p_{x+1}(t) - \mu p_x(t), & p_x(0) = 0, \quad x = 1, 2, \dots, N-1, \dots, \\ \frac{dp_N(t)}{dt} = -\mu p_N(t), & p_N(0) = 0. \end{cases} \tag{3.54}$$

We leave it as an exercise to show that the solution of Equation (3.54) is as follows:

$$\begin{cases} p_0(t) = 1 - \left[\sum_{x=1}^{N} (\lambda t)^{N-x} \frac{e^{-\lambda t}}{(N-x)!} \right], \\ p_x(t) = \frac{(\lambda t)^{N-x}}{(N-x)!} e^{-\lambda t}, \quad x = 1, 2, \dots, N. \end{cases} \tag{3.55}$$

EXERCISES

3.1 Find similar relations to (3.8)–(3.10) for a finite pure death process and consequently for a B–D process.

3.2 Show that $\Psi_n(s), 0, 1, 2, \dots, N$, satisfies the set of DDEs (Equation 3.25).

3.3 Show that Equation (3.36) is true.

3.4 Show that the solution of Equation (3.54) is Equation (3.55).

3.5 Consider a B–D process where λ_i and μ_i are given as

a.

$$\begin{cases} \lambda_i = 20, & i = 0, 1, 2, \dots, \\ \mu_i = 25, & i = 1, 2, \dots. \end{cases}$$

b.

$$\begin{cases} \lambda_i = 10, & i = 0, 1, 2, \dots, \\ \mu_i = 20, & i = 1, 2, \dots. \end{cases}$$

c.

$$\begin{cases} \lambda_i = 6, & i = 0, 1, 2, \dots, \\ \mu_i = 8, & i = 1, 2, \dots. \end{cases}$$

d.

$$\begin{cases} \lambda_i = 10, & i = 0, 1, 2, \dots, \\ \mu_i = 40, & i = 1, 2, \dots. \end{cases}$$

(i) Find the stationary probability vector.

(ii) Find the stationary probability for state 3.

3.6 Consider a B–D process with Poisson birth and exponential death distributions, where death occurs singly. Denote the birth and death rates as λ and μ, respectively, given as

a.

$$\begin{cases} \lambda = 10, \\ \mu = 11. \end{cases}$$

b.

$$\begin{cases} \lambda = 30, \\ \mu = 50. \end{cases}$$

c.

$$\begin{cases} \lambda = 2, \\ \mu = 8. \end{cases}$$

Let E_k denote the state of the system, that is, population at any time is k. Find

(i) the stationary probability vector;

(ii) the stationary probability that population is 2; and

(iii) the probability that there is no one in the system.

3.7 A small town beauty salon has two stylists, Angel and Barbara, and an additional chair for one waiting customer. If a customer arrives when there are three customers in the salon, she leaves. From the past experience, the average time between arrivals of customers is 30 min. Angel completes serving a customer in 30 min, while Barbara does it in 1 h. Angel has priority serving the customer when there is only one customer in the shop, as she is faster than Barbara. Furthermore, assume that interarrival times and service times of customers are independent exponential random variables.

a. Draw a state transition diagram with possible states and corresponding B–D rates.

b. After a long time passes, what is the probability that the shop is empty?

c. What is the probability mass function of the number of customers in the shop?

d. After a long time passes, what is the probability that an arriving customer is turned away?

e. What is the distribution of the number of customers arriving in the first 3 h?

f. What is the probability that no customer arrives in a 3-h time interval?

g. What is the mean number of arrivals in 3 h?

h. What is the distribution of time until the fifth customer arrives?

i. What is the probability that the time until the fifth customer arrives is less than 2.5 h?

3.8 Consider a walker who walks on a real line starting at 0 with moving one step forward with probability p and backward with probability q, $p + q = 1$. Let X_n describe the position of the walker after n steps.

a. What is the probability that the walker is at point 0 on the line after two steps?

b. What is the probability that the walker is at point -1 on the line after three steps?

c. What is the probability that the walker is at point 3 on the line after three steps?

d. If the walker is at point 4 after 10 steps, does the probability that it will be at point 8 after 16 steps (six more steps) depend on how it moves to point 4 within the first 10 steps?

e. Are $X_{10} - X_4$ and $X_{16} - X_{12}$ independent?

f. Are $X_{10} - X_4$ and $X_{12} - X_8$ independent?

4

STANDARD QUEUES

4.1 INTRODUCTION OF QUEUES (GENERAL BIRTH AND DEATH PROCESS)

In this book, words such as "queue," "waiting line," and "congestion" are used interchangeably. Although a queueing model could be deterministic, queueing theory is an application of stochastic processes, that is, mathematical theory and analysis of waiting lines and delays while waiting in a line or queue for obtaining service from a system.

The basic queueing models, as described later, are examples of birth and death processes. In other words, we consider the stochastic behavior of processes arising in connection with services in cases where random fluctuations occur. The idea is to predict fluctuating demands from observational data, which will ultimately enable an enterprise to provide adequate service with tolerable waiting times for its clients. The art of queueing theory is to construct a model that is simple to carry out mathematical analysis, but containing sufficient details so that its performance measures may reflect the behavior of the real system. Thus, queueing theory is considered as a branch of operations research, because the results are often used when making business decisions about the resources needed to provide a service. The reader may refer to Haghighi and Mishev (2014) for some examples. Recently, expressions such as ***Q-nomy's queue management system*** has been used to "organize waiting lines by providing visitors with an easy ordering method based on "first-in-first-out," prescheduled appointments, or more sophisticated methods" (see qnomy.com).

For development of queues, two stages have been mentioned in the literature (Frontiers in Queueing 1997): "Rise and Fall" and "Revival and Separation." The former is about ideas and methods of early queueing theory and the latter is concerned with the abstract and practical.

Delayed and Network Queues, First Edition. Aliakbar Montazer Haghighi and Dimitar P. Mishev.

A.K. Erlang (1878–1929)

Although there is a claim that queueing theory started when Napoleon asked Poisson how many cannons he would have to send to the front to have 80% confidence in getting 100 canons to where he wanted them, when he wanted them, it is almost universally acceptable that historically, at the beginning of the 20th century, A. K. Erlang (1878–1929), a famous Danish scientist of Copenhagen Telephone Company, was among the first pioneers in the development of the theory of queues with his most noted paper of 1917. Thus, he is referred to as the father of queueing theory (Brockmeyer et al. 1948).

Erlang used queueing theory for representing the behavior of telephone traffic and telephone exchanges (Syski 1960). When operations research appeared and then grew during World War II, other applications of queueing theory appeared, while telephony was yet the dominant application and most of that work utilized models that consisted of a single-server queue.

Erlang's published papers appeared during the period of 1909–1920. His idea attracted interests of mathematicians, engineers, and economists. The range of applications has grown to include not only telecommunications (such as traffic modeling, resource provisioning, and traffic management, collectively called ***teletraffic***) and computer science, but also signal processing, particle counting, stock processes, machine repairs, electronic networks, manufacturing, air traffic control, industrial engineering, management science, military logistics, the design of theme parks, factories, shops, offices, hospitals, and many other areas that involve service systems with random demands.

It took years after Erlang's publication for the next result in the field to appear. For instance, Vaulot (1927) and Pollaczek (1930) (1892–1981). Then, 2 years later the famous work of Khintchine (1932) (1894–1959) was published. It seems that the "delay systems" started to be considered by C. D. Crommelin during the period 1932–1934. Supplementary variables were first studied by Vaulot (1927).

It took many years for theoretical analysis of queueing systems to grow considerably with the advent of operations research in the late 1940s and early 1950s. The first textbook on this subject entitled *Queues, Inventories, and Maintenance* was authored by Morse in 1958.

Then, Saaty wrote his famous book entitled *Elements of Queueing Theory with Application* in 1961. In fact, 1960s was a period of intensive research in queueing theory. The number of papers published on queues appeared exponentially.

An important development was the role of ***fluctuation theory*** and ***virtual waiting time*** or ***Takács Process*** by Lajos Takács. The virtual waiting time is the period of time an imaginary customer would have to wait after having arrived at a particular observation time point t, not including his/her service time (Takács 1974).

At present, more than 40 books have been written on queues: Haghighi and Mishev (2014) is one of the most recent ones. In addition, thousands of papers on queues have been published and several journals have been dedicated to queueing research papers.

For the historic telephone model of Erlang, the underlying Markov process representing the number of customers in such systems is known as the birth and death process, which is widely used in population models. The birth–death terminology is used to represent increase and decrease of population size. The corresponding events in queueing systems are arrivals and departures. In this chapter, we present some of the important models belonging to this class.

As a final note, it should be mentioned that an occasional concern is raised about the theory of queues that it is too mathematical and not much applied. However, the criticism, although having the same merit, should not be of much concern, because, after all, queueing theory involved deeply elegant and sophisticated mathematics.

Thus, it is meaningful to say that queueing theory has revolutionized the industry with its immense applications in many areas such as city traffic, air traffic, biosciences, population studies, and health sector.

4.1.1 Mechanism, Characteristics, and Types of Queues

A basic queue (or a queueing system or queueing model) consists of ***arrival of tasks (customers)***, ***servers (processor)***, ***service discipline*** (order of service), and ***departure*** (Fig. 4.1). The basic characteristics of a queue include, but not limited to ***average queue length***, ***average busy period***, ***average service time***, ***average waiting time***, ***average sojourn time*** (time spending in the system in queue, in service, and if there is a feedback, with all feedback times) and ***throughput***. Tasks (or customers, arrivals, calls, items, or units) arrive at a service station (or channel, counter, or server) according to some probabilistic laws such as Poisson, Erlang, Yule, Palm, and recurrent. For example, in a computer system, where a programmer sitting at a terminal (arrival) requests access to a CPU (server) for the processing of a transaction (service), both the arrival time of the request for access and the amount of processing time requested

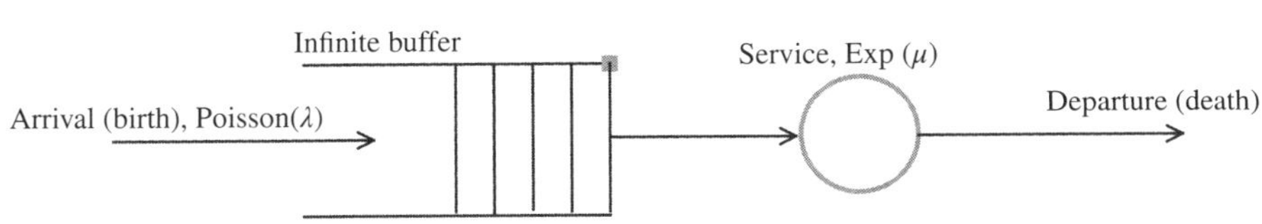

Figure 4.1 Single-server Markovian queueing process, exponential interarrival times, and exponential service times; λ = average arrival rate, μ = average service rate, $1/\mu$ = average service time.

are random. Thus, the essence of queueing theory is to consider the randomness of the arrival process and service process. This random walk and birth–death process was described in the previous chapter.

If the source of arrival is external and infinite, the system is referred to as an ***open*** queue; otherwise, the queue is termed a ***closed*** queue. A service station may be served by one or more servers. The system with one server is called a ***single-server*** queueing system and that with more servers is called a ***multi-server*** (or ***many-server***) queueing system. ***Infinite-server*** queueing system refers to the case when there is no waiting for an arriving task, that is, service starts once it arrives. Service is rendered singly or in batch according to some disciplines such as first-come first-served (FCFS), first-in-first-out (FIFO), random, priority, and last-come first-served (LCFS). The service times are assumed to be random variables following some given probabilistic laws such as constant, negative exponential, and geometric. After being served, the tasks depart from the service station (the outputs), but not necessarily from the system, as they may acquire further services with some positive probability. A system with the latter characteristic is called a queueing system with "***feedback.***" As shown by its basic mechanism, a queueing system may be modeled as a birth–death process.

The number of tasks in the system is a characteristic that will help design the system's ***waiting room*** or ***buffer***. The buffer's capacity may be finite or infinite. The system with no buffer is called a ***loss system***. This model is called the ***Erlang loss model***. The arrival process, in this model, follows a Poisson process and arrivals who find the system busy at the time of their arrival are lost. The fraction of arrival of tasks who find all the servers busy, that is, the loss probability, is calculated by the famous ***Erlang loss formula***. In telephone traffic engineering, the Erlang loss model is one of the fundamental models, where tasks are "telephone calls," the servers are "trunks," and those who are not able to attend the system calls are called the "lost" calls.

The number of tasks in the system (waiting in line and in service) is called the ***state*** of the system, the ***queue size,*** or ***queue length***. ***Waiting time*** is another basic characteristic of a queueing system that will inform of how long a task would have to wait before its service starts and how long it will take to get out of the system. The latter sometimes is called the ***sojourn time***. Finally, a third basic characteristic of a queueing system is the duration of time in which the server is busy. This characteristic is called ***busy period of the server***. A similar term is ***busy period of the system***, which means the duration of time in which the system starts to work until it stops for the first time after the start.

Other characteristics of a queue that are not basic also exist. Queueing systems with impatient tasks have a special significance for the business community. A task (customer) is said to be ***impatient*** if he/she tends to join the queue only when a short wait is expected and remain in the line if his/her wait has been very small. In general, impatience takes three forms: (1) ***balking***, that is, the task decides not to join the queue at all upon arrival; (2) ***reneging***, that is, the task is reluctant to remain in the waiting line after joining and waiting; and (3) ***jockeying*** between lines when each of a number of parallel service channels has its own queue (Gross and Harris 1985).

The notion of impatient tasks in queueing theory is shown by Haight (1957). Recently, a review on queueing systems with impatient customers was presented by Wang et al. (2010). Ancker and Gafarian (1963a) studied $M/M/1/N$ queueing system with balking and reneging, and derived its steady-state solution. Ancker and Gafarian (1963a) obtained results of a balking system (no reneging) by setting the reneging parameter to zero. Gavish and Schweitzer (1977) considered a deterministic reneging model with an additional assumption that arrivals can be labeled by their service requirement before joining the queue and arriving tasks are admitted only if their waiting time in the system does not exceed a certain fixed value.

Liau (2011) developed a queueing model to estimate business loss, which is quite difficult to estimate. Balking index and reneging rate are used in the model to represent different configurations of balking behavior and reneging behavior, respectively, for different queueing systems. Use of balking index and reneging rate provides decision makers with the capability to estimate the incurred business loss for different values of balking index, reneging rate, and service level.

A queueing system may have more than one service station, which may be set ***parallel***, in ***series*** (or in ***tandem***), or ***mixed***. In that case, each station may have its own buffer or there might be a common buffer for all. In other words, if several service stations are set in parallel, each station may have a buffer of its own and an arrival may choose a waiting line. In this case, a task may choose to move among waiting lines, that is, it may ***jockey***. On the contrary, there might be only one buffer in which all arriving tasks attend in and each will be called for service as a server becomes available.

Finally, a queueing system may be considered ***time-dependent*** or ***transient*** when all events occur as functions of time, otherwise it is said to be in ***Steady-state*** (or in stationary or equilibrium state) if it has been in operation for a "long" time, so that occurrences of events are time-independent.

Before defining some other terms, based on the aforementioned definitions, we introduce some symbolic terms. Let the random variable $\xi(t)$ denote the queue length at time t, $t \geq 0$ and also denote the probability of the (time-dependent) system being in state k by $\psi_k(t)$, that is,

$$\psi_k(t) = P\{\xi(t) = k, \quad t \geq 0\}. \tag{4.1}$$

Hence, the process $\{\xi(t) = k\}$ is an irreducible continuous-time Markov chain. As t approaches infinity, we denote the limiting value (i.e., the process has run for long enough and hence becomes time independent) by ψ_k, that is,

$$\psi_k = \lim_{t\to\infty} \psi_k(t), \tag{4.2}$$

for each state k (assuming the limits exists). In other words, the state of the system is the event $\{\xi = k\}$. In a stationary queueing system, the ratio of the average number of arrivals, λ, to the average number of services completed, μ, denoted by, ρ, is

$$\rho = \frac{\lambda}{\mu}, \tag{4.3}$$

which is called the ***traffic intensity*** (or ***utilization factor***) of the system. In other words, ρ is the probability of the server being busy. We can also express Equation (4.3) as follows:

$$\rho = \frac{\text{mean service time of a task}}{\text{mean interarrival time}}. \tag{4.4}$$

We pause to note that going through the characteristics of a simple queue such as the one presented in Figure 4.1, we can observe the analogy between a queueing system and general B–D process. In fact, we take the opportunity to show one more of these characteristics; recall that for the Poisson pure birth process, we showed that distribution of holding times was exponential, that is, also the case for the B–D process. Different methods exist to proceed with the proof. One of them is as follows.

Let θ_k denote the random variable representing a holding time of $\xi(t)$ in state k, that is, the duration for which the population size (queue length) remains in steady state before a change occurs. Thus, the question is to find the distribution of θ_k. We define $H_k(t)$ as

$$H_k(t) = P\{\theta_k \geq t, \quad k = 0, 1, 2, \ldots\}. \tag{4.5}$$

From the Markov property and other materials discussed so far, for a small time interval Δt, we obtain the following equations:

$$\begin{aligned} H_k(t + \Delta t) &= H_k(t)H_k(\Delta t) \\ &= H_k(t)[p_{k,k}(\Delta t) + o(\Delta t)] \\ &= H_k(t)[1 - \Delta t(\lambda_k + \mu_k)] + o(\Delta t) \\ &= H_k(t) - H_k(t)\Delta t(\lambda_k + \mu_k) + o(\Delta t). \end{aligned}$$

Hence,

$$\begin{aligned} \frac{H_k(t + \Delta t) - H_k(t)}{\Delta t} &= \frac{H_k(t) - H_k(t)\Delta t(\lambda_k + \mu_k) + o(\Delta t) - H_k(t)}{\Delta t} \\ &= \frac{-H_k(t)(\lambda_k + \mu_k) + o(\Delta t)}{\Delta t}. \end{aligned}$$

Passing through the limit as Δt approaches zero, we obtain

$$H_k'(t) = -(\lambda_k + \mu_k)H_k(t), \tag{4.6}$$

which is a linear first-order differential equation. Assuming that $H_k(0) = 1$, the solution of Equation (4.6) is

$$H_k(t) = e^{-t(\lambda_k + \mu_k)}, \tag{4.7}$$

from which, we observe that

$$\begin{aligned} P\{\theta_k < t\} &= 1 - P\{\theta_k \geq t\} \\ &= 1 - H_k(t) \\ &= 1 - e^{-t(\lambda_k + \mu_k)}, \quad k = 0, 1, 2, \ldots. \end{aligned} \tag{4.8}$$

In addition to the aforementioned types, there are many other types of queueing systems in the real world – industry, health services, and business. However, two basic types, $M/M/1$ and $M/M/1/K$, are considered standard and numerous research papers have been published about them and their variations. We discuss these standard queueing systems below.

David G. Kendall (1918–2007)

Standard queueing systems are symbolized, generically, by the notation A/B/C/D (introduced by Kendall in 1953). The letters represent the arrival type (for instance, deterministic, or stochastic; Markovian, or non-Markovian; or general), and service type (such as deterministic or probabilistic; Markovian, or Erlangian; or even general). The symbols also indicate the number of servers in parallel, and the capacity of the system (that is, the capacity of the buffers and the service stations). Hence, M/M/1/K means that the interarrival times are exponential (Markovian), that is, the arrival process is a Poisson process, and service times distribution is also exponential (Markovian) with a single server and finite buffer with capacity $K-1$, with total capacity K. When the capacity of the buffer is infinite, the letter K is dropped.

Among factors of interest in a queueing system, in both transient and stationary cases, are the average queue length, the average waiting time, the average sojourn time, and the length of the busy period of a server and the system.

It should be noted that not all queues are Markovian. This indicates that arrival process, service process, or both may have arbitrary distributions. In such cases, we may have systems such as $M/G/1$, $G/M/1$, $G/G/1$, or all with multiservers, finite capacity, feedback, balking, reneging, and being in tandem. Thus, we need methods of solutions of Markovian queue as well as of queueing systems that are derived by non-Markovian stochastic processes.

Among the classical methods of solutions available are: ***imbedded Markov chain method*** due to Plam (1943) and Kendall (1951), ***method of stages*** (only when the interarrival and service time probability density functions (pdfs) have Laplace transforms that are rational), ***Lindley integral equation method*** (particularly useful for $G/G/1$ system; Lindley 1952), ***functional equation method*** (used in solving tandem queues; Haghighi and Mishev 2009), and ***method of supplementary variables*** (Cox 1955 and Kendall 1953). In particular, for non-Markovian queueing system, ***random walk method***, ***combinatorial approach***, and ***Green's function method*** (Keilson 1965) exist.

4.2 REMARKS ON NON-MARKOVIAN QUEUES

As Markovian queues are extensively described in the following chapters and we have to deal with some non-Markovian queues, as well in this section, we present some remarks on nonstandard queues.

4.2.1 Takács's Waiting Time Paradox

Takács in his celebrated 1962 book "Introduction to the Theory of Queues" provides an example that shows surprising phenomena may occur in waiting time problem: Suppose that buses arrive at a particular bus station according to a homogenous recurrent process. The interarrival times of buses has a distribution function $F(x)$ with average of β unit of time. The following question arises: What is the average waiting time of a passenger arriving at time t at the station?

A quick answer to the question may be $\beta/2$, because if the length of time between arrival of two buses is β, then the average would be a half way between. However, this answer is correct only if the buses run at exactly β-time interval. Because that is not the case and the average waiting time depends on variance of the interarrival times, the correct answer is $\beta/2 + \sigma_\beta^2/(2\beta)$, where σ_β^2 is the variance of the interarrival times. If σ_β^2 is infinite, then the average waiting time of a passenger is also infinite. However, the average waiting time of a passenger is, indeed, finite! This is known as Takács's paradox.

We now express the idea that has been rigorously reviewed by Takács. Also see Haghighi and Mishev (2014).

Theorem 4.2.1 Takács Theorem Suppose that buses arrive at a bus stop according to a homogeneous recurrent process. Let the distribution of interarrival times be denoted by $F(x)$, and the average interarrival times by β. The average waiting time of a passenger arriving at time t at the station, denoted by $\overline{W}_t$, is given by

$$\overline{W}_t = \begin{cases} \dfrac{\beta}{2} + \dfrac{\sigma_\beta^2}{2\beta}, & \sigma_\beta^2 < \infty, \\ \infty, & \sigma_\beta^2 = \infty, \end{cases} \tag{4.9}$$

where σ_β^2 is the variance of the interarrival times.

Proof: Let us denote by $\tau_1, \tau_2, \ldots, \tau_n, \ldots$, the arrival epochs of buses in the time interval $[0, \infty)$. Let $W_t(x)$ denote the distribution of the waiting time of a passenger arriving at the station at time t. That is, $W_t(x)$ is the probability that the waiting time of this passenger will be less than or equal to x at time t. Thus, we can write

$$\begin{aligned} W_t(x) &= \sum_{n=1}^{\infty} P\{t < \tau_n \le t + x < \tau_{n+1}\} \\ &= \sum_{n=1}^{\infty} \int_t^{t+x} [1 - F(t + x - u)] dP(\tau_n \le u), \quad x \ge 0. \end{aligned} \tag{4.10}$$

Now, the waiting time is less than or equal to x, if at least one bus arrives in the time interval $(t, \quad t+x]$. However, this event can occur in many mutually exclusive ways: the last bus arriving in $(t, \quad t+x]$ may be the 1st, 2nd, … , nth, …. Therefore, the process of arrival being homogeneous implies that the expected number of arriving buses in the time interval $[0, \quad \infty)$, denoted by $\overline{\xi}_t$, is

$$\overline{\xi}_t = \sum_{n=1}^{\infty} P\{\tau_n \leq t\} = \frac{t}{\beta}. \tag{4.11}$$

Hence,

$$W_t(x) = \frac{1}{\beta}\int_t^{t+x} [1 - F(t+x-u)] \; du = \frac{1}{\beta}\int_0^x [1 - F(y)] \; dy, \tag{4.12}$$

which is independent of t. Thus,

$$\overline{W}_t = \begin{cases} \int_0^\infty x dW_t(x) = \frac{1}{\beta}\int_0^\infty x[1 - F(x)] \; dx = \frac{\beta^2+\sigma_\beta^2}{2\beta}, & \sigma_\beta^2 < \infty, \\ \int_0^\infty x \; dW_t(x) = \infty, & \sigma_\beta^2 = \infty, \end{cases} \tag{4.13}$$

which completes the proof.

4.2.2 Virtual Waiting Time and Takács's Integro-Differential Equation

In general, as indicated in Figure 4.1, a queueing system consists of arrivals, services, and departures, in case of an open queue and no departure in a cyclic queue. In addition to some of the stochastic processes studied in the queueing system, the unfinished work is also of interest, denoted by $U(t)$, that exists in the system at time t. Thus,

$U(t) \triangleq$ is the ***unfinished work*** in the system at time t, or
the remaining time required to empty the system of all tasks present at time t,

where the notation $\triangleq$ is to read as "***equals by definition***." When $U(t) > 0$, the system is referred to as ***busy*** and when $U(t) = 0$, it is ***idle***.

It is important to note that in case of the unfinished work, from its definition, $U(t)$ itself is independent of order of service only if the server is busy as long as a task is in the system and no task leaves the system until its service is complete. Under this condition, the system is referred to as ***work conserving system***.

When the service discipline in the queueing system under consideration is FCFS, the unfinished work of a task at time t is referred to, by Takács, as the ***virtual waiting time at time t***. Denoting by random variable $\eta(t)$, the virtual waiting time of a task at time t is the time an arriving task at time t has to spend in the queue before its work completion, in other words, how long a (virtual) task would wait in queue if it arrived at time t. One may be interested in the ***duration*** and ***location*** of the busy and idle periods. Therefore, we can study the system by observing the system as it passes through the cycle of busy and idle periods. Hence, we can find the distribution of the busy and idle periods.

We now consider the ***distribution of the virtual waiting time***. The arrival epochs of tasks in the time interval $[0, \quad \infty)$ are denoted by $\tau_1, \tau_2, \dots, \tau_n, \dots$. Let us describe the behavior of the process $\{\eta(t), \quad 0 \le t < \infty\}$. $\eta(0)$ is the initial occupation time of the server. Then, we have the following two cases:

1. If $\eta(0) = 0$, then the server is idle at time $t = 0$.
2. If $\eta(0) \neq 0$, then $\eta(0) = 0$ is the instance when the server is not busy for the first time, if no new task arrives.

In combinatorics terminology, we describe the problem as follows: We start by denoting C_n as the nth task, τ_n as the arrival epoch of $C_n, t_n = \tau_{n+1} - \tau_n$, $n = 0, 1, \dots, \tau_0 = 0$, as the time between the tasks C_n and $C_{n-1,}$ and x_n the service time for C_n.

It is assumed that $\{t_n, \quad n = 0, 1, 2, \dots,\}$ are independent, identically distributed (iid) positive random variables with distribution function $F(x) = P\{t_n \le x\}$, where

$$F(x) = \begin{cases} 1 - e^{-\lambda x}, & x \ge 0, \\ 0, & x < 0, \end{cases} \tag{4.14}$$

with $\lambda > 0$ as the mean of the Poisson arrival process. It is further assumed that the duration of the successive service times are iid positive random variables with an arbitrary distribution function $H(x)$.

We also denote durations of the busy period by $B_1, B_2, \dots$ and of the idle periods by $I_1, I_2, \dots$. It should be noted that the durations of successive idle period and busy period are independent random variables.

We assume that task C_1 arrives at τ_1 when the system is idle, that is the system is empty, and requires a work of size x_1. Thus, arrival of C_1 will terminate the previous idle period and a new busy period begins. It is clear therefore that the unfinished work at the arrival of C_1 was zero and the system's unfinished work jumps up to the size x_1. As time progresses and the server is serving C_1, the unfinished work size decreases linearly with slope of -1. If we assume the time unit as a second, t_2 second later C_2 arrives at τ_2 and will require a work of size x_2. Thus, the $\eta(t)$ will have another vertical jump, this time of size x_2. The process continues with vertical jumps and slope decreases, as the server serves the arriving tasks, until the system has served all tasks in it and becomes empty, that is it has reached the instance $\tau_1 + B_1$. At this time the busy period ends and a new idle period begins. The cycles of busy and idle periods begin and end as C_3, C_4 and other tasks enter the system. Departure instances may be found on the horizontal axis as reflection of decreased points of $\eta(t)$.

As mentioned, the process $\{\eta(t)\}$ decreases linearly with slope -1, except that it jumps at $\tau_n = t_n$ with magnitude of the jump as $x_n, n = 0, 1, 2, \dots$, and when it becomes 0, it will remain there until a jump occurs. The stochastic process described by the virtual waiting times random variable family $\{\eta(t), 0 \le t < \infty\}$ forms a ***continuous time-parameter Markov process of the mixed type***, subject to discontinuous jumps. See Kolmogorov (1931) and Feller (1936). In such a process, changes of states occur with jumps.

Let us now denote the distribution of the virtual waiting time $\eta(t)$ by $W(t,x)$, that is,

$$W(t,x) = P\{\eta(t) \leq x\}. \tag{4.15}$$

Also, let the Laplace–Stieltjes of $W(t,x)$ be denoted by $W^*(t,s)$ and of $W(0,x)$ by $\widehat{W}(s)$, that is,

$$W^*(t,s) = E\{e^{-s\eta(t)} = \int_0^\infty e^{-sx} d_x W(t,x) \tag{4.16}$$

and

$$\widehat{W}(s) = W(0,s). \tag{4.17}$$

Theorem 4.2.2 Takács Theorem The Laplace–Stieltjes transform of the virtual waiting time $\eta(t)$ is given by:

$$W^*(t,s) = e^{st-[1-\psi(s)]\lambda t} \left\{ \widehat{W}(s) - s \int_0^t e^{-su+[1-\psi(s)]\lambda u} P_0(u)\, du \right\}, \tag{4.18}$$

where $P_0(t) = P\{\eta(t) = 0\}$ can be obtained by the Laplace transform

$$\int_0^\infty e^{-st} P_0(t) dt = \frac{\widehat{W}[s + \eta(1 - \gamma(s))]}{s + \lambda - \lambda\gamma(s)}, \quad \mathcal{R}(s) > 0, \tag{4.19}$$

and $z = \gamma(s)$ is the root with smallest absolute value of:

$$z = \psi[s + \lambda(1 - z)], \quad \text{within the unit circle } |z| < 1. \tag{4.20}$$

Proof: See Takács (1962, p. 52).

It is remarkable that Takács showed the distribution of the virtual waiting time, $W(t,x)$, satisfies:

$$\frac{\partial W(t,x)}{\partial t} = \frac{\partial W(t,x)}{\partial x} - \lambda \left[W(t,x) - \int_0^x H(x-y) d_y W(t,y) \right], \quad \forall x \geq 0, t \geq 0. \tag{4.21}$$

Formula (4.21) is commonly referred to as the ***Takács's integro-differential equation***. For detail, see Takács (1962, pp. 54–55).It can be shown that except for a jump at $t = 0$, $W(t,x)$ is a continuous function of t. Thus, with that exception, $d_y W(t,y)$ in Equation (4.21) can be written as $W(t,y)dy$. It can also be shown the relation (4.21) remains to be valid if λ is replaced by $\lambda(t)$.

Now, for $M/G/1$, it is clear that the distribution of idle periods is given by (4.14). This is because, when a busy period ends, a new idle period begins. This idle period terminates as soon as a task arrives. Based on the memoryless property of the arrival process, the time until arrival of the next task follow (4.14).

For the busy periods, if $\eta(0) = 0$, then the process starts with an idle period and the length of every busy period has the same distribution, say, $G(x)$. On the other hand, if $\eta(0) \neq 0$, then the process starts with a busy period. For this case, we define the distribution of the length of the initial busy period by $\widehat{G}(x)$. It is assumed that

$$\widehat{G}(x) = \begin{cases} 1, & \eta = 0, \quad x \geq 0, \\ 0, & \eta = 0, \quad x < 0. \end{cases} \tag{4.22}$$

The distribution function of the length of other busy periods is by $G(x)$.

Define the Laplace transforms of $G(x)$ and $\widehat{G}(x)$ by $\Gamma(s)$ and $\widehat{\Gamma}(s)$, respectively, for $\mathcal{R}(s) \geq 0$.

Theorem 4.2.3 The Laplace transform of the distribution of the length of a busy period, $\Gamma(s)$, is

$$\Gamma(s) = \gamma(s), \quad \mathcal{R}(s) \geq 0, \tag{4.23}$$

where $\gamma(s)$ is the root with smallest absolute value in z of the equation

$$z = \psi[s + \lambda(1 - z)], \tag{4.24}$$

and

$$\lim_{x \to \infty} G(x) = \omega, \tag{4.25}$$

where ω is the root with smallest absolute value in z of the equation

$$z = \psi[\lambda(1 - z)]. \tag{4.26}$$

Let the average virtual waiting time $E[\eta(t)]$ be denoted by $\alpha(t)$. Then,

if $\lambda\alpha \leq 1$, then $\omega = 1$ and $G(x)$ is a proper distribution function, and

if $\lambda\alpha > 1$, then $\omega < 1$ and $G(x)$ is an improper distribution function. In this case, the busy period will be infinite with probability $1 - \omega$.

The Laplace transform $\widehat{\Gamma}(s)$ can be found from

$$\widehat{\Gamma}(s) = \widehat{W}[s + \lambda - \lambda\Gamma(s)]. \tag{4.27}$$

If $\lambda\alpha \leq 1$, then $\widehat{G}(\infty) = 1$. If, however, $\lambda\alpha > 1$ and $\eta(0) \neq 0$, then $\widehat{G}(\infty) < 1$. Finally, the distribution of general busy period, $G(x)$, can be expressed explicitly by

$$G(x) = \sum_{n=1}^{\infty} \int_0^x e^{-\lambda x} \frac{(\lambda u)^{n-1}}{n!} dH_n(u), \tag{4.28}$$

where $H_n(x)$ is the nth iterated convolution of $H(x)$ with itself. Assuming that $\eta(0) = y$, where y is a constant, then the distribution of the initial busy period, $\widehat{G}(x)$, is:

$$\widehat{G}(x) = \begin{cases} \lambda_y \sum_{n=0}^{\infty} \int_0^{x-y} e^{-\lambda(y+u)} \frac{\lambda(y+u)^{n-1}}{n!} dH_n(u), & x \geq y, \\ 0, & x < y. \end{cases} \tag{4.29}$$

Proof: See Takács (1962, p. 59).

Example 4.2.1 Reliability of Bandwidth Measurement Efficient and reliable available bandwidth measurement is yet of great concern for many applications. Recently a new class of active probing techniques has emerged based on observing an increased separation of probe packets due to local saturation of the queue at the narrow link, and do not require the knowledge of link capacities along the path. Hága et al. (2006) introduce a theoretical model of packet pair separation based on a transient solution of the Takács integro-differential equation. They show that in addition to the parameters of the fluid approximation (physical bandwidth and the average cross traffic rate) the introduction of a new parameter characterizing the granularity of the cross traffic is necessary. They determine the distribution of the output spacing δ' in a single hop packet-pair measurement, where the input spacing of the probe packets is δ. The packet-pair consists of probe packets of size p(in bits). The queue is assumed to be infinite with an outgoing physical bandwidth C (in bps) and has a FIFO policy.

The queue size just before the arrival of the first probe packet is w_1 (in bits) and changes to $w_1 + p$ due to the arrival of the probe packet. The second probe packet arrives δ time later, when the queue length is w_2. The spacing between the exit times of the probe packets is

$$\delta' = \delta + \frac{w_2 - w_1}{C}. \tag{4.30}$$

Assuming steady-state process, the distribution of the output spacing is related to the conditional probability $F(w, t|w_0)$ of having queue length w at time t given the queue length is w_0 at $t = 0$. Thus, $t = \delta, w = w_2, w_0 = w_1 + p$ and the distribution of the output spacing is

$$P(\delta'|\delta, w_1) = F[(\delta' - \delta)C + w_1, \delta|w_1 + p]. \tag{4.31}$$

In Equation (4.31), $P(\delta'|\delta, w_1)$ depends on the initial value of the queue length w_1.

In their analysis, they show that the model of the cross traffic defines an $M/G/1$ queueing system, where the time evolution of the queue length distribution is described by the Takács integro-differential equation of the form

$$\frac{\partial F(w,t)}{\partial t} = \lambda \left[\int_0^w F(w-x,t)b(x)dx - F(w,t) \right] + C\frac{\partial F(w,t)}{\partial w}, \tag{4.32}$$

where the cross traffic arrival follows a Poisson distribution with parameter of λ per second and pdf $b(x)$ of the arriving packets of size bits.

For details of this example, the reader should refer to Hága et al. (2006).

4.2.3 The Unfinished Work

We continue consideration of the $M/G/1$. Again, for such a system, if at an epoch of time t we are to characterize the complete past history of the system, then we should specify the number of tasks present at time t, say $N(t)$. Also, since the service time distribution is not necessarily of memoryless type, it is assumed an arbitrary, we need to specify the service time already received by the task in service at time t, say $X_r(t)$.

Hence, the random process $N(t)$ is a non-Markovian process. However, the vector $[N(t), X_r(t)]$ is Markovian and is appropriate for $M/G/1$. This is because, it completely characterizes all past history relevant to the future development of the system.

It is commonly happens that an arriving task finds a task in the service station being partially serviced. Let us discuss an illustrative example. Suppose taxicabs arrive at a taxi station according with a Poisson distribution with average rate of λ cars per minutes. Hence, the average time between arrivals of cars will be $1/\lambda$ minutes. Since the interarrival times are exponentially distributed and, thus, memoryless property holds, the time until the next arrival is independent of how long it has passed since the previous arrival. Thus, a new arrival passenger at the taxi station must wait $1/\lambda$ minutes, on the average, before catching a cab. Suppose passengers arrive at the taxi station randomly and one at a time. Then, the average time between arrival of the last car and the next car will be $2/\lambda$ minutes.

Now, let us consider the same problem with the interarrival times distribution of the cabs as an arbitrary distribution. Then, the problem becomes one that belongs to the ***renewal theory***. So, let A_k denotes the kth cab and it will arrive at the time epoch τ_k. We assume that interarrival times $\tau_{k+1} - \tau_k$ are iid random variables with cumulative distribution function (cdf) and pdf as $F(x)$ and $f(x)$, respectively, defined by

$$F(x) \triangleq P\{\tau_{k+1} - \tau_k \leq x\} \text{ and } f(x) \triangleq \frac{dF(x)}{dx}. \tag{4.33}$$

Suppose now that a passenger arrives at the cab station at a random time t. We also suppose that A_{n-1} and A_n denote the last cab arrived at the station before the time epoch t and the cab arrived after epoch t, respectively. Let X denote this special interarrival time and Y denote the time the arriving passenger must wait until the next cab arrives.

An analogy of the sequence of arrival epochs $\{\tau_k\}$ is the sequence of instances at which broken or failed components must be replaced by new ones. From the perspective of renewal theory, the random variable X is referred to as the ***lifetime*** of the component under consideration; Y is referred to as the ***residual life*** of the component at time t, and $X_r = X - Y$ is called the ***age*** of the component at time t. The interest is in the pdf for X and Y, assuming that the renewal process has been in operation for an arbitrary long time, such that we can find the steady-state pdfs.

In order to find the pdfs in question, let us start with denoting the probability distribution and density functions of the residual life, Y, of the component, respectively, by $\widehat{F}(x)$ and, $\widehat{f}(x)$, that is,

$$\widehat{F}(x) \triangleq P\{Y \leq x\} \text{ and } \widehat{f}(x) \triangleq \frac{d\widehat{F}(x)}{dx}. \tag{4.34}$$

Similarly, we denote the cdf and pdf of the life time, X, of the component, respectively, by $F_X(x)$ and $f_X(x)$, that is,

$$F_X(x) \triangleq P\{X \leq x\} \text{ and } f_X(x) \triangleq \frac{dF_X(x)}{dx}. \tag{4.35}$$

It is left as an exercise for the reader to derive $\widehat{f}(x)$, rigorously. What is interesting is that it turns out that distribution of X is not $F(x)$ as in Equation (4.35). It would indicate that a long interval between renewal points contains larger segments of the time axis than do the shorter intervals. Thus, it is more likely that the selected random time epoch t will fall in a long interval. Thus, the probability that an interval of length x is chosen would be proportional to the length (x) and the relative occurrence of such intervals, which is given by $f(x)dx$. In other words, for the selected interval, we have

$$f_X(x)dx = Kxf(x)dx, \tag{4.36}$$

where

$$f_X(x)dx = P\{x < X \leq x + dx\}, \tag{4.37}$$

and the left-hand side of (4.37) expresses the linear weighing with respect to interval length that includes the constant K, which can be calculated through normalizing the density function. Thus, integrating both sides of Equation (4.37) yields

$$K = \frac{1}{m_1}, \tag{4.38}$$

where

$$m_1 \triangleq E\{\tau_k - \tau_{k-1}\} \tag{4.39}$$

and is the common average time between renewals, that is, between arrivals of taxicab cars. Therefore,

$$f_X(x) = \frac{xf(x)}{m_1}. \tag{4.40}$$

Now that we have reached answer to the first objective, and we proceed with the second one, that is, the density function of residual life, $f(x)$. Suppose that $X = x$. The probability that the residual life Y does not exceed the value y is given by

$$P\{Y \leq y | X = x\} = \frac{y}{x}, \quad 0 \leq y \leq x. \tag{4.41}$$

Relation (4.41) is true because a point has been chosen randomly from within the selected interval. Thus, the point must be distributed uniformly. Hence, the joint density function of X and Y is as follows:

$$\begin{aligned} P\{y < Y \leq y + dy, \quad x < X \leq x + dx\} &= \left(\frac{dy}{x}\right)\left(\frac{xf(x)dx}{m_1}\right) \\ &= \frac{f(x)dydx}{m_1}, \quad 0 \leq y \leq x. \end{aligned} \tag{4.42}$$

Integrating Equation (4.42) with respect to x, we have

$$f(y)dy = \int_{x=y}^{\infty} \frac{f(x)dydx}{m_1}, \tag{4.43}$$

from which we obtain

$$f(y) = \frac{1 - F(y)}{m_1}. \tag{4.44}$$

Thus, the density function of Y, the residual life of the component, has been found in terms of the common distribution of interval length and its mean.

It is left as an exercise for the readers to show that the stationary pdf for age, X_r, is the same as that for Y, that is, relation (4.44).

4.3 STATIONARY *M/M/*1 QUEUEING PROCESS

*M/M/*1 stands for a single-server queueing system with Markovian arrival and service processes, possessing the Markov property, Poisson arrivals from infinite resources, Markovian service times (exponentially distributed), and infinite buffer capacity. In addition, we assume the services are rendered by a single-server according to FCFS discipline. We make additional assumptions as follows: (1) The Poisson process is with constant mean λ, that is, the interarrival times are exponentially distributed with a constant average $1/\lambda$ time units. (2) The exponential service times has a constant average rate of $1/\mu$.

In summary, the classical *M/M/*1 queueing model has a simple structure. That is,

(i) there is a single server;
(ii) service times are iid exponential random variables with mean $1/\mu$;
(iii) tasks arrive according to a Poisson process with rate λ; and
(iv) the system can hold an infinite number of tasks.

These properties guarantee that the process is Markovian in continuous time. On the other hand, it is the simplest nontrivial example of a birth-death process with a constant birth rate λ and constant death rate μ.

Among many applications of *M/M/*1, very recently, Maragatha and Srinivasan (2012) have used it for automatic teller machine (ATM), which is used to save the cost and reachability of a bank by satisfying customer needs. Since the major problem faced by these ATMs seems to be the long queue of customers, the *M/M/*1 queueing system with FCFS queue discipline to study the waiting lines of customers in ATM has been chosen as a case study for centers of particular banks. Their main purpose of this study was to develop an efficient procedure for ATM waiting line reduction. As a result, they concluded that more ATMs should be installed in the centers to reduce the traffic.

We note that Burke (1956) proved the following theorem:

Theorem 4.3.1 Burke's Theorem For $M/M/1$, $M/M/c$ or $M/M/\infty$ queueing systems in the steady state with arrival rate λ, (1) the departure process is also Poisson with the same rate λ and (2) at time t, the number of tasks is independent of the sequence of departure times prior to t.

Proof: See Burke (1956).

The description of the system with state space $S = \{0, 1, 2, \dots\}$. We analyze this system in the stationary case and find the distribution of the queue length, that is, probability that there are n tasks in the system, denoted by $\psi_n, n = 0, 1, 2, \dots$.

It can be shown that ψ_n satisfies the following system of difference equations (see Haghighi and Mishev 2013, Theorem 5.5.2.1):

$$\begin{cases} \lambda\psi_0 = \mu\psi_1, \\ (\lambda + \mu)\psi_n = \lambda\psi_{n-1} + \mu\psi_{n+1}, \quad n = 1, 2, \dots, \\ \sum_{n=0}^{\infty} \psi_n = 1. \end{cases} \tag{4.45}$$

The system (4.45) can be solved iteratively and ψ_n can be found as

$$\psi_n = (1 - \rho)\rho^n, \quad n = 0, 1, 2, \dots, \rho < 1, \tag{4.46}$$

where ρ is the traffic intensity, as in Equation (4.47).

It is important to note that when there is no task in the system, the server will be idle. Hence, ψ_0 represents the probability that the server is idle,

$$\psi_0 = P(\xi = 0) = 1 - \rho. \tag{4.47}$$

Thus, the utilization factor, ρ, the proportion of time that the system is busy, is $\rho = 1 - \psi_0$.

From Equation (4.46), the expected queue length for the stationary *M/M/*1 system, denoted by L, is

$$L \equiv E(\xi) = \sum_{n=0}^{\infty} n\psi_n = (1 - \rho)\rho \sum_{n=0}^{\infty} n\rho^{n-1} = \frac{\rho}{1 - \rho}. \tag{4.48}$$

The mean queue length for the waiting line (buffer), denoted by L_q, can be found as follows:

$$L_q = \sum_{n=0}^{\infty} (n - 1)\psi_n = \frac{\rho^2}{1 - \rho}. \tag{4.49}$$

We leave it as an exercise to show that

$$L = L_q + \rho. \tag{4.50}$$

For a stationary queue, the mean waiting time in the waiting line, denoted by W_q and in the system, denoted by W, may be found through relations with the mean queue length in the waiting line and system, respectively. These relationships are obtained using the well-known ***Little***'s formula or theorem (named after J. D. C. Little (1961), who was the first to present a formal proof). The theorem is as follows:

Theorem 4.3.2 Little's Theorem When a queueing system attains a steady state, the average number of tasks (packets, jobs, customers) in a queue is equal to the product of the average arrival rate of the tasks and the average time a task is kept in the queue. In other words,

$$L = \lambda W \tag{4.51}$$

and

$$L_q = \lambda W_q. \tag{4.52}$$

Proof: See Little (1961) or Haghighi and Mishev (2013).

Thus, for *M*/*M*/1, from Equation (4.49) to Equation (4.52), we have

$$W = \frac{1}{\mu - \lambda} \tag{4.53}$$

and

$$W_q = \frac{\lambda}{\mu(\mu - \lambda)}. \tag{4.54}$$

It is clear that the mean sojourn time can be found by adding the average service time $1/\mu$ to W_q given in Equation (4.54), that is,

$$W = W_q + \frac{1}{\mu}. \tag{4.55}$$

Denoting the length of a busy period by the random variable B, the mean busy period of the server can be obtained as

$$E(B) = \frac{1}{\mu(1 - \rho)}. \tag{4.56}$$

See Haghighi and Mishev (2013).

Example 4.3.1 Calculating Queue Measures Suppose that tasks, in this case, jobs, arrive at a computer main server for processing at the mean rate λ. The processor runs 50 programs per minute, while there are nine jobs in the system on average. Assuming that the arrival and processing distributions are Poisson and exponential, respectively, calculate (1) λ, (2) the probability that the system is empty, (3) the probability that there are four jobs in the system, (4) the mean number of jobs in the waiting line, (5) the probability of not more than five jobs in the system, (6) the average sojourn time and average waiting time in the waiting line for a job, and (7) the mean busy period of the system.

This is an example of a stationary *M*/*M*/1 queueing system. Hence, we answer the questions as follows:

1. In order to find λ, we note that $\mu = 50$ and $L = 9$. Therefore, from Equation (4.49), we have $\lambda = 45$ with $\rho = 0.9 < 1$.
2. In order to find the probability of the server being idle from Equation (4.48), we have $\psi_0 = 0.10$.

3. From Equation (4.46) we have $\psi_3 = P(\xi = 3) = (0.1)(9/10)^4 = 0.06561$.
4. From Equation (4.50), we have $L_q = 8.1$.
5. The probability of at most five jobs in the system can be found from Equation (4.46) by summing from 0 to 5, which is $P\{\xi \leq 5\} = 0.468559$.
6. From Equation (4.53), we have $W = 12$ s and $W_q = 10$ s.
7. From Equation (4.55), we have $E(B) = 0.20$.

4.4 A PARALLEL *M/M/C/K* WITH BAKING AND RENEGING

In order to show how to treat a parallel Markovian queueing system, we discuss the transient distribution of the queue size for a finite multiserver parallel Markovian queueing system with balking and reneging considered in Montazer-Haghighi et al. (1986), which is a relatively general case. Here is the model:

We consider a queueing system with a finite buffer. Tasks arrive from an infinite source one at a time according to a Poisson process with parameter λ. There are c servers in place, parallel to each other, so that an arrival may be served by any one of them. The buffer (waiting room) size in front of the service station is finite with capacity $K - c$, and together with the number of servers, this make the system capacity as K. An arriving task has a choice of server, if not all servers are busy. However, an arriving task finding the system at its capacity at its time of arrival will be lost. Service of tasks is in the order of arrivals. The service times of the tasks are *iid* random variables having exponential distribution with parameter μ.

Further, an arriving task that finds all the servers busy on arrival, but the buffer not full, may balk (i.e., decline to join the waiting line) with probability q or join the system with probability $p, q + p = 1$. Thus, $q\lambda$ is the instantaneous balking rate (i.e., $q\lambda$ is the arriving rate when all servers are busy, but the buffer is not full). In addition, after joining the queue, a task may renege; that is, it will wait only for a certain length of time for service to begin, and then depart from the system. This length of waiting time is a random variable that, we assume, has exponential distribution with parameter α. If a task balks or reneges and decides to return later, it will be considered as a new arrival, independent of its previous numbers of balking or reneging.

On the basis of the description of the model, we have a finite birth and death process that was discussed in the previous chapter (Section 3.2). Thus, in our model, we have

$$\lambda_n = \begin{cases} \lambda, & 0 \leq n \leq c-1, \\ p\lambda, & c \leq n \leq K, \end{cases} \tag{4.57}$$

and

$$\mu_n = \begin{cases} n\mu, & 0 \leq n \leq c-1, \\ c\mu + (n-c)\alpha, & c \leq n \leq K. \end{cases} \tag{4.58}$$

From the system of difference equations (3.25) and relations (4.57) and (4.58), we have the following cases:

(i) For $0 \leq i \leq c-1$:

$$\psi_n(s) = \begin{cases} \dfrac{i!}{n!}\mu^{i-n}\dfrac{T_n(s)B_{K-i}(s)}{A_K(s)}, & 0 \leq n \leq i, \\ \lambda^{n-i}\dfrac{T_i(s)B_{K-n}(s)}{A_K(s)}, & i+1 \leq n \leq c-1, \\ (p\lambda)^{n-i}\dfrac{T_i(s)B_{K-i}(s)}{A_K(s)}, & c \leq n \leq K. \end{cases} \tag{4.59}$$

(ii) For $i = c$:

$$\psi_n(s) = \begin{cases} \dfrac{c!}{n!}\mu^{c-n}\dfrac{T_n(s)B_{K-c}(s)}{A_K(s)}, & 0 \leq n \leq c-1, \\ (p\lambda)^{n-i}\dfrac{T_c(s)B_{K-n}(s)}{A_K(s)}, & c \leq n \leq K. \end{cases} \tag{4.60}$$

(iii) For $c+1 \leq i \leq K$:

$$\psi_n(s) = \begin{cases} \dfrac{c!}{n!}\mu^{c-n}\displaystyle\prod_{k=c+1}^{i}[c\mu + (k-c)\alpha]\dfrac{T_n(s)B_{K-i}(s)}{A_K(s)}, & 0 \leq n \leq i-1, \\ T_i(s)B_{K-i}(s), & n = i, \\ (p\lambda)^{n-i}\dfrac{T_i(s)B_{K-n}(s)}{A_K(s)}, & i+1 \leq n \leq K. \end{cases} \tag{4.61}$$

Thus, the Laplace transform of the distribution of the length of a busy period for this special case is given by Eqs. (4.59)–(4.61).

We leave it as exercise for the reader to use the method we have illustrated before to invert Equation (4.61) and obtain the distribution.

It should be noted that recently Kaczynski et al. (2012) have analyzed a transient parallel infinite buffer queue and offered algorithmic exact distribution of the nth customer's sojourn time.

4.5 STATIONARY *M/M/1/K* QUEUEING PROCESS

We choose the assumptions as in Section 3.4 for the $M/M/1$, except that the buffer is finite with capacity $K-1$, making the total capacity of the system K, as explained earlier. In this case, the system of stationary difference equations is

$$\begin{cases} \lambda\psi_0 = \mu\psi_1, \\ (\lambda+\mu)\psi_n = \lambda\psi_{n-1} + \mu\psi_{n+1}, \quad n = 1, 2, \ldots, K-1, \\ \mu\psi_K = \lambda\psi_{K-1}, \\ \displaystyle\sum_{n=0}^{K}\psi_n = 1. \end{cases} \tag{4.62}$$

Hence, similar to the *M/M/*1 case, the following can be shown (see Haghighi and Mishev 2013, Theorem 5.6.2):

$$\psi_n = \begin{cases} \dfrac{1-\rho}{1-\rho^{K+1}}\rho^n, & \text{if } \rho \neq 1, \\ \frac{1}{K+1}, & \text{if } \rho = 1. \end{cases} \quad n = 0, 1, \ldots, K \tag{4.63}$$

We leave it as an exercise to show that from Equation (4.63), the mean queue length, denoted by L_K, is given by

$$L_K = \begin{cases} \dfrac{\rho}{1-\rho} - \dfrac{(K+1)\rho^{K+1}}{1-\rho^{K+1}}, & \rho \neq 1, \\ \frac{K}{2}, & \rho = 1. \end{cases} \tag{4.64}$$

Since the buffer is of finite size, it is natural that some arriving tasks be lost if the system is full. The expected number of tasks attending the system *M/M/*1*/K* in a unit time is called ***effective arrival rate***. We leave it as an exercise to show that the effective arrival rate, denoted by δ, is

$$\delta = \begin{cases} \dfrac{\lambda(1-\rho^K)}{1-\rho^{K+1}}, & \rho \neq 1, \\ \dfrac{\lambda K}{K+1}, & \rho = 1. \end{cases} \tag{4.65}$$

Hence, the mean interarrival times attending the system is $1/\delta$ and the expected number of possible tasks lost due to the finiteness of the buffer, denoted by L_{loss}, is

$$L_{\text{loss}} = \lambda\psi_K = \begin{cases} \dfrac{\lambda(1-\rho)\rho^K}{1-\rho^{K+1}}, & \text{if } \rho \neq 1, \\ \dfrac{\lambda}{K+1}, & \text{if } \rho = 1. \end{cases} \tag{4.66}$$

We note that for *M/M/*1*/K*, the utilization factor, denoted by ρ_K, is given by

$$\rho_K = \frac{\delta}{\mu} = \rho(1-\psi_K) = \begin{cases} \dfrac{\rho(1-\rho^K)}{1-\rho^{K+1}}, & \text{if } \rho \neq 1, \\ \dfrac{\rho K}{K+1}, & \text{if } \rho = 1. \end{cases} \tag{4.67}$$

Furthermore, using the Little's formula, the expected waiting time in the system, denoted by W_K, is

$$W_K = \frac{L_K}{\delta}, \tag{4.68}$$

where L_K and δ are given by Eqs. (4.64) and (4.65), respectively.

4.6 BUSY PERIOD OF AN *M/M/1/K* QUEUE

The busy period analysis forms an integral part of a queueing system. It plays a vital role in understanding various operations taking place in a queueing system. The concept of busy period was introduced by Borel (1942). The busy period for the Poisson queue was studied by Plam (1943, 1947) and later followed, among others, by Karlin and McGregor (1958), Takács (1962), Rice (1962), Neuts (1964), Erlander (1965), Shanbhag (1966), Nance et al. (1972), Conolly (1971, 1974), Bunday and El-Badri (1985), Parthasarthy and Sharafali (1989), Sharma (1990), and Stadje (1995). Among others, Mohanty et al. (1993), Krishna Kumar (1996), and Tarabia (2003) discussed the busy period analysis of an *M/M/1* queue with balking.

Ammar et al. (2013) studied the transient solution for the probabilities for the density function of the busy period of an *M/M/1* queue with balking and reneging. They discussed the transient solution of the busy period of an *M*/M/1 queueing system where balking and reneging occurs if and only if the system size equals or exceeds a threshold value k. Using the generating function technique, they obtained an explicit expression for the distribution of busy period in terms of modified Bessel function is obtained. This technique is to define the generating function in a special way, which results in a simple differential equation. Applying the properties of the Bessel function in the solution of this differential equation, the probabilities can be extracted in a direct way. Calculation of the effect of the threshold in network of queues in delivering packets is of importance.

We now describe a finite-state B–D process with $K+1$ states, namely $0, 1, 2, \dots, K$, with birth and death rates λ_n and μ_n, respectively, when the process is in state n, $n = 0, 1, 2, \dots, K$. We define an m-state busy period to begin the queue process with an arrival to the system at the time when the process is in state m to the very next time when the process returns to $m-1$. If the busy period starts from the time of arrival to an empty system, then $m = 1$. The system of differential–difference equations for such a busy period, with $\psi_k(t)$ defined by Equation (4.45), is given by Mohanty et al. (1993) as follows:

$$\begin{cases} \psi'_{m-1}(t) = \mu_m \psi_m(t), \\ \psi'_m(t) = -(\lambda_m + \mu_m)\psi_m(t) + \mu_{m+1}\psi_{m+1}(t), \\ \psi'_n(t) = -(\lambda_n + \mu_n)\psi_n(t) + \mu_{n-1}\psi_{n-1}(t) + \mu_{n+1}\psi_{n+1}(t), \quad m+1 \le n \le K-1, \\ \psi'_N(t) = -\mu_K \psi_K + \lambda_{K-1}\psi_{K-1}(t), \end{cases} \tag{4.69}$$

where $\psi_m(0) = 1$. It should be noted that $\psi_{m-1}(t)$, which is equal to $\mu_m \psi_m(t)$, is the density of the distribution of the length of a busy period, which is the unknown variable.

In order to solve the system (4.69) for $\psi_{m-1}(t)$, we let the Laplace transform of $\psi_k(t)$ be denoted by $\Psi_k(s)$. Thus, the Laplace transform of (4.69), skipping the first equation since that is the term we are looking for, is as follows:

$$\begin{cases} (\lambda_m + \mu_m + s)\Psi_m(s) - \mu_{m+1}\Psi_{m+1}(s) = 1, \\ -\lambda_{n-1}\Psi_{n-1}(s) + (\lambda_n + \mu_n + s)\Psi_n(s) - \mu_{n+1}\Psi_{n+1}(s) = 0, \quad m+1 \le n \le K-1, \\ -\lambda_{K-1}\Psi_{K-1}(s) + (\mu_K + s)\Psi_K(s) = 0. \end{cases} \tag{4.70}$$

Let $D_{N,k}(s), k = m-1, m, m+1, \dots, K$, be the determinants of the $k \times k$ matrices formed at the bottom right corner of the coefficient matrices of Equation (4.70). Then, we have

$$\Psi_m(s) = \frac{D_{K,m+1}(s)}{D_{K,m}(s)}. \tag{4.71}$$

Following a similar argument in an earlier section, it can be shown that the zeros of $D_{K,m}(s)$ are distinct and negative, say $u_1, u_2, \dots, u_{K-m+1}$. Hence, the density function of the length of a busy period at time t is as follows:

$$\psi'_{m-1}(t) = \mu_m \psi_m(t) = \mu_m \sum_{k=1}^{K-m+1} \gamma_k e^{u_k t}, \tag{4.72}$$

where

$$\gamma_k = \frac{\prod_{j=1}^{K-m}(u_k - v_j)}{\prod_{j=1,j\neq k}^{K-m+1}(u_k - u_j)}, \quad k = 1, 2, \dots, K-m+1, \tag{4.73}$$

where v_j are the zeros of $D_{K,m+1}(s)$.

We now prove that the distribution of a busy period is hyperexponential. However, we first note that from Equation (38.2) of Wilkinson (1965, p. 301), γ_k given in Equation (4.73) are positive. Hence,

$$\frac{\mu_m \gamma_k}{-u_k} > 0. \tag{4.74}$$

Thus, we have to show that the sum of the terms given in Equation (4.74) is 1. For that, observe that $\Psi_m(s)$ from Equation (4.71) is similar to the one given in Equation (3.30). Thus, we can write similar partial fractions as in Equation (3.31) as follows:

$$\Psi_n(s) = \sum_{k=1}^{K-m+1} \left[\frac{D_{K,m+1}(u_k)}{\prod_{j=0,j\neq k}^{K-m+1}(u_k - u_j)} \right] \left(\frac{1}{s - u_k} \right). \tag{4.75}$$

Rewriting γ_k given in Equation (4.73) as in Equation (3.32), we have

$$\sum_{k=1}^{K-m+1} \frac{\gamma_k}{-u_k} = \sum_{k=1}^{K-m+1} \left[\frac{D_{K,m+1}(u_k)}{\prod_{j=0,j\neq k}^{K-m+1}(u_k - u_j)} \right] \left(\frac{1}{-u_k} \right). \tag{4.76}$$

Thus, from Eqs. (4.71), (4.75), and (4.76), we have

$$\sum_{k=1}^{K-m+1} \frac{\gamma_k}{-u_k} = \Psi_m(0) = \frac{D_{K,m+1}(0)}{D_{K,m}(0)}. \tag{4.77}$$

We leave it as an exercise to show that

$$D_{K,k}(0) = \prod_{i=k}^{K} \mu_{m+i}. \tag{4.78}$$

Substituting Equation (4.78) into Equation (4.77) completes the proof.

We leave it as an exercise to show that the mean and variance of a busy period are as follows:

$$\text{mean of a busy time} = \sum_{k=1}^{K-m+1} \frac{\mu_m \gamma_k}{(u_k)^2} \tag{4.79}$$

and

$$\text{varianceof a busy time} = 2\left[\sum_{k=1}^{K-m+1} \frac{\mu_m \gamma_k}{(u_k)^3}\right] - \left[\sum_{k=1}^{K-m+1} \frac{\mu_m \gamma_k}{(u_k)^2}\right]^2. \tag{4.80}$$

Example 4.6.1 Calculating Queue Length Under the conditions stated in this section, distribution of the length of a busy period when $K = 7$ and $c = 3$ is given in Table 4.1.

4.7 STATIONARY *M/M/*1 AND *M/M/*1*/K* QUEUEING PROCESSES WITH FEEDBACK

Studies of queueing systems with feedback originally started with the consideration of a single-server queue with instantaneous Bernoulli feedback by Takács (1963). Montazer-Haghighi (1976) discussed Takács' extension for multichannel and briefed the delayed-feedback single-server as a particular case of a tandem queueing model without offering a solution. Here we only present the stationary case. Thus, we consider the stationary single-server queueing system, *M/M/*1, discussed in an earlier section. After departure from the service, a task may return to the waiting line, with probability p, $0 < p < 1$, for further service and exit the system with probability of q, $q = 1 - p$. If $p = 0$, then there is no feedback and the system is a *M/M/*1. It is assumed that the return of a task is an event independent of any other event involved in the system and, in particular, independent of the number of its previous returns.

A task may return as many times as necessary before its job is complete and leave the system. Thus, the actual completion rate of service is $q\mu$, which is referred to as the ***effective job service rate.*** This is the rate that changes the state of the system. Also, the rate at which jobs feed back into the system for further work (with no change of the state) is $p\mu$. With these new rates, we can find a similar performance measure as for *M/M/*1, keeping in mind that the traffic intensity in this case is $\rho = \lambda/q\mu$.

All said applies to the *M/M/*1*/K* with feedback.

TABLE 4.1 Distribution of a Busy Period with $N = 7$ and $c = 3$

$\lambda = 1, \mu = 4, \ p = 0.5, \ \alpha = 5$

$U_{m,k}$	$k=1$	$k=2$	$k=3$	$k=4$	$k=5$	$k=6$	$k=7$
$m=1$	−3.19258	−8.08754	−13.24790	−17.25638	−22.05824	−27.48892	−34.6849
$m=2$	−6.62035	−12.94451	−17.22064	−22.05713	−27.48889	−34.66849	
$m=3$	−10.95807	−16.85530	−22.03029	−27.48785	−34.66848		

$p_{m,1}$	$m=1$	$m=2$	$m=3$
$t=0$	0.0000000	0.0000000	0.0000000
$t=1$	0.9651053	0.9989347	0.9999854
$t=2$	0.9985688	0.9999986	1.0000000
$t=3$	0.9999412	1.0000000	1.0000000
$t=4$	0.9999976	1.0000000	1.0000000
$t=5$	0.9999999	1.0000000	1.0000000

$\lambda = 1, \ \mu = 4, \ p = 0.9, \ \alpha = 5$

$U_{m,k}$	$k=1$	$k=2$	$k=3$	$k=4$	$k=5$	$k=6$	$k=7$
$m=1$	−3.18806	−7.93601	−12.57752	−17.13647	−22.19103	−28.16837	−36.40253
$m=2$	−6.52730	−12.22036	−17.09310	−22.18842	−28.16830	−36.40253	
$m=3$	−10.28033	−16.62927	−22.12370	−28.16422	−36.40249		

$p_{m,1}$	$m=1$	$m=2$	$m=3$
$t=0$	0.0000000	0.0000000	0.0000000
$t=1$	0.9650481	0.9988738	0.9999738
$t=2$	0.9985602	0.9999983	1.0000000
$t=3$	0.9999406	1.0000000	1.0000000
$t=4$	0.9999976	1.0000000	1.0000000
$t=5$	0.9999999	1.0000000	1.0000000

$\lambda = 10, \ \mu = 9, \ p = 0.9, \ \alpha = 8$

$U_{m,k}$	$k=1$	$k=2$	$k=3$	$k=4$	$k=5$	$k=6$	$k=7$
$m=1$	−3.77587	−15.40871	−26.10722	−38.66673	−53.03257	−69.47661	−91.53228
$m=2$	−8.60126	−21.58944	−35.87337	−52.02763	−69.37713	−91.53117	
$m=3$	−13.97569	−29.16635	−47.86215	−68.48946	−91.50636		

$p_{m,1}$	$m=1$	$m=2$	$m=3$
$t=0$	0.0000000	0.0000000	0.0000000
$t=1$	0.9837366	0.9998931	0.9999996
$t=2$	0.9996273	1.0000000	1.0000000
$t=3$	0.9999915	1.0000000	1.0000000
$t=4$	0.9999998	1.0000000	1.0000000

$\lambda = 10, \ \mu = 9, \ p = 0.5, \ \alpha = 8$

$U_{m,k}$	$k=1$	$k=2$	$k=3$	$k=4$	$k=5$	$k=6$	$k=7$
$m=1$	−3.94459	−18.24255	−27.76168	−39.44819	−50.93458	−61.73813	−79.93029
$m=2$	−9.75588	−24.65945	−37.47423	−49.58648	−61.59456	−79.92938	
$m=3$	−17.83933	−31.16371	−45.21683	−60.86388	−79.91626		

$p_{m,1}$	$m=1$	$m=2$	$m=3$
$t=0$	0.0000000	0.0000000	0.0000000
$t=1$	0.9857127	0.9999607	1.0000000
$t=2$	0.9997234	1.0000000	1.0000000
$t=3$	0.9999946	1.0000000	1.0000000
$t=4$	0.9999999	1.0000000	1.0000000

4.7.1 Stationary Distribution of the Sojourn Time of a Task

For a pure Poisson birth process, the time between occurrences of events (births) is the holding time in each state. We have already alluded to the fact that this holding time is a random variable that has exponential distribution with the same parameter as of the Poison. In other words, if $\tau_0 = 0, \tau_1, \tau_2, \ldots$ are the times of occurrences of events (births or arrivals) and $\theta_k = \tau_{k+1} - \tau_k, k = 0, 1, 2, \ldots$, are the holding times, then θ_k are mutually independent identically distributed random variables with distribution function as

$$P\{\theta_k \leq x, \quad k = 0, 1, 2, \ldots\} = 1 - e^{-\lambda x}, \quad x > 0. \tag{4.81}$$

In addition to the state holding time, we might be interested in the time for which a task is held in the system until it exits. This time is called the holding time or sojourn time of a task and that is what we are interested in finding for the B–D processes and, in particular, the *M/M/*1, *M/M/*1*/K*, and single-server Markovian queues with feedback. Hence, we consider the latter mentioned so that other cases may be considered as special cases.

Thus, we now consider an *M/M/*1 with feedback. Let us define the following notations:

1. $F(x)$: Interarrival times distribution, which is an exponential with parameter λ, that is,

$$F(x) = \begin{cases} 1 - e^{-\lambda x}, & \text{if } x \geq 0, \\ 0, & \text{if } x < 0. \end{cases} \tag{4.82}$$

2. $H(x)$: Service times distribution, which is an exponential with parameter μ, that is

$$H(x) = \begin{cases} 1 - e^{-\mu x}, & \text{if } x \geq 0, \\ 0, & \text{if } x < 0. \end{cases} \tag{4.83}$$

3. $\phi(s)$: The Laplace–Stieltjes transform of $H(x)$ is defined by

$$\phi(s) = \int_0^\infty e^{-sx} dH(x), \quad \Re(s) \geq 0 = \frac{\mu}{\mu + s}, \tag{4.84}$$

in our case, exponential service times distribution.

4. α_r: The rth moment of $H(x)$ is defined by

$$\alpha_r = \int_0^\infty x^r dH(x), \quad r = 0, 1, 2, \ldots, \quad \Re(s) \geq 0. \tag{4.85}$$

5. α: The average service time is defined by

$$\alpha = \alpha_1 = \int_0^\infty x dH(x) = \frac{1}{\mu}. \tag{4.86}$$

6. $H_k(x)$: The kth-iterated convolution of $H(x)$ with itself.

   For two or more independent random variables, the probability distribution of their sum is called the ***convolution*** of their individual distributions.

   The convolution of two density functions $f(t)$ and $g(t)$ of two nonnegative random variables, denoted by $(f * g)(t)$, is defined as

$$(f * g)(t) = \int_{-\infty}^{\infty} f(u)g(t-u)du. \tag{4.87}$$

   We note that

$$(f * g)(t) = (g * f)(t). \tag{4.88}$$

7. $\psi_n(t)$: Distribution of the queue size to be n at time t, that is, probability of the number of tasks in the system, including the one being service.
8. ψ_n: Stationary distribution of the queue length to be of size n.
9. $\xi(t)$: The random variable representing the queue size (the number of tasks in the system including the one being served) at time t.
10. ζ_n: The random variable representing the total time spent in the system by the nth task.
11. $\Phi(s)$: The Laplace transform of distribution of ζ_n.
12. $\zeta(x)$: The distribution function of the total service time of a task.
13. β_r: The rth moment of $\zeta(x)$ is defined by

$$\beta_r = \int_0^{\infty} x^r d\zeta(x), \quad r = 0, 1, 2, \ldots, \quad \Re(s) \geq 0. \tag{4.89}$$

14. β: The average service time is defined by

$$\beta = \beta_1 = \int_0^{\infty} x d\zeta(x). \tag{4.90}$$

15. $\zeta^*(x)$: Laplace transform of the distribution function of the total service time of a task is defined as

$$\zeta^*(s) = \int_0^{\infty} e^{-sx} d\zeta(x), \quad \Re(s) \geq 0. \tag{4.91}$$

16. $X(t)$: The time needed to complete the current service (if any) at time t. $X(t) = 0$, if $\xi(t) = 0$.

Our interests are to find

1. the distribution of the total time of service by a task, $\zeta(x)$;
2. the stationary distribution of $\xi(t)$, that is, the distribution of $\xi(t)$ when it is the same distribution for all $t \geq 0$; and
3. the stationary distribution of ζ_n, that is, stationary distribution of ζ_n, when it is the same for every $n = 0, 1, 2, \ldots$.

We leave it as an exercise to prove the following:

1. The limiting distribution of $\xi(t)$, that is, $\lim_{t\to\infty} P\{\xi(t) = n\}, n = 0, 1, 2, \ldots$, exists and is independent of the initial state if and only if $\xi(t)$ has a stationary distribution and the limiting distribution is identical with the stationary distribution.
2. The limiting distribution $\lim_{n\to\infty} P\{\zeta_n \le x\}$ exists and is independent of the initial state if and only if ζ_n has a stationary distribution and the limiting distribution is identical with the stationary distribution.

4.7.2 Distribution of the Total Time of Service by a Task

In order to answer the first question, we prove a relationship between $\zeta(x)$ and $H_k(x)$ as follows:

Theorem 4.7.1

$$\zeta(x) = q\sum_{n=1}^{\infty} p^{n-1} H_n(x). \tag{4.92}$$

Proof: The probability that a task joins the queue exactly n times is qp^{n-1}, $n = 1, 2, \ldots$, because once it joins the queue for the first time, it would have to return $n-1$ times before it departs from the system permanently. Each return having a probability p and permanent departure having probability q gives us the probability of joining the queue n times as qp^{n-1}. Now, if it does join the queue n times, then its total service time equals the sum of n mutually independent random variables, each of which has the distribution function $H(x)$. The distribution of the total service times will be the convolution of $H(x)$ with itself multiplied by the probability of joining the queue n times, and summed over all n's. This proves the theorem.

We leave as exercises to show the followings four items. We should note that the first item together with Equation (4.92) will give the specific answer to one of the questions of interest.

1. The kth-iterated convolution of an exponential cdf with parameter μ with itself is gamma distribution with parameters k and $\mu, \Gamma(k, \mu), \mu > 0, k = 1, 2, \ldots$, where pdf of gamma with parameters k and μ, denoted by $\gamma(x; k, \mu)$, is

$$\gamma(x; k, \mu) = \frac{x^{k-1} e^{-\frac{x}{\mu}}}{\mu^k \Gamma(k)}, \quad \text{for } x > 0, k, \mu > 0, \text{ and } \Gamma(k) = \int_0^\infty x^{k-1} e^{-x} dx. \tag{4.93}$$

2.

$$\zeta^*(x) = q\sum_{n=1}^{\infty} p^{n-1}[\phi(s)]^n = \frac{q\phi(s)}{1 - p\phi(s)}, \quad |p\phi(s)| < 1. \tag{4.94}$$

3.

$$\beta = \frac{\alpha}{q}. \tag{4.95}$$

4.

$$\beta_r = \alpha_r + \frac{p}{q}\sum_{i=1}^{r}\binom{r}{i}\ \alpha_i\beta_{r-i}. \tag{4.96}$$

4.7.3 Stationary Distribution of the Feedback Queue Size

We now concentrate on the second question, the distribution of the queue size. The process under consideration is what is known as the Takács' type $[F(x), H(x), p]$ model and we want to consider an equivalent process to the Takács such that tasks join the queue only once and are served in one stretch. In this case, the service time of a task will equal its total service times. Hence, the process now, according to Takács' type, will be $[F(x), \zeta(x), 0]$. With this equivalence, the distribution of the queue size for both processes $[F(x), H(x), p]$ and $[F(x), \zeta(x), 0]$ is the same.

Takács (1962a,b) showed that for the process $[F(x), \zeta(x), 0]$, a stationary distribution $\{\psi_n^*\}$ exists if and only if $\lambda\beta < 1$, where ψ_n^* is denoted as the stationary distribution of queue size for $[F(x), \zeta(x), 0]$ model. Denoting by $G^*(z)$ the generating function of ψ_n^*, in his papers Takács (1962a,b), Takács gives the solution through ***A. Y. Khintchine's formula***:

$$G^*(z) = \sum_{n=0}^{\infty}\psi_n^* z^n = \frac{(1-\lambda\beta)(1-z)\zeta[\lambda(1-z)]}{\zeta[\lambda(1-z)]-z}. \tag{4.97}$$

As a consequence of what we have discussed so far, we now state the following theorem due to Takács.

Theorem 4.7.2 Takács' Theorem If $\lambda\alpha < q$, then the process $\{\xi(t), 0 \le t < \infty\}$ has a unique distribution:

$$P\{\xi(t) = n\} = \psi_n^*, \quad n = 0, 1, 2, \ldots \tag{4.98}$$

and for $|z| \le 1$, we have

$$G^*(z) = \sum_{n=0}^{\infty}\psi_n^* z^n = \left(1 - \frac{\lambda\alpha}{q}\right)\frac{q(1-z)\zeta[\lambda(1-z)]}{(q+pz)\zeta[\lambda(1-z)]-z}. \tag{4.99}$$

If $\lambda\alpha \ge q$, then a stationary distribution does not exist.

4.7.4 Stationary Distribution of ζ_n (Sojourn Time of the nth task)

Now, in order to address the third question, let us denote by $X(t)$ the time needed to complete the current service (if any) at time t. If the system is empty at time 0, that is, $\xi(t) = 0$, then $X(t) = 0$. Takács remarked that the vector process $\{\xi(t), X(t), 0 \leq t < \infty\}$ is a Markov process . This sequence has a stationary distribution only if $\lambda\alpha < q$, and coincides with the stationary distribution of $\{\xi_n, X_n, n = 0, 1, 2, \dots\}$.

The random variables ξ_n and X_n determine ζ_n. If $\{\xi_n, X_n, n = 0, 1, 2, \dots\}$ has a stationary distribution, every $\zeta_n, n = 1, 2, \dots$, has the same distribution. Hence, we have the third theorem of Takács (1963).

Theorem 4.7.3 Takács' Theorem If $\lambda\alpha < q$, then ζ_n has a unique stationary distribution $P\{\zeta_n \leq x\}$. The Laplace transform of this distribution, denoted by $\Phi(s)$, is given by

$$\Phi(s) = q\sum_{n=1}^{\infty} p^{n-1} U_n(s, 1), \quad \Re(s) \geq 0, \tag{4.100}$$

where

$$U_1(s, z) = \psi_0\phi[s + \lambda(1 - z)] + U\{s + (1 - z), (q + pz)\phi[s + \lambda(1 - z)]\}, \tag{4.101}$$

for $Re(s) \geq 0$ and $|z| \leq 1$,

$$\psi_0 = 1 - \frac{\lambda\alpha}{q}, \tag{4.102}$$

$U(s, z)$ is defined as

$$\begin{aligned} U(s, z) &= \sum_{n=1}^{\infty} \prod_{n} \phi(s) z^n \\ &= \left(1 - \frac{\lambda\alpha}{q}\right) \frac{\lambda z(1 - z)\{\phi(s) - \phi[\lambda(1 - z)]\}}{\{z - (q + pz)\phi[\lambda(1 - z)]\}[s - \lambda(1 - z)]}, \end{aligned} \tag{4.103}$$

and

$$U_{n+1}(s, z) = \phi[s + \lambda(1 - z)]U_n\{s, (q + pz)\phi[s + \lambda(1 - z)]\}, \quad n = 1, 2, \dots. \tag{4.104}$$

Proof: See Haghighi et al. (2011).

In order to calculate the explicit expressions for the moments of the sojourn time of a task, we state the fourth theorem of Takács as follows:

Theorem 4.7.4 Takács' Theorem If $\lambda\alpha < q$, then ζ_n has a stationary distribution, then assuming α_2 is finite, we have

$$E(\zeta_n) = \frac{\lambda\alpha_2 + 2\alpha_1(1 - \lambda\alpha_1)}{2(q - \lambda\alpha_1)} \tag{4.105}$$

and assuming α_3 is finite, we have

$$E\{\zeta_n^2\} = \frac{q^2 - 2q}{6(q-\lambda\alpha_1)^2[q^2 - q(2+\lambda\alpha_1) + \lambda\alpha_1]\cdot \left\{ \begin{array}{l} 2q[6\lambda\alpha_1^3 - 6\alpha_1^2 - 6\lambda\alpha_1\alpha_2 + 3\alpha_2 + \lambda\alpha_3] \\ -[12\lambda\alpha_1^3 - 12\alpha_1^2 - 6\lambda\alpha_1\alpha_2 - 3\lambda^2\alpha_2^2] \end{array} \right\}}. \tag{4.106}$$

Proof: See Haghighi et al. (2011).

We leave it as an exercise to show that, for stationary *M*/*M*/1 with feedback, we have

$$E\{\zeta_n\} = \frac{1}{q\mu - \lambda} \tag{4.107}$$

and

$$E\{\zeta_n^2\} = \frac{2(2q-q^2)\mu}{(q\mu-\lambda)^2[(2q-q^2)\mu - (1-q)\lambda]}. \tag{4.108}$$

Of course, by letting $q = 1$, the same information will be obtained for stationary *M*/*M*/1.

4.8 QUEUES WITH BULK ARRIVALS AND BATCH SERVICE

The words ***bulk*** and ***batch*** are used interchangeably to mean a group of tasks with size of greater than 1. It is well known that batch service started with Bailey (1954), although one might say it begun with Erlang in his solution of the $M/E_k/1$ model in the early 20th century (see Brockmeyer et al. 1948) that implicitly gives the solution of the $M^k/M/1$ queue model. The notation $M/E_k/1$ stands for a queueing system with Markovian arrival and Erlang service with parameter k. The notation $M^k/M/1$ stands for a queueing system with Markovian bulk arrival, each bulk of size k, and Markovian service with a single server. The study of bulk arrival started several years after the work of Bailey (1954) (Gaver 1954). Neuts (1967) studied a general class of bulk queues with Poisson input. He defined various cases of bulk services and analyzed the general case. Kleinrock (1975) considered usual bulk service models that are equivalent to Erlang arrival systems.

Madan (2003) studied two $M^X/M^{a,\ b}/1$ models with random breakdowns. This notation indicates that tasks arrive by batches of varying sizes and they are served when there is at least a task waiting, but not exceeding b task to be served at the same time. Jain et al. (2007) considered the general bulk rule with general service distribution. Khalaf et al. (2011) studied an $M^{[X]}/G/1$ queue with Bernoulli schedule, general vacation times, random breakdowns, general delay times, and general repair times. They assumed that the repair process does not start immediately after a breakdown and there is a "delay time" waiting for repairs to start. They found the steady state solution of the system using supplementary variables technique. Haridass and Arumuganathan (2011) considered the bulk service queueing system with variant threshold policy for secondary jobs. Vijaya Laxmi and Yesuf (2011) considered an infinite buffer single-server batch service queue with single exponential working vacation policy. Gupta and Banerjee (2011) considered a Markovian queue with finite buffer and bulk service with general distribution. Singh et al. (2011) investigated a single-server bulk queueing system with state-dependent rates and second optional service. Singh et al. (2012) studied single-server state-dependent queueing systems, wherein tasks arrive in batches and follow the Poisson process with state-dependent arrival rates.

Below, we offer specific simple examples, namely *M/M/*1 with bulk arrivals.

Example 4.8.1 *M/M/*1 with Bulk Arrivals with Varying Sizes Let us consider an *M/M/*1 in which tasks arrive by batches with varying size represented by a random variable X. Hence, the system will be an $M^X/M/1$. Among many authors, Jensen et al. (2006) studied $M^X/M/1$. They assumed arrivals by bulk of size $j, j = 1, 2, \dots, m$, according to a Poisson process $\{N_j(t) : t \geq 0\}$ with mean $\lambda_j, j = 1, 2, \dots, m$, and that these Poisson processes are mutually independent. Under these assumptions, letting

$$M(t) = N_1(t) + N_2(t) + \cdots + N_m(t), \tag{4.109}$$

the superposed process $\{M(t) : t \geq 0\}$ is a Poisson process with mean $\sum_{j=1}^{m} \lambda_j = \lambda$. $M(t)$ represents the number of batch arrivals in $(0, t]$ and the mean rate of batches as λ. Denoting

$$P(X = j) = \lambda_j / \lambda, \quad j = 1, 2, \dots, m, \tag{4.110}$$

they took X to have a multinomial distribution. Assuming exponential single-server service with mean service time $1/\mu$, for $m = 2$, from Harris (1966), it is concluded that

$$P_n = P_0 \sum_{i=0}^{[n/2]} \rho_2^i \rho_1^{n-2i} \binom{n-1}{i, n-2i}, \quad n \geq 1, \tag{4.111}$$

where [.] is the greatest integer symbol,

$$\rho_j = \frac{j\lambda_j}{\mu}, \quad j = 1, 2,$$

$$P_o = 1 - \rho, \quad \rho < 1,$$

$$\rho \triangleq \sum_{j=1}^{2} \rho_j.$$

See Jensen et al. (2006, p. 354, formula 10).

Example 4.8.2 An $M^X/M/1$ with Geometric Bulk Size Distribution Again consider $M^X/M/1$. We let X have a geometric distribution, that is,

$$\alpha_x = P(X = x) = p(1-p)^{x-1}, \quad 0 < p < 1, x = 1, 2, \dots. \tag{4.112}$$

Then, Medhi (2003) showed the generating function, expected value, and distribution as

$$A(z) = \frac{pz}{1-(1-p)z}, \tag{4.113}$$

$$E(X) = \overline{\alpha} = A'(1) = \frac{1}{p}. \tag{4.114}$$

and

$$P_n = \frac{\rho\rho(1-\rho)}{[1-\rho(1-\rho)]^{n-1}}, \quad n \geq 1, \tag{4.115}$$

respectively, with $P_0 = 1 - \rho$ and $\rho = \lambda/p\nu$.

4.9 A PRIORITY QUEUE WITH BALKING AND RENEGING

Priority was mentioned above as a type of service discipline. Here is an example in brief. Details can be found in Haghighi and Mishev (2006) and Haghighi and Mishev (2014). In a particular case, they consider a system that consists of two service stations, set in parallel, as in Disney (1962). Each station consists of one server and has a waiting room, called buffer. Buffer sizes are $M_1 - 1$ and $M_2 - 1$, for stations 1 and 2, respectively. Thus, the capacities of the stations are M_1 and M_2, respectively, with total capacity of the system as $M = M_1 + M_2$. Tasks arrive from an infinite source probabilistically with Poisson distribution with mean λ to the system. The service times at each station are *iid* random variables having negative exponential distributions with mean $1/\mu_i, i = 1, 2$, where μ_is are positive real numbers. At each station, tasks will be served on FIFO bases. If an arriving task finds both servers busy at the time of its arrival, it will join the system with a constant probability p, $0 \leq p \leq 1$, that is, it will balk with probability $q = 1 - p$. While a task is waiting for service, it may decide to leave the system, that is, it may renege. The length of time a task may wait before it reneges is assumed to be a random variable having a negative exponential distribution with mean $1/\alpha_i, i = 1, 2$, where α_is are positive real numbers. Reneging cannot occur for the task in service. Thus, the average number of reneged units from the system while servers are busy will be $\alpha_1 + \alpha_2$. If a task leaves the system due to balking, reneging, or completion of its service and rejoins the system, it will be considered as a new arrival, independent of the previous balking, reneging, or being served.

Service stations are ordered such that station 1 has the first priority of receiving the arrivals. That is, if a task arrives, it must join a queue by station 1, unless there are already M_1 in that station; in that case, it will go before station 2 unless there are already M_2 in station 2. If both stations are full, then the arriving task will be considered lost. Thus, loss to the system will be due to balking, reneging, or both stations being full. A special case is chosen when the capacity of each station is one or more.

We leave it as an exercise to set up the set of difference equations for the length of the queue in the general case. However, for a particular case when $M_1 = 1$ and $M_2 \geq 3$, letting $P_{m1,m2}$ denote the stationary probability of having m_1 task in the first station and m_2 tasks in the second, the system of difference equations is as follows.

However, it should be noted that reneging is another way of servicing. Thus, in order to reduce the number in the system, either we have to have a service completed or a reneging occurred.

$$\lambda P_{0,0} = \mu_1 P_{1,0} + \mu_2 P_{0,1},$$

$$(\lambda + \mu_2)P_{0,1} = \mu_1 P_{1,1},$$

$$(\lambda + \mu_1)P_{1,0} = \lambda P_{0,0} + \mu_2 P_{1,1},$$

$$(p\lambda + \mu_1 + \mu_2)P_{1,1} = \lambda P_{0,1} + \lambda P_{1,0} + (\mu_1 + \alpha_2 + \mu_2)P_{1,2},$$

$$[p\lambda + \mu_1 + (m_2 - 1)\alpha_2 + \mu_2]P_{1,m_2}$$
$$= p\lambda P_{1,m_2-1} + (\mu_1 + m_2\alpha_2 + \mu_2)P_{1,m_2+1}, \quad 2 \le m_2 \le M_2 - 1,$$

$$(\mu_1 + (M_2 - 1)\alpha_2 + \mu_2)P_{1,M_2} = p\lambda P_{1,M_2-1},$$

$$P_{0,0} + P_{0,1} + \sum_{m_2=0}^{M_2} P_{1,m_2} = 1.$$

Steps are given in these sources to solve the system for the system queue length distribution. The following example is offered by them.

Example 4.9.1 A Priority Queues with Balking and Reneging For the model just discussed, we choose the following data:

M_1	M_2	λ	$\rho_1 = \dfrac{\lambda}{\mu_1}$	$\rho_2 = \dfrac{\lambda}{\mu_2}$	$\rho_s = \dfrac{\lambda}{\mu_1 + \mu_2}$
1	1	5	0.75	0.25	0.1875
	2	15	1	0.75	0.3333
	3	25	1.25	1	0.4688
	4	35	2	1.5	0.6667
	5	45	2.5	2	0.9375
	6	55	3	2.5	1.2000
	7	65	3.5	3	1.4583
		85	4	3.5	1.7143
		95	4.5	4	1.9687
			5		2.2222

Since $M_1 = 1$, then $\alpha_1 = 0$. We consider three cases:

1. $\mu_1 \neq \mu_2, p = 0.6, \alpha_2 = 5$,
2. $\mu_1 = \mu_2, p = 0.6, \alpha_2 = 5$, and
3. $\mu_1 = \mu_2, p = 1, \alpha_2 = 0$.

We graphically present the probabilities of the system being idle, loss to the system, stations one and two being busy, also the first and second moments of the queue

TABLE 4.2 Summary of Values for Case $M_1 = 1$ and $M_2 = 7$

$M_1 = 1,\ M_2 = 7$	% System Idle	% System Loss	% Station 1 Busy	% Station 2 Busy	Average System Loss due to Balking and reneging	Average Attended to the System
$\mu_1 = \mu_2,\ \alpha_2 = 0,\ p = 1,\ \rho = 1$	5	82	57	40	0	1
$\mu_1 = \mu_2, \alpha_2 = 5, p = 0.6, \rho = 1$	15	55	52	35	0.37	0.78
$\mu_1 \neq \mu_2, \alpha_2 = 5, p = 0.6, \rho = 1$	12	59	50	42	0.3	0.78
$\mu_1 \neq \mu_2, \alpha_2 = 5, p = 0.6, \rho = 2$	2	90	70	75	0.83	1.25

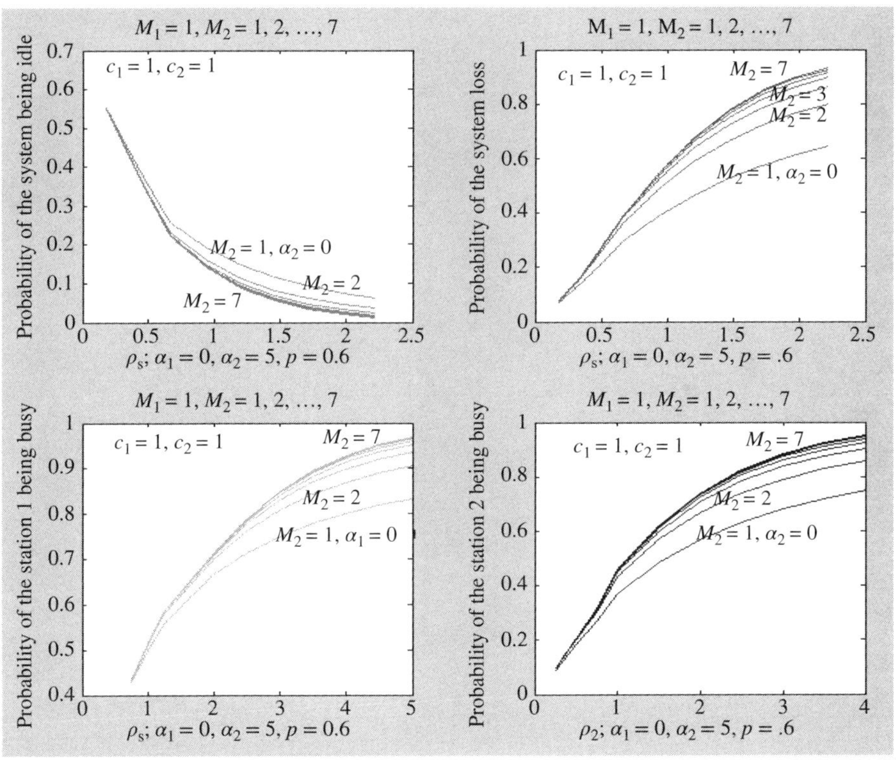

(*Continued*)

TABLE 4.2 (*Continued*)

$M_1 = 1,\ M_2 = 7$	% System Idle	% System Loss	% Station 1 Busy	% Station 2 Busy	Average System Loss due to Balking and reneging	Average Attended to the System

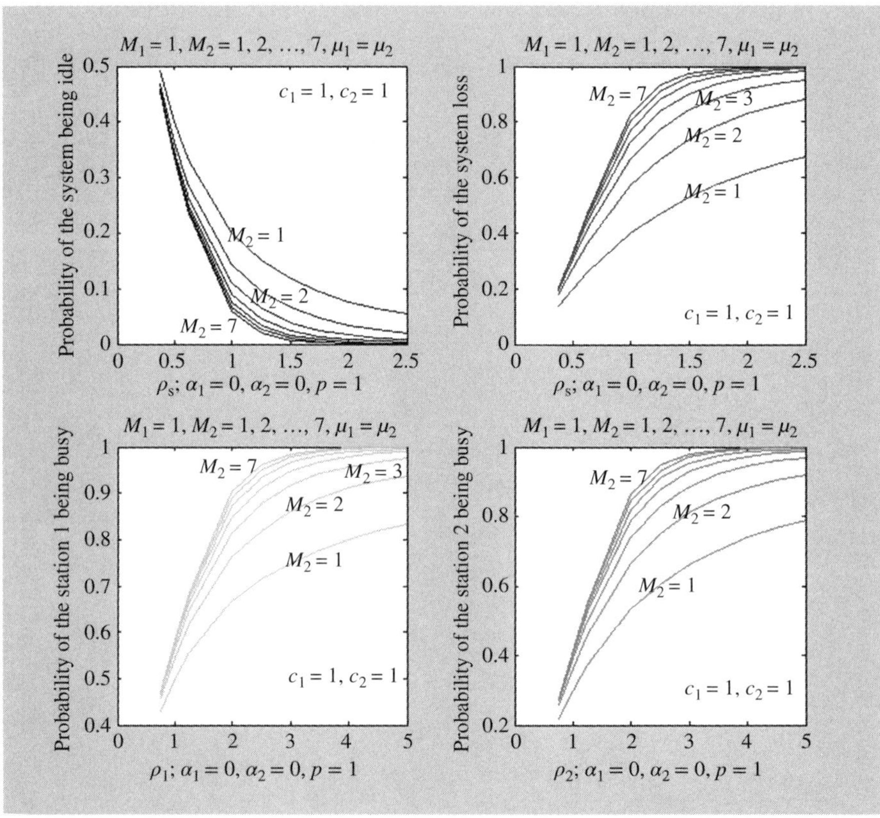

TABLE 4.2 (*Continued*)

$M_1 = 1,\ M_2 = 7$	% System Idle	% System Loss	% Station 1 Busy	% Station 2 Busy	Average System Loss due to Balking and reneging	Average Attended to the System

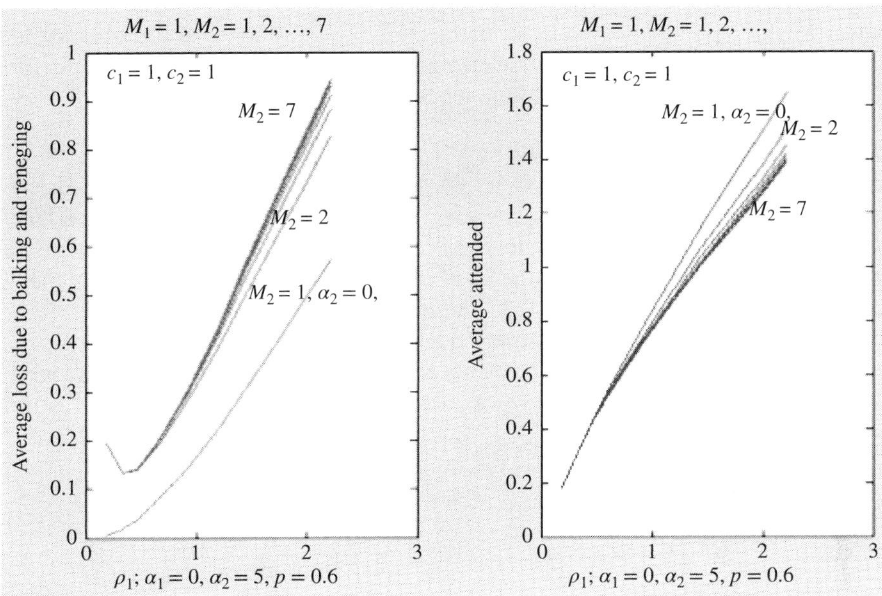

lengths of the system as well as each station, the average loss to the system due to balking and reneging. The graphs are versus ρ_1, ρ_2, and ρ_s, depending on the case for Station 1, Station 2, and the system, respectively. Values for the case $M_1 = 1$ and $M_2 = 7$ are summarized in Table 4.2.

4.10 DISCRETE TIME *M/M/*1 QUEUEING PROCESS, COMBINATORICS METHOD (LATTICE PATHS)

So far, our discussions have been focused on continuous-time models of queues, that is, cases that the time parameter t is a real number and interarrival times and service times are real-valued random variables. Although real-valued time parameter is a fact for most studies of queues, the reality is the opposite. We refer the reader to Hunter (1983, pp. 189–190) and Kobayashi (1983). Hence, as promised at the beginning of the chapter, in this section, we discuss discrete time models. That is, we segment the time axis into a sequence of time intervals (or slots) of unit duration.

It is well known that combinatorial methods provide elegant and powerful tools of analysis in the theory of queues. See, for example, Takács (1967) and Jain et al. (2007). The combinatorics of lattice paths can be applied to solve problems in the theory of queues, from classical ones like $M/M/1$ systems to open tandem systems with and without global blocking and to queueing process that are related to random walks in a quarter plane like the Flatto–Hahn systems with preemptive priorities (Böhm 2008).

We now start our discussion of the combinatorial method by defining a ***lattice point***. A lattice point is the point of intersection of two or more gridlines (Mohanty 1979). A ***lattice path*** is a path that leads from one lattice point to another. Formally, we offer the following definition (Flajolet, and Guillemin 1999).

Definition 4.10.1 A (***lattice***) ***path*** $\nu = (U_0, U_1, \ldots, U_n)$ is a sequence of points in the lattice $\mathbb{N} \times \mathbb{N}$ such that if $U_j = (x_j, y_j)$, then $x_j = j$ and $|y_{j+1} - y_j| \leq 1$. An ***edge*** (U_j, U_{j+1}) is called a***scent*** ($\underline{a}$) if $y_{j+1} - y_j = +1$, ***ascent*** ($\underline{b}$) if $y_{j+1} - y_j = -1$ and ***level step*** ($\underline{c}$) if $y_{j+1} - y_j = 0$. The quantity n is the ***length of the path,*** $o(\nu) = y_0$ is the ***initial altitude***, and $h(\nu) = y_n$ is the ***final altitude***. The quantities $\sup\{\nu\} = \max_j y_j$ and $\inf\{\nu\} = \min_j y_j$ are called ***height*** and ***depth*** of the path, respectively.

We note that some of the notations used in Definition 4.10.1 may vary depending on the authors using them. We also note that lattice paths can effectively model phenomena in physics, chemistry, and probability theory. Asymptotic enumeration of lattice paths is linked by entropy in the physical systems being modeled. Lattice paths restricted to different regions of the plane are well suited to a functional equation approach for an exact and asymptotic enumeration.

As an example of a discrete process, we discuss the ***Ballot Problem***.

4.10.1 The Basic Ballot Problem

Suppose two candidates are running for a position. In a ballot, candidate A receives x votes and candidate B receives y votes, with $x > y$. The following is a question to answer.

Question: What is the probability that candidate A is ahead of candidate B throughout the counting of votes?

On the basis of a well-known ***ballot theorem***, the answer is as follows:

Answer:

$$P\{x > y \text{ throughout the counting}\} = \frac{x-y}{x+y}. \tag{4.116}$$

Example 4.10.1 A Ballot Problem Let us suppose that in an election of two executives from among a board of five persons eligible to vote, votes have been counted and there are three votes for candidate A and 2 votes for candidate B (so $x = 3$ and $y = 2$). The 10 possibilities for the order of the votes casted are

(1) AAABB, (2) AABAB, (3) ABAAB, (4) BAAAB, (5) AABBA, (6) ABABA, (7) BAABA, (8) ABBAA, (9) BABAA, and (10) BBAAA.

To see how the counts are, obviously the choice AAABB is a right possibility since the first A counts 1 for candidate A, the second A gives the candidate A another score and makes the count as 2, and third A makes the count for candidate A as three and he wins the election with leading the voting score throughout the counting process. Another possibility is AABAB because the first A counts 1 for candidate A, the second A gives the candidate A another score and makes the count as 2, but the third records a vote for B so that A still leads with 2 against B with 1. The fourth count is for A again and put him with score of 3 and he is the winner with leading throughout. However, the choice of AABBA is not a right one, because the first two counts give the candidate A two votes in a row, but the third gives the candidate B one vote and A still with 2. The fourth vote going to B gives B 2 votes while A is yet with 2 votes and is not leading. Although, the last vote is in favor of A and makes him win, but he did not lead throughout the counting. Hence, considering all 10 possibilities, we will see that there are only two choices that satisfy the criteria and thus the probability of winning A with the condition of leading throughout is $2/1 = 1/5$, which is $(3-2)/(3+2) = 1/5$.

Proof of the ballot theorem based on a ***combinatorial approach***, ***lattice path***, is as follows: Let us represent a vote for the candidate A by a horizontal unit line segment and the candidate B by a vertical unit line segment. This way, a sequence of voting count will be represented by a lattice path from point $(0, \ 0)$ to point $(x, \ y)$. The paths are developed with the same chance, that is, they are likely events. In terms of lattice path, the answer to the question for the ballot problem is the ratio of the number of paths that do not touch the line $y = x$ except at the origin to the total number of paths. See Figure 4.2 with eight votes for candidate A and five votes for candidate B. Thus, let the number of paths that touch or cross the line $y = x$ at some point be denoted by

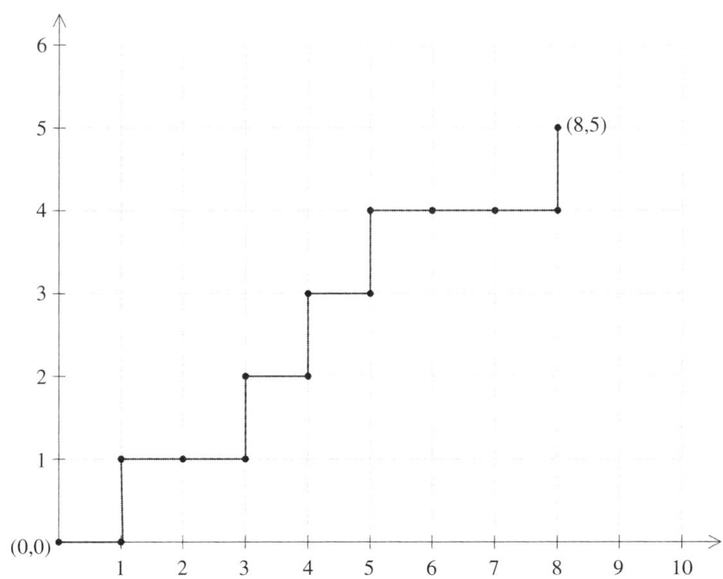

Figure 4.2 Lattice path for the ballot problem with eight and five votes for candidates A and B, respectively

n. Then, the answer to the ballot problem question is

$$P\{x > y \text{ throughout the counting}\} = 1 - \frac{n}{\binom{x+y}{y}}. \tag{4.117}$$

The number n can be calculated as

$$n = 2\binom{x+y+1}{y-1}. \tag{4.118}$$

Hence, from Equation (4.117), we have

$$P\{x > y \text{ throughout the counting}\} = 1 - \frac{n}{\binom{x+y}{y}} = 1 - \frac{2y}{x+y}, \tag{4.119}$$

which is the same as Equation (4.116). See Jain et al. (2007, p. 198).

4.10.2 Ballot Problem (based on Takács 1997)

Although Lajos Takács is well known in the field of probability and stochastic processes, particularly in queueing theory, we state a few lines about him.

Lajos Takács (1924–2015)

Emeritus Professor Lajos Takács was born on August 21, 1924, in Maglód, a small town at distance of 16 miles from Budapest. While he was a student in probability theory and mathematical statistics courses with Charles Jordán, he contributed to his class too. Hence, Jordán added this short note from Takács in an already finished manuscript of his book of probability theory.

Takács published six books (three in Hungarian and three in English language) and 225 original papers, the last was published in 1999 (Takács 1999). Starting in 1973, until recently, he has prepared 1600 pages on the theory of random fluctuation that is waiting to be published according to his terms. He was a doctoral dissertation advisor of nine students at Columbia University in New York, including the well-known Burk, and 14 at Case Western Reserve University in Cleveland, Ohio, USA, including the first author of this book.

There are papers published in the honor of Lajos Takács for his contribution, for instance, see (1) Dshalalow (1994), (2) Galambos and Gani (1994), (3) and a whole page in his honor appeared in Haghighi and Mishev (2014). He is known as one the frontiers of queueing theory. Among his various rewards, he is the winner of the 1994 John von Neumann Theory Prize.

The 8th International Conference on Lattice Path Combinatorics and Applications at California Polytechnic State University, Pomona, California, during August 17–20, 2015, honored Takács through a joint presentation by Aliakbar Montazer Haghighi and with Emeritus Professor Sri Gopal Mohanty (McMaster University, Hamilton, Ontario, Canada), as guest speakers, titled Lajos Takács: Life and Contribution to Combinatorics. It was an occasion, also, to celebrate Takács's 91st birthday. He passed away on December 4, 2015. In the Preface of AAM, Vol. 10, Issue 2, December 2015, Haghighi wrote about his life and his publications. It can be viewed at https://www.pvamu.edu/mathematics/wp-content/uploads/sites/49/01_aam_vol_10_issue_2_preface_ready_amh.pdf.

Aliakbar Montazer Haghighi

Sri Gopal Mohanty

In his last paper on Combinatorics (1997) extracted from his book of 1967, presented in honor of Sri Gopal Mohanty, he discussed the historic development of various ballot theorems, provided several proofs for these theorems, and gave some applications. As he put it, "the simplicity and the generality of the ballot theorems might explain their wide range of uses."

Theorem 4.10.1 The Classical Ballot Theorem Suppose in a ballot, candidate A receives x votes and candidate B receives y votes, with $x > \alpha y$, where α is a positive integer. Assume that all possible voting records are equiprobable. Then, the probability that throughout the counting process, the number of votes registered for candidate A is always greater than α times the number of votes registered for candidate B, denoted by $P(x, y, \alpha)$, is

$$P(x, y, \alpha) = \frac{x - \alpha y}{x + y}, \tag{4.120}$$

provided that all the possible voting records are equally probable.

Note that for $\alpha = 1$ we have Equation (4.116), as a special case.

Historically, Equation (4.116) was first published by W.A. Whitworth in 1878. However, Joseph Louis François Bertrand rediscovered it in 1887, and thus, the name of Bertrand stayed with the answer. See Feller (1968). Also, Equation (4.116) was proved in the same year by André (1887). Barbier (1887) noted that if α is a positive integer, then Equation (4.120) is true, but did not prove it. Finally, after 37 years, the ballot theorem in the form of Equation (4.120), that is, for $\alpha \geq 1$, was proved by Aeppli (1923, 1924) in his dissertation under the guidance of Professor György Pólya. The proof given below follows Aeppli's "remarkable proof," as Takács stated in his 1997 paper. Thus, Takács (1997) is the first paper in which this proof appeared. The proof is also remarkable, since it uses only a fundamental property of the arithmetic triangle. Addario-Berry and Reed (2008) present a survey of ballot theorems "old and new" and, thus, covered Takács's related papers up to 1970. They did not mention the one of 1997. This may be because the 1997 paper is extracted from his book of 1964.

Blaise Pascal (1623–1662)

Eyn Newe
Unnd wolgegründte
underweysung aller Kauffmanß Rechnung in dreyen Büchern/ mit schönen Regeln unnd fragstucken begriffen. Sunderlich was fortl unnd behendigkait in der Welschē Practica unnd Tollern gebraucht wirdt/ des gleychen fürmalß wider in Teützscher noch in Welscher sprach nie gedrückt. durch Petrum Apianū von Leyßnick/ d Astronomei zū Ingolstat Ordinariū/ verfertiget.

1
1 1
1 2 1
1 3 3 1
...

Written in 1653 and appeared posthumously in 1665

Pascal's triangle is named after the French mathematician and philosopher Blaise Pascal. His famous book is a *Treatise on the Arithmetical Triangle*. However, he was not the first to create this triangle or notice its amazing properties! The famous arrangement of numbers was known before Pascal's time in China, India, and Persia (Iran) to a very limited extent.

Omar Khayyam (1048–1122)

Persian poet and philosopher

Long before Pascal, the triangle appeared in the writings of Omar Khayyam, the great 11th-century Persian (Iranian) astronomer, poet, philosopher, and mathematician. However, the history of figure numbers goes back to 540 BC. See Edwards (2002) and references therein.

Proof: Let A and B denote votes for candidates A and B, respectively. Hence, every voting count can be represented by a sequence of $x$letters A and y letters B. Since $y < x$, the number of possible vote counts, in which A received x votes and B receives y votes is

$$\binom{x+y}{y} = \frac{(x+y)!}{x!y!}. \tag{4.121}$$

Let $n(x, \quad y, \quad \alpha)$ denote the favorable voting counts, that is, voting counts in which throughout the counting process, the number of votes registered for candidate A is always greater than α times the number of votes registered for candidate B. Then,

$$P(x, y, \alpha) = \frac{n(x, y, \alpha)}{\binom{x+y}{y}}. \tag{4.122}$$

Now, the last vote is casted for either A or B. Hence, we have

$$n(x, y, \alpha) = n(x-1, y, \alpha) + n(x, y-1, \alpha), x > \alpha y, y \geq 1, \tag{4.123}$$

which is a recursive formula.

TABLE 4.3 Calculation of $n(x, y, 1)$ for $0 \leq y \leq x \leq 10$

x	y 0	1	2	3	4	5	6	7	8	9	10
0	0	0	0	0	0	0	0	0	0	0	0
1	1	0	0	0	0	0	0	0	0	0	0
2	1	1	0	0	0	0	0	0	0	0	0
3	1	2	2	0	0	0	0	0	0	0	0
4	1	3	5	5	0	0	0	0	0	0	0
5	1	4	9	14	14	0	0	0	0	0	0
6	1	5	14	28	42	42	0	0	0	0	0
7	1	6	20	48	90	132	132	0	0	0	0
8	1	7	27	75	165	297	429	429	0	0	0
9	1	8	35	110	275	572	1001	1430	1430	0	0
10	1	9	44	154	429	1001	2002	3432	4862	4862	0

Note that

$$\begin{cases} n(x, 0, \alpha) = 1, & a \geq 1, \\ n(\alpha y, y, \alpha) = 0, & b \geq 1. \end{cases} \tag{4.124}$$

As an example, from the recursive formula (4.123), we can calculate $n(x, y, 1)$ for $0 \leq y \leq x \leq 10$, as follows:

As noted in Takács (1997), printed tables, such as Table 4.3, for values of $x, y = 0, 1, 2, 3, 4, 5,$ and 6 were already available since the 16th century. See, for example, Apianus (1527) for $x + y \leq 10$, Cardano (1570, p. 135) for $x + y \leq 12$, Mersenne (1635–1636, Libr. VII, p. 134) for $x \leq 25$, and $y \leq 12$, and Pascal (1908a, p. 446) for $x + y \leq 10$. It is also known that $B(x, \ y)$ has combinatorial interpretation. For instance, it could be considered as the number of different ways $x - 1$ letters A and y letters B can be arranged in a row.

It should also be noted that after publication of the famous ***Treaties on the Arithmetic Triangle*** in 1665 by Blaise Pascal (Pascal 1908a,b), it has become generally known that the ***binomial coefficients*** or ***figure numbers***, denoted by $B(x, y)$,

$$B(x, y) = \binom{x + y - 1}{y} = \frac{(x + y - 1)!}{(x - 1)!y!}, \quad x \geq 1, \ y \geq 0, \tag{4.125}$$

can be calculated by the following recurrence formula:

$$\begin{cases} B(x, y) = B(x - 1, y) + B(x, y - 1), & x \geq 1, \ y \geq 1, \\ B(x, 0) = 1, & x \geq 0, \\ B(0, y) = 0, & y \geq 0. \end{cases} \tag{4.126}$$

Table 4.4 shows the probability that throughout the counting number of votes registered for candidate A, x, is always greater than the number of votes registered for candidate B, y, for $x, y \leq 10$ (Table 4.5).

TABLE 4.4 Calculation of $B(x, y)$ for $x, y \leq 10$

x \ y	0	1	2	3	4	5	6	7	8	9	10
0	1	0	0	0	0	0	0	0	0	0	0
1	1	1	1	1	1	1	1	1	1	1	1
2	1	2	3	4	5	6	7	8	9	10	11
3	1	3	6	10	15	21	28	36	45	55	66
4	1	4	10	20	35	56	84	120	165	220	286
5	1	5	15	35	70	126	210	330	495	715	1001
6	1	6	21	56	126	252	462	792	1287	2002	3003
7	1	7	28	84	210	462	924	1716	3003	5005	8008
8	1	8	36	120	330	792	1716	3432	6435	11440	19448
9	1	9	45	165	495	1287	3003	6435	12870	24310	43758
10	1	10	55	220	715	2002	5005	11440	24310	48620	92378

TABLE 4.5 Probability that Throughout the Counting Number of Votes Registered for Candidate A, x, is Always Greater than the Number of Votes Registered for Candidate B, y, for $x, y \leq 10$

x \ y	0	1	2	3	4	5	6	7	8	9	10
0	0	0	0	0	0	0	0	0	0	0	0
1	1	0	0	0	0	0	0	0	0	0	0
2	1	0.333	0	0	0	0	0	0	0	0	0
3	1	0.500	0.200	0	0	0	0	0	0	0	0
4	1	0.600	0.333	0.143	0	0	0	0	0	0	0
5	1	0.667	0.429	0.250	0.111	0	0	0	0	0	0
6	1	0.714	0.500	0.333	0.200	0.91	0	0	0	0	0
7	1	0.750	0.556	0.400	0.273	0.167	0.077	0	0	0	0
8	1	0.778	0.600	0.455	0.333	0.231	0.143	0.067	0	0	0
9	1	0.800	0.636	0.500	0.385	0.286	0.200	0.125	0.059	0	0
10	1	0.818	0.667	0.539	0.429	0.333	0.250	0.177	0.111	0.53	0

It is important to note that since both $B(x, y)$ and $B(y, x)$ satisfy Equation (4.123), so will

$$n(x, y, \alpha) = B(x, y) - \alpha B(y, x), \tag{4.127}$$

when $x > \alpha y > 0$. But, Equation (4.127) satisfies the boundary conditions (4.124). Hence,

$$n(x, y, \alpha) = \binom{x + y - 1}{y} - \alpha \binom{x + y - 1}{x}, x \geq \alpha y. \tag{4.128}$$

Therefore, from Eqs. (4.122) and (4.128), we have Equation (4.120), and this proves Theorem 4.10.1.

Theorem 4.10.2 Bertrand's Ballot Theorem, Special Case of the Classic Ballot Theorem, $\alpha = 1$ We discussed the proof of the special case of the classic ballot theorem, case $\alpha = 1$ (referred to as Bertrand's ballot theorem), through lattice path in the previous section. We now discuss the proof by André (1887) from Takács (1997), although the proof originally appeared in the book of Bertrand (1889, pp. 18–20). André's proof was also in the book of Poincaré (1896, pp. 21–26).

André's reasoning is as follows: Every voting count can be considered as a sequence of x letters A and y letters B. The total number of voting counts is given by Equation (4.121). André showed that the number of unfavorable voting counts, denoted by $U(x, y)$, is

$$U(x, y) = 2\binom{x+y-1}{x}. \tag{4.129}$$

This is because the set of unfavorable voting counts can be divided into two disjoint parts: (1) the set of all the voting counts in which the first letter is B and additionally, there are x letters A and $y - 1$ letters B and (2) the set of all unfavorable voting counts in which the first letter is A and additionally, there are $x - 1$ letters A and y letters B. There is a 1–1 correspondence between the voting numbers in these two subsets, since if a voting count belongs to the second subset, then counting the letters from left to right, there is a shortest subsequence, which contains an equal number of letters A and B. The last letter in this shortest sequence must be B. Therefore, let us remove all the letters except the last B and put them at the end of the voting count in the same order that belongs to the second subset. One-to-oneness of the mapping is trivial. Hence, both subsets contain $\binom{x+y-1}{x}$ voting counts. Thus, the total number is as in Equation (4.129).

Therefore,

$$n(x, y, 1) = \binom{x+y}{y} - 2\binom{x+y-1}{x}, \quad x, y \geq 1. \tag{4.130}$$

Relation (4.130) implies (4.116).

Takács found *a generalization of the classical ballot theorem* of Bertrand, which made it possible to solve many problems in queueing theory, in the theory of dams, and in order statistics. In his 1997 book, Takács presented a generalization of the basic ballot theorem as follows:

Theorem 4.10.3 General Ballot Theorem Suppose that a box contains n cards marked with nonnegative integers $k_1, k_2, \ldots, k_n$, such that $k_1 + k_2 + \cdots + k_n = k \leq n$. Cards are drawn one at the time without replacement from the box. Denoting by v_r the number obtained at the rth drawing ($r = 1, 2, \ldots, n$), then,

$$P\{v_1 + v_2 + \cdots + v_r < r, \quad \text{for } r = 1, 2, \ldots, n\} = \frac{n-k}{n}, \tag{4.131}$$

provided that all the possible results are equally probable.

Before we present proof of the theorem, we show that Theorem 4.10.2 is a special case of Theorem 4.10.3. For that, consider a box that contains a cards marked 0 and b cards marked $\mu + 1$. All the $a + b$ cards are drawn from the box without replacement. It is assumed that all the possible results are equally probable. Further, suppose that a card marked 0 corresponds to a vote for A and a card marked $\mu + 1$ corresponds to a vote for B. If among the first r drawings there are α_r cards marked 0 and β_r cards marked $\mu + 1$, then

$$\alpha_r 0 + \beta_r(\mu + 1) < r = \alpha_r + \beta_r$$

holds if and only if $\alpha_r > \beta_r \mu$. Now,

$$\alpha_1 + \alpha_2 + \cdots + \alpha_{a+b} = a \quad \text{and} \quad \beta_1 + \beta_2 + \cdots + \beta_{a+b} = b. \tag{4.132}$$

If $a \geq b\mu$ and we let

$$n = a + b \quad \text{and} \quad k = b(\mu + 1),$$

then we obtain

$$P\{\alpha_r > \beta_r \mu, \quad r = 1, 2, \ldots, a + b\} = \frac{a - b\mu}{a + b}, \tag{4.133}$$

which is in agreement with Equation (4.120). See Takács (1961, 1962).

In (1997), using mathematical induction and that $S(1, 0) = 1$ and $S(1, 1) = 0$, Takács proved that Theorem 4.10.3 is equivalent to the following theorem:

Theorem 4.10.4 Supposed that n cards are marked with nonnegative integers $k_1, k_2, \ldots, k_n$, such that $k_1 + k_2 + \cdots + k_n = k \leq n$. Among the $n!$ permutations of the n cards, there are exactly

$$S(n, k) = (n - k)(n - 1)!$$

permutations in which the sum of the numbers on the first r cards is less than r for every $r = 1, 2, \ldots, n$.

Drawing cards without replacement implies consideration of $n!$ permutations of n cards. Theorem 4.10.3 is true if permutations are replaced by cyclic permutations and is stated as follows as a counting result.

Theorem 4.10.5 Suppose that n cards are marked with nonnegative integers $k_1, k_2, \ldots, k_n$, such that $k_1 + k_2 + \cdots + k_n = k \leq n$. Among the n cyclic permutations of the n cards, there are exactly $n - k$ integers in which the sum of the numbers on the first r cards is less than r for every $r = 1, 2, \ldots, n$.

Proof: Let

$$k_{r+n} = k_r, r = 1, 2, \ldots,$$

and set

$$\phi_r = k_1 + k_2 + \cdots + k_r, \quad r = 1, 2, \dots; \ \phi_0 = 0.$$

Define

$$\delta_r = \begin{cases} 1, \ i - \phi_i > r - \phi_r, & i > r, \\ 0, \ \text{otherwise,} \end{cases}$$

and

$$\psi_r = \inf\{i - \phi_i, i \geq r\}, \quad r = 0, 1, \dots .$$

Note that

$$\delta_r = \psi_{r+1} - \psi_r.$$

Since

$$\phi_{r+n} = \phi_r + \phi_n,$$

we have

$$\delta_{r+n} = \delta_r$$

and

$$\psi_{r+n} = \psi_r + n - k, \quad r = 0, 1, 2, \dots .$$

Therefore, among the n cyclic permutation of $\{k_1, k_2, \dots, k_n\}$, there are exactly

$$\sum_{r=1}^{n} \delta_r = \psi \sum_{r=1}^{n} \delta_r = \psi_{n+1} - \psi_1 = n - k,$$

for which the sum of the first r elements is less than r, for $r = 1, 2, \dots, n$. This completes the proof of Theorem 4.10.5.

It can be seen that Theorem 4.10.5 implies Theorem 4.10.4. Note also that Theorem 4.10.5 can be formulated in a more general way as follows:

Theorem 4.10.6 Let $\nu_1, \nu_2, \dots, \nu_n$, be interchangeable or cyclically interchangeable discrete random variables taking values as nonnegative integers. Let

$$N_r = \nu_1 + \nu_2 + \cdots + \nu_r, \quad r = 1, 2, \dots, n \ N_0 = 0.$$

We have

$$P\{N_r < r, 1 \leq r \leq n\}.$$

Among the n cyclic permutations of the n cards, there are exactly $n - k$ integers in which the sum of the numbers on the first r cards is less than r for every $r = 1, 2, \dots, n$.

4.10.3 Transient Solution of the *M/M/*1 by Lattice Path Method

The literature is rich with extensive studies of the time-dependent (or transient) solution of *M/M/*1. See, for instance, Prabhu (1965) and Cohen (1982). In addition to the standard method used, there are those that are different. For instance, Abate and Whitt (1986) developed approximation. They used exponential and hyperexponential approximation for the first two moments to show how the process approaches steady state. Leguesdron et al. (1993) used generating function and its inversion to drive the solution.

All the aforementioned publications are on continuous time. However, it seems Champernowne (1956) is the first to provide a combinatorial solution of the classical *M/M/*1 queueing process. The process is a single-server Markovian queueing system with Poisson arrival with rate λ and exponential service times with parameter μ, with infinite arriving source and infinite capacity waiting room. Thus, letting random variables X_t and Y_t, respectively, represent the number of arrivals and departures (after completion of services) during time interval $(0,\ t]$, we have

$$P\{X_t = x\} = e^{-\lambda t}\frac{(\lambda t)^x}{x!}, \quad x = 0, 1, 2, \dots, \tag{4.134}$$

and

$$P\{Y_t = y\} = e^{-\mu t}\frac{(\mu t)^y}{y!}, \quad y = 0, 1, 2, \dots. \tag{4.135}$$

We now view the system as a birth and death process within the time interval $(0,\ t]$. Each event in the process is either an arrival into the buffer or departure from the service due to completion of service, with, respective, rates $\frac{\lambda}{\lambda+\mu}$ and $\frac{\mu}{\lambda+\mu}$. We let the number of tasks in the system, after occurrence of m events, be denoted by ϕ_m, $m = 0, 1, 2, \dots$. Thus, $\{\phi_m\}$ becomes a restricted (bounded) random walk with, respective, transition probabilities (a birth and death or a step forward and backward) as follows:

$$P\{\phi_{m+1} = z + 1 \mid \phi_m = z\} = \frac{\lambda}{\lambda + \mu} \tag{4.136}$$

and

$$P\{\phi_{m+1} = z + 1 \mid \phi_m = z\} = \frac{\mu}{\lambda + \mu}. \tag{4.137}$$

Thus, assuming $X_t = x$ and $Y_t = y$, a one-to-one correspondence can be established between the unrestricted (not bounded) random walk and the *M/M/*1 queueing process through lattice path representation. (See Jain et al. 2007). For that, we represent an arrival by a horizontal unit and departure by a vertical unit. Suppose the system start working when there are i tasks in the system. For instance, if $x = 8, y = 7$, and $i = 2$, then Figure 4.3 represents the lattice paths corresponding to the random walk and its associated queueing process:

$$P\{x > y \text{ throughout the counting}\} = 1 - \frac{m}{\binom{x+y}{y}}. \tag{4.138}$$

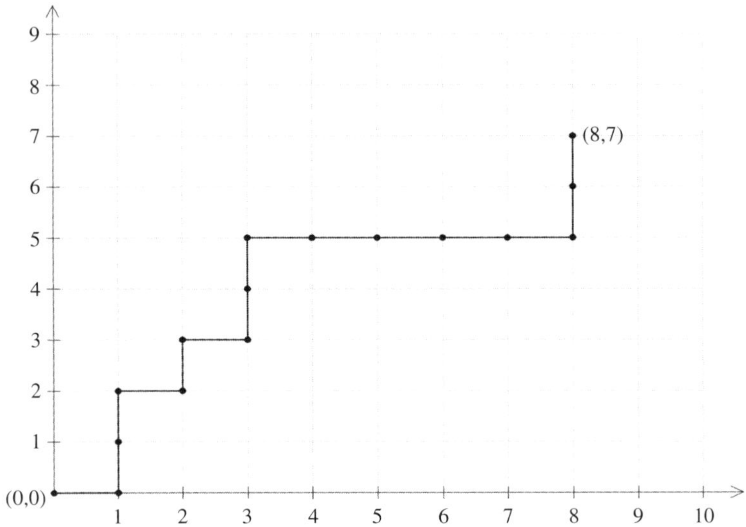

Figure 4.3 Lattice path representation of random walk.

It is a known property of the *M*/*M*/1 queue with continuous time that the transient probability distribution of the number of tasks, say m, in the system at time t, denoted by $\phi_m(t)$, starting with i tasks at $t = 0$, is

$$\phi_m(t) = e^{-(\lambda+\mu)t}\left[\rho^{\frac{m-i}{2}} I_{m-i}(2\sqrt{\lambda\mu t}) + \rho^{\frac{m-i-1}{2}} I_{m+i+1}(2\sqrt{\lambda\mu t}) + (1-\rho)\rho^m \sum_{j=m+i+2}^{\infty} \rho^{-\frac{j}{2}} I_j(2\sqrt{\lambda\mu t})\right], \quad (4.139)$$

where $\rho = \frac{\lambda}{\mu}$, I_r is the modified Bessel function of the first type given by

$$I_r(x) = \sum_{m=0}^{\infty} \frac{1}{m!\Gamma(m+r+1)}\left(\frac{x}{2}\right)^{2m+r}, \quad (4.140)$$

and $\Gamma(u)$ is the gamma function given by

$$\Gamma(u) = \int_0^{\infty} t^{u-1} e^{-t} dt, \quad (4.141)$$

with

$$\Gamma(n) = (n-1)!, \quad (4.142)$$

where n is a positive integer.

Now, for the discrete-time case under consideration, the assumptions are

1. no more than one task may arrive or complete service in any given slot;
2. events mentioned in (1) are mutually exclusive;
3. events mentioned in (1) in different slots are independent;
4. for the events mentioned in (1), the probability of an arrival in a slot is α; and
5. for the events mentioned in (1), the probability of a departure in a slot is

$$\begin{cases} \delta, & \text{if the system is not empty,} \\ 0, & \text{otherwise.} \end{cases}$$

On the basis of the aforementioned assumptions, the probability of no event occurs, that is, no arrival and no departure in a slot is

$$\begin{cases} 1-\alpha-\delta, & \text{if the system is not empty,} \\ 1-\alpha, & \text{otherwise.} \end{cases}$$

Thus, time instances $1, 2, \ldots$, indicate the end of intervals of unit duration. We let the iid random variables $X_n, n = 0, 1, 2, \ldots$, represent the nth increments in the system. Then, the queueing system under consideration may be associated with a random walk $\{S_n, n = 0, 1, 2, \ldots\}$, where

$$S_0 = 0 \text{ and } S_n = X_1 + X_2 + \cdots + X_n, \quad n = 1, 2, \ldots .$$

Now, if we assume that initially there are i tasks in the system, the problem of determining the probability of having m tasks in the system at the time epoch n, denoted by $P\{X_n = m \,|\, X_0 = i\}$, referred to as the transient solution of the queueing system, is to consider the random walk with a barrier at state $-i$ such that the walk moves forward $(+1)$, backward (-1), or stay put (0) such that

if $S_{n-1} > -i$, then

$$\begin{aligned} P\{X_n = +1\} &= p, \\ P\{X_n = -1\} &= q, \\ P\{X_n = 0\} &= 1-p-q, \end{aligned}$$

and

if $S_{n-1} = -i$, then

$$\begin{aligned} P\{X_n = +1\} &= p, \\ P\{X_n = 0\} &= \gamma. \end{aligned}$$

Thus, if the increments $X_n = 1, 0, -1$, are represented by a horizontal unit, diagonal unit, and vertical unit, respectively, the random walk can be represented by a lattice path with diagonal steps starting from the origin and encountering a barrier at line $x = y - i$. At the barrier, the path may move horizontally or diagonally. See Mohanty and Panny (1990).

For $\phi_m(t)$ with x, y, and m as used above, we have

$$\phi_m(t) = \sum_x \sum_y P\{Z(t) = m \mid X_t = x, Y_t = y\} P\{X_t = x\} P\{Y_t = y\}. \tag{4.143}$$

Thus, given $X_t = x$, $Y_t = y$, and that any realization of the random walk has the probability

$$\left(\frac{\lambda}{\lambda+\mu}\right)^x \left(\frac{\mu}{\lambda+\mu}\right)^y,$$

we have

$$\begin{aligned} P\{Z(t) = m | X_t = x, Y_t = y\} &= \frac{r\left(\frac{\lambda}{\lambda+\mu}\right)^x \left(\frac{\mu}{\lambda+\mu}\right)^y}{\binom{x+y}{y}\left(\frac{\lambda}{\lambda+\mu}\right)^x \left(\frac{\mu}{\lambda+\mu}\right)^y} \\ &= \frac{r}{\binom{x+y}{y}}, \end{aligned} \tag{4.144}$$

where r is the number of random walk paths corresponding to the event $Z(t) = m$. Jain et al. (1990, p. 200) show that for paths that do not touch the line $y = x + i$, that is, $m = x + i - y$ and the number of paths is the total number of paths minus the number of paths that touch or cross the line $y = x + i$. Thus,

$$r = \binom{x+y}{y} - \binom{x+y}{y-i}, \tag{4.145}$$

where $m = x + i - y$. In a case in which paths touch or cross the line $y = x + i$, we have

$$m = x + i - y + k, \quad k = 0, 1, 2, \ldots, y - i,$$

where k is found by the line $y = x + i + k$, which is touched but not crossed by random walk paths. Thus, the number of paths that touch but do not cross the line $y = x + i + k$, $\quad k = 0, 1, \ldots, y - i$, is

$$\binom{x+y}{y-i-k} - \binom{x+y}{y-i-k-1}, \quad k = 0, 1, 2, \ldots, y - i. \tag{4.146}$$

Thus, from the last relations, we have

$$\phi_m(t) = e^{-(\lambda+\mu)t}\left[\sum_x\sum_y\left(\frac{1}{x!y!} - \frac{1}{(x+i)!(y-i)!}\right)\lambda^x\mu^y t^{x+y}\right.$$
$$+\sum_x\sum_y\sum_{k=0}^{y-i}\left(\frac{1}{(x+i+k+1)!(y-i-k)!}\right.$$
$$\left.\left.-\frac{1}{(x+i+k+1)!(y-i-k)!}\right)\lambda^x\mu^y t^{x+y}\right], \tag{4.147}$$

where the first and second sums are subject to $m = x + i - y$ and $m = x + i - y + k$, $k = 0, 1, \ldots, y - i$, respectively. Therefore, the combinatorial transient solution of the *M*/*M*/1 system is

$$\phi_m(t) = e^{-(\lambda+\mu)t}\left[\left(\sum_{y=0}^{\infty}\frac{1}{y!(y+m-i)!} - \sum_{y=i}^{\infty}\frac{1}{(y-i)!(y+m)!}\right)\lambda^{y+m-i}\mu^y t^{x+y}\right.$$
$$+\sum_{y+i}^{\infty}\sum_{k=0}^{y-i}\left(\frac{1}{(y-i-k-1)!(y+m)!} - \frac{1}{(y-i-k-1)!(y+m+1)!}\right)$$
$$\left.\times\,\lambda^{y+m-i-k}\mu^y t^{2y+m-i-k}\right]. \tag{4.148}$$

We refer the reader to Jain et al. (1990, p. 201) to see that Eqs. (4.139) and (4.148) are equivalent.

4.11 STATIONARY *M/M/C* QUEUEING PROCESS

A queueing system in which its service station consists of more than 1, say c, servers set in parallel that they work independently of each other, is referred to as a ***parallel queueing system***. Such a system is sometimes referred to as a ***c-server queueing system, many-server queueing system***, ***multichannel*** queueing ***system***, ***multiprocessor system***, ***multiserver queueing system***, or other similar names.

Transient behavior of an $M/M/c$ queueing processes was first considered by Karlin and McGregor (1958). They used the ***spectral theory method*** for the differential equations of single birth and death processes. The method requires determining a set of polynomials and a spectral density function on these polynomials. After 2 years, Saaty (1960) offered the solution in a direct manner and obtained an explicit expression for the Laplace transform of the transient queue size for the two-server case. It is commonly accepted that general models may not possess explicit closed-form solutions. Jackson and Henderson (1966) suggested a relatively simple method for transient queues. Nonetheless, inversion of the Laplace transforms involved in such models may be an extremely difficult task. We, in this section, consider only the stationary-process case.

4.11.1 A Stationary Multiserver Queue

Let us consider an $M/M/c$ queueing system, that is, Poisson arrival of density λ, exponential service time with parameter μ for each one of c servers, and FCFS service discipline. Letting $\tau_0, \tau_1, \ldots, \tau_n$ be the epochs of arrivals with $\tau_0 = 0$, that is, starting time is set at zero, the interarrival times $\tau_n - \tau_{n-1}, \quad n = 1, 2, \ldots$, are independent identically distributed positive random variables with a common distribution function, $F(x)$, such that

$$F(x) = 1 - e^{-\lambda x}, \quad x \geq 0. \tag{4.149}$$

Suppose that the c service stations, with one server in each, are set in parallel. Furthermore, we suppose that the service times are independent identically distributed positive random variables with a common distribution function, $H(x)$, such that

$$H(x) = 1 - e^{-\mu x}, \quad x \geq 0, \tag{4.150}$$

independent of the arrival processes. Finally, we say that the system is in state E_n at time t if there are n tasks in the system, that is, in the waiting line and at all stations.

Let us denote the number of tasks in the system at time t by $\xi(t)$. We also denote the stationary distribution of the queue length by $P_n, n = 0, 1, 2, \ldots$, that is,

$$\lim_{t\to\infty} P\{\xi(t) = n\} = P_n, \quad n = 0, 1, 2, \ldots . \tag{4.151}$$

Similar to the case $M/M/1$ and its system of balance equations (4.45), we can write the system of balance equations for stationary $M/M/c$ as follows:

$$\begin{cases} \lambda P_0 = \mu P_1 \\ (\lambda + n\mu)P_n = \lambda P_{n-1} + (n+1)\mu P_{n+1}, & 1 \leq n \leq c, \\ (\lambda + c\mu)P_n = \lambda P_{n-1} + c\mu P_{n+1}, & n \geq c, \end{cases} \tag{4.152}$$

with $P_i(0) = \delta_{ij}$, where δ_{ij} is the Kronecker delta, which is defined in Equation (1.143).

The system (4.151) can easily be solved by recursive method. The unique solution can be obtained based on Haghighi and Mishev (2013, Theorem 5.9.2.1) as follows:

If $\lambda < c\mu$,

$$P_n = \begin{cases} \dfrac{\left(\dfrac{\lambda}{\mu}\right)^n}{n!} P_0, & \text{if } 1 \leq n \leq c, \\ \dfrac{\left(\dfrac{\lambda}{c\mu}\right)^n c^c}{c!} P_0, & \text{if,} \quad n \geq c, \end{cases} \tag{4.153}$$

where P_0 is given by

$$P_0 = \frac{1}{\sum_{n=0}^{c-1}\left[\frac{\left(\frac{\lambda}{\mu}\right)^n}{n!}\right] + \frac{\left(\frac{\lambda}{\mu}\right)^c}{c!\left(1-\frac{\lambda}{c\mu}\right)}}. \tag{4.154}$$

If $\lambda > c\mu$, then stationary distribution does not exist and $P_n = 0,\ \ \forall n$.

We note that $\rho = \frac{\lambda}{c\mu}$ is the traffic intensity for the system $M/M/c$. From the intensity condition for the existence of the stationary distribution, that is, $\rho = \frac{\lambda}{c\mu} < 1$, we can find the minimum number of servers during the stationary period as $c > \frac{\lambda}{\mu}$. Also, when $n \geq c$, the system $M/M/c$ may be regarded as an $M/M/1$ with service rate $c\mu$. Further, a task has to wait for service only if all servers are busy. The probability of such an event is $\sum_{n=m}^{\infty} P_n$. Thus, we have

$$P\{\text{a task waits for service}\} = \frac{\left(\frac{\lambda}{\mu}\right)^c}{c!} \times \frac{1}{1-\frac{\lambda}{c\mu}} \times \frac{1}{\sum_{n=0}^{c-1}\left[\frac{\left(\frac{\lambda}{\mu}\right)^n}{n!}\right] + \frac{\left(\frac{\lambda}{\mu}\right)^c}{c!\left(1-\frac{\lambda}{c\mu}\right)}}. \tag{4.155}$$

Chan and Lin (2003) presented a method of analysis of waiting time distribution for the $M/M/c$ queueing system. They showed that the conditional waiting time follows an Erlang distribution with rate $c\mu$. The authors obtained an explicit closed-form solution by means of the probability density function of the Erlang distribution. They have also shown that the waiting time distribution can be obtained from Takács's waiting time distribution for the $G/M/c$ queue as a special case. This reveals some insight into the significance of Takács's more general, but rather complex, result.

We should also remark that in case the system includes a feedback factor, that is, a task will depart from the system with probability q and returns to further service with probability $p, p + q = 1$, then service rate may be considered as $q\mu$ and thus Eqs. (4.153) and (4.154) will become as follows:

If $\lambda < cq\mu$,

$$P_n = \begin{cases} \dfrac{\left(\dfrac{\lambda}{q\mu}\right)}{n!} P_0, & 1 \leq n \leq c, \\ \dfrac{c^c\left(\dfrac{\lambda}{cq\mu}\right)^n}{c!} P_0, & n \geq c, \end{cases} \tag{4.156}$$

where P_0 is given by

$$P_0 = \frac{1}{\left[\sum_{n=0}^{c-1} \frac{\left(\frac{\lambda}{q\mu}\right)^n}{n!} + \frac{\left(\frac{\lambda}{q\mu}\right)^c}{c!\left(1 - \frac{\lambda}{cq\mu}\right)}\right]}. \tag{4.157}$$

If $\lambda > cq\mu$, then stationary distribution does not exist and $P_n = 0$, for every n.

EXERCISES

4.1 Derive $\widehat{f}(x)$ defined in Equation (4.34).

4.2 Show that the stationary pdf for age, X_r, is the same as for Y, that is, relations (4.44).

4.3 Prove the relation (4.50).

4.4 Invert Equation (4.61) and obtain the distribution.

4.5 Prove the relation (4.64).

4.6 Show that the effective arrival rate for the stationary $M/M/1/K$ process follows Equation (4.65).

4.7 Prove the relation (4.78).

4.8 Show that the mean and variance of a busy period for the stationary $M/M/1/K$ process are given by Eqs. (4.79) and (4.80), respectively.

4.9 For the stationary $M/M/1/K$ process, prove the following:

a. The limiting distribution of $\xi(t)$, that is, $\lim_{t\to\infty} P\{\xi(t) = n\}, n = 0, 1, 2, \ldots$, exists and is independent of the initial state if and only if $\xi(t)$ has a stationary distribution and the limiting distribution is identical with the stationary distribution.

b. The limiting distribution $\lim_{n\to\infty} P\{\zeta_n \leq x\}$ exists and is independent of the initial state only if ζ_n has a stationary distribution and the limiting distribution is identical with the stationary distribution.

4.10 Prove the relations (4.93)–(4.96).

4.11 For the stationary $M/M/1$ process with feedback, show that relations (4.107) and (4.108) hold.

4.12 For a priority queue with balking and reneging in Section 4.9, set up the set of difference equations for the length of the queue.

4.13 Consider an $M/M/1$ queueing system with mean arrival and service rates λ and μ, respectively, and FCFS service discipline. A task, on its arrival, joins the queue with probability 1, if the number of tasks in the system is less than a threshold value k, otherwise, it will join the queue with probability p and balk with probability $1-p$. After joining the queue, a task will wait for a certain length of time T for service to begin. If it has not begun by then, it will leave the queue without getting service. The time T is a random variable with exponential distribution of parameter α. The arrival and departure of the impatient tasks without service are independent events.

a. Find the average reneging rate of tasks.

b. Denote the probability of the queue length (waiting and being served) being n at time t by $P_n(t)$. Assume that the system starts with 1 task in the it, that is, $P_n(0) = \delta_{1,n}$, where $\delta_{1,\ n}$ is the Kronecker's delta and 0 is the absorbing state. Write the system of differential difference equation for this queueing system.

4.14 Consider an $M/M/1$ queue described as a bank with one teller and a lobby with a capacity of 10 clients. Assume that clients arrive at the rate $\lambda = 0.4$ per minute and are served at the rate of $\mu = 0.5$. Find:

a. average number of clients waiting in the lobby;

b. average time a client spends in the bank;

c. average time a client has to wait for service;

d. the probability that there is no client in the bank;

e. the probability that there are k clients in the bank; and

f. the probability of number of clients lost due to the capacity limit.

4.15 Consider an $M/E_k/1$ queue in which tasks arrive according to the Poisson process with rate λ and the service times follow an iid Erlang distributions of order k with mean service time $1/\mu$. Denote the number of tasks in the system at time t (including the one in service) by $N(t)$. Also, denote by $J(t)$ the remaining service phases of the task in service at time t. Assume that the system starts serving at $t = 0$. This assumption implies that $J(0) = k$.. Define the first passage time τ_j as

$$\tau_j \triangleq \inf\{t > 0,\ \ N(t) = 0 \,|\, N(0) = j,\ \ J(0) = k\}.$$

Note that τ_j is the length of the busy period that starts with j tasks in the system. Also, define the following:

$$Q_{n,i}^{(j)}(t) \triangleq P\{N(t) = n,\ \ J(t) = i,\ \ \tau_j > t \mid N(0) = j,\ \ J(0) = k\},$$

$$n, j \geq 1,\ \ 1 \leq i \leq k,\ \ t > 0,$$

$$Q_0^{(j)}(t) \triangleq P\{\tau_j < t \mid N(0) = j,\ \ J(0) = k\},\ j \geq 1,\ \ t > 0.$$

For the continuous Markov chain $\{N(t),\ J(t)\}$ with absorbing barrier at $N(\tau_j) = 0$, write the differential difference system of equations.

5

QUEUES WITH DELAY

5.1 INTRODUCTION

The service interruptions are an unavoidable phenomenon in many real-life situations. In most of the studies of queuing systems, it is assumed that the server is available on a permanent basis as long as at least a task is present in the waiting line, but service station never fails. However, these assumptions are practically unrealistic. In practice, we often meet a case where the server may fail and can be repaired. Applications of such models may be found in the areas of computer communication network and flexible manufacturing system.

The word "***delay***" in queuing theory has been used with different meanings on different occasions, including the waiting time. In this book, "waiting time" is actually waiting time. The term "delay" in this book is used to mean the additional time to the service time and waiting in line that a task may have to spend in the system before completion of its service. Feedback time is not a delay time, and we refer to it as the feedback time. However, there are terms that use "delay," but they are actually waiting time. For instance, in a normal situation, arrivals enter a system at a buffer and wait for service. If the server is busy, then an arrival has to wait for service. This is what we mean by standard "***waiting time***." Another example is in computer network, wherein ***queuing delay*** is the time the packet is in a waiting line before transmission starts. The time depends on the number of packets ahead of it in the queue and the rate they are removed from the queue. Yet another example of a standard waiting time is ***processing delay*** in computer network, which is the time from the arrival of a packet until it is assigned to a queue of an output link for transmission. This is what we refer to as the ***sojourn time***. If the packet size is fixed, the time is a constant. However, "real delay" may have different forms. Below, we list some types of delays as examples.

Delayed and Network Queues, First Edition. Aliakbar Montazer Haghighi and Dimitar P. Mishev.

1. ***Delay in Voice Over IP* Services**. Delay has been a source of problems with telephone networks for many years. Before, the major source of delay was satellite connections, typically 700 ms per satellite hop. However, significant delay can now occur, even on local connections, because of the use of voice over IP (VoIP) service and cellular technology. This type of delay may have a great impact on conversational quality and how to define acceptable levels of delay in service level definitions or agreements. In a voice conversation, a call participant is aware of the time taken for the remote user to react. This time includes the round trip delay experienced by the voice signal and the reaction time of the remote user. The user, however, may not be aware of any asymmetry that may be present in the time taken by the outgoing versus incoming voice signal.

   Round trip delay refers to the total delay for the sending and receiving direction combined. ***One-way delay*** refers to the delay in either the sending or receiving direction. ***Symmetric one-way delay*** refers to the delay in the sending or receiving direction with the assumption that they are equal (i.e., half the round trip delay). Within a VoIP service, delay has three key elements: (1) ***IP Network Delay***, which is generally the round trip delay for an IP packet within the IP network; (2) ***End System Delay***, which is the sum of the ***encoding*** (the time taken to accumulate and encode a voice packet, and will always be at least as large as the packet size, typically 10–30 ms), ***Decoding*** (the time taken to decode and start to play out a received voice packet this may be quite short), ***jitter buffer*** (IP phones and gateways use jitter buffers to remove short-term packet delay variations from the received packet stream by inserting some additional delay that allows the system to reinsert late-arriving packets in the proper order before they are decoded and played out) and other data handling delays that occur within a typical IP phone or gateway; and (3) ***External Delay***.

2. ***Warm-up Time*** period before starting service, ***breakdown***, ***working vacation***, and ***break*** or ***vacation***, constitute "delay" in arrival, service, feedback, and other system's activities.

3. If a task is allowed to return to the service station after completion of its service, but it is ***rerouting*** through a filter, the time spend for filtration constitutes "delay" for us.

4. ***Transmission Delay*** is the time between the transmission of the first bit and last bit of the packet. If the packet size is fixed, the time is a constant.

5. ***Propagation Delay*** is the time for signals to traverse the link. In other words, in computer networks, propagation delay is the time taken by the head of the signal to travel from the sender to the receiver. It can be computed as the ratio of the link length to the propagation speed over the specific medium. The propagation delay is a constant, if the length and propagation speed of the link are fixed. It is independent of the traffic in the network.

6. ***Latency*** in a signal transmission process is the sum of delays at links. The delay at a link is the sum of four components, if the retransmission is not

considered. These components are queuing delay, processing delay, transmission delay, and propagation delay.

7. If the system imposes a minimum number of tasks in the buffer before service starts, waiting for the ***threshold*** constitute a "delay."
8. The ***(k, K)-batch policy*** is a varying batch size policy, that is, either arrival or service process requires a minimum of k and maximum of K tasks in a batch before it starts operating. See Haghighi and Mishev (2016b), for example. In their papers, they considered the busy period and queue length for system with (k, K) policy, that is, a system with delayed feedback.
9. The ***N-policy*** or ***threshold policy*** is a rule specifying the beginning of a busy period. That is, once the server starts an idle period, it waits for the number of tasks in the buffer to accumulate to N, $(N \geq 1)$, before it begins to serve. See Dshalalow (1997). One of the reasons to apply N-policy is to minimize server switch-overs, that is, usually followed by start-ups. However, this prolongs the idle period and, thus, delays the service, that is, increases the sojourn time of a task. Thus, the time for which a task has to wait until the number N has been reached is called as "delay." Hence, the optimal value of N is an objective.

   The concept of optimal N-policy was introduced by Yadin and Naor (1963) for $M/M/1$ queuing system without start-up time. They also referred to N-policy as ***removable server or servers***. As an example, see Jain et al. (2012), who looked at an $M^X/G/1$ queue with Bernoulli feedback under N-policy.

   Here are two applications of the N-policy: (1) Variable vacation time, that is, vacation time plus the elapse time for accumulation of N tasks. In other words, as soon as all tasks in the system are served, the server goes on a single vacation and returns when there are at least N tasks accumulated in the system. (2) In a machine repair system, the N-policy is one of the cost-effective measures, as during the accumulation of failed units, the repairman may perform ancillary duty.
10. The ***F-policy*** is a policy to control arrivals to a queuing system. For instance, in a multicomponent machining system, the F-policy does not allow a task to enter the system when the system has reached its capacity, say K, until enough tasks in the system have been served so that the number of tasks in the system decreases to a threshold number F, $0 \leq F < K \leq 1$.

    In the real-life day-to-day operations, such as production, communication, transportation, and manufacturing systems, the F-policy and N-policy are being used as cost-effective approaches. The N-policy queuing system investigates the most common issue of controlling service, while the F-policy queuing system deals with the most common issue of controlling arrivals.
11. The ***(M, N)-policy*** is a policy that the server starts servicing only when the number of tasks in the buffer becomes N and remains busy as long as there are at least $M \geq 0$ tasks waiting. We will, in due course, present a paper on this policy.
12. In a queuing system with vacation time, the server may not be available for a period of time due to many reasons such as waiting for the server maintenance,

serving another queue, scanning for new job in a typical communication system, or just taking a break time. This period of unavailability of server is referred to as ***vacation***.

13. Tasks attending many queuing systems may be served repeatedly for various reasons. When the service of a task is unsatisfactory, service may be repeated as many times as necessary until a successful service is completed. Such queuing systems arise in the stochastic modeling of many real-life situations. For instance, in data transmission, a packet transmitted from the source to the destination may be returned and it may continue until the packet is finally transmitted. In the ***retrial setup***, after completion of each service, the server will remain idle in the system until the arrival of the next ***primary*** or ***retrial task***. Server's idle time is reduced by the introduction of search of ***orbital tasks*** immediately after a service completion.

    A ***retrial queuing model*** is characterized by the feature that an arriving task finds the server busy, it is ejected or rejected by the system, and hence, neither joins the waiting line nor leaves the system immediately forever, but it goes to some virtual space, called as ***orbit***, and tries again and again after some random time to obtain service. Such queuing systems arise in many communication protocols, local area networks, wireless and computer networks, and daily-life situations. A simple example is a person making a phone call. If the line is busy, then he/she is not able to queue but dials again after some time until connection is made.

    A retrial queue is similar to any ordinary queuing system in the sense that there is an arrival process and one or more servers. The elemental differences are first the tasks that enter during a down or busy period of the server or servers may reattempt service at some random time in the future, and second a buffer, which is known as a primary waiting line, in the context of retrial queues is not mandatory. Instead of an ordinary buffer in an ordinary queuing system, an orbit or a retrial orbit is used to which tasks proceed after an unsuccessful attempt obtaining service, and from which it retries service according to a given probabilistic or deterministic policy.

    Neuts and Ramalhoto (1984) were the first to introduce a classical queue with search for ***orbital tasks*** immediately on termination of a service. Because of the practical importance of telecommunication, wireless and other types of network technologies, an ample literature on this type of queue has emerged over the past several decades. For a general survey of retrial queues and summary of many results, we refer the readers to the following references as examples: Falin (1990), Walrand (1991), Choi and Kim (1998), Falin and Artalejo (1998), Libman and Orda (2002), Gharbi and Ioualalen (2006), Ebrahimi (2006), Bouchentouf and Sakhi (2015), Rajadurai et al. (2016) and Bouchentouf et al. (2015).

In this chapter, we discuss different models covering various types of delays defined above. We start with a model that was published in 2008 and also appeared in a book written by the authors of this book. Since that book contains a complete detail of the paper, here we outline the model as it is an interesting delay model.

5.2 A QUEUING SYSTEM WITH DELAYED SERVICE

In this section, we consider a single-server Poisson queue with delayed service. The delay in this model is caused by tardiness in starting the service after the task has entered the service station.

Aliakbar M. Haghighi

Stefanka Chukova

Dimitar Mishev

This model has been studied by Haghighi, et al. (2008) for the first time and has been discussed in detail in Haghighi and Mishev (2014). The authors proved that the model presented is a non-Markovian one and is an $M/G/1$ queuing system. Then, using $M/G/1$, they obtained many valuable and interesting results. The delay time τ is approximated by an Erlang-distributed random variable with a large number of stages. Thus, the authors are able to write the system of differential difference equations for the new model and obtain the distribution of the queue length in both steady-state and transient cases. Here, we only present highlights of the paper.

Here is the model: Consider an $M/G/1$ queuing system, a non-Markovian queue, with a mean arrival rate λ. The processing time of a task consists of two independent parts: the delay time and service time. The delay time may be a ***deterministic (fixed)*** and ***stochastic*** random variable, in our case, exponential with parameter μ_τ. The authors provide an explicit closed-form distribution for the length of a busy period. The delay time will be approximated by an Erlang-distributed random variable with a large number of stages. The authors also obtain an explicit closed-form steady-state distribution for the queue length.

For the nth task, $n = 1, 2, \ldots$, we denote its processing time by Y_n, delayed time, τ, and service time by X_n. Thus, $Y_n = X_n + \tau$. We assume that $\{Y_n\}$ is a sequence of identically distributed, mutually independent (*iid*), positive random variables with distribution $K(t)$. We also denote by $H(t)$ the distribution of the sequence $\{X_n\}$, where $H(t)$ is exponential with parameter μ_X. Let X and Y represent cumulative distributions of $\{X_n\}$ and $\{Y_n\}$, respectively, that is,

$$H(t) = P(X \leq t) = 1 - e^{-\mu_X t}, \quad t > \tau, \tag{5.1}$$

and

$$K(t) = P(Y \leq t) = P(X + \tau \leq t), \tag{5.2}$$

respectively.

It should be noted that since a service cannot start before the delay period has elapsed, probability of completion of processing of a task before τ is zero. Hence, $Y \geq \tau$, for processing time.

The probability density function, *pdf*, of Y, denoted by $k(t)$, when $\tau_x \neq \mu_\tau$, is

$$k(t) = c_1 \mu_X e^{-\mu_X t} + c_2 \mu_\tau e^{-\mu_\tau t}, \tag{5.3}$$

with $c_1 + c_2 = 1$ (see Cox and Miller 1984, p. 257). On the contrary, if $\mu_X = \mu_\tau = \mu$, then completion of a process will require a task to undergo two stages with mean $\mu_X + \mu_\tau = 2\mu$. Under the aforementioned situation, the system will be an $M/E_2/1$ one with parameter $\frac{1}{2\mu}$ (see Medhi (2003, p. 178)).

We remind the reader that E_2 stands for Erlang distribution with shape parameter equal to 2. (See Chapter 1, Eqs. (1.68) and (1.69)). The deterministic delay time will be approximated by an Erlang-distributed random variable with a large number of stages.

In case of the exponential delay time, since Y is a sum of two *iid* exponential random variables, the Laplace transform of its distribution, denoted by $K^*(s)$, is

$$K^*(s) = \left(\frac{\mu_X}{\mu_X + s}\right)\left(\frac{\mu_\tau}{\mu_\tau + s}\right). \tag{5.4}$$

In the stationary case, we can readily have expected measures of effectiveness such as the mean waiting time in the line, W_q, mean waiting time in the system, W, mean waiting queue length, L_q, and mean queue length, L, as

$$W_q = \frac{3\rho}{2\mu(1-\rho)}, \quad W = \frac{L}{\lambda} = W_q + \frac{2}{\mu}, \tag{5.5}$$

$$L_q = \lambda W_q = \frac{3\rho^2}{2\mu(1-\rho)}, \text{ and } L = L_q + \rho, \tag{5.6}$$

where $\mu = \mu_X = \mu_\tau$ and $\rho = \lambda/\mu$. (Gross and Harris 1998, pp. 133–134). In case of a deterministic (fixed) delay time, distribution of the processing time, Y, denoted by $K(t)$, under the assumption of $\mu = \mu_X = \mu_\tau$, is given by

$$K(t) = \begin{cases} 1 - e^{-\mu(t-\tau)}, & t \geq 0, \\ 0, & \text{otherwise.} \end{cases} \tag{5.7}$$

In other words, $K(t)$ is the shifted exponential distribution (Hogg and Tanis 1993) with parameters (μ, τ). The mean and variance of the processing time are, respectively,

$$E(Y) = \frac{1}{\mu} + \tau \text{ and } \sigma_Y^2 = \frac{1}{\mu^2}. \tag{5.8}$$

Now, let the random variable $\xi(t)$ represent the number of tasks in the system at time t, including the one being processed. Let the probability of $n (n \geq 0)$ tasks in the system at time t be denoted by $P_n(t)$, that is,

$$P_n(t) = P\{\xi(t) = n\}. \tag{5.9}$$

Let also

$$P_n = \lim_{t\to\infty} P_n(t) \tag{5.10}$$

be the steady-state probability of having n tasks in the system. Further, let $P_0(t)$ represent the probability that the system is empty. In addition, let the utilization factors or traffic intensities, in this case, be

$$\rho_X = \frac{\lambda}{\mu}, \quad \text{and} \quad \rho_Y = \lambda\left(\frac{1}{\mu} + \tau\right). \tag{5.11}$$

Then, in case of deterministic delay time, if $\rho_Y < 1$, the steady-state distribution of the queue length in the system exists, which is given by

$$P_0 = 1 - \rho_Y, \tag{5.12}$$

$$P_n = (1-\rho_Y)d_n, \quad n = 1,2,\dots, \tag{5.13}$$

where

$$d_n = \begin{cases} \dfrac{(\lambda\tau)^{n-1}(\lambda\tau - n)}{n!} - \displaystyle\sum_{i=1}^{n} b_i d_{n-i}, & n = 1,2,\dots, \\ 1, & n = 0, \end{cases} \tag{5.14}$$

and

$$\begin{cases} b_1 = \lambda\tau - \left(1 + \dfrac{\lambda}{\mu}\right) e^{\lambda\tau}, \\ b_2 = \dfrac{(\lambda\tau)^2}{2} + \dfrac{\lambda}{\mu} e^{\lambda\tau}, \\ b_i = \dfrac{(\lambda\tau)^i}{i!}, \quad i = 3,4,\dots. \end{cases} \tag{5.15}$$

The mean stationary queue length, denoted by L_D, is

$$L_D = \frac{2\rho_Y(1-\lambda\tau) + \lambda^2\tau^2}{2(1-\rho_Y)}. \tag{5.16}$$

It is well known that

$$E_r = X_1 + X_2 + \cdots + X_r, \tag{5.17}$$

where, as we mentioned earlier, $\{X_i\}$ is a sequence of *iid* random variables with exponential distribution with parameter r/τ, that is,

$$P\{X_i \le t\} = 1 - e^{-\frac{r}{\tau}t}, \tag{5.18}$$

with

$$E(X_i) = \frac{\tau}{r} \text{ and } \operatorname{Var}(X_i) = \frac{\tau^2}{r^2}. \tag{5.19}$$

Thus,

$$E(E_r) = r\frac{\tau}{r} = \tau, \operatorname{Var}(E_r) = r\frac{\tau^2}{r^2} = \frac{\tau^2}{r} \tag{5.20}$$

and

$$\lim_{r\to\infty} \operatorname{Var}(E_r) = 0. \tag{5.21}$$

Therefore, we have

$$Y = \tau + X \approx \widehat{Y} = E_r + X. \tag{5.22}$$

In other words, tasks arrive in the system and eventually enter the service station one at a time, where the in-service requires passing through $r+1$ stages. Times for these stages are $r+1$ independent random variables, the first r of which are identically exponentially distributed with parameter r/τ and the last one is exponentially distributed with parameter μ.

The busy period of the system is defined as the interval of time during which the system is continuously busy serving tasks. Let us also denote the *CDF* of a busy period, T_{bp}, by $B(t)$, that is, $B(t) = P(T_{\text{bp}} \leq t)$. In general, it is possible that a busy period starts at the arrival of a task not necessarily to an idle system, for instance, when there are $m-1, m \leq 1$, tasks in the system, and ends the next time when the system returns to $m-1$, that is, the time interval $t_2 - t_1 \equiv T_{\text{bp}}$ between two consecutive instances t_1 and t_2 such that $\xi(t_1) - \xi(t_2) = m - 1$. This case is referred to an m-fold busy period. Thus, $\xi(t) \geq m, \forall t \in (t_1, t_2)$. Here, however, we assume that a busy period starts when the system is empty. Then, for the exponential delay time, it can be shown, in Haghighi et al. (2008), the following:

1. if $\tau < \frac{1}{\lambda} - \frac{1}{\mu}$, the cumulative distribution of a busy period T_{bp} is given by

$$B(t) = \sum_{n=1}^{\left[\frac{t}{\tau}\right]} \frac{\lambda^{n-1}\mu^n e^{\mu\tau n}}{n!} \sum_{k=0}^{n-1} \frac{(n+k-1)!}{k!(n-k-1)!} \frac{(-n\tau)^{n-k-1}}{(\lambda+\mu)^{n+k}} \times \sum_{i=0}^{n+k-1} \frac{1}{i!}[e^{-(\lambda+\mu)n\tau}[n\tau(\lambda+\mu)]^i - e^{-(\lambda+\mu)t}[t(\lambda+\mu)^i]], \tag{5.23}$$

where $\left[\frac{t}{\tau}\right]$ is the greatest integer less than or equal to $\frac{t}{\tau}$,

2.
$$B(\infty) = 1, \tag{5.24}$$

and

3.
$$E(T_{\text{bp}}) = \begin{cases} \dfrac{1+\mu\tau}{\mu - \lambda(1+\mu\tau)}, & \rho_Y < 1, \\ \infty, & \rho_Y \geq 1. \end{cases} \tag{5.25}$$

It should be noted that the length of a busy period is independent of the queue discipline. Indeed, this is the fact that Takács (1962) has "brilliantly" (as Kleinrock (1975) puts it) used to find the distribution of a busy period.

Example 5.2.1 We, finally, present this example from Haghighi et al. (2008) and Haghighi and Mishev (2014). The computer software Matlab has been used in this example. We choose different values of λ, μ, τ, and t to examine the behavior of the distribution of the busy period as well as the distribution of the queue length given in Figures 5.1–5.8. From the numerical examples, we observe that when the values of τ is "very close" to its upper limit $(1/\lambda) - (1/\mu)$, the distribution of the queue length converges very slowly to one.

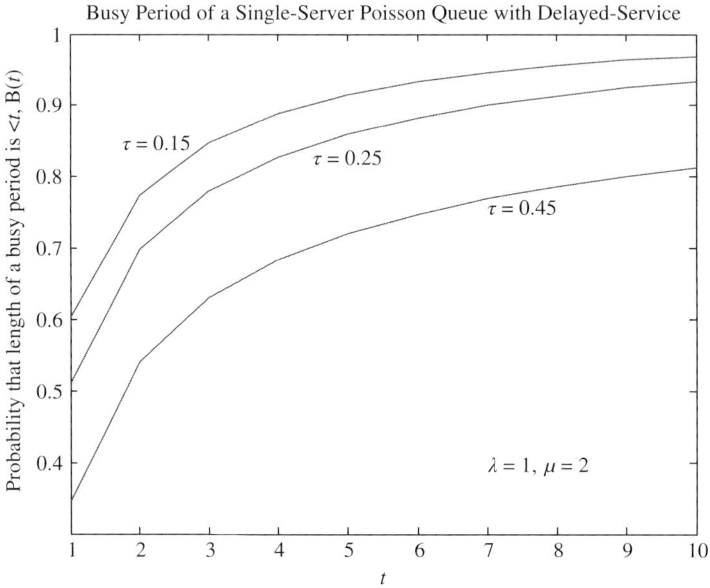

Figure 5.1 Probability of busy period approaching 1 as time becomes large; $\lambda = 1$, $\mu = 2$.

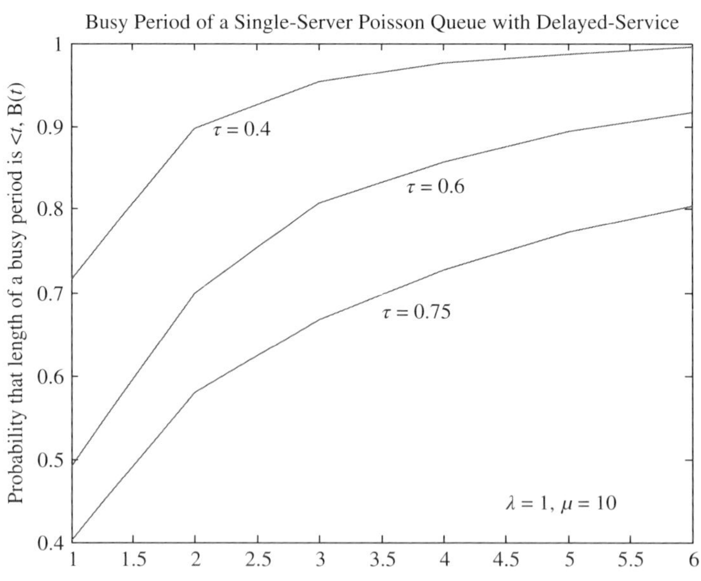

Figure 5.2 Probability of busy period approaching 1 as time becomes large; $\lambda = 1$, $\mu = 10$.

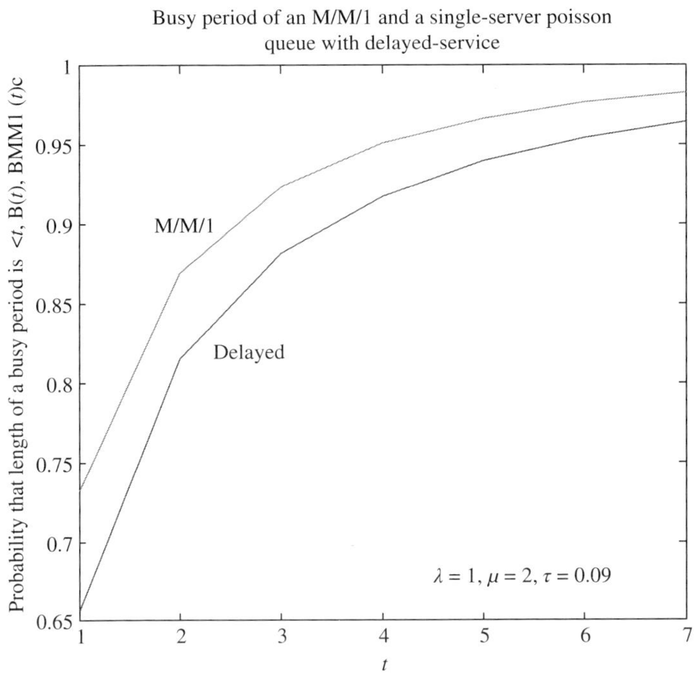

Figure 5.3 Difference between busy period of model discussed and standard *M/M/*1; $\lambda = 1$, $\mu = 2$, $\tau = 0.09$.

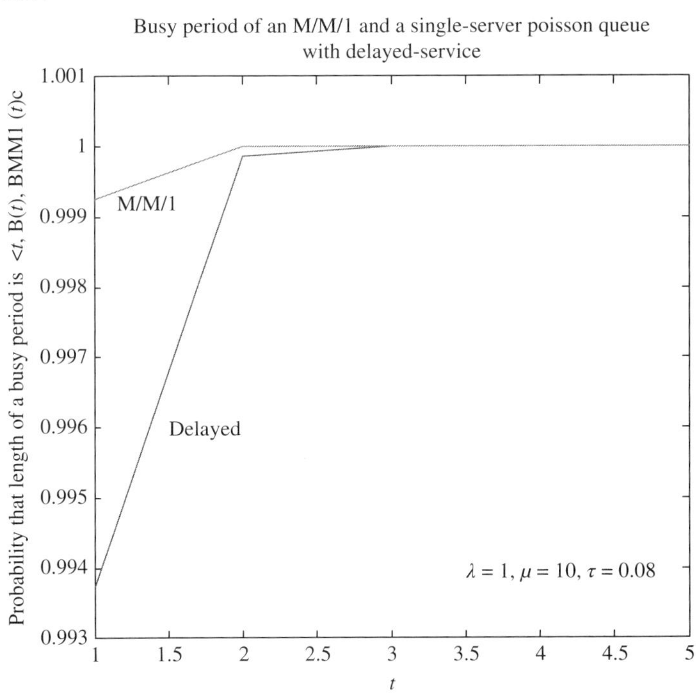

Figure 5.4 Difference between busy period of model discussed and standard *M/M/*1; $\lambda = 1$, $\mu = 10$, $\tau = 0.08$.

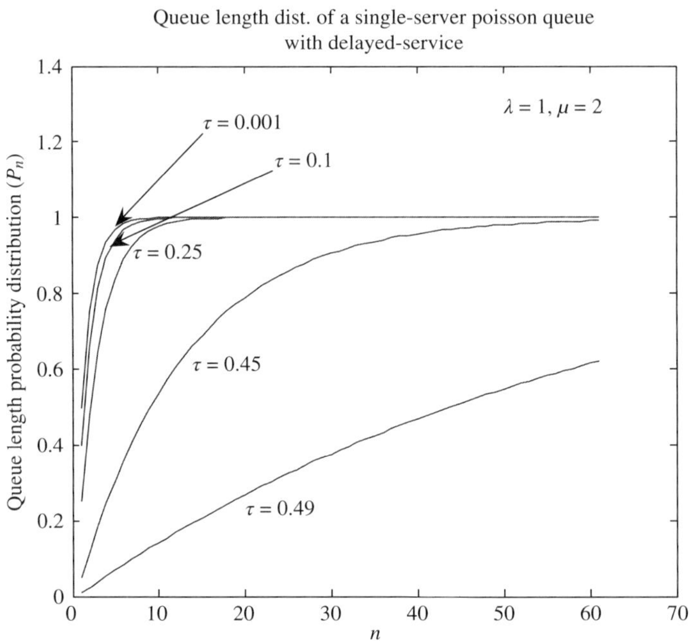

Figure 5.5 Cumulative probabilities approaching 1, regardless of values of τ; $\lambda = 1$, $\mu = 2$.

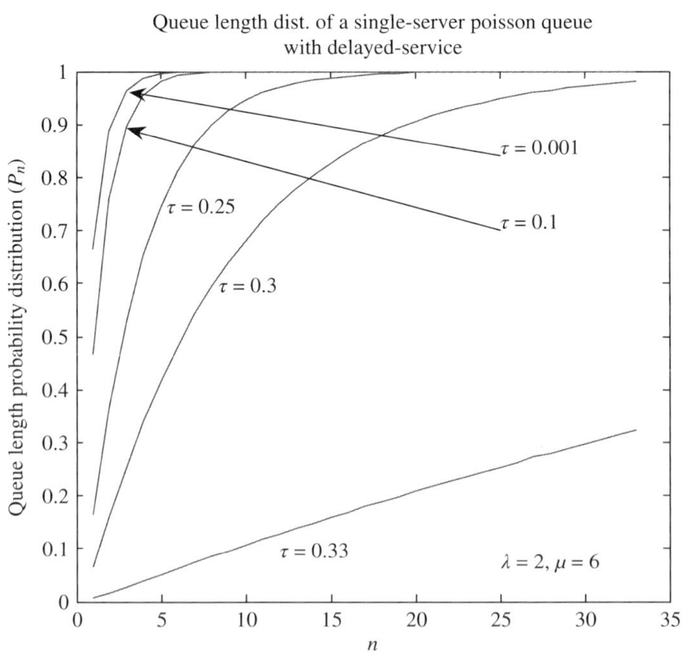

Figure 5.6 Cumulative probabilities approaching 1, regardless of values of τ; $\lambda = 2$, $\mu = 6$.

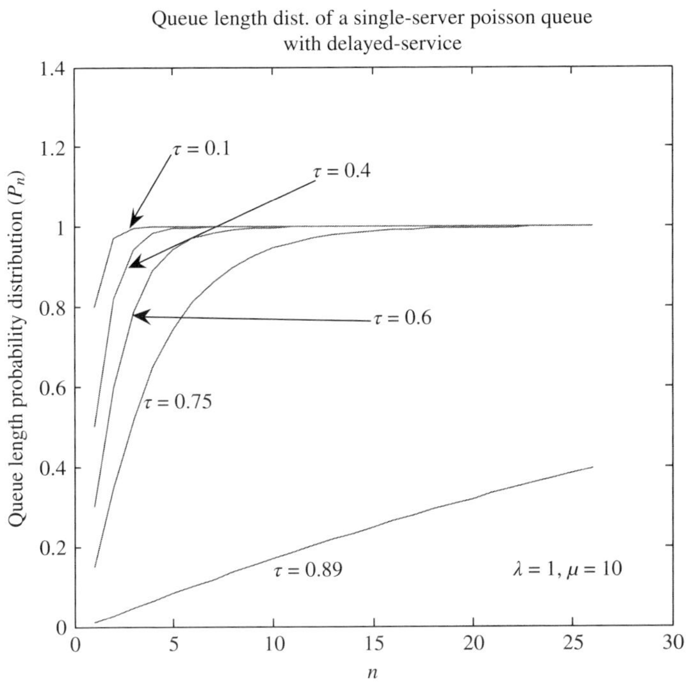

Figure 5.7 Cumulative probabilities approaching 1, regardless of values of τ; $\lambda = 1$, $\mu = 10$.

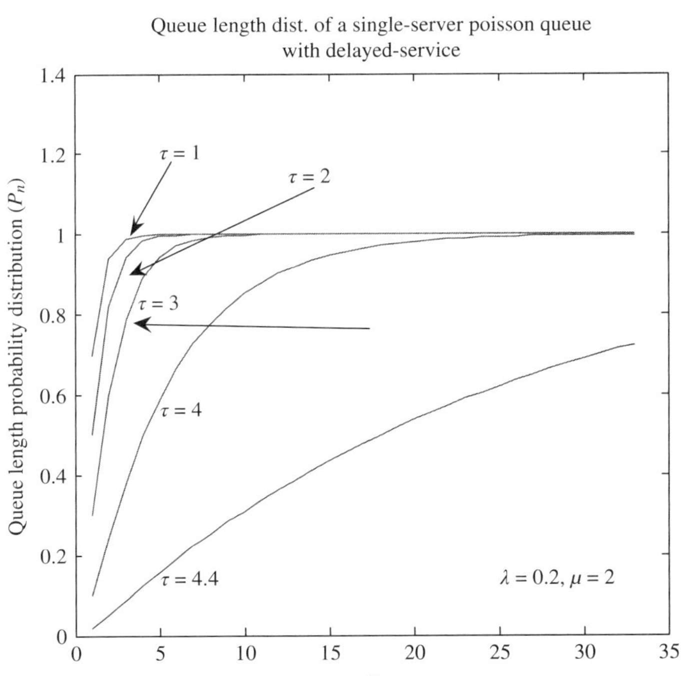

Figure 5.8 Cumulative probabilities approaching 1, regardless of values of τ; $\lambda = 0.02$, $\mu = 2$.

In order to evaluate the distribution of the busy period, when $t = 10$, and $\tau = 0.1$, would require nearly 2 million rounds of operations involving expressions such as 100^{100}, e^{100}, and 100!. Our experience shows that the mathematical software Matlab that, for instance, can handle a limited number of values of $\left[\frac{t}{\tau}\right]$, which will impose some constraints on the choice of λ, μ, and τ. On the contrary, a large delay time will require a small arrival and large service rates for the steady state to exist. This may not be desirable and, indeed, practically not acceptable. On the basis of the above constraints, the values of λ, μ, and τ, for this example, have been chosen.

As noted from Figures 5.1 and 5.2, the busy period probabilities are approaching 1 as time increases. Comparative figures have been illustrated to show the difference between the busy period of the model discussed (on the graph denoted by $B(t)$), and the one for the standard *M*/*M*/1 (on the graph denoted by $BMM(t)$), Figures 5.3 and 5.4. As can be seen for values of τ very close to zero, *M*/*M*/1 is a good approximation for the model. It seems (apparent from our observations) that there exists a threshold for τ, say τ^*, such that if $\tau > \tau^*$, the two systems are significantly different. The difference between the two systems is identified by a prespecified performance diagnostics. In Figures 5.4 and 5.5, for instance, 0.09 and 0.08 are closed to the threshold.

For the queue length distribution, Figures 5.5–5.8, cumulative probabilities are approaching 1, as expected. The size of τ in this case is of no concern.

5.3 AN *M*/*G*/1 QUEUE WITH SERVER BREAKDOWN AND WITH MULTIPLE WORKING VACATION

S. Pazhani Bala Murugan

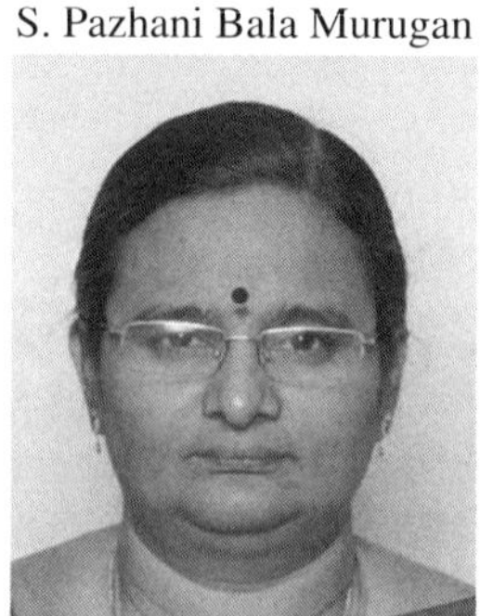

Tmt. K Santhi

In this section, we present a model discussed in Murugan and Santhi (2015). The model deals with the steady-state behavior of an $M/G/1$ queuing system that is subject to ***server breakdown***, in which the server takes multiple working vacations. In other words, the system may break down at random and repair time is assumed arbitrary. Further, just after completion of a task's service, the server may take a multiple working vacation. However, during the vacation times, the server works with different service times rather than completely stopping the service. Both the regular service times and service times in a vacation period are assumed to be generally distributed random variables. Because of the complexity of the model, rather than explicit queue length distribution, the authors find the probability generating function (pgf) of the number and the mean number in the system and discuss some particular cases. We will discuss this paper with the same notations as they appear in the paper.

5.3.1 Mathematical Formulation of the Model

Tasks arrive singly according to a Poisson distribution with arrival rate $\lambda(\lambda > 0)$ to a system and are served with first-come-first-serve (FCFS) discipline. The service times are assumed to be presented by random variables that are generally distributed with pdf and cdf as $s_b(x)$ and $S_b(x)$, respectively, with $S_b^*(\theta)$ as the Laplace–Stieltjes transform (***LST***) of $S_b(x)$. As soon as the system becomes empty at a service completion instant, the server starts a working vacation. Durations of the vacation times are assumed random variables distributed exponentially with parameter η. At return from vacation, if there are tasks in the system, the server will start a new busy period, otherwise, he/she goes back to continue his/her working vacation period. This is what is referred to as ***multiple working vacation***s. During a working vacation period, the server provides the service with pdf and cdf as $s_v(x)$ and $S_v(x)$, respectively, and LST $S_v^*(\theta)$.

It is assumed that the system may break down at random. Breakdown of the server may occur whether the server is on a working vacation or not. Breakdowns are assumed to occur according to a Poisson distribution with mean breakdown rate $\alpha_1(> 0)$ if the server is on a working vacation and $\alpha_2(> 0)$ if the server is not on a working vacation. Further, it is assumed that once the system breaks down, the task whose service is disrupted moves to the head of the waiting line (in the buffer) and the system enters a repair mode immediately. The repair times are assumed to be random variables with general pdf and cdf as $s_{r_1}(x)$ and $S_{r_1}(x)$, respectively, if the server is on working vacation and $s_{r_2}(x)$ and $S_{r_2}(x)$, respectively, if the server is not on working vacation with $S_{r_1}^*(\theta)$ and $S_{r_2}^*(\theta)$ as the LST of each cdf, respectively.

5.3.2 Steady-State Mean Number of Tasks in the System

We pause to note that if a stochastic process is not regenerative, ***supplementary variables*** are used to make it so. Supplementary variables are used to include in a non-Markovian process to make it Markovian. These variables were first introduced by Cox (1995). Alfa and Rao (2000) showed that the use of remaining or elapsed times as supplementary variables leads to the notion of forward or backward Chapman–Kolmogorov equations, respectively. The technique of supplementary variables is also a method to solve for the stationary distribution of an $M/G/1$

queue. The idea is to assign to every nonexponential governing random variable a supplementary variable that fixes either the elapse time or the remaining time to this random variable. The supplementary variable makes the process a Markovian, incorporating all necessary information to supplementary variables to predict the future of the process knowing only its current state, the Markov property. For instance, Wang and Ke (2000) provided a recursive method and used the supplementary variable method to develop the stationary probability distributions for the *M/G/1/K* queue. In addition, Ke and Wang (2002) considered the *N*-policy *G/M/1/K* queuing system based on the supplementary variable method.

Murugan and Santhi (2015) used the supplementary variable technique to find the distribution of the queue length. That is, from the joint distribution of the queue length and the remaining service time of the task in service, if the server is busy, or the remaining service time of the task, if the server is on working vacation. However, finding the distribution for the model under consideration, if not impossible, is extremely difficult. Thus, the authors only find the mean queue size in the stationary case. For that, they find the pgf of the queue size.

We now present the process of calculating the mean queue length. Let $N(t)$ denote the system size at time t. Denoting the "working vacation" by WV, in order to define a bivariate Markov process, we define the following ***supplementary random variables***:

$$\begin{cases} S_b^0(t) : \text{the remaining service time in not WV period,} \\ S_v^0(t) : \text{the remaining service time in WV period,} \\ S_{r_1}^0(t) : \text{the remaining repair time in not WV period, and} \\ S_{r_2}^0(t) : \text{the remaining repair time in WV period,} \end{cases} \tag{5.26}$$

with their respective LST, with parameter θ, as

$$\begin{cases} \mathcal{L}\{S_b^0(t)\} = S_b^{0*}(\theta), \\ \mathcal{L}\{S_v^0(t)\} = S_v^{0*}(\theta), \\ \mathcal{L}\{S_{r_1}^0(t)\} = S_{r_1}^{0*}(\theta), \text{ and} \\ \mathcal{L}\{S_{r_2}^0(t)\} = S_{r_2}^{0*}(\theta). \end{cases} \tag{5.27}$$

We also define the random variable $Y(t)$ as

$$Y(t) = \begin{cases} 0, & \text{if the server is idle in WV at time } t, \\ 1, & \text{if the server is busy on not WV period at time } t, \\ 2, & \text{if the server is busy on WV period at time } t, \\ 3, & \text{if the server is waiting for repair during not WV period at time } t, \text{ and} \\ 4, & \text{if the server is waiting for repair during WV period at time } t. \end{cases} \tag{5.28}$$

We now define the ***bivariate Markov process*** $\{N(t), \delta(t); t \geq 0\}$, where

$$\delta(t) = \begin{cases} S_b^0(t), & Y(t) = 1, \\ S_v^0(t), & Y(t) = 2, \\ S_{r_1}^0(t), & Y(t) = 3, \\ S_{r_2}^0(t), & Y(t) = 4. \end{cases} \tag{5.29}$$

In order to set up the system of difference equations and its solution, we define the following limiting probabilities:

$$\begin{cases} Q_0 = \lim\limits_{t\to\infty} P\{N(t) = 0, \quad Y(t) = 0\} \\ P_n(x) = \lim\limits_{t\to\infty} P\{N(t) = n, \quad Y(t) = 1, \quad x < S_b^0(t) \leq x + dx\}, \quad n \geq 1, \\ Q_n(x) = \lim\limits_{t\to\infty} P\{N(t) = n, \quad Y(t) = 2, \quad x < S_v^0(t) \leq x + dx\}, \quad n \geq 1, \\ R_{1,n}(x) = \lim\limits_{t\to\infty} P\{N(t) = n, \quad Y(t) = 3, \quad x < S_{r_1}^0(t) \leq x + dx\}, \quad n \geq 1, \\ R_{2,n}(x) = \lim\limits_{t\to\infty} P\{N(t) = n, \quad Y(t) = 4, \quad x < S_{r_2}^0(t) \leq x + dx\}, \quad n \geq 1, \end{cases} \tag{5.30}$$

with their respective LST, with parameter θ, as

$$\begin{cases} \mathcal{L}\{Q_0(x)\} = Q_0^*(\theta), \ \ \mathcal{L}\{Q_n(x)\} = Q_0^*(\theta), \ \ \mathcal{L}\{P_0(x)\} = P_0^*(\theta), \\ \mathcal{L}\{P_n(x)\} = P_n^*(\theta), \ \ \mathcal{L}\{R_{1,n}(x)\} = R_{1,n}^*(\theta), \ \ \mathcal{L}\{R_{2,n}(x)\} = R_{2,n}^*(\theta). \end{cases} \tag{5.31}$$

It should be noted that the equations in (5.30) represent the steady-state probabilities. For instance, $Q_n(x)$ represents the steady-state probability that there is at least one task in the system, including the one in the service station, when the server is busy in the working vacation period, where the remaining service time for the task in working vacation period, that is, $S_v^0(t)$, lies between x and $x + dx$.

Thus, we have

$$\lambda Q_0 = P_1(0) + Q_1(0), \tag{5.32a}$$

$$-Q_1'(x) = -(\lambda + \alpha_2 + \eta)Q_1(x) + Q_2(0)s_v(x) + \lambda Q_0 s_v(x) + R_{2,1}s_v(x), \tag{5.32b}$$

$$-Q_n'(x) = -(\lambda + \alpha_2 + \eta)Q_n(x) + Q_{n+1}(0)s_v(x) + \lambda Q_{n-1}s_v(x) + R_{2,n}(0)s_v(x), n \geq 2, \tag{5.32c}$$

$$-P_1'(x) = -(\lambda + \alpha_1)P_1(x) + P_2(0)s_b(x) + \eta s_b(x)\int_0^\infty Q_1(y)\,dy + R_{1,1}(0)s_b(x), \tag{5.32d}$$

$$\begin{aligned} -P_n'(x) = {} & -(\lambda + \alpha_1)P_n(x) + P_{n+1}(0)s_b(x) + \eta s_b(x)\int_0^\infty Q_n(y)dy \\ & + \lambda P_{n-1}(x) + R_{1,n}(0)s_b(x), \quad n \geq 2, \end{aligned} \tag{5.32e}$$

$$-R_{2,1}'(x) = -(\lambda + \eta)R_{2,1}(x) + \alpha_2 s_{r_2}(x)\int_0^\infty Q_1(x)\,dx, \tag{5.32f}$$

$$-R'_{2,n}(x) = -(\lambda+\eta)R_{2,n}(x) + \lambda R_{2,n-1}(x)\alpha_2 s_{r_2}(x)\int_0^\infty Q_n(x)dx, \quad n \geq 2, \tag{5.32g}$$

$$-R'_{1,1}(x) = -\lambda R_{1,1}(x) + \alpha_1 s_{r_1}(x)\int_0^\infty P_1(x)\,dx + \eta s_{r_1}(x)\int_0^\infty R_{2,1}(y)\,dy, \tag{5.32h}$$

$$-R'_{1,n}(x) = -\lambda R_{1,n}(x) + \lambda R_{1,n-1}(x) + \alpha_1 s_{r_1}(x)\int_0^\infty P_n(x)dx + \eta s_{r_1}(x)\int_0^\infty R_{2,n}(y)dy,$$
$$n \geq 2. \tag{5.32i}$$

We define the following pgf of the probabilities:

$$\begin{cases} Q(z,0) = \sum_{n=1}^\infty Q_n(0)z^n, \quad Q^*(z,0) = \sum_{n=1}^\infty Q_n^*(0)z^n, \quad Q^*(z,\theta) = \sum_{n=1}^\infty Q_n^*(\theta)z^n, \\ P(z,0) = \sum_{n=1}^\infty P_n(0)z^n, \quad P*(z,0) = \sum_{n=1}^\infty P_n^*(0)z^n, \; P*(z,\theta) = \sum_{n=1}^\infty P_n^*(\theta)z^n, \\ R_i(z,0) = \sum_{n=1}^\infty R_{i,n}^*(0)z^n, \; R_i^*(z,0) = \sum_{n=1}^\infty R_{i,n}^*(\theta)z^n, \quad R_i^*(z,\theta) = \sum_{n=1}^\infty R_{i,n}^*(\theta)z^n, \quad i = 1,2. \end{cases} \tag{5.33}$$

Now, applying the LST on the system (5.32b)–(5.32i), we have

$$\theta Q_1^*(\theta) - Q_1(0) = (\lambda+\alpha_2+\eta)Q_1^*(\theta) - Q_2(0)S_v^*(s) - \lambda Q_0 S_v^*(s) - R_{2,1}(0)S_v^*(\theta), \tag{5.34a}$$

$$\begin{aligned}\theta Q_n^*(\theta) - Q_n(0) &= (\lambda+\alpha_2+\eta)Q_n^*(\theta) - Q_{n+1}(0)S_v^*(\theta) \\ &\quad - \lambda Q_{n-1}^*(\theta) - R_{2,n}(0)S_v(\theta), \quad n \geq 2,\end{aligned} \tag{5.34b}$$

$$\theta P_1^*(\theta) - P_1(0) = -(\lambda+\alpha_1)P_1^*(\theta) - P_2^*(0)S_b(\theta) - \eta S_b^*(\theta)Q_1^*(0) - R_{1,1}(0)S_b^*(\theta), \tag{5.34c}$$

$$\begin{aligned}\theta P_n^*(\theta) - P_n(0) &= (\lambda+\alpha_1)P_n^*(\theta) - P_{n+1}(0)S_b^*(\theta) - \eta S_b^*(\theta)Q_n^*(0) - \lambda P_{n-1}^*(\theta) \\ &\quad - R_{1,n}(0)S_b^*(\theta), \quad n \geq 2,\end{aligned} \tag{5.34d}$$

$$\theta R_{2,1}^*(\theta) - R_{2,1}(0) = (\lambda+\eta)R_{2,1}^*(\theta) - \alpha_2 S_{r_2}^*(\theta)Q_1^*(0), \tag{5.34e}$$

$$\theta R_{2,n}^*(\theta) - R_{2,n}(0) = (\lambda+\eta)R_{2,n}^*(\theta) - \lambda R_{2,n-1}^*(\theta) - \alpha_2 S_{r_2}^*(\theta)Q_n^*(0), \quad n \geq 2, \tag{5.34f}$$

$$\theta R_{1,1}^*(\theta) - R_{1,1}(0) = \lambda R_{1,1}^*(\theta) - \alpha_1 S_{r_1}^*(\theta)P_1^*(\theta) - \eta S_{r_1}^*(\theta)R_{2,1}(0), \tag{5.34g}$$

$$\begin{aligned}\theta R_{1,n}^*(\theta) - R_{1,n}(0) &= \lambda R_{1,n}^*(\theta) - \lambda R_{1,n-1}^*(\theta) - \alpha_1 S_{r_1}^*(\theta)P_n^*(\theta) \\ &\quad - \nu S_{r_1}^*(\theta)R_{2,n}^*(0), \quad n \geq 2.\end{aligned} \tag{5.34h}$$

Multiplying Equation (5.34b) by powers n of z and summing over values of n, from 2 to ∞, and multiplying Equation (5.34a) by n of z, adding the results, we obtain the following:

$$[\theta-(\lambda-\lambda z+\alpha_2+\eta)]Q^*(z,\theta)$$
$$=\left[\frac{z-\theta_v^*(\theta)}{z}\right]Q(z,0)-S_v^*(\theta)[\lambda z Q_0+R_2(z,0)-Q_1(0)]. \tag{5.35}$$

Relation (5.35) is the Laplace transform of the generating function of the queue size. However, in order to find the mean queue size, we have to find the derivative of the generating function when z is 1. In other words, we are to consider the generating function at the boundary of its region of existence, that is, $z=1$.

Durations of the vacation times are assumed to be random variables distributed exponentially with parameter η.

It should be noted that at the boundary of the region $|z|\le 1$, that is, when $z=1$, if there is no breakdown (i.e., $\alpha_2=0$) and if the vacation is just to be started or just completed (i.e., $\eta=0$), then the value of $[\theta-(\lambda-\lambda z+\alpha_2+\eta)]$ is 0. That is, when the Laplace transform's parameter θ reaches the value $(\lambda-\lambda z+\alpha_2+\eta)$, we obtain the boundary value through $Q(z,0)$ from (5.35).

Thus, letting $\theta=\lambda-\lambda z+\alpha_2+\eta$, we have

$$Q(z,0)=\frac{zS_v^*(\lambda-\lambda z+\alpha_2+\eta)[\lambda z Q_0+R_2(z,0)-Q_1(0)]}{z-S_\mu^*(\lambda-\lambda z+\alpha_2+\eta)}. \tag{5.36}$$

Multiplying (5.34f) by z^n and summing over values of n, from 2 to ∞, yields

$$[\theta-(\lambda-\lambda z+\eta)]R_2^*(z,\theta)=R_2(z,0)-\alpha_2 S_{r_2}^*(\theta)Q^*(z,0). \tag{5.37}$$

As mentioned above, letting $\theta=\lambda-\lambda z+\eta$, from (5.37) we have

$$R_2(z,0)=\alpha_2 S_{r_2}^*(\lambda-\lambda z+\eta)Q^*(z,0). \tag{5.38}$$

Substituting (5.38) into (5.37) and letting $\theta=0$, we obtain

$$R_2^*(z,0)=\frac{\alpha_2 Q^*(z,0)[1-S_{r_2}^*(\lambda-\lambda z+\eta)]}{\lambda-\lambda z+\eta}. \tag{5.39}$$

By substituting (5.39) into (5.37), we obtain

$$Q(z,0)=\frac{zS_v^*(\lambda-\lambda z+\alpha_2+\eta)[\lambda z Q_0+\alpha_2 S_{r_2}^*(\lambda-\lambda z+\eta)Q^*(z,0)-Q_1(0)]}{z-S_v^*(\lambda-\lambda z+\alpha_2+\eta)}. \tag{5.40}$$

Again, substituting (5.38) and (5.40) into (5.35) yields

$$[\theta-(\lambda-\lambda z+\alpha_2+\eta)]Q^*(z,\theta)$$
$$=\frac{z[S_v^*(\lambda-\lambda z+\alpha_2+\eta)-S_v^*(\theta)][\alpha_2 S_{r_2}^*(\lambda-\lambda z+\eta)Q^*(z,0)+\lambda z Q_0-Q_1(0)]}{z-S_v^*(\lambda-\lambda z+\alpha_2+\eta)}. \tag{5.41}$$

Finally, note that $Q(z,0)$ is one of the boundary conditions. Thus, in order to solve the algebraic equations obtained as a result of application of the LST on the differential difference equations governing the system, we are to find the value of $Q(z,0)$. Here, the component 0 represents the case that either the service is yet to be started or it is just completed (i.e., no remaining service). See Medhi (2009, p 17). Hence, by substituting $\theta = 0$ in (5.41), we have

$$Q^*(z,0) = \frac{z[1 - S_v^*(\lambda - \lambda z + \alpha_2 + \eta)][\lambda z Q_0 - Q_1(0)]}{D}, \tag{5.42}$$

where

$$\begin{aligned} D = {} & (\lambda - \lambda z + \alpha_2 + \eta)[z - S_v^*(\lambda - \lambda z + \alpha_2 + \eta)] \\ & - \alpha_2 z(1 - S_v^*[\lambda - \lambda z + \alpha_2 + \eta)S_{r_2}^*(\lambda - \lambda z + \eta]. \end{aligned} \tag{5.43}$$

The denominator of (5.41) has a unique root, say z_1, in the interval (0, 1), (why?). Thus,

$$Q_1(0) = \lambda z_1 Q_0. \tag{5.44}$$

By substituting (5.44) into (5.41), we obtain

$$Q^*(z,0) = \frac{\lambda z(z - z_1)[1 - S_v^*(\lambda - \lambda z + \alpha_2 + \eta)]Q_0}{D_1}, \tag{5.45}$$

where

$$\begin{aligned} D_1 = {} & (\lambda - \lambda z + \alpha_2 + \eta)[z - S_v^*(\lambda - \lambda z + \alpha_2 + \eta)] \\ & - \alpha_2 z(1 - S_v^*(\lambda - \lambda z + \alpha_2 + \eta)S_{r_2}^*(\lambda - \lambda z + \eta). \end{aligned} \tag{5.46}$$

Also, substituting (5.46) into (5.39), we obtain

$$R_2^*(z,0) = \frac{Q_0\alpha_2[1 - S_{r_2}^*(\lambda - \lambda z + \eta)]\lambda z(z - z_1)1 - S_v^*(\lambda - \lambda z + \alpha_2 + \eta)]}{D_2}, \tag{5.47}$$

where

$$\begin{aligned} D_2 = {} & (\lambda - \lambda z + \eta)\{(\lambda - \lambda z + \alpha_2 + \eta)[z - S_v^*(\lambda - \lambda z + \alpha_2 + \eta)] \\ & - \alpha_2 z(1 - S_v^*(\lambda - \lambda z + \alpha_2 + \eta)S_{r_2}^*(\lambda - \lambda z + \eta)\}. \end{aligned} \tag{5.48}$$

Multiplying (5.34d) by z^n and summing over values of n, from 2 to ∞, then adding to term obtained by z times (5.34c), yields

$$\begin{aligned} & [\theta - (\lambda - \lambda z + \alpha_1]P^*(z,\theta) \\ & = \left[\frac{z - [S_b^*(\theta)]}{z}\right] P(z,0) - S_b^*(\theta)[\eta Q^*(z,0) + R_1(z,0) - P_1(0)]. \end{aligned} \tag{5.49}$$

Again, by letting $\theta = \lambda - \lambda z + \alpha_1$ and the fact that $\lambda(1 - z_1)Q_0 = P_1(0)$, (5.49) can be rewritten as

$$P(z,0) = \frac{zS_b^*(\lambda - \lambda z + \alpha_1)[\eta Q * (z,0) + R_1(z,0) - \lambda(1 - z_1)Q_0]}{z - S_b^*(\lambda - \lambda z + \alpha_1)}. \tag{5.50}$$

Multiplying (5.34h) by z^n and summing over values of n, from 2 to ∞, then adding to the term obtained by z times (5.34g), yields

$$[\theta - (\lambda - \lambda z)]R_1^*(z,\theta) = R_1(z,0) - \alpha_1 S_{r_1}^*(\theta)P^*(z,0) - \eta S_{r_1}^*(\theta)R_2^*(z,0). \tag{5.51}$$

Again, letting $\theta = \lambda - \lambda z$ in (5.51), we obtain

$$R_1(z,0) = S_{r_1}^*(\lambda - \lambda z)[\alpha_1 P * (z,0) + \eta R_2(z,0)]. \tag{5.52}$$

Substituting (5.52) into (5.51) with letting $\theta = 0$, yields

$$R_1^*(z,0) = \frac{[1 - S_{r_1}^*(\lambda - \lambda z)][\alpha_1 + P^*(z,0) + \eta R_2^*(z,0)]}{\lambda - \lambda z}. \tag{5.53}$$

Now substituting (5.39), (5.42), (5.50), (5.52), $\theta = 0$, and $\lambda(1 - z_1)Q_0 = P_1(0)$ into (5.48), we have the following:

$$P^*(z,0) = \frac{N_3}{D_3}, \tag{5.54}$$

where

$$\begin{aligned} N_3 = {} & Q_0\lambda z[1 - S_b^*(\lambda - \lambda z + \alpha_1)] \\ & \times \{\eta z(z - z_1)(1 - S_v^*(\lambda - \lambda z + \alpha_2 + \eta)] \\ & \times [(\lambda - \lambda z + \eta) + \alpha_2 S_{r_1}^*(\lambda - \lambda z)[1 - S_{r_2}^*(\lambda - \lambda z + \eta)]] \\ & - (\lambda - \lambda z + \eta)(1 - z_1)\{(\lambda - \lambda z + \alpha_2 + \eta)[z - S_v^*(\lambda - \lambda z + \alpha_2 + \eta)] \\ & - \alpha_2 z[1 - S_v^*(\lambda - \lambda z + \alpha_2 + \eta)]S_{r_2}^*(\lambda - \lambda z + \eta)]\}. \end{aligned} \tag{5.55}$$

$$\begin{aligned} D_3 = {} & (\lambda - \lambda z + \eta)\{(\lambda - \lambda z + \alpha_1)[z - S_b^*(\lambda - \lambda z + \alpha_1)] \\ & - z\alpha_1[1 - S_b^*(\lambda - \lambda z + \alpha_2)]S_{r_1}^*(\lambda - \lambda z)\} \\ & \times \{(\lambda - \lambda z + \alpha_2 + \eta)[z - S_v^*(\lambda - \lambda z + \alpha_2 + \eta)] \\ & - \alpha_2 z[1 - S_v^*(\lambda - \lambda z + \alpha_2 + \eta)]S_{r_2}^*(\lambda - \lambda z + \eta)\}. \end{aligned} \tag{5.56}$$

Substituting (5.47) and (5.54) into (5.53), we obtain

$$R_1^*(z,0) = \frac{N_4}{D_4}, \tag{5.57}$$

where

$$\begin{aligned}
N_4 = & Q_0[1 - S^*_{r_1}(\lambda - \lambda z)] \\
& \times \{\alpha_1 \lambda z[1 - S^*_b(\lambda - \lambda z + \alpha_1)] \\
& \times \{\eta z(z - z_1)[1 - S^*_v(\lambda - \lambda z + \alpha_2 + \eta)] \\
& \times [(\lambda - \lambda z + \eta) + \alpha_2 S^*_{r_1}(\lambda - \lambda z)[1 - S^*_{r_2}(\lambda - \lambda z + \eta)]] \\
& - (\lambda - \lambda z + \eta)(1 - z_1)[(\lambda - \lambda z + \alpha_2 + \eta)[z - S^*_v(\lambda - \lambda z + \alpha_2 + \eta)] \\
& - \alpha_2 z[1 - S^*_v(\lambda - \lambda z + \alpha_2 + \eta)]S^*_{r_2}(\lambda - \lambda z + \eta)]]\} \\
& + \alpha_2 \eta \lambda z(z - z_1)[1 - S^*_v(\lambda - \lambda z + \alpha_2 + \eta)][1 - S^*_{r_2}(\lambda - \lambda z + \eta)]. \\
& \times \{(\lambda - \lambda z + \alpha_1)[z - S^*_b(\lambda - \lambda z + \alpha_1)] \\
& - \alpha_1 z[1 - S^*_b(\lambda - \lambda z + \alpha_1 + \eta)]S^*_{r_1}(\lambda - \lambda z)\}\}. \qquad (5.58)
\end{aligned}$$

$$\begin{aligned}
D_4 = & (\lambda - \lambda z)(\lambda - \lambda z + \eta) \\
& \times \{(\lambda - \lambda z + \alpha_1)[z - S^*_b(\lambda - \lambda z + \alpha_1)] \\
& - \alpha_1 z[1 - S^*_b(\lambda - \lambda z + \alpha_1)]S^*_{r_1}(\lambda - \lambda z)\} \\
& \times \{(\lambda - \lambda z + \alpha_2 + \eta)[z - S^*_v(\lambda - \lambda z + \alpha_2 + \eta)] \\
& - \alpha_2 z[1 - S^*_v(\lambda - \lambda z + \alpha_2 + \eta)]S^*_{r_2}(\lambda - \lambda z + \eta)\}. \qquad (5.59)
\end{aligned}$$

Let us now define the pgf of the number of tasks in the system when the server is not on working vacation period and on working vacation by $P_B(z)$ and $P_V(z)$, respectively, where

$$P_B(z) = P^*(z, 0) + R^*_1(z, 0) \qquad (5.60)$$

and

$$P_V(z) = Q^*(z, 0) + R^*_2(z, 0) + Q_0. \qquad (5.61)$$

Hence, the pgf of the number of tasks in the system will be

$$P(z) = P_B(z) + P_V(z). \qquad (5.62)$$

It should be noted that the only unknown in (5.52) is Q_0. This unknown may be determined by using the L'Hôpital's rule, $z = 1$, and the normalizing factor $P(1) = 1$, as follows:

$$Q_0 = \frac{N_5}{D_5}, \qquad (5.63)$$

where

$$\begin{aligned}
N_5 = & \eta\{\alpha_1 S^*_b(\alpha_1) - \lambda[1 - S^*_b(\alpha_1)][1 + \alpha_1 E(S_{r_1})]\} \\
& \times [1 - S^*_v(\eta + \alpha_2)]\{\eta + \alpha_2[1 - S^*_v(\eta)]\} \qquad (5.64)
\end{aligned}$$

and

$$-\eta\lambda(1-z_1)\{[1-S_b^*(\alpha_1)]S_v^*(\eta+\alpha_2)\{\alpha_2+\eta[1+\alpha_1 E(S_{r_1}^*)]\}$$
$$+\alpha_1\alpha_2 E(S_{r_1})\{S_v^*(\eta+\alpha_2)-S_b^*(\alpha_1)\{1-S_{r_2}^*(\eta)[1-S_v^*(\eta+\alpha_2)]\}\}]\}, \quad (5.65)$$

where $E(S_{r_1})$ is the mean repair time in not working vacation period (regular service period).

Now, in order to find the mean queue length in the system, we denote the mean number of tasks in the system in the steady-state case during the working vacation and not working vacation periods by L_v and L_b, respectively. Thus, we have

$$L_v = \frac{d}{dz}P_V(z)\bigg|_{z=1} = \frac{d}{dz}\left[\frac{A(z)}{D(z)}+\frac{B(z)}{(\lambda-\lambda z+\eta)D(z)}\right]Q_0\bigg|_{z=1}, \quad (5.66)$$

where $P_V(z)$ is given in (5.64), D is given in (5.43) and $A(z)$ and $B(z)$ are, respectively, given by

$$A(z)=\lambda z(z-z_1)[1-S_v^*(\lambda-\lambda z+\eta+\alpha_2)], \quad (5.67)$$

and

$$B(z)=\alpha_2\lambda z(z-z_1)[1-S_{r_2}^*(\lambda-\lambda z+\eta)][1-S_v^*(\lambda-\lambda z+\alpha_2+\eta)]. \quad (5.68)$$

Thus, from (5.66), we will have the following:

$$L_v = Q_0\left\{\frac{D(1)A'(1)-A(1)D'(1)}{[D(1)]^2}+\frac{\eta[D(1)B'(1)-B(1)D'(1)+\lambda B(1)D(1)]}{[\eta D(1)]^2}\right\}, \quad (5.69)$$

where

$$A(1)=\lambda(1-z_1)[1-S_v^*(\eta+\alpha_2)], \quad (5.70)$$

$$A'(1)=[1-S_v^*(\eta+\alpha_2)][\lambda+\lambda(1-z_1)]+\lambda^2(1-z_1)S_v'(\eta+\alpha_2), \quad (5.71)$$

$$D(1)=[1-S_v^*(\eta+\alpha_2)]\{\eta+\alpha_2[1-S_v^*(\eta+\alpha_2)]\}, \quad (5.72)$$

$$B(1)=\alpha_2\lambda(1-z_1)[1-S_{r_2}^*(\eta)][1-S_v^*(\eta+\alpha_2)], \quad (5.73)$$

and

$$B'(1)=\alpha_2[\lambda(1-z_1)[1-S_v^*(\eta+\alpha_2)].$$
$$\times[1-S_{r_2}^*(\eta)\lambda S_{r_2}^*(\eta)]+\lambda[1-S_{r_2}^*(\eta)]$$
$$\times S_v'(\eta+\alpha_2)]+\lambda\{1-S_{r_2}^*(\eta)[1-S_v^*(\eta+\alpha_2)]\}\}]. \quad (5.74)$$

We also have

$$L_b = \frac{d}{dz}P_B(z)\bigg|_{z=1}$$
$$= \frac{d}{dz}\left[\frac{N_6(z)N_7(z)}{D_2(z)D_6(z)} + \frac{N_8(z)N_9(z)}{D_2(z)D_6(z)(\lambda - \lambda z)}\right] Q_0\bigg|_{z=1}, \tag{5.75}$$

where $P_B(z)$ is given in (5.65),

$$N_6(z) = \lambda z[1 - S_b^*(\lambda - \lambda z + \eta + \alpha_1)], \tag{5.76}$$

$$\begin{aligned} N_7(z) = {} & \eta z(z - z_1)[1 - S_v^*(\lambda - \lambda z + \alpha_2 + \eta)] \\ & \times \{[\lambda - \lambda z + \eta + \alpha_2 S_{r_1}^*(\lambda - \lambda z)[1 - S_{r_2}^*(\lambda - \lambda z + \eta)]\} \\ & - (\lambda - \lambda z + \eta)(1 - z_1)[(\lambda - \lambda z + \alpha_2 + \eta)[z - S_v^*(\lambda - \lambda z + \eta + \alpha_2)] \\ & - \alpha_2 S_{r_2}^*(\lambda - \lambda z + \eta)[1 - S_v^*(\lambda - \lambda z + \eta + \alpha_2)]], \end{aligned} \tag{5.77}$$

$$N_8(z) = 1 - S_{r_1}^*(\lambda - \lambda z), \tag{5.78}$$

$$\begin{aligned} N_9(z) = {} & \alpha_1 \lambda z[1 - S_b^*(\lambda - \lambda z + \alpha_1)] \\ & \times \{\eta z(z - z_1)[1 - S_v^*(\lambda - \lambda z + \eta + \alpha_2)] \\ & \times \{\lambda - \lambda z + \eta + \alpha_2 S_{r_1}^*(\lambda - \lambda z)[1 - S_{r_2}^*(\lambda - \lambda z + \eta)]\} \\ & - (\lambda - \lambda z + \eta)(1 - z_1) \\ & \times [(\lambda - \lambda z + \alpha_2 + \eta)[z - S_v^*(\lambda - \lambda z + \eta + \alpha_2)] \\ & - \alpha_2 z[1 - S_v^*(\lambda - \lambda z + \eta + \alpha_2)]S_{r_2}^*(\lambda - \lambda z + \eta)]\} \\ & + \alpha_2 \eta \lambda z(z - z_1)[1 - S_v^*(\lambda - \lambda z + \alpha_2)][1 - S_{r_2}^*(\lambda - \lambda z + \eta)] \\ & \times \{(\lambda - \lambda z + \alpha_1)[z - S_b^*(\lambda - \lambda z + \alpha_1)] \\ & - \alpha_1 z[1 - S_b^*(\lambda - \lambda z + \alpha_1)]S_{r-1}^*(\lambda - \lambda z)\}, \end{aligned} \tag{5.79}$$

$D_2(z)$ is given in (5.48) and

$$\begin{aligned} D_6 = {} & \{(\lambda - \lambda z + \alpha_2 + \eta)[z - S_v^*(\lambda - \lambda z + \alpha_2 + \eta)] \\ & - \alpha_2 z[1 - S_v^*(\lambda - \lambda z + \alpha_2 + \eta)]S_{r_2}^*(\lambda - \lambda z + \eta)\}. \end{aligned} \tag{5.80}$$

Using Little's formula, the mean waiting times for both cases of tasks in the system during working vacation period and regular service period can be, respectively, calculated as follows:

$$W_v = \frac{L_v}{\lambda} \text{ and } W_b = \frac{L_b}{\lambda}. \tag{5.81}$$

We leave it as an exercise for the reader to find the evaluated derivatives at $z = 1$ or refer to Murugan and Santhi (2015) to obtain a simplified expression for L_b.

5.3.3 A Special Case

As a particular case, if the system does not experience any breakdown, that is, $\alpha_1 = \alpha_2 = 0$, then from (5.62), we have

$$P(z) = \frac{N_{11}}{(\lambda - \lambda z)(\lambda - \lambda z + \eta)[z - S_b^*(\lambda - \lambda z)][z - S_v^*(\lambda - \lambda z + \eta)]}, \tag{5.82}$$

where

$$\begin{aligned} N_{11} = Q_0\{ & \lambda z[1 - S_b^*(\lambda - \lambda z)][nz(z - z_1)[1 - S_v^*(\lambda - \lambda z + \eta)] \\ & - (1 - z_1))(\lambda - \lambda z + \eta)[z - S_v^*(\lambda - \lambda z + \eta)]] \\ & + (\lambda - \lambda z)S_b^*(\lambda - \lambda z)[\lambda z(z - z_1)[1 - S_v^*(\lambda - \lambda z + \eta)] \\ & + (\lambda - \lambda z + \eta)[z - S_v^*(\lambda - \lambda z + \eta)]\}, \end{aligned} \tag{5.83}$$

$$Q_0 = \frac{1 - p_b}{\dfrac{\lambda - \lambda z + \eta}{\eta} - \dfrac{p_b(1 - z_1)S_v^*(\eta)}{1 - S_v^*(\eta)}}, \tag{5.84}$$

and

$$p_b = \lambda E(S_b). \tag{5.85}$$

Example 5.3.1 Let the distribution of both service times (regular and working vacation) be exponential and assume the following facts:

$$S_b^*(\alpha_1) = \frac{\mu_b}{\alpha_1 + \mu_b}, \quad S_v^*(\eta + \alpha_2) = \frac{\mu_v}{\eta + \alpha_2 + \mu_v}, \quad S_{r_2}^*(\eta) = \frac{\mu_{r_2}}{\eta + \mu_{r_2}},$$

$$S_b^{*'}(\alpha_1) = -\frac{\mu_b}{(\alpha_1 + \mu_b)^2}, \quad S_v^{*'}(\eta + \alpha_2) = -\frac{\mu_v}{(\eta + \alpha_2 + \mu_v)}, \quad S_{r_2}^{*'}(\eta) = -\frac{\mu_{r_2}}{(\eta + \mu_{r_2})^2},$$

$$E(S_{r_1}) = \frac{1}{\mu_{r_1}}, \quad E(S_{r_1}^2) = \frac{2}{\mu_{r_1}^2},$$

and fixed values as follows:

$$z_1 = 0.6, \quad \mu_v = 6, \quad \mu_b = 15, \quad \mu_{r_1} = 2, \quad \mu_{r_2} = 5, \quad \alpha_1 = 2, \quad \alpha_2 = 1,$$
$$\lambda = 3.1, 3.2, 3.3, 3.4, 3.5, \text{ and } \eta = 3.1, 3.3, 3.5, 3.7, 3.9.$$

We calculate the corresponding values of L_b and W_b for multiple working vacations and list them in Tables 5.1 and 5.2. See Murugan and Santhi (2015).

From Tables 5.1 and 5.2, it can be seen that as λ increases, both L_b and W_b increase for various values of η.

We now choose the following values to calculate L_v and W_v for multiple working vacations and list them in Tables 5.3 and 5.4. See Murugan and Santhi (2015).

$$z_1 = 0.8, \quad \mu_v = 5, \quad \mu_b = 11, \quad \mu_{r_1} = 2, \quad \mu_{r_2} = 4, \quad \alpha_1 = 1, \quad \alpha_2 = 1,$$
$$\lambda = 1.5, 1.6, 1.7, 1.8, 1.9, \quad \text{and} \quad \eta = 2.10, 2.15, 2.20, 2.25, 2.30.$$

From Tables 5.3 and 5.4, it can be seen that, as in the previous case, as λ increases, both L_v and W_v increase for various values of η.

TABLE 5.1 Arrival Rate, λ, versus Mean System Size, L_b, in Regular Service Period

λ	η 3.1	3.3	3.5	3.7	3.9
3.1	0.484350	0.475244	0.461146	0.461146	0.455593
3.2	0.566134	0.553773	0.543432	0.534666	0.527152
3.3	0.654603	0.638586	0.625199	0.613862	0.604151
3.4	0.750153	0.730055	0.713273	0.699071	0.686915
3.5	0.853213	0.828583	0.808033	0.790658	0.775796

TABLE 5.2 Arrival Rate, λ, versus Mean Waiting Time, W_b, in Regular Service Period

λ			η		
	3.1	3.3	3.5	3.7	3.9
3.1	0.156242	0.153304	0.150844	0.148757	0.146966
3.2	0.176917	0.173054	0.169823	0.167083	0.164735
3.3	0.198364	0.193511	0.189454	0.186019	0.183076
3.4	0.220633	0.214722	0.209786	0.205609	0.202034
3.5	0.243775	0.236738	0.230867	0.225902	0.221656

TABLE 5.3 Arrival Rate, λ, versus Mean System Size, L_v in Regular Service Period

λ			η		
	2.10	2.15	2.20	2.25	2.30
1.5	0.310922	0.310464	0.309687	0.308635	0.307349
1.6	0.343920	0.342170	0.340202	0.338051	0.335746
1.7	0.377454	0.374326	0.371091	0.367771	0.364388
1.8	0.411382	0.406801	0.402232	0.397684	0.393166
1.9	0.445566	0.439468	0.433507	0.421979	0.421979

TABLE 5.4 Arrival Rate, λ, versus Mean Waiting Time, W_v in Regular Service Period

λ	η 2.10	2.15	2.20	2.25	2.30
1.5	0.207281	0.206976	0.206458	0.205757	0.204899
1.6	0.214950	0.213856	0.212626	0.211282	0.209842
1.7	0.222032	0.220192	0.218289	0.216336	0.214346
1.8	0.228545	0.226000	0.223462	0.220936	0.218426
1.9	0.234509	0.231299	0.228162	0.225094	0.222094

5.4 A BULK QUEUING SYSTEM UNDER N-POLICY WITH BILEVEL SERVICE DELAY DISCIPLINE AND START-UP TIME

In this section, we discuss a queuing system studied by Muh (1993) that in addition to a delay due to an N-Policy, arrival process is with bulks and service is rendered by batches, the system is with a compound Poisson input, bilevel generally distributed service times and delay discipline, start-up time, and a fixed accumulation level with control operating policy. It is assumed that when the queue length falls below a predefined level $r(r \geq 1)$, the system, with server capacity R, immediately stops service until the queue length reaches or exceeds the second predefined accumulation level $N(\geq r)$. This could be a case in computer communication networks. In such a case, it is common that the system stops servicing when the total number of tasks in the input buffer falls below a preassigned level r, which is less than or equal to the server capacity R. The service resumes when at least r tasks are accumulated in the system. This is the case referred to as ***bilevel service delay discipline***, that is, the level r and the server capacity R, or $(\boldsymbol{r}, \boldsymbol{R})$***–quorum***.

Note the meaning of the three integers r, R, and N:

R is the server's capacity level,
r is the minimum batch size, that is, the control level, and
N is the control operating policy that can be adjusted to optimize system performance, that is, the integer at which the server changes from idle to busy state when queue length equals or exceeds N.

Depending on the situation, N can be either selected from between levels r and R, or made greater than R. In the case of a very intense input, arriving tasks can be grouped within small intervals of time. This is what is referred to as a bulk input.

The model described is a generalization of some existing classes of N-policy $M/G/1$ models. The author of the paper finds explicit formulas for this stationary distribution, as we will see below. Some numerical examples will be offered to demonstrate the workability of the obtained results. Thus, the purpose of this section is to find the pgf of the stationary distribution of the queuing process with two cases, $N \leq R$ and $N \geq R$. Hence, three different models under N-Policy, are considered. They are:

1. model with $r \leq N \leq R$,
2. model with $r \leq R \leq N$, and
3. modification of model (1), that is, tasks are allowed to arrive during the start-up time period in excess of the queue length under the condition of bulk size limit for service.

5.4.1 Analysis of the Model

After describing the Muh (1993) model above, we now consider its analysis. The model consists of an infinite-sized buffer and a single server. The system is with bulk arrivals and batch services. Let $\tau_n, n = 1, 2, \ldots$, denote the arrival epochs of bulks of tasks according to a Poisson process with mean rate λ bulk of tasks per unit time. The bulk sizes are represented by iid random variables $X_n, n = 1, 2, \ldots$, with mean rate α. Sequences of time at which batches of tasks complete their services are denoted by $\nu_0 = 0, \nu_1, \nu_2, \ldots$. It is assumed that the service times and batch sizes to be served are independent of the queue length. We also denote by σ_n the service time of the nth batch of tasks. We further denote by $Q(t),\ t \geq 0$, the number of tasks in the system at time $t \geq 0$ with state space $\Psi = \{0, 1, 2, \ldots\}$. We define the embedded process

$$Q_n = Q(\nu_n+) = Q(\nu_n), \quad n = 1, 2, \ldots. \tag{5.86}$$

Let

$$S_k = X_0 + X_1 + X_2 + \cdots + X_k, \quad X_0 = Q_0 \tag{5.87}$$

and define the common pgf of X_n by $A(z)$, that is,

$$A(z) = E(z^{X_n}), \quad n = 1, 2, \ldots. \tag{5.88}$$

Let the so-called ***random index*** with which S_k first reaches or exceeds level N after the moment of time ν_n at which the total number of tasks in the system is Q_n be denoted by ξ_n, where

$$\xi_n = \inf\{n \geq 0; S_n \geq N\}. \tag{5.89}$$

Thus, τ_{ξ_n} is the ***first passage time*** of the queue to reach or exceed N after ν_n, and S_{ξ_n} gives the total number of tasks in the system at time instant τ_{ξ_n}.

At time $\tau_n + 0$, the sever starts its $(n + 1)$st service and carries a batch of tasks of size $\min\{Q_n,\ R\}$, if at least r tasks are available; otherwise, the server will stay idle until the queue length Q_n reaches or exceeds the integer N for the first time.

We note two cases:

1. If $Q_n \geq r$, then $\nu_{n+1} - \nu_n$ will be the service time of the $(n + 1)$st batch of tasks. In this case, it is assumed that the service time is a random number, say σ_{n+1}, with general distribution function B and finite mean β.
2. If $Q_n < r$, then the server waits as long as necessary for the number of tasks in the queue reaches at least N. Then, the server resumes service at the instant that the queue for the first time reaches or exceeds N. In this case, the system enters the start-up mode, which lasts ξ_{n+1} with arbitrary distribution D and

finite mean δ, followed by $(n+1)$st service. Thus, a batch of size $\min\{S_{\xi_n}, \ R\}$ will be serviced during the time σ_{n+1} of service after start-up time ξ_{n+1}. Hence, $\nu_{n+1} - \nu_n$ is the sum of the server waiting time $\tau_{\xi_n} - \tau_n$, the actual service time σ_{n+1}, and the start-up time ξ_{n+1}.

In Model 1, not all new arrivals are accepted during the start-up servicing batch, sometimes referred to as the ***artificial start-up service policy***. In Model 2, new arrivals during the start-up time have no effect on the start-up servicing batch since $S_{\xi_n} \geq R$. In both Models 1 and 2, when the server starts to serve the $(n+1)$st batch of tasks, its load, denoted by $L_{n+1}(Q_n)$, could be defined as

$$L_{n+1}(Q_n) = \begin{cases} \min\{S_{\xi_n}, R\}, & Q_n < r, \\ \min\{Q_n, R\}, & Q_n \geq r. \end{cases} \tag{5.90}$$

A modified version of Model 1, which might be a more realistic service policy, accepts new arrivals during the start-up period to the start-up servicing batch, excluding those in excess of R. In this case, when the server starts to serve the $(n+1)$st batch of tasks, its load could be defined as

$$L_{n+1}(Q_n) = \begin{cases} \min\{S_{\xi_n} + \omega_{n+1}, R\}, & Q_n < r, \\ \min\{Q_n, R\}, & Q_n \geq r, \end{cases} \tag{5.91}$$

where $\omega_n = \omega(s_n)$ is the number of tasks that arrived during the start-up time s_n.

Let $V_n = V(\sigma_n)$ denote the number of tasks that arrive during the service time σ_n. Let us also denote $g^+ = \sup\{g, 0\}$. Then, it can be shown that the value of $Q_n, n = 1, 2, \ldots$, satisfies the following recursive relations:

Model 1. $r \leq N \leq R$ (service does not include the number of tasks arrived during the a start-up time):

$$Q_{n+1} = \begin{cases} (S_{\xi_n} - R)^+ + V_{n+1} + \omega_{n+1}, & Q_n < r, \\ (Q_n - R)^+ + V_{n+1}, & Q_n \geq r. \end{cases} \tag{5.92}$$

Model 2. $r \leq R \leq N$:

$$Q_{n+1} = \begin{cases} S_{\xi_n} - R + V_{n+1} + \omega_{n+1}, & Q_n < r, \\ (Q_n - R)^+ + V_{n+1}, & Q_n \geq r. \end{cases} \tag{5.93}$$

Model 3. $r \leq N \leq R$ (server may take some tasks for service, who arrived during the a start-up time):

$$Q_{n+1} = \begin{cases} (S_{\xi_n} + \omega_{n+1} - R)^+ + V_{n+1} + \omega_{n+1}, & Q_n < r, \\ (Q_n - R)^+ + V_{n+1}, & Q_n \geq r. \end{cases} \tag{5.94}$$

It should be noted that the way models are defined in (5.92)–(5.94) are state-dependent and all three have (r, R)–quorum (i.e., they follow the minimum required service batch size and the server capacity) and N-policy service discipline.

We also note that the availability of the start-up time is a form of the general state dependency service time policy (see Abolnikov and Dshalalow 1992 for details).

Example 5.4.1 Consider the Model 1, $r \leq N \leq R$, in which it is understood that as long as the queue length is less than r, the system will be idle. In order to make the idle period minimal, we need to calculate the probability of the system turn-off, which is the sum of the probability of queue length from 0 to $r-1$. We then need to compare the sum obtained for various values of r and N. For the steady state, we should select r and N in such a way that the probability of the turn-off time is the smallest.

Let us choose the bulk arrival distribution to be geometric with parameter $p = 0.3$. Let us also choose the service time distribution be exponential with rate $b = 0.2$, that is, $\mu = 1/b = 5$. Choose $R = 6$, and $r, N = 10, 1, 2, 3, 4, 5$. Under these assumptions, we have

$$A(z) = \frac{pz}{1-(1-p)z} \tag{5.95}$$

and the LST of the service distribution is

$$\beta(s) = \frac{\mu}{\mu + s}. \tag{5.96}$$

Thus,

$$\beta(\lambda - \lambda A(z)) = \frac{1}{1 + \rho - \rho A(z)}, \tag{5.97}$$

where

$$\rho = \frac{\lambda}{\mu} \tag{5.98}$$

is the system intensity.

5.5 INTERRELATIONSHIP BETWEEN N-POLICY $M/G/1/K$ AND F-POLICY $G/M/1/K$ QUEUES WITH START-UP TIME

The server start-up corresponds to the preparatory work of the server before starting service, as it is the case in some real-life situations. Several authors have studied queuing systems with start-up time focusing mainly on the $M/G/1$ queue. It seems Baker (1973) is the first to derive the N-policy $M/M/1$ queue with an exponential start-up time. Other researchers continued and extended Baker's work. For example, Gupta (1995) was the first to offer analytic solution of the F-policy $M/M/1/K$ queuing system with exponential start-up time and Kuo et al. (2011) that we will discuss their work below. Although much studies of types discussed have been done on Markovian processes, it is extremely difficult, if not impossible, to develop the interrelationship between the N-policy and F-Policy for non-Markovian queues having general interarrival times or general service times.

In this section, we discuss two finite single-server finite-buffer queuing models by Kuo et al. (2011) with both N-policy and F-policy. The main goals of this model are to:

1. find an algorithmic distribution of the queue length for the stationary process *N*-policy *M/G/1/K* queuing system with start-up time to develop the interrelationship between *N*-policy *M/G/1/K* queuing system and the *F*-policy *G/M/1/K* queuing system, and
2. demonstrate that the *N*-policy *M/G/1/K* queuing system efficiently gives algorithmic solution to the *F*-policy *G/M/1/K* queuing system.

In order to reach their goals, the authors apply recursive method, using the supplementary variable technique and treating the supplementary variable as the remaining service time, to develop the solution algorithm for the *N*-policy *M/G/1/K* queue with start-up time. They showed that the *N*-policy *M/G/1/K* queuing system with start-up time has been effectively used to derive the algorithmic solution for the *F*-policy *G/M/1/K* queuing system with start-up time. The authors, finally, illustrate the interrelationship between the *N*-policy and *F*-policy queuing systems analytically for three-stage Erlang (service or interarrival) time distribution. We discuss this model below.

5.5.1 *N*-Policy *M/G/1/K* Queuing System with Exponential Start-up Time

As we mentioned earlier, Yadin and Naor (1963) introduced the concept of *N*-policy. The *N*-policy dictates that the server is turned on when N or more tasks are in the system and is turned off only when the system is empty. After the server is off, it may not operate until N tasks have accumulated in the system. However, the *F*-policy dictates that when the number of tasks in the system reached the system's capacity, K, $K > N$, that is, the system is full, no arrival is allowed until enough tasks have been serviced, so that the number of tasks in the system has decreased to a threshold F, $0 \leq F \leq K - 1$. At that time, the server requires to take an exponential start-up time to start allowing tasks to arrive in the system. This process repeats itself, that is, after arrival permit, the system operates normally until the number of tasks in the system reaches its capacity again, and at which time the above process is repeated. As we mentioned earlier, Gupta (1995) offered analytic solution of the *F*-policy *M/M/1/K* queuing system with exponential start-up time.

Kuo et al. (2011)'s model assumptions are as follows:

1. Interarrival times are iid random variables, denoted by A.
2. The remaining interarrival times are random variables denoted by V.
3. The random variable A has a general distribution function denoted by $A(v)$, with pdf denoted by $a(v)$, $v \geq 0$, and the mean interarrival time a_1.
4. Service times are iid random variables, denoted by S.
5. The remaining service times is a random variable denoted by U.
6. The cumulative distribution function of the random variable S is denoted by $S(u)$, with its pdf denoted by $s(u)$, $u \geq 0$, and the mean service time s_1.
7. The service process is independent of the arrival process.
8. Arriving tasks form a single waiting line.

9. Service discipline is based on the FCFS.
10. The server serves one task at a time.
11. Upon arrival, if a task finds the server busy, it has to wait in the queue until the server is free.

Define the following notations:

$P_{0,0}(t)$: The probability of no task in the system at time t when the server is turned off

$P_{0,n}(t)$: The probability of n task in the system at time t when the server is turned off, where $n = 1, 2, \ldots, K$, and

$P_{1,n}(t)$: The probability of n tasks in the system at time t when the server is turned on, where $n = 1, 2, \ldots, K$,

The states of the system at time t are given by

$N(t)$: The number of tasks in the ystem at time t, and

$U(t)$: The remaining service time for the task being served at time t.

Define

$$P_{1,n}(u, t) = P\{N(t) = n, u < U(t) \leq u + du\}, u \geq 0, n = 1, 2, \ldots, K \tag{5.99}$$

and

$$P_{1,n}(t) = \int_0^\infty P_{1,n}(u, t)du, = n, \quad n = 1, 2, \ldots, K. \tag{5.100}$$

For the steady-state cases, define

$$\begin{aligned}
P_{0,0} &= \lim_{t\to\infty} P_{0,0}(t),\\
P_{0,n} &= \lim_{t\to\infty} P_{0,n}(t), \quad n = 1, 2, \ldots, K,\\
P_{1,n} &= \lim_{t\to\infty} P_{1,n}(t), \quad n = 1, 2, \ldots, K,\\
P_{1,n}(u) &= \lim_{t\to\infty} P_{1,n}(u, t), \quad n = 1, 2, \ldots, K, \text{ and}\\
P_{0,n}(u) &= P_{0,n}s(u), \quad n = 1, 2, \ldots, K,
\end{aligned}$$

On the basis of the notations and definitions above, for the stationary N-policy $M/G/1/K$ queuing system with exponential start-up under consideration, we have the following system of difference and differential difference equations:

$$\begin{cases}
\lambda P_{0,0} = P_{1,1}(0), & \\
\lambda P_{0,n} = \lambda P_{0,n-1}, & 1 \leq n \leq N-1,\\
(\lambda + \gamma)P_{0,n} = \lambda P_{0,n-1}, & N \leq n \leq K-1,\\
\gamma P_{0,K} = \lambda P_{0,K-1}, & n = K.
\end{cases} \tag{5.101}$$

$$\begin{cases} -\dfrac{d}{du}P_{1,1}(u) = -\lambda P_{1,1}(u) + P_{1,2}(0)s(u), & n = 1, \\ -\dfrac{d}{du}P_{1,n}(u,t) = -\lambda P_{1,n}(u) + \lambda P_{1,n-1}(u) + P_{1,n+1}(0)s(u), & 2 \le n \le N-1, \\ -\dfrac{d}{du}P_{1,n}(u) = -\lambda P_{1,n}(u) + \lambda P_{1,n-1}(u) + \gamma P_{0,n}s(u) + P_{1,n+1}(0)s(u), & N \le n \le K-1, \\ -\dfrac{d}{du}P_{1,K}(u) = \lambda P_{1,K-1}(u) + \gamma P_{0,K}s(u), & n = K. \end{cases} \tag{5.102}$$

We now define the following Laplace transforms:

$$\mathcal{L}S(u) \equiv S*(\theta), \quad \mathcal{L}P_{0,n}(u) \equiv P^*_{0,n}(\theta), \quad \mathcal{L}P_{1,n}(u) \equiv P^*_{1,n}(\theta),$$

$$P_{0,n} = P^*_{0,n}(0), \quad P_{1,n}(u) = P^*_{1,n}(0),$$

$$\int_0^\infty e^{-\theta u}\frac{\partial}{\partial u}P_{0,n}(u)du = \theta P^*_{0,n}(\theta) - P_{0,n}(0), \text{ and}$$

$$\int_0^\infty e^{-\theta u}\frac{\partial}{\partial u}P_{1,n}(u)du = \theta P^*_{1,n}(\theta) - P_{1,n}(0).$$

We further denote the lth-order derivative of $S^*(\theta)$ and $a^*(\theta)$ with respect to θ by $S^{*(l)}(\theta)$ and $a^{*(l)}(\theta)$, respectively.

Thus, applying LST on (5.102), using definitions above, we have the following:

$$\begin{cases} (\lambda-\theta)P^*_{1,1}(\theta) = P_{1,2}(0)S^*(\theta) - P_{1,1}(0), & n = 1, \\ (\lambda-\theta)P^*_{1,n}(\theta) = \lambda P^*_{1,n-1}(\theta) + P_{1,n+1}(0)S^*(\theta) - \lambda P_{1,n}(\theta), & 2 \le n \le N-1, \\ (\lambda-\theta)P^*_{1,n}(\theta) = \lambda P^*_{1,n-1}(\theta) + \gamma P_{0,n}S^*(\theta) + P_{1,n+1}(0)S*(\theta) - P_{1,n}(0), & \\ & F+1 \le n \le K-2, \\ -\theta P^*_{1,K}(\theta) = \lambda P^*_{1,K-1}(\theta) + \gamma P_{0,K}S^*(\theta) - P_{1,K}(0), & n = K. \end{cases} \tag{5.103}$$

In order to solve the system (5.103), recursive method is to be used. First, to find $P^*_{0,n}(0)$ and $P^*_{1,n}(0)$, we algorithmically find $P_{0,n}$, $1 \le n \le K$. From (5.101), we obtain

$$P_{0,n} = \phi_n P_{0,0}, \quad 1 \le n \le K, \tag{5.104}$$

where

$$\phi_n = \begin{cases} 1, & 1 \le n \le N-1, \\ \left(\dfrac{\lambda}{\lambda+\gamma}\right)^{n-N+1}, & N \le n \le K-1, \\ \dfrac{\lambda^{K-N+1}}{\gamma(\lambda+\gamma)^{K-N}}, & n = K. \end{cases} \tag{5.105}$$

Thus, from (5.104) and (5.105), the unknowns $P_{0,1}, P_{0,2}, \dots, P_{0,K}$ can be found in terms of $P_{0,0}$. Then, from the third equation of (5.103) and (5.104), and by setting $\theta = \lambda$ in the first three equations in (5.103), we obtain

$$P_{1,2}(0) = \frac{P_{1,1}(0)}{S^*(\lambda)}, \quad n = 1, \tag{5.106}$$

$$P_{1,n+1}(0) = \frac{P_{1,n}(0) - \gamma\phi_{n,N}\phi_n P_{0,0} S^*(\lambda) - \lambda P^*_{1,n-1}(\lambda)}{S^*(\lambda)}, \quad 2 \le n \le K-1, \quad (5.107)$$

where

$$\phi_{n,N} = \begin{cases} 1, & N \le n \le K-1, \\ 0, & \text{otherwise.} \end{cases} \quad (5.108)$$

Once again, using (5.104) in the third equation of (5.103), differentiating the first three equations of (5.103) $l-1$ times with respect to θ and setting $\theta = \lambda$, we find $P^*_{1,n-1}(\lambda)$, $1 \le n \le K-1$, as follows

$$P^{*(l-1)}_{1,1}(\lambda) = -\frac{S^{*(l)}(\lambda)}{l}[P_{1,2}(0)], \quad n = l, l = 1, \dots, K-1, \quad (5.109)$$

$$P^{*(l-1)}_{1,n}(\lambda) = -\frac{1}{l}[P_{1,n+1}(0)S^{*(l)}(\lambda) + \gamma\phi_{n,N}\phi_n P_{0,0} S^{*(l)}(\lambda) + \lambda P^{*(l)}_{1,n-1}(\lambda)],$$
$$2 \le n \le K-1, l = 1, \dots, K-n, \quad (5.110)$$

where

$$l_n = \begin{cases} -\dfrac{(-\lambda)^n S^{*(n)}(\lambda)}{n! S^*(\lambda)}, & 1 \le n \le K-2, \\ 0, & \text{otherwise.} \end{cases} \quad (5.111)$$

Now substituting (5.110) into (5.107), we obtain

$$P_{1,n}(0) = \frac{1}{S^*(\lambda)} P_{1,n-1}(0) + \sum_{i=2}^{n-1} l_{n-1} P_{1,i}(0)$$
$$+ \gamma \left(\phi_{n-2,N} \sum_{i=N}^{n-2} l_{n-i-1}\phi_i - \phi_{n-1,N}\phi_{n-1} \right) P_{0,0}, \quad 3 \le n \le K. \quad (5.112)$$

We leave it as an exercise for the reader to show that with more manipulation and application of recursive method, we have

$$P_{1,1}(0) = A(1)P_{0,0}, \quad n = 1, \quad (5.113)$$

and

$$P_{1,n}(0) = \sum_{i=2}^{n} \Psi_{n-i} A(i) P_{0,0}, \quad 2 \le n \le K, \quad (5.114)$$

where

$$\Psi_n = \begin{cases} 1, & n = 0, \\ \displaystyle\sum_{1 \le k \le n} \sum_{\substack{\tau_1 + \cdots + \tau_k = n \\ \tau_{1,\dots,\tau_k \in \{1,\dots,n\}}}} k_{\tau_1}, \dots, k_{\tau_k}, & n = 1, \dots, K-2, \\ 0, & \text{otherwise,} \end{cases} \quad (5.115)$$

$$k_n = \begin{cases} -\dfrac{1}{S^*(\lambda)} + l_1, & n = 1, \\ l_n, & n = 2, 3, \ldots, K-2, \\ 0, & \text{otherwise}, \end{cases} \tag{5.116}$$

(e.g., $\Psi_4 = k_4 + k_3k_1 + k_2k_2 + k_1k_3 + k_1k_1k_2 + k_1k_2k_1 + k_2k_1k_1 + k_1k_1k_1k_1 = k_4 + 2k_3k_1 + k_2^2 + 3k_1^2k_2 + k_1^4$,) and

$$A(n) = \begin{cases} \lambda, & n = 1, \\ \dfrac{\lambda}{S^*(\lambda)_1}, & n = 2, \\ \gamma \left(\phi_{n-2,N} \displaystyle\sum_{i=N}^{n-2} l_{n-i-1}\phi_i - \phi_{n-1,N}\phi_{n-1} \right), & 3 \leq n \leq K. \end{cases} \tag{5.117}$$

Furthermore, the reader can show as an exercise that

$$P^*_{1,n}(0) = \frac{1}{\lambda} \left[P_{1,n+1}(0) - P_{1,1}(0) + \gamma\phi_{n,N} \sum_{i=N}^{n} \phi_i P_{0,0} \right], \quad 1 \leq n \leq K-1. \tag{5.118}$$

Knowing $P_{1,n}(0), 1 \leq n \leq K$, from (5.118), we can recursively find $P^*_{1,n}(0), 1 \leq n \leq K-1$. $P^*_{1,K}(0)$ can be found as

$$P^*_{1,K}(0) = -\left[\sum_{i=1}^{K-1} P_{1,i} + \gamma S^{*(1)}(0) \sum_{i=N}^{K} \phi_n P_{0,0} + +S^{*(1)}(0) \sum_{i=2}^{K} P_{1,i}(0) \right]. \tag{5.119}$$

$P_{0,0}$ can be found from

$$\sum_{i=0}^{K} P_{0,i} + \sum_{i=1}^{K} P_{1,i} = 1. \tag{5.120}$$

In order to define the steps for algorithmic solution, we remind the reader that N is the threshold and K is the maximum capacity of the system. Then,

Step	Action
1	Compute $\phi_n, n = 1, 2, \ldots, K$, using (5.105)
2	Compute $P^*_{0,n}(0), n = 1, 2, \ldots, K$, in terms $P_{0,0}$, using (5.104)
3	Compute l_n and $k_n, n = 1, 2, \ldots, K-2$, in terms $P_{0,0}$, using (5.111) and (5.116)
4	Compute $\Psi_n, n = 1, 2, \ldots, K-2$, using (5.115)
5	Compute $A(n), n = 1, 2, \ldots, K$, using (5.117)
6	Compute $P_{1,n}(0)$ and $k_n, n = 1, 2, \ldots, K$, in terms $P_{0,0}$, using (5.113) and (5.114)
7	Compute $P^*_{1,n}(0), n = 1, 2, \ldots, K-1$, in terms $P_{0,0}$, using (5.118)
8	Compute $P^*_{1,n}(0), n = K$, in terms $P_{0,0}$, using (5.119)
9	Compute $P_{0,0}$, using (5.120)
10	Compute $P^*_{0,n}(0), n = 1, 2, \ldots, K$, from Steps 2 and 9
11	Compute $P^*_{1,n}(0), n = 1, 2, \ldots, K$, from Steps 7 and 8
12	Obtain steady-state probabilities of the *F*-policy *G/M/1/K* queue by changing the index of $P^*_{0,n}(0)$ and $P^*_{1,n}(0)$ from n to $K-n$.

Example 5.5.1 *N*-Policy *M*/*E*$_3$/1 Queue This is an example for the *N*-policy *M*/*G*/1/*K* queue with exponential start-up time for a three-stage Erlang service distribution. Thus, we suppose a queuing system is made of three independent and identical exponential stages, each with mean of $1/(3\lambda)$. We assume that $\gamma = 1/\mu, N = 2$, and $K = 3$. Thus, we have

$$S^*(\theta) = \left(\frac{3\mu}{3\mu + \theta}\right)^3. \tag{5.121}$$

Now, we follow the steps laid down above:

Step 1. Computing ϕ_n, for $n = 1, 2$, and 3, using (5.105), and letting

$$\eta = \frac{3\lambda}{3\lambda + \mu}, \tag{5.122}$$

we have

$$\phi_1 = 1, \phi_2 = \frac{\eta}{3 - 2\eta}, \text{ and } \phi_3 = \frac{\eta^2}{3(1-\eta)(3-2\eta)}.$$

Step 2. Computing $P^*_{0,n}(0)$, for $n = 1, 2,$ and 3, using (5.104) in terms of $P_{0,0}$, we have

$$P^*_{0,1}(0) = \phi_1 P_{0,0} = P_{0,0},$$

$$P^*_{0,2}(0) = \phi_2 P_{0,0} = \frac{\eta}{3 - 2\eta} P_{0,0}, \text{ and}$$

$$P^*_{0,3}(0) = \phi_3 P_{0,0} = \frac{\eta^2}{3(1-\eta)(3-2\eta)} P_{0,0}.$$

Step 3. Compute l_n and k_n, for $n = 1, 2,$ and 3, using (5.111) and (5.116). Then, letting

$$\sigma = \frac{\lambda}{3\mu}, \tag{5.123}$$

we have

$$l_1 = \frac{-3\sigma}{1 + \sigma} \text{ and } k_1 = \frac{1 + \sigma + 6\sigma^2 + 4\sigma^3 + \sigma^4}{1 + \sigma}.$$

Step 4. Computing Ψ_n, for $n = 0,$ and 1, using (5.115), we have

$$\Psi_0 = 1 \text{ and } \Psi_1 = \frac{1 + \sigma + 6\sigma^2 + 4\sigma^3 + \sigma^4}{1 + \sigma}.$$

Step 5. Computing $A(n)$, for $n = 1, 2,$ and 3, using (5.117), we have

$$A(1) = \lambda, A(2) = \lambda(1 + \sigma)^3, \text{ and } A(3) = \frac{-3\lambda(1 - \eta)}{3 - 2\eta}.$$

Step 6. Computing $P_{1,n}(0)$, for $n = 1, 2,$ and 3, using (5.113) and (5.114) in terms of $P_{0,0}$, we have

$$P_{1,1}(0) = A(1)P_{0,0} = \lambda P_{0,0},$$

$$P_{1,2}(0) = \Psi_0 A(2)P_{0,0} = \lambda(1+\sigma)P_{0,0},$$

$$P_{1,3}(0) = [\Psi_1 A(2) + \Psi_0 A(3)]P_{0,0}$$

$$= \frac{\lambda(3-2\eta)(1+\sigma)^2(1+\sigma+6\sigma^2+4\sigma^3+\sigma^4) - 3\lambda(1-\eta)}{3-2\eta}P_{0,0}.$$

Step 7. Computing $P^*_{1,n}(0)$, for $n = 1$ and 2, using (5.118) in terms of $P_{0,0}$, we have

$$P^*_{1,1}(0) = \sigma P^*_{1,1}(0) = \sigma(3+3\sigma+\sigma^2)P_{0,0} \text{ and}$$

$$P^*_{1,2}(0) = \sigma(3+9\sigma+17\sigma^2+15\sigma^3+6\sigma^4+\sigma^5)P_{0,0}.$$

Step 8. Computing $P^*_{1,n}(0)$, for $n = 3$, using (5.119) in terms of $P_{0,0}$, we have

$$P^*_3(0) = \frac{1}{3-2\eta}\{\sigma[\eta(3-12\sigma-36\sigma^2-78\sigma^3-78\sigma^4-34\sigma^5-6\sigma^6)$$
$$+\sigma(18+54\sigma+117\sigma^2+117\sigma^3+51\sigma^4+9\sigma^5)]\}P_{0,0}.$$

Step 9. Computing $P_{0,0}$ using (5.120), we have

$$P_{0,0} = 3(1-\eta)(3-2\eta)[18-\eta(27-10\eta)$$
$$+27\sigma(2+6\sigma+12\sigma^2+18\sigma^3+15\sigma^4+6\sigma^5+5\sigma^6)$$
$$-9\sigma\eta(27+30\sigma+60\sigma^2+270\sigma^3+75\sigma^4+30\sigma^5+5\sigma^6)$$
$$+9\sigma\eta^2(3+12\sigma+24\sigma^2+36\sigma^3+30\sigma^4+12\sigma^5+2\sigma^6)]^{-1}.$$

Steps 10–12 can now be achieved.

5.5.2 *F*-Policy *G/E/1/K* Queuing System with Exponential Start-up Time

In this part, we use the algorithmic solution of the *N*-policy *M/G/1/K* with exponential start-up time to find the algorithmic solution of the *F*-policy *G/M/1/K* with exponential start-up time. In order to proceed, we note that all parameters in the algorithmic solution for *N*-policy *M/G/1/K* with exponential start-up time remain the same for *F*-policy *G/M/1/K* except the following three that will change as follows: (1) threshold N to *K–F*, (2) interarrival rate λ to service rate μ, and (3) service time random variable *S* to interarrival time random variable *A*. In summary, we list the parameters and symbols used in Table 5.5.

TABLE 5.5 Interrelationship between the *N*-Policy and *F*-Policy Problems

Term	*N*-Policy *M/G/1/K*				*F*-Policy *G/M/1/K*
Capacity	K	$\Rightarrow$	K		K
Threshold	N	$\Rightarrow$	K-F		F
Arrival times	λ	$\Rightarrow$	μ		A
Service time	S	$\Rightarrow$	A		μ
Start-up time	γ	$\Rightarrow$	γ		γ
Probabilities	$P^*_{0,n}(0)$, $n = 1, 2, \ldots, K$	$\Rightarrow$	$P^*_{0,K-n}(0)$	$=$	$P^*_{0,n}(0)$, $n = 1, 2, \ldots, K$
	$P^*_{0,n}(0)$, $n = 1, 2, \ldots, K$	$\Rightarrow$	$P^*_{1,K-n}(0)$	$=$	$P^*_{1,n}(0)$, $n = 1, 2, \ldots, K$

Thus, the steps for F-policy $G/M/1/K$ are as follows:

Step 1. Compute ϕ_n, for $n = 1, 2, \ldots, K$, where

$$\phi_n = \begin{cases} 1, & 1 \le n \le K - F - 1, \\ \left(\dfrac{\mu}{\mu + \gamma}\right)^{n-K+F+1}, & K - F \le n \le K - 1, \\ \dfrac{\mu}{\gamma}\left(\dfrac{\mu}{\mu + \gamma}\right)^{F}, & n = K. \end{cases} \tag{5.124}$$

Step 2. Compute ϕ_n, for $n = 1, 2, \ldots K$, using (5.125)

$$P^*_{0,n}(0) = \phi_n P_{0,0}, \quad 1 \le n \le K. \tag{5.125}$$

Step 3. Compute l_n and k_n, for $n = 1, 2, \ldots, K - 2$, using (5.126) and (5.127)

$$l_n = \begin{cases} -\dfrac{(-\mu)^n a *^{(n)}(\mu)}{n! a * (\mu)}, & 1 \le n \le K - 2, \\ 0, & \text{otherwise.} \end{cases} \tag{5.126}$$

and

$$k_n = \begin{cases} -\dfrac{1}{a * (\mu)} + l_1, & n = 1, \\ l_n, & 2 \le n \le K - 2, \\ 0, & \text{otherwise.} \end{cases} \tag{5.127}$$

Step 4. Compute Ψ_n, for $n = 0, \ldots, K - 2$, using (5.115).

Step 5. Compute $A(n)$, for $n = 1, \ldots, K$, using (5.128)

$$A(n) = \begin{cases} \mu, & n = 1, \\ \dfrac{\mu}{a * (\mu)}, & n = 2, \\ \gamma\left(\phi_{n-2,K-F} \displaystyle\sum_{i=K-F}^{n-2} l_{n-i-1}\phi_i - \phi_{n-1,K-F}\phi_{n-1}\right), & 3 \le n \le K, \end{cases} \tag{5.128}$$

where

$$\phi_{n,K-F} = \begin{cases} 1, & K-F \leq n \leq K-1, \\ 0, & \text{otherwise.} \end{cases} \tag{5.129}$$

Step 6. Compute $P_{1,n}(0)$, for $n = 1, 2, \ldots K$, using (5.130) and (5.131) in terms of $P_{0,0}$:

$$P_{1,1}(0) = A(1)P_{0,0} \tag{5.130}$$

and

$$P_{1,n}(0) = \sum_{i=2}^{n} \psi_{n-i}A(i)P_{0,0}, \quad 2 \leq n \leq K. \tag{5.131}$$

Step 7. Compute $P^*_{1,n}(0)$, for $n = 1, \ldots, K-1$, using (5.118) in terms of $P_{0,0}$:

$$P^*_{1,n}(0) = \frac{1}{\mu}\left[P_{1,n+1}(0) - P_{1,1}(0) + \gamma\phi_{n,K-F}\sum_{i=K-F}^{n}\phi_i P_{0,0}\right], \quad 1 \leq n \leq K-1. \tag{5.132}$$

Step 8. Compute $P^*_{1,n}(0)$, for $n = K$, using (5.133) in terms of $P_{0,0}$:

$$P^*_{1,K}(0) = -\left[\sum_{i=1}^{K-1} P_{1,i} + \gamma a*^{(1)}(0)\sum_{i=K-F}^{K}\phi_n P_{0,0} + +a*^{(1)}(0)\sum_{i=2}^{K} P_{1,i}(0)\right]. \tag{5.133}$$

Step 9. Compute $P_{0,0}$ using (5.120).

Steps 10–12 can now be achieved.

Example 5.5.2 Interrelationship between the *N*-Policy *M/G/1/K* and *F*-Policy *G/M/1/K*, E_3/*M*/1/*K* We now go through an illustrative example for a three-stage Erlang interarrival time distribution to show how the interrelationship between the *N*-policy *M/G/1/K* with exponential start-up time and the *F*-policy *G/M/1/K* with exponential start-up time. The three-stage Erlang distribution consists of three independent and identical exponential stages, each with mean $1/3\mu$. Set the mean service time $s_1 = /\mu$, $N = 2$ and $K = 3$. Again, we use an algorithmic method to solve the steady-state probabilities of F-policy $E_3/M/1/K$ queue. Before we start, we change $S^*(\theta) = [3\mu/(3\mu+\theta)]^3$ to $a^*(\theta) = [3\lambda/(3\lambda+\theta)]^3$.

Step 1. Computing ϕ_n, for $n = 1, 2,$ and 3, using (5.124), letting

$$\gamma = \frac{3\mu}{3\mu+\gamma}, \tag{5.134}$$

we have

$$\phi_1 = 1, \phi_2 = \frac{\gamma}{3-2\gamma}, \text{ and } \phi_3 = \frac{\gamma^2}{3(1-\gamma)(3-2\gamma)}.$$

Step 2. Computing $P^*_{0,n}(0)$, for $n = 1, 2,$ and 3, using (5.125) in terms of $P_{0,0}$, we have

$$P^*_{0,1}(0) = \phi_1 P_{0,0} = P_{0,0},$$

$$P^*_{0,2}(0) = \phi_2 P_{0,0} = \frac{\gamma}{3 - 2\gamma} P_{0,0}, \text{ and}$$

$$P^*_{0,3}(0) = \phi_3 P_{0,0} = \frac{\gamma^2}{3(1-\gamma)(3-2\gamma)} P_{0,0}.$$

Step 3. Compute l_n and k_n, for $n = 1$, using (5.126) and (5.127). Then, letting

$$\tau = \frac{\mu}{3\lambda}, \tag{5.135}$$

we have

$$l_1 = \frac{-3\tau}{1+\tau} \text{ and } k_1 = \frac{1 + \tau + 6\tau^2 + 4\tau^3 + \tau^4}{1+\tau}.$$

Step 4. Computing Ψ_n, for $n = 0$ and 1 and 1, using (5.115), we have

$$\Psi_0 = 1 \text{ and } \Psi_1 = \frac{1 + \tau + 6\tau^2 + 4\tau^3 + \tau^4}{1+\tau}.$$

Step 5. Computing $A(n)$, for $n = 1, 2,$ and 3, using (5.128) we have

$$A(1) = \mu, \ A(2) = \mu(1+\tau)^3, \text{ and } A(3) = \frac{-3\mu(1-\gamma)}{3-2\gamma}.$$

Step 6. Computing $P_{1,n}(0)$, for $n = 1, 2,$ and 3, using (5.130) and (5.131) in terms of $P_{0,0}$, we have

$$\begin{aligned}
P_{1,1}(0) &= A(1)P_{0,0} = \mu P_{0,0},\\
P_{1,2}(0) &= \Psi_0 A(2) P_{0,0} = \mu(1+\tau)P_{0,0},\\
P_{1,3}(0) &= [\Psi_1 A(2) + \Psi_0 A(3)] \ P_{0,0}\\
&= \frac{\mu(3-2\gamma)(1+\tau)^2(1+\tau+6\tau^2+4\tau^3+\tau^4) - 3\mu(1-\gamma)}{3-2\gamma} P_{0,0}.
\end{aligned}$$

Step 7. Computing $P^*_{1,n}(0)$, for $n = 1$ and 2, using (5.132) in terms of $P_{0,0}$, we have

$$\begin{aligned}
P^*_{1,1}(0) &= \tau(3 + 3\tau + \tau^2)P_{0,0} \text{ and}\\
P^*_{1,2}(0) &= \tau(3 + 9\tau + 17\tau^2 + 15\tau^3 + 6\tau^4 + \tau^5)P_{0,0}.
\end{aligned}$$

Step 8. Computing $P^*_{1,n}(0)$, for $n = 3$, using (5.133) in terms of $P_{0,0}$, we have

$$\begin{aligned}
P^*_{1,3}(0) = \frac{1}{3-2\gamma}\{\tau[\gamma(3 - 12\tau - 36\tau^2 - 78\tau^3 - 78\tau^4 - 34\tau^5 - 6\tau^6)\\
+ \tau(18 + 54\tau + 117\tau^2 + 117\tau^3 + 51\tau^4 + 9\tau^5)]\}P_{0,0}.
\end{aligned}$$

Step 9. Computing $P_{0,0}$ using (5.134), we have

$$\begin{aligned}P_{0,0} = {} & 3(1-\gamma)(3-2\gamma)[18-\gamma(27-10\gamma) \\ & + 27\tau(2+6\tau+12\tau^2+18\tau^3+15\tau^4+6\tau^5+5\tau^6) \\ & - 9\tau\gamma(27+30\tau+60\tau^2+270\tau^3+75\tau^4+30\tau^5+5\tau^6) \\ & + 9\tau\gamma^2(3+12\tau+24\tau^2+36\tau^3+30\tau^4+12\tau^5+2\tau^6)]^{-1}.\end{aligned}$$

Steps 10–12 can now be achieved.

5.6 A TRANSIENT $M/M/1$ QUEUE UNDER (M, N)-POLICY, LATTICE PATH METHOD

In this section, we present a model that involves the (M, N)-policy. This is a model addressed by Böhm and Mohanty (1993). Here is the model: consider an $M/M/1$ queuing system, in which the (M, N)-policy is applied. That is, consider a single-server Poisson arrival and exponential service times, in which the server starts servicing only when the number of tasks in the buffer becomes N and remains busy as long as there are at least $M \geq 0$ tasks waiting in line.

Note that the case $M = 0$, discrete-time analogue of this model and by using combinatorics, the method has been studied by Bohm and Mohanty (1990).

The goal of the model we are discussing is to find the transient distribution of the number of tasks in the system with combinatorial method. The authors showed that the transient solution of the standard $M/M/1$ queuing system is a special case of results obtained in this study.

The authors begin the study of this model by discretizing the time, as they did in Mohanty and Panny (1990), and implementing lattice path with diagonal steps in the plane. In order to count such paths, they showed how a bijective mapping can be constructed, which can be encountered directly by standard methods of lattice path combinatorics. Hence, the time axis is segmented into a sequence of time slots of unit length.

In addition to the assumptions that define the (M, N)-policy mentioned above, the following further assumptions are part of the makeup of the model to be considered:

1. Tasks may arrive one at a time.
2. No more than one task may finish service in a given slot.
3. Arrivals and services are mutually exclusive.
4. Events in different slots are independent.
5. The probability of an arrival in a slot is α.
6. The probability of a departure is 0 or γ, depending upon whether the server is idle or busy, respectively.
7. The queue length remains unchanged in a slot with probability $\delta = 1 - \alpha$ and $\beta = 1 - \alpha - y$ during idle periods and busy periods, respectively.

8. Whenever the system becomes idle, the server remains idle until the Nth new arrival; the arrival of the Nth task terminates the idle period and service starts immediately.
9. Initially, there are m tasks waiting for service.
10. The number of tasks in the queue during an idle period is always at least M, but less than N.
11. The number of tasks in the queue during a busy period is always greater than M.

Note that the model under these assumptions seems to be equivalent to the geometric/geometric/1 queuing system given in Hunter (1983, Chapter 9), as the authors acknowledge.

Now, let ω_i and σ_i represent the ith idle and ith busy period, respectively. Define S_n as follows:

$$S_n = \begin{cases} \omega_i, & \text{represents the event that the server is idle at time } n \text{ for the } i\text{th time}, \\ \sigma_i, & \text{represents the event that the server is busy at time } n \text{ for the } i\text{th time}, \\ \sigma, & \text{represents the event that the server is busy at time } n. \end{cases}$$

Let Q_n denote the number of tasks waiting at time n. For simplicity, the authors have chosen their goals as to find

$$P\{Q_n = k + M, S_n = \sigma_i | Q_0 = m, S_0 = \sigma_0\}, \quad k > 0 \tag{5.136}$$

and

$$P\{Q_n = k + M, S_n = \sigma | Q_0 = m, S_0 = \sigma_0\}, \quad k > 0. \tag{5.137}$$

All other cases, such as the server being idle at time n and at time 0, can be treated with the same method as in special cases.

5.6.1 Solution in Discrete Time

The method of lattice path is as follows: a lattice path with n steps is associated with a queuing process up to time n, where

(i) an arrival is represented by a horizontal line unit,
(ii) a departure is represented by a vertical line unit, and
(iii) no change by a diagonal line unit.

For counting lattice paths with diagonal steps, see Mohanty (1979).

There are four types of steps corresponding to the associated probabilities, namely $\alpha-, \gamma-, \beta-$, and $\delta-$. We denote the number of these types of steps by n_1, n_2, n_3, and n_4, respectively. Then,

$$n_1 + n_2 + n_3 + n_4 = n \text{ and } n_1 + m - n_2 = k + M. \tag{5.138}$$

Thus, the model may be looked at as a random walk with reflecting barrier having a further restriction due to the (M, N)-policy. However, enumeration of such paths is complicated by the fact that the diagonal steps have different weights, depending upon whether the server is idle or busy. In order to circumvent these difficulties, the authors discard all diagonal steps. The resulting path, which the authors call the ***skeleton path***, still has all the relevant information. In order to count the path, cut out and concatenate i path segments corresponding to $\omega_1, \omega_2, \ldots, \omega_i$ in that order. Then, join to it the concatenation of path segments corresponding to $\sigma_0, \sigma_1, \ldots, \sigma_i$. It is left as an exercise for the reader to show that the resulting new path is in one-to-one correspondence to the original skeleton path. Every transformed path starts with a segment from the point $(0, 0)$ to the point $((N-M)i,\ 0)$ and then reaches $(n_1,\ n_2)$ without touching the line $y = x + m - M$, but touching or crossing the line $y = x - N + m$. The number of such paths is denoted by $S_i(n_1,\ n_2\,|\,\overline{m - M,\ \ -N + m})$.

Two cases are to be considered: $k \le N - M$ and $k \ge N - M$.

1. Case $k \le N - M$. In this case, the second condition is clearly satisfied since for $i \ge 0$ and $k \le N - M$, we have

$$S_i(n_1, n_2|\overline{m - M, -N + m}) = \binom{n_1 + n_2 - (N-M)i}{n_1 - (N-M)i} - \binom{n_1 + n_2 - (N-M)i}{n_1 + m - M}. \tag{5.139}$$

   Relation (5.139) gives the number of paths from $((N-M)i,\ 0)$ to $(n_1,\ n_2)$, not touching the line $y = x + m - M$.

2. Case $k \ge N - M$. For $i \ge 1$, subtract from the number of paths which do not touch the line $y = x + m - M$ the number of paths which do not touch the line $y = x + m - M$ and obtain

$$S_i(n_1, n_2|\overline{m - M, -N + m}) = \binom{n_1 + n_2 - (N-M)i}{n_1 + m - N} - \binom{n_1 + n_2 - (N-M)i}{n_1 + m - M}. \tag{5.140}$$

We are to reintroduce diagonal steps. Hence, "the balls into cells" method technique of Goodman and Narayana (1969) is to be used. It should be observed that in every transformed skeleton path, there are $n_1 + n_2 - (N-M)i + 1$ and $(N-M)i$ lattice points, where we can insert n_3 of β–type and n_4 of δ–type steps, respectively. The number of ways of doing this is

$$\binom{n_1 + n_2 + n_3 - (N-M)i}{n_3} - \binom{n_4 - (N-M)i - 1}{n_4}. \tag{5.141}$$

Hence, we have the following three cases:

(i) For $i \ge 1$ and $k \le N - M$, we have

$$P\{Q_n = k + M, S_n = \sigma_i | Q_0 = m, S_0 = \sigma_0\}$$

$$= \sum_{\substack{n_1+n_2+n_3+n_4=n,\\ n_1-n_2=k+M-m}} \left[\binom{n_1+n_2-(N-M)i}{n_1-(N-M)i} - \binom{n_1+n_2-(N-M)i}{n_1+m-M}\right]\cdot \times \binom{n_1+n_2+n_3-(N-M)i}{n_3}\binom{n_4-(N-M)i-1}{n_4} \times \alpha^{n_1}\gamma^{n_2}\beta^{n_3}\delta^{n_4}. \tag{5.142}$$

(ii) For $i \geq 1$ and $k \geq N-M$, we have

$$P\{Q_n = k+M, S_n = \sigma_i | Q_0 = m, S_0 = \sigma_0\} = \sum_{\substack{n_1+n_2+n_3+n_4=n,\\ n_1-n_2=k+M-m}} \left[\binom{n_1+n_2-(N-M)i}{n_1+m-N} - \binom{n_1+n_2-(N-M)i}{n_1+m-M}\right]\cdot \times \binom{n_1+n_2+n_3-(N-M)i}{n_3}\binom{n_4-(N-M)i-1}{n_4} \times \alpha^{n_1}\gamma^{n_2}\beta^{n_3}\delta^{n_4}. \tag{5.143}$$

(iii) For $i = 0$, we have

$$P\{Q_n = k+M, S_n = \sigma_0 | Q_0 = m, S_0 = \sigma_0\} = \sum_{\substack{n_1+n_2+n_3+n_4=n,\\ n_1-n_2=k+M-m}} \left[\binom{n_1+n_2}{n_1} - \binom{n_1+n_2}{n_1+m-M}\right]\cdot \times \binom{n_1+n_2+n_3}{n_3} \times \alpha^{n_1}\gamma^{n_2}\beta^{n_3}. \tag{5.144}$$

Relations (5.142)–(5.144) can be simplified as

(i) For $i \geq 1$ and $k \leq N-M$, we have

$$P\{Q_n = k+M, S_n = \sigma_i | Q_0 = m, S_0 = \sigma_0\}$$

$$\times \binom{n_1+n_2+n_3}{n_3} \times \alpha^{n_1}\gamma^{n_2}\beta^{n_3}.$$

$$
\begin{aligned}
&= \sum_{\substack{n_1+n_2+n_3=n-n_4 \\ n_1-n_2=k+M-m}} \frac{(n_1+n_2+n_3-(N-M)i)!}{n_1-(N-M)i)!n_2!n_3!} \times \alpha^{n_1}\gamma^{n_2}\beta^{n_3}\delta^{n_4} \\
&- \sum_{n_4=0}^{n-(N-M)i} \binom{n_4+(N-M)i-1}{(N-M)i-1} \delta^{n_4} \\
&\times \sum_{\substack{n_1+n_2+n_3=n-n_4 \\ n_1-n_2=k+M-m}} \frac{(n_1+n_2+n_3-(N-M)i)!}{(n_1+m)!(n_2-(N-M)i)!-m)!n_3!} \\
&\times \alpha^{n_1}\gamma^{n_2}\beta^{n_3}\delta^{n_4}.
\end{aligned}
\tag{5.145}
$$

Using the trinomial coefficient, defined as

$$
\binom{n;\beta,\gamma}{n+k} = [v^{n+k}](\alpha v^2+\beta v+\gamma)^n = \sum_{\substack{n_1+n_2+n_3=n \\ n_1-n_2=k}} \frac{(n!}{n_1!n_2!n_3!}\alpha^{n_1}\gamma^{n_2}\beta^{n_3},
\tag{5.146}
$$

and

$$
\rho = \frac{\alpha}{\gamma},
\tag{5.147}
$$

we have

$$
\begin{aligned}
&P\{Q_n=k+M, S_n=\sigma_i|Q_0=m, S_0=\sigma_0\} \\
&= \rho^{-m+M} \sum_{r=(N-M)i}^{n} \binom{r-1}{(N-M)i-1} \gamma^{(N-M)i}(1-\alpha)^{r-(N-M)i} \\
&\times \left[\rho^k \binom{n-r;\alpha,\beta,\gamma}{n-r+m-k-M+(N-M)i}\right. \\
&\left. - \binom{n-r;\alpha,\beta,\gamma}{n-r+m+k-M+(N-M)i}\right].
\end{aligned}
\tag{5.148}
$$

(ii) For $i \geq 1$ and $k \geq N-M$, we have

$$
\begin{aligned}
&P\{Q_n=k+M, S_n=\sigma_i|Q_0=m, S_0=\sigma_0\} \\
&= \rho^{-m+M} \sum_{r=(N-M)i}^{n} \binom{r-1}{(N-M)i-1} \gamma^{(N-M)i}(1-\alpha)^{r-(N-M)i} \\
&\times \left[\rho^{N-M} \binom{n-r;\alpha,\beta,\gamma}{n-r+m+k+M+(N-M)i-2N}\right. \\
&\left. - \binom{n-r;\alpha,\beta,\gamma}{n-r+m+k-M+(N-M)i}\right].
\end{aligned}
\tag{5.149}
$$

(iii) For $i = 0$, we have

$$P\{Q_n = k + M, S_n = \sigma_0 | Q_0 = m, S_0 = \sigma_0\}$$
$$= \binom{n; \alpha, \beta, \gamma}{n - m + k + M} - \rho^{-m} \binom{n; \alpha, \beta, \gamma}{n - m + k - M}. \tag{5.150}$$

This completes the first goal, relation (5.136). It should be noted that if $M = 0$, then (5.148)–(5.150) match with those in Böhm and Mohanty (1990).

We now address the second goal, relation (5.137). In this case, Eqs (5.148) and (5.149) should be summed over i. Indeed, these two relations are built by terms of the following types:

$$A_n(a) = \sum_{i\geq 1} \sum_{l=Qi}^{n} \binom{l-1}{Qi-1} \gamma^{Qi}(1-\alpha)^{l-Qi} \binom{n-l; \alpha, \beta, \gamma}{n-l+a+Qi}, \tag{5.151}$$

where a is an integer, possibly negative, and $Q = N - M$. Let us set

$$T = \alpha\nu + \beta + \frac{\gamma}{\nu}. \tag{5.152}$$

Then, (5.16) can be rewritten as

$$A_n(a) = [\nu^a] \sum_{i\geq 1} \sum_{l=Qi}^{n} \binom{l-1}{Qi-1} \left(\frac{\gamma}{1-\alpha}\right)^{Qi} \nu^{-Qi} T^{n-1} (1-\alpha)^l$$
$$= [\nu^a] \sum_{i\geq 1} \left(\frac{\gamma}{\nu(1-\alpha)}\right)^{Qi} \sum_{l=Qi}^{n} \binom{l-1}{Qi-1} \left(\frac{1-\alpha}{T}\right)^i. \tag{5.153}$$

Now let

$$z = \frac{1-\alpha}{T}. \tag{5.154}$$

Using (5.154), differentiating a finite geometric series $Q - 1$ times, and the incomplete beta function (see, for instance, Abramowitz and Stegun 1972), the second sum in (5.153) can be done as follows:

$$\sum_{l=Qi}^{n} \binom{l-1}{Qi-1} \left(\frac{1-\alpha}{T}\right)^i = \sum_{l=Qi}^{n} \binom{l-1}{Qi-1} z^l$$
$$= \frac{z^{Qi}}{(1-z)^{Qi}} \left(1 - \sum_{j=0}^{Qi-1} \binom{n}{j} z^{n-j} (1-z)^j\right)$$
$$= n \frac{z^{Qi}}{(1-z)^{Qi}} \binom{n-1}{Qi-1} \int_0^{1-z} t^{Qi-1} (1-t)^{n-Qi} dt. \tag{5.155}$$

Hence, from (5.153) and (5.155), we have

$$A_n(a) = [v^a]nT^n \times \int_0^{1-z} (1-t)^{n-1} \sum_{i\geq 1} \binom{n-1}{Qi-1} \left(\frac{\gamma z}{v(1-\alpha)(1-z)}\right)^{Qi} \times t^{Qi-1}(1-t)^{-Qi+1}dt. \tag{5.156}$$

By multisectioning the series in (5.156), we have

$$\begin{aligned} A_n(a) &= [v^a]\frac{nT^n}{Q}\int_0^{1-z}(1-t)^{n-1}\frac{\gamma z}{v(1-\alpha)(1-z)} \\ &\times \sum_{j=1}^{Q} e^{\frac{2\pi ji}{Q}} \left(1+\frac{\gamma z t e^{\frac{2\pi ji}{Q}}}{v(1-\alpha)(1-z)(1-t)}\right)^{n-1} dt \\ &= [v^a]\frac{nT^n}{Q}\sum_{j=1}^{Q} e^{\frac{2\pi ji}{Q}} \frac{\gamma z}{[v(1-\alpha)(1-z)]^n} \\ &\times \int_0^{1-z}\left\{v(1-\alpha)(1-z)+t\left[\gamma z e^{\frac{2\pi iji}{Q}} - v(1-\alpha)(1-z)\right]\right\}^{n-1} dt. \end{aligned} \tag{5.157}$$

Conducting the integral, whose integrand is a binomial and resubstituting z, we obtain the following:

$$\begin{aligned} A_n(a) &= [v^a]\frac{\rho^{-1}}{Q}\sum_{j=1}^{Q}\frac{e^{\frac{2\pi ji}{Q}}T^n}{v^2-\rho^{-1}v+\rho^{-1}\left(1-e^{\frac{2\pi ji}{Q}}\right)} \\ &- [v^a]\frac{\rho^{-1}}{Q}\sum_{j=1} Q\frac{e^{\frac{2\pi ji}{Q}}v^{-n}\left(v(1-\alpha)+\gamma e^{\frac{2\pi ji}{Q}}\right)^n}{v^2-\rho^{-1}+\rho^{-1}\left(1-e^{\frac{2\pi ji}{Q}}\right)}. \end{aligned} \tag{5.158}$$

The roots of the denominator of (5.158), denoted by k_{1j} and k_{2j}, respectively, are

$$k_{1j} = \frac{1}{2}\rho^{-1}\left(1+\sqrt{1-4\rho\left(1-e^{\frac{2\pi ji}{Q}}\right)}\right) \text{ and } k_{2j} = \frac{1}{2}\rho^{-1}\left(1-\sqrt{1-4\rho\left(1-e^{\frac{2\pi ji}{Q}}\right)}\right), \tag{5.159}$$

with $k_{1Q} = \rho^{-1}$ and $k_{2Q} = 0$. With some algebra such as expanding $A_n(a)$ in partial fraction and extracting the coefficient of v^a, we have

$$A_n(a) = \frac{1}{Q}\left(\rho^{a+1} - \sum_{v\geq 0}\binom{n;\alpha,\beta,\gamma}{n+a-v+1}\rho^{v}\right.$$
$$+\frac{\rho^{-1}}{Q}\sum_{j=1}^{Q-1}\frac{e^{\frac{2\pi ji}{Q}}}{k_{2j}-k_{1j}}\sum_{v\geq 0}\binom{n;\alpha,\beta,\gamma}{n+a-v}(k_{1j}^{-v-1}-k_{2j}^{-v-1})$$
$$\left.-k_{2j}^{-a-1}\left(1-\alpha+\gamma k_{1j}^{-1}e^{\frac{2\pi ji}{Q}}\right)^{n}\right). \quad (5.160)$$

We summarize the forgoing in the following theorem:

Theorem 5.6.1

1. If $k \leq N - M$, then we have

$$P\{Q_n = k + M, S_n = \sigma Q_0 = m, S_0 = \sigma_0\}$$
$$= \binom{n;\alpha,\beta,\gamma}{n-m+k+M} - \rho^{-m+M}\binom{n;\alpha,\beta,\gamma}{n+m+k-M}.$$
$$+\rho^{k-m+M}A_n(m-k-M) - \rho^{-m+M}A_n(m+k-M). \quad (5.161)$$

2. If $k \geq N - M$, we have

$$P\{Q_n = k + M, S_n = \sigma | Q_0 = m, S_0 = \sigma_0\}$$
$$= \binom{n;\alpha,\beta,\gamma}{n-m+k+M} - \rho^{-m+M}\binom{n;\alpha,\beta,\gamma}{n+m+k-M}.$$
$$+\rho^{N-m}A_n(m+k+M-2N) - \rho^{-m+M}A_n(m+k-M). \quad (5.162)$$

5.6.2 Solution in Continuous Time

A Poisson-type limiting method is to be used for the transient solution in continuous time. That is, as $n \to \infty$ and $\alpha, \gamma \to 0$, then $n\alpha \to \lambda$ and $n\gamma \to \mu$. These limits will be used in this section.

From (5.161), we see that in $A_n(a)$, the only terms that depend on n are the generalized trinomial coefficients and the following terms:

$$\left(1-\alpha+\gamma k_{ij}^{-1}e^{\frac{2\pi ji}{Q}}\right)^{n} = \left(1-\frac{1}{n}\left(\lambda-\mu k_{ij}^{-1}e^{\frac{2\pi ji}{Q}}\right)\right)^{n}, \quad i = 1,\ 2. \quad (5.163)$$

Mohanty and Panny (1990) show the following limit is true:

$$\lim \binom{n; \alpha, \beta, \gamma}{n+a} = e^{-(\lambda+\mu)} \rho^{\frac{a}{2}} I_a(2\sqrt{\lambda\mu}), \tag{5.164}$$

where $I_a(2\sqrt{\lambda\mu})$ denotes the modified Bessel function of order a. From (5.163), we have

$$\lim \left(1 - \frac{1}{n}\left(\lambda - \mu k_{ij}^{-1} e^{\frac{2\pi ji}{Q}}\right)\right)^n = e^{-(\lambda - \mu k_{ij}^{-1} e^{\frac{2\pi ji}{Q}}}, \quad i = 1,\ 2. \tag{5.165}$$

We let

$$\lim A_n(a) = A(a). \tag{5.166}$$

Thus, denoting by $\mathcal{R}$ the real part of z, we have

$$\begin{aligned} A(a) = \frac{1}{Q}\Bigg(& \rho^{a+1} e^{-(\lambda+\mu)t} \sum_{v\geq 0} \rho^{\frac{a+v+1}{2}} I_{a-v+1}(2t\sqrt{\lambda\mu})\Bigg) \\ &+ \frac{\rho^{-1}}{Q}\mathcal{R}\left[\sum_{j=1}^{Q-1} \frac{e^{\frac{2\pi ji}{Q}} e^{-(\lambda+\mu)t}}{k_{2j} - k_{1j}} \sum_{v\geq 0} \rho^{\frac{a-v}{2}} I_{a-v}(2t\sqrt{\lambda\mu})(k_{1j}^{-v-1} - k_{2j}^{-v-1})\right] \\ &- \frac{\rho^{-1}}{Q}\mathcal{R}\left\{\sum_{j=1}^{Q-1} \frac{e^{\frac{2\pi ji}{Q}} e^{-(\lambda+\mu)t}}{k_{2j} - k_{1j}} \left[k_{1j}^{-a-1} e^{-\left(\lambda - \mu k_{1j}^{-1} e^{\frac{2\pi ji}{Q}}\right)t} - k_{2j}^{-a-1} e^{-\left(\lambda - \mu k_{1j}^{-1} e^{\frac{2\pi ji}{Q}}\right)t}\right]\right\}. \end{aligned} \tag{5.167}$$

Hence, we have the following theorem:

Theorem 5.6.2

1. If $k \leq N - M$, then we have

$$\begin{aligned} P\{Q_t &= k + M, S_t = \sigma | Q_0 = m, S_0 = \sigma_0\} \\ &= \rho^{(k-m+M)/2} e^{-(\lambda+\mu)t}[I_{k-m+M}(2t\sqrt{\lambda\mu}) - I_{k+m+M}(2t\sqrt{\lambda\mu})] \\ &\quad + \rho^{k-m+M} A(m - k - M) - \rho^{-m+M} A(k + m - M). \end{aligned} \tag{5.168}$$

2. If $k \geq N - M$, we have

$$\begin{aligned} P\{Q_t &= k + M, S_t = \sigma | Q_0 = m, S_0 = \sigma_0\} \\ &= \rho^{(k-m+M)/2} e^{-(\lambda+\mu)t}[I_{k-m+M}(2t\sqrt{\lambda\mu}) - I_{k+m-M}(2t\sqrt{\lambda\mu})] \\ &\quad + \rho^{N-m} A(m + k + M - 2N) - \rho^{-m+M} A(m + k - M). \end{aligned} \tag{5.169}$$

The authors prove that when $M = 0$ and $N = 1$, Theorem 5.6.1 gives the transient solution of $M/M/1$ for $k \geq 1$ tasks waiting at time n. In the proof, they use

$$\sum_{\nu \geq 0} \binom{n; \alpha, \beta, \gamma}{\nu} = 1. \tag{5.170}$$

5.7 STATIONARY *M/M/*1 QUEUING PROCESS WITH DELAYED FEEDBACK

A ***delayed-feedback queue*** appears to have been first discussed in Nakamura (1971) followed by Hannibalsson and Disney (1977), Fayolle (1979), and subsequently by Foley and Disney (1983). Haghighi et al. (2011) considered an $M/M/1$ queue with random delayed feedback with an additional feature, that is, a splitting device set after the service station that a serviced task might choose to go through. This model appeared in Haghighi and Mishev (2013). Here, in this section, we study a special case of this model, and in a later section, we will consider a generalization of it.

Thus, we consider the single-processor model illustrated in Figure 5.9.

The model consists of two types of inputs: external arrivals and inputs from feedbacks. External tasks arrive according to a Poisson distribution with parameter λ to a waiting buffer with infinite capacity in front of a single server. The service distribution is exponential with parameter μ. We refer to this server and its buffer as the ***main service station***. After a task leaves the main service station, it undergoes one of the following two modifications:

1. leaves the system with probability $q_o, 0 \leq q_o \leq 1$; or
2. returns to the end of waiting buffer for further service with probability p_f, through the ***delay station***, which consists of an infinite-capacity buffer and single delay processor.

The movement from the delay station to the tail of the queue, that is, the main service station is singly with exponential distribution and parameter ν. After processing

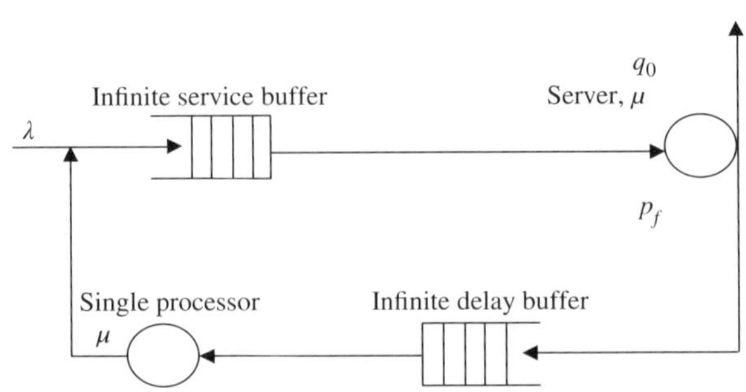

Figure 5.9 Single processor with splitting and delayed feedback

at the delay station is completed, the processed task joins the queue tail of the main service station. We assume that the capacities of both buffers (one in the main service station and the other in the delay station) are infinite. Hence, no blocking is possible. It is assumed that return is an event independent of any other event involved and, in particular, independent of the number of its returns. Discipline in the main service station is in the order of arrivals.

Hence, we have a tandem queue with two infinite buffers, Poison external arrival, single servers at each station, exponential service distribution with parameter μ at the main service station and ν at the delay station.

5.7.1 Distribution of the Queue Length

Let the random variables $\xi_1(t)$ and $\xi_2(t)$ represent the number of tasks, at time t, in the main service station, including the one being served and at the delay station, including those being processed, respectively. Further, we let the joint probability of $m, m \geq 0$, tasks at the main service station, including the one being served, and $n, n \geq 0$, tasks at the delay station, including the one being processed, at time t be denoted by $\Psi_{m,n}(t)$, that is,

$$\Psi_{m,n}(t) = P\{\xi_1(t) = m, \xi_2(t) = n\}. \tag{5.171}$$

Thus, $P\{\xi_1(t) = m, \xi_2(t) = n\}$ is an irreducible Markov chain. We further let

$$\Psi_{m,n}(t) = \lim_{t\to\infty} \Psi_{m,n}(t) \tag{5.172}$$

to be the steady-state probability of having m tasks in the main service station and n tasks in the delay station. It is known that our system is not Markovian. See, for example, Brémaud (1978).

The intention here is to find the steady-state queue length, that is, $\Psi_{m,n}$. Let us denote by λ_1 and λ_2 as the average arrival rates to the service station and delay station, respectively. Then, since when the system is busy, the arrival rates in and out of delay station are the same; from Figure 5.9, it can be seen that

$$\begin{cases} \lambda_1 = \lambda + \lambda_2, \\ \lambda_2 = \lambda_1 p_f, \end{cases} \tag{5.173}$$

from which we have

$$\lambda_1 = \frac{\lambda}{1-p_f} \quad \text{and} \quad \lambda_2 = \frac{\lambda p_f}{1-p_f}. \tag{5.174}$$

Theorem 5.7.1 shows the (theoretical) existence of the solution. For the solution of the functional equation mentioned in the theorem, we refer the readers to Haghighi et al. (2011). However, we will use the theorem to find the expected values of the queue length.

Theorem 5.7.1 Let

$$a = q_o\mu, b = p_f\mu, \quad \text{and} \quad d = q_o. \tag{5.175}$$

If $d > 0$, $\frac{\lambda_1}{\mu} < 1$, and $\frac{\lambda_2}{\nu} < 1$, then the stationary joint distribution function of the number of tasks in each station exists and can be obtained by the coefficient of Maclaurin series expansion of solution of the functional equation for the generating function of the distribution function

$$A(w,z)G_1(w,z) = B_1(w,z)G_1(0,z) + B_2(w,z)G_1(w,0), \tag{5.176}$$

where A, B_1, B_2 and $G_1(w,z)$ are defined as

$$A(w,z) = bz^2 + [a + \lambda w^2 - (\lambda + \mu + \nu)w]z + \nu w^2, \tag{5.177}$$

$$B_1(w,z) = bz^2 + (a - \mu w)z, \tag{5.178}$$

$$B_2(w,z) = \upsilon\omega^2 - \nu wz, \tag{5.179}$$

and

$$G_1(w,z) = \sum_{m=0}^{\infty}\sum_{n=0}^{\infty} \Psi_{m,n} w^m z^n, \quad |z| < 1, \quad |w| < 1 \tag{5.180}$$

is the generating function of the joint distribution function of the number of tasks in each station. The index of G indicates the value of k, which in this case is 1. Furthermore, expected number of tasks at each station is found using the algorithm given in the following proof.

Proof: From the description of the model, the system of stationary balance difference equations in this case is as follows:

(i)

$$\lambda\Psi_{0,0} = a\Psi_{1,0}, \tag{5.181}$$

(ii)

$$(\lambda + \nu)\Psi_{0,1} = a\Psi_{1,1} + b\Psi_{1,0},$$

(iii)

$$(\lambda + \nu)\Psi_{0,n} = a\Psi_{1,n} + b\Psi_{1,n-1}, \quad n \geq 2,$$

(iv)

$$(\lambda + \mu)\Psi_{m,0} = \lambda\Psi_{m-1,0} + a\Psi_{m+1,0} + \nu\Psi_{m-1,1}, \quad m \geq 1,$$

(v)

$$(\lambda + \mu + \nu)\Psi_{m,1} = \lambda\Psi_{m-1,1} + a\Psi_{m+1,1} + b\Psi_{m+1,0} + \nu\Psi_{m-1,2}, \quad m \geq 1,$$

(vi)

$$(\lambda + \mu + \nu)\Psi_{m,n} = \lambda\Psi_{m-1,n} + a\Psi_{m+1,n} + b\Psi_{m+1,n-1} + \nu\Psi_{m-1,n+1}, \quad m \geq 1,$$
$$n \geq 1,$$

and

(vii)

$$\sum_{m=0}^{\infty}\sum_{n=0}^{\infty}\Psi_{m,n} = 1.$$

It should be noted that by letting $q_o = 1$, the system of equations (5.181) reduces to that of *M/M/*1. Note also that the system (5.181) is similar to the one by Kleinrock and Gail (1996) with the exception of the delayed distribution. Furthermore, the system is similar to the one by Jackson (1957) with the exception of allowing exit in deciding to feedback.

We define

$$G_1(w,z) = \sum_{m=0}^{\infty}\sum_{n=0}^{\infty}\Psi_{m,n}w^m z^n \tag{5.182}$$

as the generating function of the joint stationary probability distribution $\Psi_{m,n}$. Let

$$F_m(z) = \sum_{n=0}^{\infty}\Psi_{m,n}z^n, \quad m = 1, 2, \dots . \tag{5.183}$$

Then, the generating function defined in (5.182) can be rewritten as

$$G_1(w,z) = \sum_{m=0}^{\infty}F_m(z)w^m. \tag{5.184}$$

Now apply (5.183) on (5.181) and, thus, from (iii) of system (5.181) we have

$$(\lambda+\nu)[F_0(z) - \Psi_{0,0} - z\Psi_{0,1}] = a[F_1(z) - \Psi_{1,0} - z\Psi_{1,1}] + bz[F_1(z) - \Psi_{1,0}]$$

or

$$(\lambda+\nu)F_0(z) - (a+bz)F_1(z) = \nu\Psi_{0,0}. \tag{5.185}$$

Now from (vi) of (5.181), after some algebra, we have

$$\begin{aligned}
&(\lambda+\mu+\nu)[F_m(z) - \Psi_{m,0} - z\Psi_{m,1}]\\
&\quad = \lambda[F_{m-1}(z) - \Psi_{m-1,0} - z\Psi_{m-1,1}] + a[F_{m+1}(z) - \Psi_{m+1,0} - z\Psi_{m+1,1}]\\
&\quad + bz[F_{m+1}(z) - \Psi_{m+1,0}] + \nu F_{m+1}(z)\\
&\quad + \frac{\nu}{z}[F_{m-1}(z) - \Psi_{m-1,0} - z\Psi_{m-1,1} - z^2\Psi_{m-1,2}], \quad m \ge 1.
\end{aligned} \tag{5.186}$$

Relation (5.186) can be reduced to

$$\left(\lambda + \frac{\nu}{z}\right)F_{m-1}(z) - (\lambda+\mu+\nu)F_m(z) + (a+bz)F_{m+1}(z) = \frac{\nu}{z}\Psi_{m-1,0} - \nu\Psi_{m,0},$$
$$m \ge 1. \tag{5.187}$$

Summing (5.187) over m gives

$$\left[\left(\lambda+\frac{\nu}{z}\right)w-(\lambda+\mu+\nu)+\frac{1}{w}(a+bz)\right]G_1(w,z)$$
$$=\left[\frac{1}{w}(a+bz)-\mu\right]F_0(z)+\left(\frac{\nu w}{z}-\nu\right)(\Psi_{0,0}+\Psi_{1,0}w+\Psi_{2,0}w^2+\cdots) \quad (5.188)$$

Since $G_1(0,z)=\sum_{n=0}^{\infty}\Psi_{0,n}z^n=F_0(z)$ and $G_1(w,0)=\sum_{m=0}^{\infty}\Psi_{m,0}w^m$, Equation (5.188) can be rewritten as

$$\left[\left(\lambda+\frac{\nu}{z}\right)w-(\lambda+\mu+\nu)+\frac{1}{w}(a+bz)\right]G_1(w,z)$$
$$=\left[\frac{1}{w}(a+bz)-\mu\right]G_1(0,z)+\left(\frac{\nu w}{z}-\nu\right)G_1(w,0) \quad (5.189)$$

or

$$\{bz^2+[q+\lambda w^2-(\lambda+\mu+\nu)w]z+\nu w^2\}G_1(w,z)$$
$$=[bz^2+[a-\mu w)z]G_1(0,z)+(\nu w^2-\nu wz)G_1(w,0), \quad (5.190)$$

from which we have (5.177).

In order to prove the remaining part of the theorem, we substitute $z=1$ in (4.4.8)–(4.4.10) and use the fact that $a+b=\mu$, after some algebra, we have

$$(\mu-\lambda w-\nu w)G_1(w,1)=\mu G_1(0,1)-\nu wG_1(w,0) \quad (5.191)$$

Now substituting $w=1$ in (5.191), we have

$$\mu G_1(0,1)-\nu G_1(1,0)=\mu-\lambda-\nu \quad (5.192)$$

In a similar manner, substituting $w=1$ in (5.177)–(5.179), and more algebra, we have

$$\begin{cases}\mu G_1(0,1)-\nu G_1(1,0)=\mu-\lambda-\nu,\\ bG_1(0,1)-\nu G_1(1,0)=b-\nu.\end{cases} \quad (5.193)$$

System (5.193) is a nonhomogeneous one of two equations with two unknowns. We solve it by the elimination method and obtain its solutions as

$$G_1(0,1)=1-\frac{\lambda_1}{\mu} \text{ and } G_1(1,0)=1-\frac{\lambda_2}{\nu}. \quad (5.194)$$

Since $0<G_1(0,\ 1)<1$ and $0<G_1(0,\ 1)<1$, from (5.194), we have $0<\frac{\lambda_1}{\mu}<1$ and $0<\frac{\lambda_2}{\nu}<1$. Now, from the assumption of the theorem, $\frac{\lambda}{d\mu}=\frac{\lambda_1}{\mu}$, and, hence, $d>0$. Thus, all conditions of Theorem 5.7.1 are satisfied.

The joint probability distribution could be obtained from the Maclaurin expansion of (5.180). On the contrary, substituting 1 for each w and z, separately, in (5.180) and evaluating the derivative with respect to the other variable at 1 gives the expected number of tasks in each station. This completes proof of Theorem 5.7.1.

5.7.2 Mean Queue Length and Waiting Time

Since analytic mean queue length and waiting time for the model is almost impossible, we offer an algorithm to find them. A numerical example is offered to illustrate workability of the algorithm. From the statement of Theorem 5.7.1, recall that $a = q_o\mu, b = p_f\mu, d = q_o, d > 0, \frac{\lambda_1}{\mu} < 1$, and $\frac{\lambda_2}{\nu} < 1$. Also, from the description of the model, recall that $q_o + p_f = 1$.

Before defining the steps, we need a brief discussion of $G(0, z)$. Let $\zeta(w)$ be an algebraic curve, say L, defined over the complex plane by $A(w, \zeta(w)) = 0$ such that

$$B_1(w, \zeta(w))G(0, \zeta(w)) + B_2(w, \zeta(w))G(w, 0) = 0 \tag{5.195}$$

Similarly, let $\omega(z)$ be an algebraic curve defined over the complex plane by $A(w(z), z) = 0$ such that

$$B_1(\omega(z), z)G(0, z) + B_2(\omega(z), z)G(\omega(z), 0) = 0. \tag{5.196}$$

Now from (5.196), we have

$$G(w, 0) = -\frac{B_1(\omega(z), z)}{B_2(\omega(z), z)}G(0, z), \omega \in [\omega_1^*, \omega_2^*], z \in L, \tag{5.197}$$

where $[\omega_1^*, \omega_2^*]$ is a cut on the real axis. Thus, $G(0, z), G(w, 0)$, and $G(w, z)$ can be found.

Since $G(w, 0)$ is real on the cut $[\omega_1^*, \omega_2^*]$, it implies that the right-hand side of (5.197) is real on the cut. We wish to move away from the cut $[\omega_1^*, \omega_2^*]$ and get on the contour L, and particularly on the unit circle L_o. This can be done by ***conformal mapping*** as we will see below.

For $z \in L$, let $\Psi(z)$ be a complex function of complex variable z, defined by

$$\Psi(z) = u(z) + iv(z) \equiv G(0, z). \tag{5.198}$$

Also, let $B(z)$ be a complex function of complex variable z, defined by

$$B(z) = b(z) + ia(z) \equiv -\frac{B_1(\omega(z), z)}{B_2(\omega(z), z)}, \tag{5.199}$$

where $b(z) = \Re(B(z))$ and $a(z) = \Im(B(z))$. Then, relation (5.197) can be rewritten as

$$G(w, 0) = B(z)\Psi(z) = [b(z)u(z) - a(z)v(z)] + i[a(z)u(z) + b(z)v(z)], z \in L. \tag{5.200}$$

Thus, for t such that $w_1^* \le t \le w_2^*$, we must have

$$\Im[B(z)\Psi(z)] = a(z)u(z) + b(z)v(z) = 0, \text{for } z \in L. \tag{5.201}$$

Relation (5.201) is a ***homogenous Hilbert boundary value problem***, see Gakhov (1966), or Problem H_0, with index χ that needs to be solved. The solution will be the unknown function $G(0, z)$ analytic in and continuous on L. It is well known that the Problem H_0 has a unique solution only if $\chi = 0$, as in our case, see Gakhov (1966), Muskhelishvili (1992), or Smirnov (1964).

Now, dividing the boundary conditions (5.201) by $\sqrt{[a(z)]^2 + [b(z)]^2}$, it will be reduced to the case when $[a(z)]^2 + [b(z)]^2 = 1$, where

$$a_1(z) = \frac{a(z)}{\sqrt{[a(z)]^2 + [b(z)]^2}} \quad \text{and} \quad b_1(z) = \frac{b(z)}{\sqrt{[a(z)]^2 + [b(z)]^2}}. \tag{5.202}$$

Thus, (5.201) can be rewritten as

$$\Im[B_0(z)\Psi(z)] = a_1(z)u(z) + b_1(z)v(z) = 0, \quad \text{for } z \in L, \tag{5.203}$$

where

$$B_0(z) = \frac{B(z)}{\sqrt{[a(z)]^2 + [b(z)]^2}} \quad \text{and } B(z) \equiv -\frac{B_1(\omega(z), z)}{B_2(\omega(z), z)}, \quad \text{for } z \in L. \tag{5.204}$$

The standard way to solve this type of boundary value problem is to transform the boundary condition (5.204) on the unit circle [see Muskhelishvili (1992) and Cohen and Boxma (1983)] and then transform back on the region L^+. Thus, we introduce the conformal mapping $t = f(z) : L^+ \to L_o^+$ onto its inverse $z = f^{-1}(t) : L_o^+ \to L^+$, where the $^+$ sign indicate the region interior of the curve. Using these mappings, we can reduce the Riemann–Hilbert problem on L to the same on the unit circle. Thus, here is the ***New Problem H$_0$***: Determine a function $\Psi_0(t)$ such that

$$\Im[\widetilde{B}_0(t)\Psi_0(t)] = \widetilde{a}_1(t)u(t) + \widetilde{b}_1(t)v(t) = 0, \quad \text{for } t \in L_o, \tag{5.205}$$

where

$$\widetilde{B}_0(t) = B_0(f^{-1}(t)), \tag{5.206}$$

$$\widetilde{a}_1(t) = a_1(f^{-1}(t)) = \frac{a_1(f^{-1}(t))}{\sqrt{[a_1(f^{-1}(t))]^2 + [b(f^{-1}(t))]^2}}, \tag{5.207}$$

and

$$\widetilde{b}_1(t) = b_1(f^{-1}(t)) = \frac{b(f^{-1}(t))}{\sqrt{[a(f^{-1}(t))]^2 + [b(f^{-1}(t))]^2}}. \tag{5.208}$$

Hence, the solution of (5.203) is

$$\Psi(z) = \Psi_0(f(z)), \qquad z \in L^+. \tag{5.209}$$

Therefore, the solution of (5.205) is

$$\Psi_0(t) = ce^{\sigma(t)}, \quad t \text{ is within the unit circle}, \tag{5.210}$$

where

$$\sigma(t) = \frac{1}{2\pi i}\int_L \frac{1}{s}\left\{\tan^{-1}\left(\frac{\widetilde{b}_1(s)}{\widetilde{a}_1(s)}\right)\right\}\frac{s+t}{s-t}ds, \tag{5.211}$$

and $\widetilde{a}_1(t)$ and $\widetilde{b}_1(t)$ are given by (5.207) and (5.208), respectively. See Cohen and Boxma (1983, p. 58).

Algorithm

Step 1. **Choosing Data.** Choose probabilities such that all conditions mentioned above are satisfied. Also, choose values for parameters so that $\lambda > 0$, $\mu \geq \frac{\lambda}{d} > 0$, and $\nu \geq \frac{\lambda(1-d)}{d} > 0$. If we let $r = \frac{b}{\mu}$, then $0 < \nu < \frac{\lambda r}{1-r}$.

Step 2. From Equation (5.175), solve $A(w, z) = 0$ for w. This gives two roots as functions of z, say $w_1(z)$ and $w_2(z)$.

Step 3. For z on the unit circle, choose the unique root $w_i(z), i = 1$ or 2, of $A(w, z) = 0$ such that $|w_i(z)| \leq 1$. For that, choose $w_0 = \frac{1}{2} + \frac{\sqrt{3}}{2}i$ so that $|z| = 1$. Then, try the values of roots mentioned in Step 2 at this z-value, choose the one with absolute value less than or equal to 1 and call it $\omega(z)$.

Step 4. From Equation (5.199), find $b(z) = \Re(B(z))$ and $a(z) = \Im(B(z))$.

Step 5. From Step 4 and (5.202), find $a_1(z)$ and $b_1(z)$.

Step 6. Find $\theta(\nu_k)$, that is, the last iteration of the following process, see Step 6.4 below. In order to find $\theta(\nu_k)$, we do as follows [for more details, see Cohen and Boxma (1983, IV.1.3)]:

Step 6.1. Solve

$$A(w, z) = (\lambda z + \nu)w^2 - [(\lambda + \mu + \nu)z]w + [az + bz^2] = 0,$$

for w, say, that gives

$$w_{1,2} = \frac{(\lambda + \mu + \nu)z \pm \sqrt{D(z)}}{2(\lambda z + \nu)}, \tag{5.212}$$

where $D(z) = (\lambda + \mu + \nu)^2 z^2 - 4(\lambda z + \nu)(az + bz^2)$. From (5.212), we have

$$w_1 + w_2 = \frac{\lambda + \mu + \nu}{\lambda z + \nu} = 2\Re(\omega(z)), \tag{5.213}$$

and

$$w_1 w_2 = \frac{az + bz^2}{\lambda z + \nu} = |\omega(z)|^2, \tag{5.214}$$

It should be noted that $D(z)$ must be negative since (5.214) shows that the roots are complex conjugates.

Now, solving (5.214) for z, we obtain

$$z = \frac{2\nu\mathfrak{R}(\omega(z))}{\lambda + \mu + \nu - 2\lambda\mathfrak{R}(\omega(z))} = \frac{2\nu\delta}{\lambda + \mu + \nu - 2\lambda\delta}, \qquad (5.215)$$

where $\delta = \mathfrak{R}(\omega(z))$. Substituting z from (5.215) into (5.214), we obtain

$$H(\delta) \equiv |\omega(z)|^2 = \frac{b}{\lambda} - \left(\frac{2\delta\nu}{\lambda + \mu + \nu - 2\delta\lambda}\right) + \left(\frac{a}{\lambda} - \frac{b\nu}{\lambda^2}\right)$$
$$- \left(\frac{a\nu}{\lambda} - \frac{b\nu^2}{\lambda^2}\right)\left(\frac{2\delta\lambda\nu}{\lambda + \mu + \nu - 2\delta\lambda} + \nu\right)^{-1}. \qquad (5.216)$$

It is clear from (5.216) that $H(\delta) > 0$.

Now, we draw a circle with an embedded triangle ΔOAB as follows:

$$\overline{OA} = \mathfrak{R}(\omega(z)) = \delta, \overline{OB} = |\omega(z)|,$$

and from

$$\Delta OAB : \cos\ (\theta_0(\nu_k)) = \overline{OA}/\overline{OB}.$$

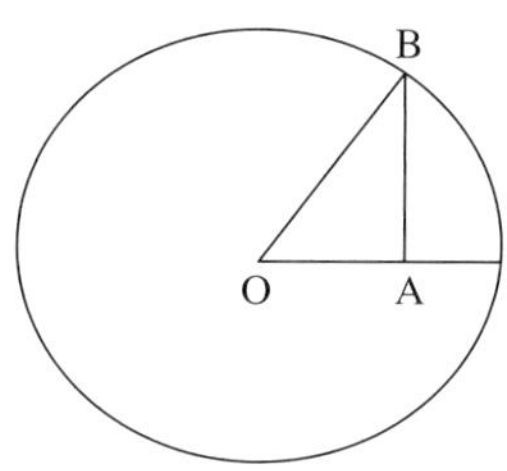

From ΔOAB and (5.216), we have

$$\cos(\theta_0(\nu_k)) = \frac{\delta}{|\omega(z)|} = \frac{\delta}{\sqrt{H(\delta)}}. \qquad (5.217)$$

Therefore,

$$\delta - \cos(\theta_0(\nu_k))\sqrt{H(\delta)} = 0, \qquad (5.218)$$

where δ depends on $\theta_0(\nu_k)$, that is, $\delta = \delta(\theta_0(\nu_k)), k = 0, 1, \ldots, K$.

Step 6.2. Set $\theta_0(\nu_k) = \nu_k = \frac{2\pi k}{K}, k = 0, 1, \ldots, K$. Using Newton's method, solve Equation (5.179), that is,

$$\delta(\theta_0(\nu_k)) - \cos(\theta_0(\nu_k))\sqrt{H(\delta(\theta_0(\nu_k)))} = 0, \qquad (5.219)$$

for $\delta(\theta_0(\nu_k)), k = 0, 1, \ldots, K$.

Note that Step 6.2 could alternatively be done (as we do in the numerical example) using the system of equations (5.220) below and solving for three unknowns $\delta = \Re(\omega(z)), |\omega(z)|$, and z.

$$\begin{cases} 2\Re(\omega(z)) = \dfrac{\lambda + \mu + \nu)z}{\lambda z + \nu}, \\ |\omega(z)|^2 = \dfrac{az + bz^2}{\lambda z + \nu}, \\ \dfrac{\Re(\omega(z))}{|\omega(z)|} = \cos(\theta_0(\nu_k)), \quad k = 0, 1, \ldots, K. \end{cases} \tag{5.220}$$

In order to solve (5.220), we do as Step 6.2. Alternative is given below [for more details, see Cohen and Boxma (1983, IV.1.3)].

Alternative: Set $\theta_0(\nu_k) = \nu_k = \frac{2\pi k}{K}, k = 0, 1, \ldots, K$, should be divisible by 4. Using Newton's method, solve Equation (5.221) below for $\delta(\theta_0(\nu_k)), k = 0, 1, \ldots, K$:

$$\sqrt{\delta(\theta_0(\nu_k))} - \cos(\theta_0(\nu_k))\sqrt{H(\delta(\theta_0(\nu_k)))} = 0, \quad k = 1, 2, \ldots, K, \tag{5.221}$$

where

$$H_1(\delta) = \frac{2}{\lambda + \mu + \nu}\left[a + b\left(\frac{2\delta\nu}{\lambda + \mu + \nu - 2\lambda\delta}\right)\right] \tag{5.222}$$

and $\delta \equiv \Re(\omega(z))$.

We should point out that the proper choice of a starting point is very important in the Newton's method, since it is possible that the method fails to converge.

For values of z on the cut, we have to find

$$\Re(\omega(z)) = \delta = \frac{(\lambda + \mu + \nu)z}{2(\lambda z + \nu)}$$

at the end points of the cut. Then, choose the initial value for the Newton's method a value between the two values of δ just found. This will assure the Newton's approximation to result in positive values.

Step 6.3. Using the Trapezoidal rule, find $\theta_1(\nu_k)$, for each $k = 0, 1, \ldots, K$ from

$$\theta_1(\nu_k) = \nu_k - \int_0^{2\pi} \ln[\rho(\theta_0(\tau))] \cot\left[\frac{1}{2}(\tau - \nu_k)\right] d\tau, \tag{5.223}$$

where $\rho(\theta_0(\tau)) = \frac{\delta(\theta_0(\tau))}{\cos\ (\theta_0(\tau))}$.

The integrand in (5.223) has a singularity at $\tau = \nu_k$. In order to deal with this singularity, we can rewrite the integrand as

$$\{\ln[\rho(\theta_0(\tau))] - \ln[\rho(\theta_0(\tau))]\} \cot\left[\frac{1}{2}(\tau - \nu_k)\right]$$
$$+ \ln[\rho(\theta_0(\nu_k))] \cot\left[\frac{1}{2}(\tau - \nu_k)\right]. \qquad (5.224)$$

The integral of the last term equals zero; the value of the first integrand at $\tau = \nu_k$ can easily be determined by the L'Hôpital's rule. Therefore, the resulting integral can be evaluated using a standard numerical integration procedure, that is, the trapezoidal rule. Thus, we can rewrite (5.223) as

$$\theta_1(\nu_k) = \nu_k - \int_0^{2\pi} \{\ln[\rho(\theta_0(\tau))]$$
$$- \ln[\rho(\theta_0(\nu_k))]\} \cot\left(\frac{\tau - \nu_k}{2}\right) d\tau. \qquad (5.225)$$

In applying the trapezoidal rule, denote the integrand of the integral in (5.223) by $h(\tau)$, that is,

$$h(\tau) = \{\ln[\rho(\theta_0(\tau))] - \ln[\rho(\theta_0(\nu_k))]\} \cot\left(\frac{\tau - \nu_k}{2}\right). \qquad (5.226)$$

Then,

$$\int_0^{2\pi} h(\tau)d\tau \approx \frac{\pi}{K}[h(\nu_0) + 2h(\nu_1) + \cdots + 2h(\nu_k)$$
$$+ \cdots + 2h(\nu_{K-1}) + h(\nu_K)], \qquad (5.227)$$

where $\nu_k = \frac{2\pi k}{K}$, $k = 0, 1, \ldots, K$. To find $h(\nu_k)$, we can rewrite $h(\tau)$ as

$$h(\tau) = \frac{\ln\left[\dfrac{\rho(\theta_0(\tau))}{\rho(\theta_0(\nu_k))}\right]}{\tan\left(\dfrac{\tau - \nu_k}{2}\right)}. \qquad (5.228)$$

However,

$$\lim_{r\to\nu} \frac{\ln\left[\dfrac{\rho(\theta_0(\tau))}{\rho(\theta_0(\nu_k))}\right]}{\tan\left(\dfrac{\tau - \nu_k}{2}\right)} = \frac{2}{\rho(\theta_0(\nu_k))} \frac{d}{d\tau}\rho(\theta_0(\tau))\bigg|_{r=\nu}. \qquad (5.229)$$

Therefore, for this case, we have

$$h(\nu_k) = \frac{2}{\rho(\theta_0(\nu_k))} \frac{d}{d\tau}\rho(\theta_0(\tau))\bigg|_{r=\nu}. \qquad (5.230)$$

Step 6.4. For $n = 1, 2, \ldots$, repeat Steps 6.1–6.3 to find $\theta_n(\nu_k)$ and $\theta_{n+1}(\nu_k)$, $k = 0, 1, \ldots, K$, until condition $\max\limits_{0 \le n \le N-1} |\theta_{n+1}(\nu_k) - \theta_n(\nu_k)| < 10^{-5}$. is satisfied. This condition may be revised in practice. The last iteration in this process will be $\theta(\nu_k), k = 0, 1, \ldots, K$.

Step 7. Find $f(1)$ from $f^{-1}(t) = te^{\frac{1}{2\pi}\int_0^{2\pi} \ln[\rho(\theta(u))] \frac{e^{i\nu}+2}{e^{i\nu}-2} d\nu}$, $|t| < 1$. For that, use the Newton's method and solve $f^{-1}(\eta) = 1$ in $[0, 1]$ Thus, $f(1) = \eta$. Ensure that $\eta < 1$.

Step 8. Find $f'(1)$. For that, use the Trapezoidal rule for the integral on the right-hand side of

$$f'(1) = \frac{1}{\frac{d}{dt}[f^{-1}(t)]\Big|_{t=\eta}}.$$

Step 9. From $f^{-1}(e^{i\nu k}) = \frac{\delta(\theta(\nu_k))}{\cos(\theta(\nu_k))} e^{i\theta(\nu_k)}, k = 0, 1, \ldots, K-1$, find $f^{-1}(e^{i\nu k})$.

Step 10. Using Trapezoidal rule, find $\frac{d}{dz}G(0, z)\Big|_{z=1}$ from

$$\frac{d}{dz}G(0, z)\Bigg|_{z=1} = G(0, 1)\frac{i}{\pi}\int_0^{2\pi} \left\{\tan^{-1}\left[\frac{b_1(f^{-1}(e^{i\nu}))}{a_1(f^{-1}(e^{i\nu}))}\right]\right\} \frac{f'(1)}{[e^{i\nu} - f(1)]^2} d\nu.$$

Step 11. Differentiating both sides of (5.228) with respect to z, setting $z = 1$ and solving for $\frac{d}{dz}G(1, z)\Big|_{z=1}$, we obtain

$$\frac{\partial}{\partial z}G(1, z)|_{z=1} = \frac{1}{\nu - b}\left[\frac{\lambda b}{a} - b\frac{\partial}{\partial z}G(0, z)|_{z=1}\right]. \tag{5.231}$$

It should be noted that since expected values for the model must be nonnegative, in the numerical evaluation, we should have one of the following two cases:

1.
$$\nu - b > 0 \text{ and } \frac{\lambda b}{a} - b\frac{\partial}{\partial z}G(1, \ z)|_{z=1} > 0,$$

or

2.
$$\nu - b > 0 \text{ and } \frac{\lambda b}{a} - b\frac{\partial}{\partial z}G(0, z)|_{z=1} < 0.$$

Step 12. Find $\frac{\partial}{\partial z}G(w, 1)|_{w=1}$ from

$$\frac{\partial}{\partial w}G(w, 1)|_{w=1} = \frac{1}{a - \lambda}\left(\lambda + a\frac{\partial}{\partial z}G(0, z)|_{z=1}\right) - \frac{\partial}{\partial z}G(1, z)|_{z=1}. \tag{5.232}$$

Here is how we obtained (5.232). We put $w = z = \zeta$ in (5.177). For this choice of w, $B_2(w, z) = 0$. Hence, (5.177) reduces to

$$A(\zeta,\zeta)G(\zeta,\zeta) = B_1(\zeta,\zeta)G(0,\zeta). \tag{5.233}$$

Differentiating both sides of (5.233) with respect to ζ, solving for $\frac{dG(\zeta,\zeta)}{d\zeta}$, $\frac{dG(\zeta,\zeta)}{d\zeta} = \frac{\partial G}{\partial w} + \frac{\partial G}{\partial z}$, simplifying, and applying L'Hôpital's rule as $\zeta \to 1$, we obtain

$$\frac{\partial}{\partial w}G(w,1)\bigg|_{w=1} + \frac{\partial}{\partial z}G(1,z)\bigg|_{z=1} = \frac{1}{a-\lambda}\left(\lambda + a\frac{\partial}{\partial z}G(0,z)\bigg|_{z=1}\right). \tag{5.234}$$

This is a relationship between the two expected values, from which (5.231) follows.

Step 13. Calculate the average waiting time at each station using $W_i = \frac{L_i}{\lambda_i}, i = 1, 2$.

Example 5.7.1 This example is to illustrate how the algorithmic steps work. We choose the number of iterations for approximations of integrals by trapezoidal rule as $K=30$, for approximations of derivatives by Newton's method as $N=10$, and the number of implementation of the approximation as $R=10$. We also use the following notations and conditions:

1.
$$a = q_o\mu, b = p_f\mu, a + b = \mu, d = q_o > 0.$$

2.
$$\lambda > 0, \mu \geq \frac{\lambda}{d}, \quad \text{and} \quad \nu \geq \frac{\lambda(1-d)}{d}.$$

3. λ_1 and λ_2 are as found in (4.3.4), $\frac{\lambda_1}{\mu} < 1$ and $\frac{\lambda_2}{\nu} < 1$.
4. $0 < G_1(0,1) < 1$ and $0 < G_1(1,0) < 1$.

In order to check the validity of the result, not only we compare the special cases with those that exist in the literature, but we will also simulate the system and compare the analytical and simulated results.

For different values of the model parameters, we obtain the following table that includes the expected values of the queue lengths at each station (including those in service and weighting for departure from the feedback buffer).

Remark on Special Cases

1. If $q_o = 1$ and $\nu = 0$, then the system (5.181) reduces to the system for stationary $M/M/1$ queue. In this case, if X represents the number of tasks in the system, then the expected value of the queue length E_{mml} is

$$E_{\text{mml}}(X) = \frac{\rho_{\text{mml}}}{1-\rho_{\text{mml}}}, \tag{5.235}$$

TABLE 5.6 $q_s = 0, q_o = 0.6, p_f = 0.4, a = q_o\mu, b = p_f\mu, d = 0.6$, **and** $r = b/\mu = 0$.

λ	μ	ν	λ_1	ρ_1	λ_2	ρ_2	ρ ρ_T	$E(X_1)$ $E_T(X_1)$	$E(X_2)$ $E_T(X_2)$
0.605	1.05	29	1.0083	0.96032	0.40333	0.013908	0.97423	22.2370	0.0460570
							0.96032	24.2000	0.0080137
1.000	2.00	30	1.6667	0.83333	0.66667	0.022222	0.85556	4.1987	0.1144200
							0.83333	5.0000	0.0111110
2.000	3.50	40	3.3333	0.95238	1.33330	0.033333	0.98571	14.2620	0.3910200
							0.95238	20.0000	0.0190480
10.000	17.50	200	16.6670	0.95238	6.66670	0.033333	0.98571	12.8370	1.8166000
							0.95238	20.0000	0.0190480
20.000	37.00	400	33.3330	0.90090	13.33300	0.033333	0.93423	7.7132	0.8332100
							0.90090	9.0909	0.0180180
30.000	53.00	500	50.0000	0.94340	20.00000	0.040000	0.98340	12.7940	2.1312000
							0.94340	16.6670	0.0226420

where $\rho_{\text{mml}} = \frac{\lambda}{\mu}$.

2. If $p_s = 0$ in the system (5.181), then the system described reduces to a stationary *M/M/*1 queue with random delayed feedback with a single return facility discussed in Kleinrock and Gail (1996). In this case, the expected values of the queue lengths for the stations are

$$E_T(S_1) = \frac{\rho_T}{1 - \rho_T}, \quad \text{and } E_T(S_2) = \frac{p_f \mu}{\nu}\rho_T, \tag{5.236}$$

where $\rho_T = \frac{\lambda}{q_o\mu}$. Note that the expected value of number of tasks in the first station is the same as the expected queue length for *M/M/*1 without feedback, that is, (5.236). See Table 5.6 for numerical values and comparison.

Note It seems that the system does not respond well when traffic intensities at each station is very low. In addition, due to stability of the system, the service and feedback rates were chosen depending upon the external arrival rate. This can be seen quite consistent throughout the numerical example. It should also be noted that $q_o > 0$ and denoting

$$\rho_1 = \frac{\lambda}{q_o\mu} \text{ and } \rho_2 = \frac{(1 - q_o)\lambda}{q_o\nu}, \tag{5.237}$$

we should have

$$\frac{\rho_1}{\rho_2} > 1. \tag{5.238}$$

Rewriting (5.225) as

$$\left.\frac{\partial}{\partial z}G(1, z)\right|_{z=1} = \frac{1}{\left(\frac{\rho_1}{\rho_2} - 1\right)}\left\{\rho_1\left(\frac{1}{q_o} - 1\right) - \left.\frac{\partial}{\partial z}G(0, z)\right|_{z=1}\right\} \tag{5.239}$$

and (5.234) as

$$\frac{\partial}{\partial w}G(w,1)\bigg|_{w=1} + \frac{\partial}{\partial z}G(1,z)\bigg|_{z=1} = \frac{1}{1-\rho_1}\left(\rho_1 + \frac{\partial}{\partial z}G(0,z)\bigg|_{z=1}\right). \tag{5.240}$$

All relations (5.238)–(5.240) must be satisfied and both (5.239) and (5.240) must be positive for the expected values to make sense and, thus, will help choosing the parameters λ, μ, and ν.

5.8 SINGLE-SERVER QUEUE WITH UNRELIABLE SERVER AND BREAKDOWNS WITH AN OPTIONAL SECOND SERVICE

So far, in this chapter, we have seen queues with various delay types. We now discuss another model that is very different from the others we have discussed. This model is a single-server Poisson arrival with general service times (*M/G/*1), but with a ***second optional service***, ***unreliable server***, and ***server breakdown***.

K. C. Madan

Madan (2000) studied an *M/G/*1 queue with second optional service. In the model, the arrival process is a Poisson process. All tasks demand the first "essential" service that follows a general distribution, while only some of them demand the second "optional" service that follows exponential distribution. For instance, in a barber's shop, all customers may need a haircut, but only a part of them may need hair coloring after the haircut. As another example, in a university, all undergraduate students must take the university's core courses, but only part of the student body must take courses for each major discipline. Using a supplementary variable method, we find the Laplace transform of generating function of the time-dependent pgfs of the queue length as the steady-state distribution.

Wang (2004) studied the same model Madan (2000) studied with an additional and more realistic feature, that is, the server is unreliable and the system is subject to server breakdown. He obtained the transient and steady-state solutions for both queuing and reliability measures. The ***delay*** in this system occurs due to the time it will take to repair the server after its breakdown. This type of model has been investigated by many authors including Cao and Cheng (1982) and Li et al. (1997). One of the motivations for considering such a model is the fast-expanding area of tele services, which prominently include telephone call centers and the emerging Internet-based market. See Koole and Mandelbaum (2002).We now discuss the *M*/*G*/1 model of Wang (2004).

The model consists of a server with two types of services: (1) ***essential*** and (2) ***optional***. The essential service is needed for all tasks arriving to the system, while optional service is only for those who need such a secondary service. Assumptions are as follows:

1. Tasks arrive according to a Poisson process with mean λ.
2. The essential service time is a random variable, denoted by X, with cdf and pdf as $B(x)$ and $b(x)$, respectively, with mean $1/\mu_1$ and hazard rate function $\mu_1(x)$.
3. As soon as a task completed its essential service, it will move on for the optional service with probability r, or leaves the system with probability $1-r$. If it chooses the optional service, commence of this service is immediate.
4. The optional service times are assumed to be exponential with mean service time $1/\mu_2$.
5. The lifetime of the server for the essential service has exponential distribution with mean $1/\alpha_1$. However, for the optional service, the server fails exponentially with mean α_2.
6. The server may break down while servicing a task. If that happens, the server is sent for repair. The task that was receiving service before server breakdown has to wait for the server to return from the repair service and complete its remaining service. The service time is cumulative.
7. The repair times of both service phases, essential and optional, are arbitrarily distributed with probability distribution functions $G_1(x)$ and $G_2(x)$, respectively, with $g_k(x), 1/\beta_k$, and $\beta_k(x), k = 1, 2$, as the corresponding pdfs, means, and hazard rates, respectively.
8. Various stochastic processes involved in the system are independent of each other.

Let $N(t)$be the number of tasks in the system at time t. Supplementary variables will be used to make the queuing process a Markovian one. Thus, we let $X(t)$ represent the elapsed service time of the task currently being served at time t and $Y(t)$ represent the elapsed repair time of the breakdown of the server at time t. Let us define the following state probabilities at time t:

1. $Q(t)$ the probability that the server is idle at time t.
2. $P_n^{(1)}(t, x)dx$: the joint probability of the events that at time t there are $n (n \geq 0)$ tasks in the queue excluding the one being provided the essential service, the server is up, and a task is being served with the elapsed service time between x and $x + dx$.

3. $P_n^{(2)}(t)$ the joint probability of events that at time t there are $n(n \geq 0)$ tasks in the queue excluding the one being provided the optional service.
4. $R_n^{(1)}(t,x,y)dy$: the joint probability of events that at time t there are $n(n \geq 0)$ tasks in the queue excluding the one being provided the essential service, the elapsed service time for the task under service is equal to x, and the server is being repaired with the elapsed repair time between y and y and $y + dy$.
5. $R_n^{(2)}(t,y)dy$: the joint probability of events that at time t there are $n(n \geq 0)$ in the queue excluding the one being provided the optional service, and the server is being repaired with the elapsed repair time between y and y and $y + dy$.

Thus, the state of the system at any time t can be characterized by random variables $N(t)$, $X(t)$, and $Y(t)$. Now, let us consider transitions of the process between time t and $t + \Delta t$. Then, we can derive the system of Kolmogorov forward equations for $n = 0, 1, 2, \ldots$, by letting $\Delta t =$ zero, as follows:

$$\left(\frac{d}{dt} + \lambda\right) Q(t) = \mu_2 P_0^{(2)}(t) + (1 - r) \int_0^\infty P_0^{(1)}(t,x)\mu_1(x)dx, \tag{5.241}$$

$$\left[\frac{\partial}{\partial t} + \frac{\partial}{\partial x} + \mu_1(x) + \lambda + \alpha_1\right] P_n^{(1)}(t,x) = \lambda P_{n-1}^{(1)}(t,x) + \int_0^\infty R_n^{(1)}(t,x,y)\beta_1(y)dy, \tag{5.242}$$

$$\left[\frac{\partial}{\partial t} + \frac{\partial}{\partial y} + \beta_1(y) + \lambda\right] R_n^{(1)}(t,x,y) = \lambda R_{n-1}^{(1)}(t,x,y), \tag{5.243}$$

$$\left(\frac{d}{dt} + \lambda + \alpha_2 + \mu_2\right) P_n^{(2)}(t) = \lambda P_{n-1}^{(2)}(t) + r \int_0^\infty R_n^{(1)}(t,x)\mu_1(x)dx + \int_0^\infty R_n^{(2)}(t,y)\beta_2(y)dy, \tag{5.244}$$

$$\left[\frac{\partial}{\partial t} + \frac{\partial}{\partial y} + \beta_2(y) + \lambda\right] R_n^{(2)}(t,y) = \lambda R_{n-1}^{(2)}(t,y). \tag{5.245}$$

For the sake of illustration, let us see how (5.242) is written, for example. Suppose that at the time $t + \Delta t$ there are n tasks in the waiting line. Also, suppose that the server is up and the time taken by the task being served is x. This situation can only arise during the interval $(t, t + \Delta t)$ from the following states:

(i) no change of state at all, with transition probability $1 - \lambda\Delta t - \alpha_1\Delta t - \mu(x)\Delta t$;
(ii) from $P_{n-1}^{(1)}(t - \Delta t)$ by the arrival of a new task which joins the queue, with transition probability $\lambda\Delta t$; or
(iii) from $R_n^{(1)}(t, x, \Delta t)$ by the completion of the repair of the server, with transition probability $\beta_1(y)\Delta t$.

Then, equating the probability of the corresponding transitions, transposing, and passing to the limit, Equation (5.242) is obtained.

We need to solve the system of equations (5.241)–(5.245) subject to the boundary conditions (5.246)–(5.249) below. The conditions are obtained by letting one of the supplementary variables starts from 0:

$$P_n^{(1)}(t,0) = (1-r)\int_0^\infty P_{n+1}^{(1)}(t,x)\mu_1(x)dx + \mu_2 P_{n+1}^{(2)}(t), n \geq 1, \tag{5.246}$$

$$P_0^{(1)}(t,0) = (1-r)\int_0^\infty P_1^{(1)}(t,x)\mu_1(x)dx + \mu_2 P_1^{(2)}(t) + \lambda Q(t), \tag{5.247}$$

$$R_n^{(1)}(t,x,0) = \alpha_1 P_n^{(1)}(t,x), \tag{5.248}$$

$$R_n^{(2)}(t,0) = \alpha_2 P_n^{(2)}(t). \tag{5.249}$$

As mentioned how the conditions have been obtained, let us consider (5.246), for example. For $n > 0$, either an essential service of a task is completed and the task leaves the system with probability $1 - r$, and a new task enters the essential service; or an optional service of a task is completed and a new task enters the essential service.

Hence, now the system of equations (5.241)–(5.249) should be solved together with the following normalizing equation:

$$Q(t) + \sum_{n=0}^{\infty} \left\{ P_n^{(2)}(t) + \int_0^\infty P_n^{(1)}(t,x)dx + \int_0^\infty R_n^{(2)}(t,y)dy \right.$$
$$\left. + \int_0^\infty \int_0^\infty R_n^{(1)}(t,x,y)dxdy \right\} = 1, \tag{5.250}$$

and the initial values $Q(0) = 1$, where $P_{-1}^{(1)}(t,x) \triangleq 0, P_{-1}^{(2)}(t) \triangleq 0, R_{-1}^{(1)}(t,x,y) \triangleq 0,$ $P_{-1}^{(2)}(t,y) \triangleq 0$, for any fixed t, x, y.

In order to solve the system, we define the following generating functions:

$$P_q^{(1)}(t,x,z) = \sum_{n=0}^{\infty} P_n^{(1)}(t,x)z^n, P_q^{(2)}(t,x,z) = \sum_{n=0}^{\infty} P_n^{(2)}(t,x)z^n, \tag{5.251}$$

$$R_q^{(1)}(t,x,y,z) = \sum_{n=0}^{\infty} R_n^{(1)}(t,x,y)z^n, \text{ and } R_q^{(2)}(t,y,z) = \sum_{n=0}^{\infty} R_n^{(2)}(t,y)z^n. \tag{5.252}$$

Denoting the Laplace transforms of a function $f(t)$ by $\widetilde{f}(s)$, we have

$$\phi_i(s,z) \equiv s + \lambda - \lambda z + \alpha_i - \alpha_i \widetilde{g}_i(s + \lambda - \lambda z), \quad i = 1, 2. \tag{5.253}$$

It is left as an exercise for the reader to show that the following LST and moment generating functions of the state probabilities are given as follows:

$$\widetilde{Q}(s) = \frac{1}{s + \lambda - \lambda z_s}, \tag{5.254}$$

$$\widetilde{P}_q^{(1)}(s,x,z) = \frac{[\phi_2(s,z)+\mu_2][(s+\lambda-\lambda z)\widetilde{Q}(s)-1]e^{[-\phi_1(s,z)x][1-B_1(x)]}}{[\widetilde{b}(\phi_1(s,z))-z]\phi_2(s,z)+\mu_2]-r\widetilde{b}(\phi_1(s,z)\phi_2(s,z)}, \tag{5.255}$$

$$\widetilde{P}_q^{(2)}(s,z) = \frac{r\widetilde{b}[\phi_1(s,z)][(s+\lambda-\lambda z)\widetilde{Q}(s)-1]}{[\widetilde{b}(\phi_1(s,z))-z][\phi_2(s,z)+\mu_2]-r\widetilde{b}(\phi_1(s,z)\phi_2(s,z)}, \tag{5.256}$$

$$\widetilde{R}_q^{(2)}(s,y,z) = \alpha_2\widetilde{P}_q^{(2)}(s,z)e^{-(s+\lambda-\lambda z)y}[1-G_2(y)], \tag{5.257}$$

$$\widetilde{R}_q^{(1)}(s,x,y,z) = \alpha_2\widetilde{P}_q^{(1)}(s,x,z)e^{-(s+\lambda-\lambda z)y}[1-G_1(y)], \tag{5.258}$$

where z_s is the root of the equation

$$x = \widetilde{b}[\phi_1(s,z)] - \frac{r\phi_2(s,x)\widetilde{b}[\phi_1(s,x)]}{\phi_2(s,x)+\mu_2}, \quad |z|, Re(s) > 0. \tag{5.259}$$

Denoting the pgf of the number of tasks in the queue, irrespective of the type of service being provided, by

$$\widetilde{P}_q(s,z) = \widetilde{P}_q^{(1)}(s,z) + \widetilde{P}_q^{(2)}(s,z) + \widetilde{R}_q^{(1)}(s,z) + \widetilde{R}_q^{(2)}(s,z), \tag{5.260}$$

the author proves the following:

$$\widetilde{P}_q(s,z) = \frac{r\phi_2(s,z)\widetilde{b}\phi_1(s,z) + [\phi_2(s,z)+\mu_2][1-\widetilde{b}\phi_1(s,z)]\left[\widetilde{Q}(s) - \frac{1}{s+\lambda-\lambda z}\right]}{[\phi_2(s,z)+\mu_2][\widetilde{b}\phi_1(s,z)-z] - r\phi_2(s,z)\widetilde{b}\phi_1(s,z)}. \tag{5.261}$$

Multiplying both sides of (5.261) by s, passing the limit as s approaches zero, applying the well-known Tauberian property:

$$\lim_{s\to 0} s\widetilde{f}(s) = \lim_{t\to\infty} f(t),$$

after some simplifications, we have

$$P_q(z) = \lim_{s\to 0} s\widetilde{P}_q(s,z) = \frac{[r\phi_2(0,z)\widetilde{b}\phi_1(0,z)] + [\phi_2(0,z)+\mu_2][1-\widetilde{b}\phi_1(0,z)]Q}{[\phi_2(0,z)+\mu_2][\widetilde{b}\phi_1(0,z)-z] - r\phi_2(0,z)\widetilde{b}\phi_1(0,z)}. \tag{5.262}$$

Now, letting $z = 1$ in (4.22), applying L'Hôpital's rule, after some simplification, since

$$Q + P_q(1) = 1, \tag{5.263}$$

and mandating by stability condition that

$$\rho_1\left(1+\frac{\alpha_1}{\beta_1}\right) - r\rho_2\left(1+\frac{\alpha_2}{\beta_2}\right) < 1, \tag{5.264}$$

where

$$\rho_1 = \frac{\lambda}{\mu_1} \text{ and } \rho_2 = \frac{\lambda}{\mu_2}, \tag{5.265}$$

Then, the steady-state probability that the server is idle is

$$Q = 1 - \rho_1 \left(1 + \frac{\alpha_1}{\beta_1}\right) - r\rho_2 \left(1 + \frac{\alpha_2}{\beta_2}\right). \tag{5.266}$$

Through direct calculations, we can find the steady-state probability that the server is busy as

$$P = \lim_{z\to 1}\lim_{s\to 0} s[\widetilde{P}_q^{(1)}(s,z) + \widetilde{P}_q^{(2)}(s,z)] = \rho_1 + r\rho_2 \tag{5.267}$$

and the steady-state probability that the server is under repair as

$$R = \lim_{z\to 1}\lim_{s\to 0} s[\widetilde{R}_q^{(1)}(s,z) + \widetilde{R}_q^{(2)}(s,z)] = \rho_1 \frac{\alpha_1}{\beta_1} + r\rho_2 \frac{\alpha_2}{\beta_2}. \tag{5.268}$$

As a theorem, the author proves that the LST of the moment generating function

$$\sum_{n=0}^{\infty} P\{N(t) = n\} z^n \tag{5.269}$$

is given by

$$\widetilde{P}_q(s,z) = \frac{r\phi_2(s,z)\widetilde{b\phi}_1(s,z)] + [\phi_2(s,z) + \mu_2][1 - \widetilde{b\phi}_1(s,z)]}{[\phi_2(s,z) + \mu_2][\widetilde{b\phi}_1(s,z) - z] - r\phi_2(s,z)\widetilde{b\phi}_1(s,z)} \times \left[\frac{1}{s + \lambda - \lambda z_s} - \frac{1}{s + \lambda - \lambda z}\right]. \tag{5.270}$$

Denoting the expected number of failures of the server in the essential service and the optional service up to time t by $M_1(t)$ and $M_2(t)$, respectively, using the fact that Shi (1985) showed

$$\widetilde{M}_i(s) = \sum_{n=0}^{\infty}(s) \int_0^{\infty} \alpha_i \widetilde{P}_n^{(i)}(s,x)dx = \alpha_i \widetilde{P}_q^{(i)}(s,1), \quad i = 1,2, \tag{5.271}$$

given that the system is initially empty and the server is idle, we have the LSTs of $M_1(t)$ and $M_2(t)$ as follows:

$$\widetilde{M}_1(s) = \frac{\alpha_1[\phi_2(s,1) + \mu_2][s\widetilde{Q}(s) - 1]}{[\widetilde{b\phi}_1(s,1) - 1][\phi_2(s,1) + \mu_2] - r\phi_2(s,1)\widetilde{b\phi}_1(s,1)} \times \frac{1 - \widetilde{b\phi}_1(s,1)}{\phi_1(s,1)}, \tag{5.272}$$

and

$$\widetilde{M}_2(s) = \frac{\alpha_2 r\widetilde{b\phi}_1(s,1)[s\widetilde{Q}(s) - 1]}{[\widetilde{b\phi}_1(s,1) - 1][\phi_2(s,1) + \mu_2] - r\phi_2(s,1)\widetilde{b\phi}_1(s,1)}. \tag{5.273}$$

The author, further, showed that the steady-state failure frequency of the server, denoted by M_f, is given by

$$M_f = \alpha_1\rho_1 + r\alpha_2\rho_2. \tag{5.274}$$

Denoting by τ the time to the first failure of the server, then the ***reliability function of the server*** is

$$R(t) = P\{\tau > t\}. \tag{5.275}$$

Then, the author proved the following theorem:

Theorem 5.8.1 The Laplace transform of $R(t)$ is given by

$$\widetilde{R}(s) = \frac{1}{s + \lambda - \lambda\omega(s)} + \frac{r(s + \alpha_2)\widetilde{b}(s + \alpha_1) + [1 - \widetilde{b}(s + \alpha_1)](s + \alpha_2 + \mu_2)}{(s + \alpha_1)\{(s + \alpha_2 + \mu_2)[\widetilde{b}(s + \alpha_1)] - r(s + \alpha_2)\widetilde{b}(s + \alpha_1)\}}$$
$$\times \left[\frac{s}{s + \lambda - \lambda\omega(s)} - 1\right], \tag{5.276}$$

where $\omega(s)$ is the root of the following equation:

$$x = \widetilde{b}(s + \lambda - \lambda x + \alpha_1)\left[1 - \frac{r(s + \lambda - \lambda x + \alpha_2)}{(s + \lambda - \lambda x + \mu_2)}\right], \tag{5.277}$$

Inside $|z| = 1$, and $Re(s) > 0$.

Denoting the mean time to the first failure of the server by MTTFF, since

$$\text{MTTFF} = \int_0^\infty R(t)dt = \widetilde{R}(0) \text{ and } \lim_{s\to 0} s\widetilde{Q}(s) = Q, \tag{5.278}$$

it follows from Theorem 5.8.1 that

$$\text{MTTFF} = \widetilde{Q}(0) + \frac{(\rho_1 + r\rho_2)\{r\alpha_1\widetilde{b}(\alpha_1) + (\alpha_2 + \mu_2)[1 - \widetilde{b}(\alpha_1)]\}}{\alpha_1[r\alpha_2\widetilde{b}(\alpha_1) + (\alpha_2 + \mu_2)[1 - \widetilde{b}(\alpha_1)]]}. \tag{5.279}$$

Example 5.8.1 The Impact of Failure Rates The author offers this example to demonstrate the impact of failure rates $\alpha_i, i = 1, 2$, on some main performance measures of the system. Let us assume

$$\lambda = 0.75, r = 0.2, \mu_1 = 1.25, \mu_2 = 1, \beta_1 = 0.15, \beta_2 = 0.1$$

and $\alpha_i, i = 1, 2$, may change from 0 to 0.08.

Accordingly, computation results of the steady-state server availability A and failure frequency M_f are summarized in Table 5.7. Clearly, high values of $\alpha_i, i = 1, 2$, will result in low server availability and high failure frequency.

TABLE 5.7 Influence of the Reliability Factors α_1 and α_2

		$\lambda = 0.75, r = 0.2, \mu_1 = 1.25, \mu_2 = 1$			
α_1	α_2	β_1	β_2	A	M_f
0	0			1	0
0.05	0	0.15	0.10	0.800	0.0300
0.05	0.02	0.15	0.10	0.770	0.0330
0.05	0.04	0.15	0.10	0.740	0.0360
0.05	0.06	0.15	0.10	0.710	0.0390
0.05	0.08	0.15	0.10	0.680	0.0420
0	0.05	0.15	0.10	0.925	0.0075
0.02	0.05	0.15	0.10	0.845	0.0195
0.04	0.05	0.15	0.10	0.765	0.0315
0.06	0.05	0.15	0.10	0.685	0.0435
0.08	0.05	0.15	0.10	0.605	0.0555
0.08	0.08	0.15	0.10	0.520	0.0600

5.9 A BULK ARRIVAL RETRIAL QUEUE WITH UNRELIABLE SERVER

The study of retrial queue in queuing theory has been focused by many authors because of its wide applicability. For instance, in modeling magnetic disk memory system, cellular mobile networks, computer networks, and local area networks with nonpersistent CSMA/CD protocols, star topology, random access protocols, and multiple access protocols. For articles in retrial queues, the reader may refer to Yang and Templeton (1987), Falin(1990), and Falin and Templeton (1997).

P. Rajadurai

M. C. Saravanarajan

In this section, we will discuss a paper by Rajadurai et al. (2015). The authors have considered queuing model $M^{[x]}/(G_1, G_2)/1$. Positive and negative tasks have recently appeared in the queuing literature. The ***positive tasks*** arrive to the system and get their services in normal manner. The ***negative tasks***, on the contrary, arrive into the system only at the service time of positive tasks.

These tasks neither join the queue nor get any service. ***A negative task will remove and destroy a positive task in service, and then the positive task loses its service and leaves the system***. Queues with negative tasks are referred to as ***G-queues***. These types of queues have extensive applications in computers, communication networks, and manufacturing system. Krishnakumar et al. (2013) studied a model with the concept of *M*/*G*/1 feedback retrial queuing system with negative tasks. Ke and Chang (2009a) have discussed modified vacation policy for *M*/*G*/1 retrial queue with balking and feedback. The concept of balking (tasks decide not to join the line at all if the server is unavailable upon arrival) was first studied by Haight in 1957. There are many situations where the tasks may be impatient, such as impatient telephone switchboard tasks and the hospital emergency rooms handling critical patients, web access, including call centers and computer systems. Ke (2007) studied the $M^{[X]}/G/1$ queue with variant vacations and balking. Authors such as Wang and Li (2009) and Gao and Wang (2014) discussed the concept balking.

Here, the authors discuss the steady-state analysis of a batch arrival feedback retrial queue with two types of service and negative tasks. Any arriving batch of positive tasks finds the server is free, one of the tasks from the batch enters into the service area and the rest of them join into the orbit. The negative task is arriving during the service time of a positive task, will remove the positive task, and the interrupted positive task either enters the orbit or leaves the system. If the orbit is empty at the service completion of each type of service, the server takes at most J vacations until at least one task is received in the orbit when the server returns from a vacation. The busy server may break down at any instance and the service channel will fail for a short interval of time.

In a vacation queuing system, the server may not be available for a period of time due to many reasons such as being checked for maintenance, working at other queues, scanning for new work (a typical aspect of many communication systems), or simply taking a break. This period of time during which the server is unavailable for primary tasks is referred as a vacation. Krishnakumar and Arivudainambi (2002) have investigated a single-server retrial queue with Bernoulli schedule, where the random decision whether to take a vacation or not is allowed only at instances when the system is not empty (and a service or vacation has just been completed). If the system is empty, the server must take a vacation that is the assumption for their model. Ke and Chang (2009) examined a batch retrial model with J vacations in which if an orbit becomes empty, the server takes at most J vacations repeatedly until at least one task appears in the orbit upon returning from a vacation. By applying the supplementary variable technique, system characteristics are derived. Later, Ke and Chang (2009a), Chen et al. (2010), and Rajadurai et al. (2014) discussed the different types of queuing model with J vacation queuing models.

It seems the work of authors we are discussing is the first that contains the combination batch arrival retrial queue, two types of service, G-queues, balking, feedback, modified vacation (at most J vacations), and breakdowns. The mathematical results and theory of queues of this model provide to serve a specific and convincing application in the transfer model of an E-mail system. In Simple Mail Transfer Protocol (SMTP), mail system is used to deliver the messages between mail servers.

5.9.1 The Model

The system consists of a batch arrival feedback retrial with two types of services, negative tasks under modified vacation policy, where the server is subject to failure, breakdown, and repair. The detailed description of model is given as follows:

1. ***Arrival Process:*** Positive tasks arrive in batches according to a compound Poisson process with rate λ. We denote the number of tasks belonging to the kth arriving batch by $X_k, k = 1, 2, \ldots$. Let us assume that the random variables $X_k, k = 1, 2, \ldots$. have a common distribution:

$$P\{X_k = n\} = x_n, \quad n = 1, 2, \ldots, \tag{5.280}$$

   and pgf $X(z)$, that is

$$X(z) = \sum_{k=0}^{\infty} x_n z^n. \quad |z| \leq 1. \tag{5.281}$$

   We further denote by $X^{[k]}$ the kth factorial moment of $X(z), k = 1, 2$.
2. ***Retrial Process:*** There is no buffer at in front of the service station. Hence, if an arriving batch finds the server free, one of the tasks from the batch begins its service and the rest join an orbit. However, if an arriving batch of tasks finds the server being busy, on vacation or in breakdown, the arriving batch either leaves the service area with probability $1-b$ or joins the pool of blocked tasks, that is, the orbit with probability b. Inter-retrial times have an arbitrary distribution $R(t)$ with corresponding LST $R^*(\theta)$.
3. ***Service Process:*** A single server that provides the two types of services. If any batch of arriving positive tasks finds the server free, then one of the tasks from the batch is allowed to start the first-type service (FTS) with probability p_1 or second-type service (STS) with probability $p_2, p_1 + p_2 = 1$, and others join the orbit. It is assumed that the ith, $i = 1, 2$, type service time random variable, denoted by S_i, follows a general distribution function $S_i(t)$ with its LST denoted by $S_i^*(\theta), i = 1, 2$.
4. ***Starting Failure Repair Process:*** Any arriving batch of positive tasks that finds the server free, service of only the task at the head of the batch is allowed to start, that is, the server is turned on, the other tasks leave the service area and join the orbit. The service discipline for the tasks in the orbit is first-retry success, first-service (FRSFS). Returning task finds the server being free (retry successfully), the task must start (turn on) the server. The start-up time of server could be negligible. Moreover, the server may fail with probability $\overline{\alpha} = 1 - \alpha$. If the server started successfully, the task will receive service immediately, otherwise, the server's repair starts immediately and the task must leave

the service area and make a retrial later. In other words, the probability of successful commencement of service is α for a new and returning task.

Note that the repair time of the failure server is of random length H with distribution function $H(t)$, LST of $H^*(\theta)$, and finite kth moment of $h^{(k)}$, $k = 1, 2$.

5. ***Feedback Rule:*** After completion of type 1 service, type 2 service for each positive task, the unsatisfied positive tasks may rejoin the orbit as a feedback task for receiving another regular service with probability $p, 0 \le p \le 1$, or may leave the system with probability $q, p + q = 1$.
6. ***Negative Arrival Process:*** Negative tasks arrive from outside the system according to a Poisson process with rate δ. These tasks arrive only during service times of positive tasks. Once a negative task arrived into the system, it will remove the positive task in service. The interrupted positive task either enters the orbit with probability θ $(0 \le \theta \le 1)$ or leaves the system forever with probability $1 - \theta$.
7. ***Vacation Process:*** When the orbit becomes empty, the server will leave for a vacation of random length V. If no task appears in the orbit when the server returns from a vacation, it leaves again for another vacation with the same length. Such pattern continues until it returns from a vacation to find at least one task found in the orbit or it has already taken J vacations. If the orbit is empty at the end of the Jth vacation, the server remains idle for new arrivals in the system. At a vacation completion epoch, the orbit is nonempty and the server waits for the tasks in the orbit or for a new arrival. It is assumed that vacation time V has distribution function $V(t)$, LST $V^*(\theta)$, and finite kth moment $v^{(k)}, k = 1, 2$.
8. ***Breakdown Process:*** While the server is busy with any type of service, it may break down at any time. The breakdowns follow exogenous Poisson processes with rates α_1 for FTS and α_2 for STS, which are referred to by the authors as some sort of disaster during FTS and STS periods, respectively.
9. ***Repair Process:*** As soon as the server breaks down, it is sent for repair, and service provided to the primary tasks stops until the repair is completed. The task that was receiving service before occurrence of server breakdown will have to wait for the duration of repair time at which time it will receive the remaining of its service. The repair time is denoted by G_1 for FTS and G_2 for STS. These random variables are generally distributed, denoted by $G_i(t), i = 1, 2$, LST$G_i^*(\theta)$ and finite kth moment $g_i^{(k)}, i, k = 1, 2$.
10. Various stochastic processes involved in the system are assumed to be independent of each other.

The model as described above has potential application in transferring E-mails. In SMTP, mail system is used to deliver the messages between mail servers. When a mail transfer program contacts a server on a remote machine, it forms a transmission control protocol (TCP) connection over which it communicates. Once the connection is in place, the two programs follow SMTP that allows the sender to identify it, specify a recipient, and transfer an E-mail message. For receiving a group of messages, client applications usually use either the post office protocol (POP) or the Internet message access protocol (IMAP) to access their mail box accounts on a mail server. Typically, contacting a group of messages arrive at the mail server following the Poisson stream.

5.9.2 Model Analysis

We now develop the steady-state difference differential equations for the retrial system by treating the elapsed retrial times, elapsed service times, elapsed vacation times, and elapsed repair times as supplementary variables. Then, we derive the pgf for the server states, the pgf for the number of tasks in the system, and orbit by using the supplementary variable method.

In the steady state, it is assumed that

$$\begin{cases} R(0) = 0, \quad R(\infty) = 1, \\ S_i(0) = 0, \quad S_i(\infty) = 1, \\ V_j(0) = 0, \quad V_j(\infty) = 1, \\ H(0) = 0, \text{ and } \quad H(\infty) = 1, \quad \text{for} \quad i = 1, 2, j = 1, 2, \cdots, J, \\ \qquad\qquad\qquad\qquad \text{are continuous at } x = 0 \text{ and} \\ G_i(0) = 0, \quad G_i(\infty) = 1, \text{ and } i = 1, 2, j = 1, 2, \cdots, J, \\ \qquad\qquad\qquad\qquad \text{are continuous at } y = 0. \end{cases} \tag{5.282}$$

Hence, for $i = 1, 2$, the functions $\alpha(x)$, $\mu_i(x)$, $\gamma(x)$, $\eta(x)$, and $\xi_i(y)$ are the conditional completion rates for retrial, service on both types, on vacation and repair on both types, respectively. These rates are as follows:

$$\alpha(x)dx = \frac{dR(x)}{1 - R(x)}, \mu_i(x)dx = \frac{dS_i(x)}{1 - S_i(x)}, \gamma(x)dx = \frac{dV(x)}{1 - V(x)}, \eta(x)dx = \frac{dH(x)}{1 - H(x)},$$

$$\text{and } \xi_i(y)dx = \frac{dG_i(y)}{1 - G_i(y)}. \tag{5.283}$$

In addition, we denote the following:

$R^0(t)$, the elapsed retrial times,
$S_i^0(t)$, service times on both types,
$V_j^0(t)$, vacation times,
$H^0(t)$, repair times on starting failure server at time t, and
$G_i^0(t)$, repair times at time t on both types, for $i = 1, 2$, and $j = 1, 2, \ldots, J + 3$.

We also note that the states of the system at time t can be described by the bivariate Markov process $\{C(t), N(t), t \geq 0\}$, where $C(t)$ denotes the server state $0, 1, 2, \ldots, J + 3$, depending on the server is idle, busy on FTS or STS, repair on starting failure server, repair on FTS or STS and first vacation, second vacation, ..., Jth, vacation; and $N(t)$ denotes the number of tasks in the orbit.

Here are some special cases:

(i) If $C(t) = 0$, and $N(t) > 0$, $R^0(t)$ represents the elapsed retrial time.
(ii) If $C(t) = 1$ and $N(t) \geq 0$, then $S_i^0(t)$, corresponding to the elapsed time of the task being served on FTS (STS), for $i = 1, 2$.

(iii) If $C(t) = 2$ and $N(t) \geq 0$, then $H^0(t)$, corresponding to the elapsed time of the failure server being repaired.
(iv) If $C(t) = 3$ and $N(t) \geq 0$, then $G_i^0(t)$, corresponding to the elapsed time of the server being repaired on FTS (STS) $i = 1, 2$.
(v) If $C(t) = 3$ and $N(t) \geq 0$, then $V_j^0(t)$, corresponding to the elapsed 1st vacation time.
(vi) If $C(t) = 4$ and $N(t) \geq 0$, then $V_j^0(t)$, corresponding to the elapsed jth vacation time.

Let $t_1, t_2, \ldots t_n$ denote the epochs at each either type 1 or 2 service completion occurs, that is, either a vacation period or repair period ends. The sequence of random vectors $Z_n = \{C(t_n+), N(t_n+)\}$ forms a Markov chain, which is embedded in the retrial queuing system. The authors showed that $\{Z_n, n \in \mathbb{N}\}$ is ergodic only if $\rho < 1$. Thus, the system will be stable when

$$\rho = X^{[1]}[1 - R^*(\lambda)] + \overline{\omega} \tag{5.284}$$

and

$$\overline{\omega} = \alpha[\theta - (\theta - p)[p_1 S_1^*(\delta) + p_2 S_2^*(\delta)]] + \overline{\alpha}(\lambda b X^{[1]} h^{(1)} + 1) \\ + \frac{\alpha \lambda b X^{[1]}}{\delta}[p_1(1 + \alpha_1 g_1^{(1)})[1 - S_1^*(\delta)] + p_2(1 + \alpha_2 g_2^{(1)})[1 - S_2^*(\delta)]]. \tag{5.285}$$

For the process $\{N(t), t \geq 0\}$, we define the probabilities

$$P_0(t) = P\{C(t) = 0, N(t) = 0\} \tag{5.286}$$

and the pdfs for $i = 1, 2, t \geq 0, (x, y) \geq 0$, and $n \geq 0$ as follows:

$$\begin{cases} P_n(x,t)dx = P\{C(t) = 0, N(t) = n, x \leq R^0(t) < x + dx\}, \\ \Pi_{i,n}(x,t)dx = P\{C(t) = 1, N(t) = n, x \leq S_i^0(t) < x + dx\}, \\ Q_n(x,t)dx = P\{C(t)=2, N(t) = n, x \leq H^0(t) < x + dx\}, \\ R_{i,n}(x,y,t)dx = P\left\{ C(t) = 3, N(t) = n, y \leq G_i^0(t) < y + \frac{dy}{S_i^0(t)=x} \right\}, \\ \Omega_{j,n}(x,t)dx = P\{C(t) = j + 3, N(t) = n, x \leq V_j^0(t) < x + dx\}, 1 \leq j \leq J. \end{cases} \tag{5.287}$$

The following probabilities are used in subsequent sections:

$P_0(t)$: the probability that the system is empty at time t.

$P_n(x, t)$: the probability that at time t there are exactly n tasks in the orbit and the elapsed retrial time of the test task undergoing retrial lying in between x and $x + dx$.

$P_{i,n}(x,t), i = 1,2$: the probability that at time t there are exactly n tasks in the orbit and the elapsed service time of the test task undergoing service lying in between x and $x + dx$ in their respective types.

$Q_n(x,t)$: the probability that at time t there are exactly n tasks in the orbit and the elapsed repair time of server lying in between x and $x + dx$ on the failure server.

$R_{i,n}(x,y,t), i = 1,2$: the probability that at time t there are exactly n tasks in the orbit, the elapsed service time of the test task undergoing service is x and the elapsed repair time of server lying in between y and $y + dy$ in their respective types.

$\Omega_{j,n}(x,t), j = 1,2,\ldots,J$: the probability that at time t there are exactly n tasks in the orbit and the elapsed vacation time of the vacation lying in between x and $x + dx$.

It is assumed that the stability condition is fulfilled and so that for $t \geq 0, x \geq 0, n \geq 1, i = 1,2,$ and, $j = 1,2,\ldots,J$, we write

$$\begin{cases} P_0 = \lim_{t\to\infty} P_0(t), \\ P_n(x) = \lim_{t\to\infty} P_n(x,t), \\ \Pi_{i,n}(x) = \lim_{t\to\infty} \Pi_{i,n}(x,t), \\ Q_n(x) = \lim_{t\to\infty} P_n(x,t), \\ \Omega_{j,n}(x) = \lim_{t\to\infty} \Pi_{i,n}(x,t), \quad \text{and} \\ R_{i,n}(x,y) = \lim_{t\to\infty} R_{i,n}(x,y,t). \end{cases} \tag{5.288}$$

Thus, using the ***supplementary variable*** method, we obtain the system of equations governing the steady-state behavior of the system as follows:

$$\lambda b P_0 = \int_0^\infty \Omega_{J,0}(x)\gamma(x)dx, \tag{5.289}$$

$$\frac{dP_n(x)}{dx} + [\lambda + a(x)]P_n(x) = 0, \quad n \geq 1, \tag{5.290}$$

$$\frac{d\Pi_{i,0}(x)}{dx} + [\lambda + \delta + \alpha_i + \mu_i(x)]\Pi_{i,0}(x) = (\lambda)(1-b)\Pi_{i,0}(x) + \int_0^\infty \xi_i(y)R_{i,0}(x,y)dy,$$
$$n = 0, i = 1,2, \tag{5.291}$$

$$\frac{d\Pi_{i,n}(x)}{dx} + [\lambda + \delta + \alpha_i + \mu_i(x)]\Pi_{i,n}(x) = (\lambda)(1-b)\Pi_{i,n}(x) + \lambda b \sum_{k=1}^{n} \chi_k \Pi_{i,n-k}(x)$$
$$+ \int_0^\infty \xi_i(y)R_{i,n}(x,y)dy, \quad n \geq 1, i = 1,2, \tag{5.292}$$

$$\frac{dQ_0(x)}{dx} + [\lambda + \eta(x)]Q_0(x) = \lambda(1-b)Q_{i,0}(x), \quad n = 0, \tag{5.293}$$

$$\frac{dQ_n(x)}{dx} + [\lambda + \eta(x)]Q_n(x) = \lambda(1-b)Q_{i,n}(x) + \lambda b \sum_{k=1}^{n} \chi_k Q_{i,n-k}(x),$$
$$n \geq 1, \tag{5.294}$$

$$\frac{dR_{i,0}(x,y)}{dy} + [\lambda + \xi_i(y)]R_{i,0}(x,y) = \lambda(1-b)R_{i,0}(x,y), \quad n=0, i=1,2, \tag{5.295}$$

$$\frac{dR_{i,n}(x,y)}{dy} + [\lambda + \xi_i(y)]R_{i,n}(x,y) = \lambda(1-b)R_{i,n}(x,y), +\lambda\sum_{k=1}^{n}\chi_k R_{i,n-k}(x,y),$$
$$n \geq 1, i=1,2, \tag{5.296}$$

$$\frac{d\Omega_{j,0}(x)}{dx} + [\lambda + \gamma(x)]\Omega_{j,0}(x) = \lambda(1-b)\Omega_{j,0}(x), \quad n=0, j=1,2,\ldots,J, \tag{5.297}$$

$$\frac{d\Omega_{j,n}(x)}{dx} + [\lambda + \gamma(x)]\Omega_{j,n}(x) = \lambda(1-b)\Omega_{j,n}(x) + \lambda b\sum_{k=1}^{n}\chi_k\Omega_{j,n-k}(x),$$
$$n \geq 1, j=1,2,\ldots,J. \tag{5.298}$$

At $x=0$ and $y=0$, the steady-state boundary conditions are as follows:

$$\begin{aligned} P_n(0) &= \sum_{j=1}^{J}\int_0^\infty \Omega_{j,n}(x)\gamma(x)dx + \int_0^\infty Q_n(x)\eta(x)dx \\ &+ \theta\delta\left(\int_0^\infty \Pi_{1,n-1}(x)dx + \int_0^\infty \Pi_{2,n-1}(x)dx\right) \\ &+ (1-\theta)\delta\left(\int_0^\infty \Pi_{1,n}(x)dx + \int_0^\infty \Pi_{2,n}(x)dx\right) \\ &+ p\left(\int_0^\infty \Pi_{1,n-1}(x)\mu_1(x)dx + \int_0^\infty \Pi_{2,n-1}(x)\mu_2 dx\right), \quad n \geq 1, \end{aligned} \tag{5.299}$$

$$\Pi_{i,0}(0) = \alpha p_i\left(\int_0^\infty P_1(x)a(x)dx + \lambda b\chi_1 P_0\right), \quad n=0, i=1,2, \tag{5.300}$$

$$\Pi_{i,n}(0) = \alpha p_i\left(\int_0^\infty P_{n+1}(x)a(x)dx + \lambda\sum_{k=1}^{n}\chi_k\int_0^\infty P_{n-k+1}(x)dx + \lambda b\chi_{n+1}P_0\right),$$
$$n \geq 1, i=1,2, \tag{5.301}$$

$$Q_n(0) = \overline{\alpha}\left(\int_0^\infty P_n(x)a(x)dx + \lambda\sum_{k=1}^{n}\chi_k\int_0^\infty P_{n-k}(x)dx + \lambda b\chi_n P_0,\right), \quad n \geq 2, \tag{5.302}$$

$$\Omega_{1,n}(0) = q\left(\int_0^\infty \Pi_{1,0}(x)\mu_1(x)dx + \int_0^\infty \Pi_{2,0}(x)\mu_2(x)dx\right)$$

$$+ (1-\theta)\delta \left(\int_0^\infty \Pi_{1,0}(x)dx + \int_0^\infty \Pi_{2,0}(x)dx \right), \quad n = 0, \tag{5.303}$$

$$\Omega_{j,n}(0) = \begin{cases} \int_0^\infty \Omega_{j-1,n}(x)\gamma(x)dx, & n = 0, j = 2, 3, \dots, J, \\ 0 & n \geq 1, \end{cases} \tag{5.304}$$

$$P_{i,n}(x, 0) = \alpha_i \Pi_{i,n}(x), \quad n \geq 1, i = 1, 2. \tag{5.305}$$

The normalizing condition is

$$P_0 + \sum_{j=1}^{J} \left(\int_0^\infty P_n(x)dx + \int_0^\infty Q_n(x)dx \right) + \sum_{n=0}^{\infty} \sum_{j=1}^{J} \int_0^\infty \Omega_{j,n}(x)dx$$

$$+ \sum_{n=0}^{\infty} \sum_{i=1}^{2} \left(\int_0^\infty \Pi_{i,n}(x)dx + \int_0^\infty \int_0^\infty R_{i,n}(x)dxdy \right) = 1. \tag{5.306}$$

5.9.3 Steady-State System Analysis

Using pgf method, we will solve the system discussed above and obtain the steady-state solution of the retrial system. Thus, as in (5.281), we define the generating functions for $|z| \leq 1$ and $i = 1, 2$, as follows:

$$P(x, z) = \sum_{n=1}^{\infty} P_n(x)z^n, \quad P(0, z) = \sum_{n=1}^{\infty} P_n(0)z^n, \quad \Pi_i(0, z) = \sum_{n=0}^{\infty} \Pi_{i,n}(0)z^n,$$

$$Q(x, z) = \sum_{n=1}^{\infty} Q_n(x)z^n, \quad Q(0, z) = \sum_{n=1}^{\infty} Q_n(0)z^n, \quad \Omega_j(x, z) = \sum_{n=1}^{\infty} \Omega_{j,n}(x)z^n,$$

$$\Omega_j(0, z) = \sum_{n=0}^{\infty} \Omega_{j,n}(0)z^n, \quad R(x, y, z) = \sum_{n=0}^{\infty} R_{i,n}(x, y)z^n,$$

$$R(x, 0, z) = \sum_{n=0}^{\infty} R_{i,n}(x, 0)z^n, \quad X(z) = \sum_{n=1}^{\infty} \chi_n z^n. \tag{5.307}$$

Now, multiplying the steady-state equations and steady-state boundary conditions (5.289)–(5.305) by z^n and summing over $n, n = 0, 1, 2, \dots,$ for $i = 1, 2$ and $j = 1, 2, \dots, J$, we obtain

$$\frac{\partial P(x, z)}{\partial x} + [\lambda + a(x)]P(x, z) = 0, \tag{5.308}$$

$$\frac{\partial \Pi_i(x, z)}{\partial x} + \{\lambda b[(1 - X(z)] + \delta + \alpha_i + \mu_i(x)\}\Pi_i(x, z) = \int_0^\infty \xi_i(y)R_i(x, y, z)dy, \tag{5.309}$$

$$\frac{\partial Q(x,z)}{\partial x} + \{[\lambda b(1 - X(z)] + \eta(x)\} Q(x,z) = 0, \tag{5.310}$$

$$\frac{\partial R_i(x,y,z)}{\partial y} + \{[\lambda b(1 - X(z)] + \xi_i(y)\} R_i(x,y,z) = 0, \tag{5.311}$$

$$\frac{\partial \Omega_j(x,z)}{\partial x} + \{[\lambda b(1 - X(z)] + \gamma(x)\} Q_j(x,z) = 0, \tag{5.312}$$

$$\begin{aligned} P(0,z) = {} & \sum_{n=1}^{\infty} \int_0^{\infty} \Omega_j(x,z)\gamma(x)dx + \delta(1 - \theta + \theta z) \\ & \times \left(\int_0^{\infty} \Pi_1(x,z)dx + \int_0^{\infty} \Pi_2(x,z)dx \right) + \int_0^{\infty} Q(x,z)\eta(x)dx \\ & + (pz + q) \left(\int_0^{\infty} \Pi_1(x,z)\mu_1 dx + \int_0^{\infty} \Pi_2(x,z)dx R_{i,n}(x)\mu_2 dx \right) \\ & - \sum_{j=1}^{J} \Omega_{j,0}(0) - \lambda b P_0, \quad n \geq 1, \end{aligned} \tag{5.313}$$

$$\Pi_i(0,z) = \alpha p_i \left[\frac{1}{z} \int_0^{\infty} P(x,z)a(x)dx + \frac{\lambda X(z)}{z} \int_0^{\infty} P(x,z)dx + \frac{\lambda X(z) P_0}{2} \right], \quad i = 1, 2, \tag{5.314}$$

$$Q(0,z) = \overline{\alpha} \left[\int_0^{\infty} P(x,z)a(x)dx + \lambda X(z) \int_0^{\infty} P(x,z)dx + \lambda b X(z) P_0 \right], \tag{5.315}$$

$$R_i(x,0,z) = \alpha_i \Pi_i(x,z). \tag{5.316}$$

Solving the partial differential equations (5.308)–(5.312) yields the following:

$$P(x,z) = P(0,z)[1 - R(x)]e^{-\lambda x}, \tag{5.317}$$

$$\Pi_i(x,z) = \Pi_i(0,z)[1 - S_i(x)]e^{-A_i(z)x}, \quad i = 1, 2, \tag{5.318}$$

$$Q(x,z) = Q(0,z)[1 - H(x)]e^{-b(z)x}, \tag{5.319}$$

$$\Omega_j(x,z) = \Omega_j(0,z)[1 - V(x)]e^{-b(z)x}, \quad j = 1, 2, \ldots, J, \tag{5.320}$$

$$R_i(x,y,z) = R_i(x,0,z)[1 - G_i(y)]e^{-b(z)y}, \quad i = 1, 2, \tag{5.321}$$

where

$$A_i(z) = \lambda b[1 - X(z)] + \delta + \alpha_i[1 - G_i^*[b(z)]] \text{ and } b(z) = \lambda b[1 - X(z)]. \tag{5.322}$$

From (5.297) we have

$$\Omega_{j,0}(x) = \Omega_{j,0}(0)[1 - V(x)]e^{-\lambda b x}, \quad j = 1, 2, \ldots, J. \tag{5.323}$$

Multiplying Equation (5.323) by $\gamma(x)$ for $j = J$ and integrating with respect to x, from 0 to ∞, then from (5.289), we have,

$$\Omega_{J,0}(0) = \frac{\lambda b P_0}{V^*(\lambda b)}. \tag{5.324}$$

From Equation (5.324) and solving (5.304) and (5.323) over the range $j = J - 1$, $J - 2, \dots, 1$, after some simplifications, we have

$$\Omega_{j,0}(0) = \frac{\lambda b P_0}{[V^*(\lambda b)]^{J-j+1}}, \quad j = 1, 2, \dots, J - 1. \tag{5.325}$$

From (5.304), (5.323), and (5.325), we obtain

$$\Omega_j(0, z) = \frac{\lambda b P_0}{[V^*(\lambda b)]^{J-j+1}}, \quad j = 1, 2, \dots, J. \tag{5.326}$$

Integrating Equation (5.323) from 0 to ∞ and using (5.324) and (5.325), we finally have

$$\Omega_{j,0}(0, z) = \frac{P_0[1 - V^*(\lambda b)]}{[V^*(\lambda b)]^{J-j+1}}, \quad j = 1, 2, \dots, J - 1. \tag{5.327}$$

It should be noted that, $\Omega_{j,0}$ represents the steady-state probability that no task appears while the server is on the jth vacation.

Now, let us define Ω_0 as the steady-state probability that no task appears in the system while the server is on vacation. Then,

$$\Omega_0 = \frac{P_0\{1 - [V^*(\lambda)]^J\}}{b[V^*(\lambda)]^J}. \tag{5.328}$$

Inserting Eqs. (5.318)–(5.320) and (5.326) in (5.327) yields

$$\begin{aligned} P(0, z) = {} & \lambda b P_0[N(z) - 1] + Q(0, z)H^*[b(z)] \\ & + (pz + q)\{\Pi_1(0, z)S_1^*[A_1(z)] + \Pi_2(0, z)S_1^*[A_1(z)]\} \\ & + \frac{\delta(1 - \theta + \theta z)}{A_1(z)A_2(z)} \left\{A_2(z)\pi_1(0, z)\left[1 - S_1^*[A_1(z)]\right]\right. \\ & \left. + A_1(z)\pi_2(0, z)\left[1 - S_2^*[A_2(z)]\right]\right\}, \end{aligned} \tag{5.329}$$

where

$$N(z) = \frac{1 - [V^*(\lambda b)]^J}{[V^*(\lambda b)]^J[1 - V^*(\lambda b)]}[V^*(\lambda b[1 - X(z)) - 1]. \tag{5.330}$$

Inserting Eqs. (5.317), (5.318) in (5.314), after some manipulations, we obtain

$$\Pi_i(0, z) = \alpha p_i \left\{ \frac{P(0, z)}{z}[R^*(\lambda) + X(z)[1 - R^*(\lambda)]] + \frac{\lambda b X(z)}{z} P_0 \right\}, \quad i = 1, 2. \tag{5.331}$$

Inserting Eqs (5.317), (5.318) in (5.315), after some manipulations, we obtain

$$Q(0,z) = \overline{\alpha}\{P(0,z)[R^*(\lambda) + X(z)[1 - R^*(\lambda)]] + \lambda bX(z)P_0\}. \tag{5.332}$$

Using Equation (5.318) in Equation (5.316), we have

$$R_i(x,0,z) = \alpha_i\{\Pi_i(0,z)[1 - S_i(x)]e^{-A_i(z)x}\}, \quad i = 1,2. \tag{5.333}$$

Using (5.331) and (5.332) in (5.329), we have

$$P(0,z) = \frac{N_r(z)}{D_r(z)}, \tag{5.334}$$

where

$$\begin{aligned} N_r(z) = \lambda bP_0\{(z[N(z) - 1] &\\ + X(z)[\alpha(pz+q)[p_1S_1^*(A_1(z)) + p_2S_2^*(A_2(z))] + \overline{\alpha}zH^*(b(z))])A_1(z)A_2(z) &\\ + X(z)[\alpha\delta(1-\theta+\theta z)[p_1A_2(z)[1 - S_1^*(A_1(z))] + p_2A_1(z)[1 - S_2^*(A_2(z))]]\}. & \end{aligned} \tag{5.335}$$

$$\begin{aligned} D_r(z) = \{z - [\alpha(pz+q)(p_1S_1^*(A_1(z)) + p_2S_2^*(A_2(z))) + \overline{\alpha}zH^*(b(z))] &\\ \times [R^*(\lambda) + X(z)[1 - R^*(\lambda)]]\}A_1(z)A_2(z) &\\ - \{R^*(\lambda) + X(z)[1 - R^*(\lambda)]\} &\\ \times \{\alpha\delta(1-\theta+\theta z)[p_1A_2(z)[1 - S_1^*(A_1(z))] + p_2A_1(z)[1 - S_2^*(A_2(z))]]\}. & \end{aligned} \tag{5.336}$$

Substituting (5.334) into (5.331) yields

$$\Pi_i(0,z) = \frac{\alpha\lambda bP_0p_i\{[N(z) - 1][R^*(\lambda) + X(z)[1 - R^*(\lambda)]] + X(z)\}A_1(z)A_2(z)}{D_r(z)}, \quad i = 1,2. \tag{5.337}$$

Substituting (5.334) into (5.332) yields

$$Q(0,z) = \frac{\overline{\alpha}z\lambda bP_0\{[N(z) - 1][R^*(\lambda) + X(z)[1 - R^*(\lambda)]] + X(z)\}A_1(z)A_2(z)}{D_r(z)}. \tag{5.338}$$

Substituting (5.337) into (5.333) yields

$$R_i(x,0,z) = \frac{\alpha\alpha_i\lambda bP_0p_i\{[N(z) - 1][R^*(\lambda) + X(z)[1 - R^*(\lambda)] + X(z)\}A_1(z)A_2(z)}{D_r(z)}, \quad i = 1,2. \tag{5.339}$$

Thus, the pgfs $P(x,z), \Pi_1(x,z), H(x,z), R_1(x,yz)$, and $R_2(x,y,z)$ can be obtained by using (5.334)–(5.339) in (5.317)–(5.321).

The marginal orbit size distributions due to the system state of the server are found below:

Theorem 5.9.1 When $\rho < 1$, and $D_r(z)$ is given in (5.336), for $i = 1,2$, the joint steady-state distributions of the number of tasks in the system when server is idle, busy on both types, on vacation, under repair on starting failure server and under repair on both types are given by the following:

$$P(z) = \frac{N_{rP}(z)}{D_r(z)}, \tag{5.340}$$

where

$$\begin{aligned} N_{rP}(z) = b[1-R^*(\lambda)]P_0\{&z[N(z)-1] \\ &+ X(z)[\alpha(pz+q)[p_1S_1^*(A_1(z)) + p_2S_2^*(A_2(z))] + \overline{\alpha}zH^*(b(z))]\}A_1(z)A_2(z) \\ &+ X(z)[\alpha\delta(1-\theta+\theta z)[p_1A_2(z)[1-S_1^*(A_1(z))] + p_2A_1(z)[1-S_2^*(A_2(z))]], \end{aligned} \tag{5.341}$$

$$\Pi_1(z) = \frac{N_{r\Pi_1}(z)}{D_r(z)}, \tag{5.342}$$

where

$$\begin{aligned} N_{r\Pi_1}(z) = {}& \alpha\lambda bP_0p_1[1-S_1^*(A_1(z))] \\ & \times \{[[N(z)-1][R^*(\lambda) + X(z)[1-R^*(\lambda)]] + X(z)]A_2(z)\}, \end{aligned} \tag{5.343}$$

$$\Pi_2(z) = \frac{N_{r\Pi_2}(z)}{D_r(z)}, \tag{5.344}$$

where

$$\begin{aligned} N_{r\Pi_2}(z) = {}& \alpha\lambda bP_0p_2[1-S_2^*(A_2(z))] \\ & \times \{[[N(z)-1][R^*(\lambda) + X(z)[1-R^*(\lambda)]] + X(z)]A_1(z)\}, \end{aligned} \tag{5.345}$$

$$Q(z) = \frac{N_{r_Q}(z)}{b(z)D_r(z)}, \tag{5.346}$$

where

$$\begin{aligned} N_{r_Q}(z) = {}& \overline{\alpha}z\lambda bP_0[1-H^*(b(z))] \\ & \{[[N(z)-1][R^*(\lambda) + X(z)[1-R^*(\lambda)]] + X(z)]A_1(z)A_2(z)\}, \end{aligned} \tag{5.347}$$

$$R_1(z) = \frac{N_{rR_1}(z)}{b(z)D_r(z)}, \tag{5.348}$$

where

$$N_{rR_1} = \alpha\alpha_1\lambda bP_0p_1\{[N(z)-1][R^*(\lambda)+X(z)[1-R^*(\lambda)]+X(z)\}$$
$$\times A_2(z)\{1-S_1^*(A_1(z))][1-G_1^*(b(z)), \quad (5.349)$$

$$R_2(z) = \frac{N_{rR_2}(z)}{b(z)D_r(z)}, \quad (5.350)$$

where

$$N_{rR_2} = \alpha\alpha_2\lambda bP_0P_1\{[N(z)-1][R^*(\lambda)+X(z)[1-R^*(\lambda)]+X(z)\}$$
$$\times A_1(z)\{1-S_2^*(A_2(z))][1-G_2^*(b(z)), \quad (5.351)$$

$$\Omega_j(z) = \frac{P_0\{[V^*(\lambda b(1-X(z)))]-1\}}{[X(z)-1][V^*(\lambda b)]^{J-j+1}}, \quad j=1,2,\ldots,J, \quad (5.352)$$

where

$$P_0 = \frac{1}{\beta}[1-X^{[1]}[1-R^*(\lambda)]-\overline{\omega}], \quad (5.353)$$

where

$$\beta = X^{[1]}[1-R^*(\lambda)](b-1)$$
$$\left(\frac{N'(1)}{X^{[1]}}+\overline{\alpha}\lambda bh^{(1)}+\frac{\alpha\lambda b}{\delta}(p_1[1-\alpha_1g_1^{(1)}][1-S_1^*(\delta)])+(p_2[1-\alpha_2g_2^{(1)}]\right.$$
$$\left.[1-S_2^*(\delta)])+1\right)$$
$$-\alpha\{\theta-1-(\theta-p)[p_1S_1^*(\delta)+p_2S_2^*(\delta)]\}\left\{\frac{N'(1)}{X^{[1]}}+b(1-R^*(\lambda)]+1\right\}, \quad (5.354)$$

$$N'(1) = \frac{\{1-[V^*(\lambda b)]^J\}\lambda bX^{[1]}v^{(1)}}{[V^*(\lambda b)]^J[1-V^*(\lambda b)]}, \quad (5.355)$$

$$A_i(z) = b(z)+\delta+\alpha_i[1-G_1^*(b(z))], \quad (5.356)$$

and

$$b(z) = \lambda b[1-X(z)]. \quad (5.357)$$

Proof: Integrating Eqs (5.317)–(5.320) with respect to x and define the partial pgfs for $i=1,2$, as

$$P(z) = \int_0^\infty P(x,z)dx, \qquad \Pi_i(z) = \int_0^\infty \Pi_i(x,z)dx,$$
$$Q(z) = \int_0^\infty Q(x,z)dx, \qquad \Omega_j(z) = \int_0^\infty \Omega_j(x,z)dx. \quad (5.358)$$

Integrating Equation (5.321) with respect to x and y and define the partial pgfs for $i = 1, 2$, as

$$R_i(x,z) = \int_0^{\infty} R_i(x,y,z)dy \quad \text{and} \quad R_i(z) = \int_0^{\infty} R_i(x,z)dx. \tag{5.359}$$

Since the only unknown is P_0, the probability that the server is idle when there is no task in the orbit, and it can be determined using the normalizing condition $j = 1, 2, \ldots, J$. Thus, by setting $z = 1$ in (5.340)–(5.352) and applying L'Hôpital's rule whenever necessary and we get

$$P_0 + P(1) + \Pi_1(1) + \Pi_2(1) + Q(1) + R_1(1) + R_2(1) + \sum_{j=1}^{J} \Omega_j(1) = 1. \tag{5.360}$$

Theorem 5.9.2 When $\rho < 1$, that is, the stability condition, the steady-state pgf of the number of tasks in the system, denoted by $K_s(z)$, and of the number of tasks in the orbit, denoted by $K_0(z)$, are as follows:

$$K_s(z) = P_0 \left[\frac{N_{r1}(z)}{D_{r1}(z)} + \frac{N_{r2}(z)}{D_{r1}(z)} + \frac{N_{r3}(z)}{D_{r1}(z)} \right], \tag{5.361}$$

$$K_0(z) = P_0 \left[\frac{N_{r4}(z)}{D_{r1}(z)} + \frac{N_{r2}(z)}{D_{r1}(z)} + \frac{N_{r3}(z)}{D_{r1}(z)} \right], \tag{5.362}$$

where P_0 is given in (5.353) and

$$\begin{aligned} N_{r1}(z) &= z\{\{\alpha[1 - [p_1 S_1^*(A_1(z)) + p_2 S_2^*(A_2(z))] + \overline{\alpha} z[1 - H^*(b(z))]]\}A_1(z)A_2(z) \\ &\quad - \{\alpha\delta[p_1 A_2(z)\{1 - S_1^*(A_1(z))] + p_2 A_1(z)\{1 - S_2^*(A_2(z))\}]\} \\ &\quad \times \{[N(z) - 1][R^*(\lambda) + X(z)[1 - R^*(\lambda)]] + X(z)\}\}, \end{aligned} \tag{5.363}$$

$$\begin{aligned} N_{r2}(z) &= -N(z)\{zA_1(z)A_2(z) \\ &\quad - [\alpha(pz + q)[p_1 S_1^*(A_1(z)) + p_2 S_2^*(A_2(z))] + \overline{\alpha} z H^*(b(z))] \\ &\quad \times [R^*(\lambda) + X(z)[1 - R^*(\lambda)]] - [R^*(\lambda) + X(z)[1 - R^*(\lambda)]] \\ &\quad \times [\alpha\delta(1 - \theta + \theta z)\{p_1 A_2(z)[1 - S_1^*(A_1(z))] + p_2 A_1(z)[1 - S_2^*(A_2(z))]\}]\}, \end{aligned} \tag{5.364}$$

$$\begin{aligned} N_{r3}(z) &= [1 - X(z)]\{zA_1(z)A_2(z)[1 + b[1 - R^*(\lambda)][N(z) - 1]] \\ &\quad - [R^*(\lambda) + X(z)(1 - b)[1 - R^*(\lambda)]][\{\alpha(pz + q)[p_1 S_1^*(A_1(z)) \\ &\quad + p_2 S_2^*(A_2(z))\} + \overline{\alpha} z H^*(b(z))]A_1(z)A_2(z) \\ &\quad + [\alpha\delta(1 - \theta + \theta z)\{p_1 A_2(z)[1 - S_1^*(A_1(z))] + p_2 A_1(z)[1 - S_2^*(A_2(z))]]\}, \end{aligned} \tag{5.365}$$

$$\begin{aligned} N_{r4}(z) &= \{\alpha[1 - [p_1 S_1^*(A_1(z)) + p_2 S_2^*(A_2(z))] + \overline{\alpha} z[1 - H^*(b(z))]]\} A_1(z) A_2(z) \\ &\quad - \{\alpha\delta[p_1 A_2(z)\{1 - S_1^*(A_1(z))] + p_2 A_1(z)\{1 - S_2^*(A_2(z))\}]\} \\ &\quad \times \{[N(z) - 1][R^*(\lambda) + X(z)[1 - R^*(\lambda)]] + X(z)\}, \end{aligned} \tag{5.366}$$

$$\begin{aligned} D_{r1}(z) &= [1 - X(z)]\{zA_1(z)A_2(z) - [\alpha(pz + q)(p_1 S_1^*(A_1(z)) + p_2 S_2^*(A_2(z))) \\ &\quad + \overline{\alpha} z H^*(b(z))][R^*(\lambda) + X(z)[1 - R^*(\lambda)]]\} A_1(z) A_2(z) \\ &\quad - \{R^*(\lambda) + X(z)[1 - R^*(\lambda)]\} \\ &\quad \times \{\alpha\delta(1 - \theta + \theta z)[p_1 A_2(z)[1 - S_1^*(A_1(z))] + p_2 A_1(z)[1 - S_2^*(A_2(z))]]\}. \end{aligned} \tag{5.367}$$

Proof: Relations (5.361) and (5.362) can be obtained by using the following equations:

$$K_s(z) = P_0 + P(z) + z\left\{ Q(z) + \sum_{i=1}^{2} [\Pi_i(z) + R_i(z)] \right\} + \sum_{j=1}^{J} \Omega_j(z) \tag{5.368}$$

and

$$K_0(z) = P_0 + P(z) + Q(z) + \sum_{i=1}^{2} [\Pi_i(z) + R_i(z)] \sum_{j=1}^{J} \Omega_j(z). \tag{5.369}$$

Substituting (5.340)–(5.352) into the above results and direct calculation proof will be completed.

5.9.4 Performance Measures

In this section, we derive system performance measures and later, we will numerically validate these results.

It should be noted that that (5.353) gives the steady-state probability that the server is idle, but available in the system. It follows from (5.340)–(5.352) that the probabilities of the server state are as given in Theorem 5.9.3 below, proof of which is a result of direct calculations:

Theorem 5.9.3 Under the stability condition $\rho < 1$, the following probabilities hold true:

$$P = P(1) = \frac{1}{\beta}\{b[1 - R^*(\lambda)][N'(1) + X^{[1]} - 1 + \overline{\omega}]\}, \tag{5.370}$$

where P denotes the steady-state probability that the server is idle during the retrial time,

$$\Pi_i = \Pi_i(1) = \frac{1}{\beta}\left[\frac{1 - S_i^*(\delta)}{\delta}\right]\{\alpha\lambda b p_i[N'(1) + X^{[1]} - 1 + \overline{\omega}]\}, \quad i = 1, 2, \tag{5.371}$$

where Π_1 and Π_2 denote the steady-state probability that the server is busy on the ith, $i = 1, 2$, type service with positive task,

$$F_{\text{loss}} = \delta(1-\theta)(\Pi_1 + \Pi_2) = \frac{1}{\beta}[\alpha\lambda b(1-\theta)[N'(1) + X^{[1]}R^*(\lambda)] \times \{p_1[1 - S_1^*(\delta)] + p_2[1 - S_2^*(\delta)]\}], \tag{5.372}$$

where F_{Loss} denotes the frequency of the task loss due to arrival of negative tasks,

$$Q = Q(1) = \frac{1}{\beta}\{\overline{\alpha}\lambda b p_1 h^{(1)}[N'(1) + X^{[1]}R^*(\lambda)]\}, \tag{5.373}$$

where Q denotes the steady-state probability that the server is on starting failure,

$$\Omega = \sum_{j=1}^{J} \Omega_j(1) = \frac{1}{\beta}\left\{\frac{N'(1)}{X^{[1]}}[1 - X^{[1]}[1 - R^*(\lambda)] - \overline{\omega}]\right\}, \tag{5.374}$$

where Ω denotes the steady-state probability that the server is on vacation,

$$R_i = R_i(1) = \frac{1}{\beta}\left[\frac{1 - S_i^*(\delta)}{\delta}\right]\{\alpha\lambda b p_i g_i^{(1)}[N'(1) + X^{[1]}R^*(\lambda)]\}, \quad i = 1, 2, \tag{5.375}$$

where R_1 and R_2 denote the steady-state probability that the server is under repair time on the ith type service.

Theorem 5.9.4 Let L_s, L_q, W_s, and W_q denote the mean of the following quantities: (1) number of tasks in the system, (2) the number of tasks in the orbit, (3) the time a task spends in the system, and (4) the time a task spends in the orbit, respectively. Then under the stability condition, $\rho < 1$, we have

$$L_q = P_0\left\{\frac{N_{r'''q}(1)D_{r''q}(1) - N_{r''q}(1)D_{r'''q}(1)}{3[D_{r''q}(1)]^2}\right\}, \tag{5.376}$$

where

$$\begin{aligned} N_{r''q}(1) = &-2\delta^2\{[1 - R^*(\lambda)]\{N'(1)[b - X^{[1]}] + (X^{[1]})^2(b-1)[1 - \overline{\alpha}\lambda b h^{(1)}]\} \\ &+ \{N'(1) + X^{[1]}[1 - [1 - R^*(\lambda)]]\} \\ &[1 - \theta + (\theta - p)[p_1 S_1^*(\delta) + p_2 S_2^*(\delta)]]\} \\ &- 2\alpha\delta\{[p_1 A_1'[1 - S_1^*(\delta)] + p_2 A_2'(1)[1 - S_2^*(\delta)]\}, \end{aligned} \tag{5.377}$$

$$\begin{aligned} N_{r'''q}(1) = &-6\alpha X^{[1]}(1-b)[1 - R^*(\lambda)]A_1'(1)A_2'(1)\{1 - [p_1 S_1^*(\delta) + [p_2 S_2^*(\delta)]]\} \\ &-3\delta^2 N''(1)\{1 - X^{[1]}(1-b)[1 - R^*(\lambda)] - \alpha[\theta - (\theta - p) \times (p_1 S_1^*(\delta) + p_2 S_2^*(\delta))]\} \end{aligned}$$

$$
\begin{aligned}
&- 3\delta^2 N'(1)\{[1 - R^*(\lambda)][\alpha - (1 - b)](2X^{[1]} + X^{[2]}) - 2\alpha(p + X^{[1]}[1 - R^*(\lambda)]) \\
&(p_1 S_1^*(\delta)A_1'(1) + p_2 S_2^*(\delta)A_2'(1) \\
&- \alpha[1 - R^*(\lambda)][p_1 S_1^*(\delta) + p_1 S_1^*(\delta)][2pX^{[1]} + X^{[2]}]\} \\
&- 3\delta N'(1)\{2[A_1'(1) + A_2'(1)]\{(b - \overline{\alpha}X^{[1]})[1 - R^*(\lambda)] - \alpha(p + X^{[1]}[1 - R^*(\lambda)] \\
&[p_1 S_1^*(\delta) + p_1 S_1^*(\delta)]\} \\
&+ 2\alpha X^{[1]}[1 - R^*(\lambda)][p_1 A_1'[1 - S_1^*(\delta)] + p_2 A_2'[1 - S_1^*(\delta)]]\} \\
&+ 6\alpha\delta\theta X^{[1]}\{1 - b[1 - R^*(\lambda)]\}([p_1 A_1'(1)[1 - S_1^*(\delta)] + p_2 A_2'[1 - S_2^*(\delta)]) \\
&+ 6\alpha\delta X^{[1]}\{1 - b[1 - R^*(\lambda)]\}(A_1'(1) + A_2'(1) - A_1'(1)A_2'(1))[p_1 S_1^*(\delta) + p_2 S_2^*(\delta)] \\
&- 2\delta[A_1'(1) + A_2'(1)]\{3X^{[1]}(1 - b)[1 - R^*(\lambda)] - 1\} \\
&+ 3\delta X^{[1]}(1 - b)[1 - R^*(\lambda)] \times \{(X^{[1]} + \alpha[p_1 S_1^*(\delta)A_1' + p_2 S_2^*(\delta)A_2'] + \overline{\alpha}bh^{(1)}X^{[1]}) \\
&- \alpha[A_1''(1) + A_2''(1)](1 - [p_1 S_1^*(\delta) + p_2 S_2^*(\delta)])\} \\
&+ 3\delta[1 - R^*(\lambda)][X^{[2]} + 2X^{[1]}(1 - b)] \times (p_1 A_1'(1)[1 - S_1^*(\delta)] + p_2 A_2'(1)[1 - S_2^*(\delta)]) \\
&+ 3\delta^2\{\alpha(\theta - (\theta - p)[p_1 S_1^*(\delta) + p_1 S_1^*(\delta)]) + \overline{\alpha}\lambda\delta^2 bh^{(1)}X^{[1]}\} \times \{2X^{[1]}(1 - b) \\
&[1 - R^*(\lambda)] + X^{[2]}(1 - b[1 - R^*(\lambda)]\} \\
&+ 3\delta^2 X^{[1]}(1 - b)[1 - R^*(\lambda)]\{2X^{[2]} + \overline{\alpha}(2X^{[1]} + \lambda bX^{[1]}[2h^{(1)} + \lambda bX^{[1]}h^{(2)}] + \lambda bX^{[2]})\} \\
&- 3\delta^2(1 - b[1 - R^*(\lambda)])\{X^{[2]} + 2X^{[1]}(\theta - p)[p_1 S_1^*(\delta)A_1'(1) + p_2 S_2^*(\delta)A_2'(1)]\} \\
&- 3\delta^2 R^*(\lambda)X^{[1]}\{\alpha[p_1(S_1^{*''}(\delta)[A_1'(1)]^2 + S_1^{*'}(\delta)A_1''(1) + p_1 S_1^{*''}(\delta)[A_2'(1)]^2 \\
&+ S_2^{*'}(\delta)A_2''(1)] + \overline{\alpha}\lambda\delta^2 X^{[2]}bh^{(1)}\}. \qquad (5.378)
\end{aligned}
$$

$$
D_{r''}(1) = -2X^{[1]}\{1 - X^{[1]}[1 - R^*(\lambda)] - \overline{\omega}\}, \qquad (5.379)
$$

$$
\begin{aligned}
D_{r'''}(1) = {} & 3X^{[1]}\alpha\delta^2 p_1[S_1^{*'}(\delta)[A_1'(1)]^2 + S_1^{*'}(\delta)A_1''(1)] \\
&+ 3X^{[1]}\alpha\delta^2 p_2[S_2^{*''}(\delta)[A_2'(1)]^2 + S_2^{*'}(\delta)A_2''(1)] \\
&+ 6\alpha\delta^2 X^{[1]}(p + X^{[1]}[1 - R^*(\lambda)])(p_1 S_1^*(\delta)A_1'(1) + p_2 S_2^*(\delta)A_2'(1)) \\
&\times 3\delta^2\alpha X^{[1]}[1 - R^*(\lambda)](p_1 S_1^*(\delta) + p_2 S_2^*(\delta))(2pX^{[1]} + X^{[2]}) \\
&+ 3\delta^2\overline{\alpha}X^{[1]}\{\lambda bX^{[1]}(2h^{(1)} + 2h^{(1)} + \lambda bX^{[1]}h^{(2)}) \\
&+ X^{[2]}(\lambda bh^{(1)} + [1 - R^*(\lambda)]) + 2X^{[1]}[1 - R^*(\lambda)](1 + \lambda bX^{[1]}h^{(1)})\} \\
&- 6\delta X^{[1]}[A_1'(1) + A_2'(1)]\{1 - \alpha[p_1 S_1^*(\delta)A_1' + p_2 S_2^*(\delta)A_2' \\
&+ [p + X^{[1]}[1 - R^*(\lambda)]][p_1 S_1^*(\delta)A_1' + p_2 S_2^*(\delta)A_2']] - \overline{\alpha}(\lambda bX^{[1]}h^{(1)} \\
&+ X^{[1]}[1 - R^*(\lambda)] + 1)\}
\end{aligned}
$$

$$- 3\alpha X^{[1]}\{1 - [p_1 S_1^*(\delta) + p_2 S_2^*(\delta)][\delta A_1''(1) + A_2''(1) + 2A_1'(1) + A_2'(1)]\}$$
$$- 3X^{[2]}\{1 - X^{[1]}[1 - R^*(\lambda)] - \overline{\omega}\}. \tag{5.380}$$

$$L_s = P_0 \left[\frac{N_{r'''s}(1) - N_{r''q}(1)}{3[D_{r''q}(1)]^2} \right], \tag{5.381}$$

where

$$N_{r'''s}(1) = N_{r'''q}(1) + 6\delta^2 \{N'(1) + X^{[1]} R^*(\lambda)]\} \tag{5.382}$$
$$\times \left\{ \frac{\alpha}{\delta}(p_1 A_1'(1)[1 - S_1^*(\delta)] + p_2 A_2'(1)[1 - S_2^*(\delta)]) - \overline{\alpha}\lambda b X^{[1]} h^{(1)} \right\}.$$

$$W_s = \frac{L_s}{\lambda E(X)} \text{ and } W_q = \frac{L_q}{\lambda E(X)}, \tag{5.383}$$

where

$$N''(1) = \frac{(\lambda b)^2 E(V^2)\{1 - [V^*(\lambda b)]^J\}}{[1 - V^*(\lambda b)][V^*(\lambda b)]^J}, \tag{5.384}$$

$$S_i^{*'}(\delta) = \int_0^\infty x e^{-\delta x} dS_i(x) \text{ and } S_i^{*''}(\delta) = \int_0^\infty x^2 e^{-\delta x} dS_i(x), \quad i - 1, 2, \tag{5.385}$$

$$A_i'(1) = \lambda b X^{[1]}(1 + \alpha_i g_i^{(1)}), \quad i = 1, 2, \tag{5.386}$$

$$A_1''(1) = \lambda b X^{[2]}(1 + \alpha_1 g_1^{(1)}) + \alpha_1 g_1^{(2)} (\lambda b X^{[1]})^2 \tag{5.387}$$

and

$$A_1''(2) = \lambda b X^{[2]}(1 + \alpha_2 g_2^{(1)}) + \alpha_2 g_2^{(2)} (\lambda b X^{[1]})^2. \tag{5.388}$$

Proof: The mean number of tasks in the orbit, L_q, under steady-state condition, is obtained by differentiating (5.362) with respect to z and evaluating at $z = 1$. Thus, we have:

$$L_q = K_0'(1) = \lim_{z \to 1} \frac{d}{dz} K_0(z). \tag{5.389}$$

The mean number of tasks in the system, L_s, under steady-state condition is obtained by differentiating (5.361) with respect to z and evaluating at $z = 1$, Thus, we have

$$L_s = K_s'(1) = \lim_{z \to 1} \frac{d}{dz} K_s(z). \tag{5.390}$$

The average time a task spends in the system, W_s, and the average time a task spends in the queue, W_q, are found by using the Little's formula:

$$W_s = \frac{L_s}{\lambda} \text{ and } W_q = \frac{L_q}{\lambda}. \tag{5.391}$$

5.9.5 Numerical Illustration

In this section, a numerical illustration is presented, in which the authors have used MATLAB, to illustrate the effect of various parameters in the system performance measures. For retrial times, service times, vacation times, and repair times, three types of distributions, namely exponential, Erlangian, and hyperexponential, have been considered. (These distributions are defined in Chapter 1.) It is also assumed that tasks arrive singly, so that $X^{[1]} = 1$ and $X^{[2]} = 0$.

The following values have been chosen for the needed parameters such that they satisfy the stability condition. They are

μ_1	μ_2	ξ_1	ξ_2	α_1	α_2	γ	c
8	10	6	8	2	1	5	0.8

The computed steady-state characteristics are as follows:

P:	Probability that the server is idle during the retrial time
Π_1	Probability that the server is busy on the first-type service with positive customer
Π_2:	Probability that the server is busy on the second-type service with positive customer
F_{Loss}:	Frequency of customer loss due to arrival of negative customers
Q:	Probability that the server is on starting failure
Ω:	Probability that the server is on vacation
R_1:	Probability that the server is under repair time on the first-type service
R_2:	Probability that the server is under repair time on the second-type service
P_0:	Probability that the server is idle
L_q:	Mean orbit size

Table 5.8 shows that when negative arrival rate, δ, increases, then the probability that server is idle, P_0, increases, the mean orbit size, L_q, increases, and the probability that frequency of the task loss due to arrival of negative task server, F_{Loss}, also increases.

TABLE 5.8 Effect of Negative Arrival Probability, δ, on $P_0 L_q$ and F_{Loss}

Retrial Distribution	*Exponential*			*Erlang Two Stage*			*Hyperexponential*		
δ	P_0	L_q	F_{Loss}	P_0	L_q	F_{Loss}	P_0	L_q	F_{Loss}
				Negative arrival rate					
4.00	0.7988	1.2789	0.0591	0.6885	1.4015	0.0891	0.7908	1.2158	0.0657
5.00	0.8161	1.6095	0.0670	0.7084	1.8551	0.0974	0.8080	1.6184	0.0737
6.00	0.8278	2.0110	0.0736	0.7219	2.5200	0.1038	0.8197	2.2086	0.0802
7.00	0.8363	2.5341	0.0791	0.7317	3.5330	0.1088	0.8281	3.0987	0.0856
8.00	0.8427	3.2411	0.0838	0.7391	5.0698	0.1127	0.8344	4.4235	0.0902

TABLE 5.9 Effect of Service Loss Probability, θ, on P_0, L_q, and P

Retrial Distribution	*Exponential*			*Erlang Two Stage*			*Hyperexponential*		
θ	P_0	L_q	P	P_0	L_q	P	P_0	L_q	P
				Service loss probability					
0.10	0.7261	0.6403	0.0401	0.6202	0.6584	0.0701	0.7204	0.5891	0.0460
0.20	0.7184	0.6667	0.0425	0.6030	0.7149	0.0776	0.7112	0.6209	0.0492
0.30	0.7101	0.6949	0.0450	0.5834	0.7782	0.0861	0.7013	0.6551	0.0526
0.40	0.7012	0.7249	0.0477	0.5611	0.8494	0.0958	0.6905	0.6919	0.0564
0.50	0.6916	0.7570	0.0507	0.5353	0.9298	0.1070	0.6787	0.7316	0.0605

TABLE 5.10 Effect of Failure Probability, $\overline{\alpha}$, on P_0, L_q, and P

Repair Distribution	*Exponential*			*Erlang Two Stage*			*Hyperexponential*		
α	P_0	L_q	P	P_0	L_q	P	P_0	L_q	P
				Successful probability					
0.30	0.6470	1.0183	0.1365	0.4467	4.2181	0.2511	0.6358	1.3797	0.1604
0.40	0.7158	0.6791	0.0963	0.5816	1.6358	0.1711	0.7091	0.8181	0.1123
0.50	0.7549	0.6032	0.0735	0.6550	1.0030	0.1278	0.7502	0.6611	0.0854
0.60	0.7800	0.5913	0.0589	0.7008	0.7807	0.1007	0.776	0.6095	0.0681
0.70	0.7975	0.5984	0.0487	0.7322	0.6900	0.0821	0.7949	0.5937	0.0561

Table 5.9 shows that when service loss probability, θ, increases, then the probability that server is idle P_0, decreases, the mean orbit size, L_q, increases, and the probability that the server is idle during retrial time, P, also increases.

Table 5.10 shows that when the successful arrival probability, α, increases, the probability that server is idle, P_0, increases, then the mean orbit size, L_q, decreases, and the probability that the server is idle during retrial time, P, also decreases.

Table 5.11 shows that when the number of vacations, J, increases, the probability that server is idle, P_0, decreases, then the probability that server is idle during retrial time, P, increases, and the probability that the server is on vacation, Ω, also increases.

Table 5.12 shows that when repair rate on FTS, ξ_1, increases, the probability that the server is idle, P_0, increases, then the mean orbit size, L_q, decreases, and the probability that the server is under repair on FTS, R_1, also decreases.

For the effect of the parameters $\lambda, \delta, p, \alpha, \theta, p_1, \gamma$, and ξ on the system performance measures in two-dimensional graphs are drawn in Figures 5.10 and 5.11.

Figure 5.10 shows that the mean orbit size, L_q, increases for increasing value of the negative arrival rate, δ.

Figure 5.11 shows that the idle probability, P_0, decreases with the increase of the service loss probability, θ.

TABLE 5.11 Effect of Number of Vacations, J, on P_0, P, and Ω.

Vacation Distribution	*Exponential*			*Erlang Two Stage*			*Hyperexponential*		
J	P_0	P	Ω	P_0	P	Ω	P_0	P	Ω
	Number of vacations								
2.00	0.3961	0.1383	0.0126	0.2095	0.2465	0.0275	0.3851	0.1600	0.0103
3.00	0.3845	0.1390	0.0248	0.1975	0.2494	0.0401	0.3754	0.1608	0.0205
4.00	0.3732	0.1396	0.0367	0.1862	0.2522	0.0519	0.3658	0.1615	0.0305
5.00	0.3622	0.1402	0.0482	0.1755	0.2548	0.0631	0.3564	0.1622	0.0403
6.00	0.3516	0.1409	0.0594	0.1654	0.2572	0.0736	0.3471	0.1630	0.0500

TABLE 5.12 Effect of Repair rate on FTS, ξ_1, on P_0, L_q, and R_1

Repair Distribution	*Exponential*			*Erlang Two Stage*			*Hyperexponential*		
ξ_1	P_0	L_q	R_1	P_0	L_q	R_1	P_0	L_q	R_1
	Repair rate on FTS								
6.00	0.4081	1.9758	0.0105	0.2222	2.9702	0.0335	0.3950	2.0571	0.0099
7.00	0.4151	1.9615	0.0090	0.2342	2.9775	0.0287	0.4011	2.0492	0.0085
8.00	0.4204	1.9509	0.0079	0.2432	2.9828	0.0251	0.4056	2.0434	0.0074
9.00	0.4245	1.9427	0.0070	0.2502	2.9869	0.0223	0.4091	2.0389	0.0065

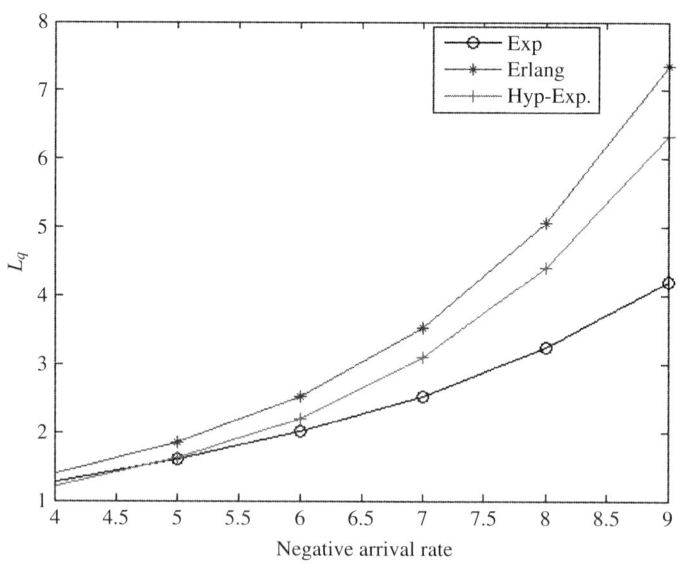

Figure 5.10 The mean orbit size, L_q versus δ.

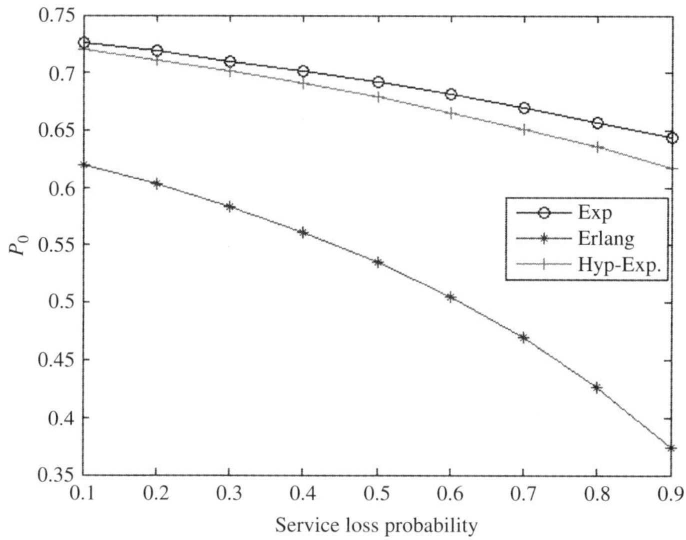

Figure 5.11 The idle probability decreases for the increasing the service loss probability; P_0 versus θ.

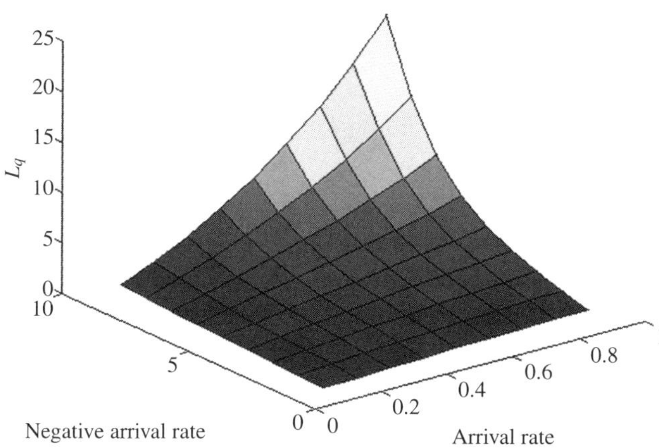

Figure 5.12 The surface displays an upward trend; L_q versus λ and δ.

Figures 5.12–5.15 illustrate three-dimensional graphs. In Figure 5.12, the surface displays an upward trend as expected for increasing the value of the arrival rate, λ, and negative arrival rate, δ, against the mean orbit size, L_q. The mean orbit size, L_q, decreases for increasing value of the feedback probability, p, and balking probability, b, in Figure 5.13. The surface displays an upward trend as expected for increasing the value of the successful service probability, α, and vacation rate, γ, against the idle probability, P_0, in Figure 5.14. In Figure 5.15, the mean orbit size, L_q, decreases for increasing value of the first-type probability, p_1, and repair rate on FTS, ξ_1.

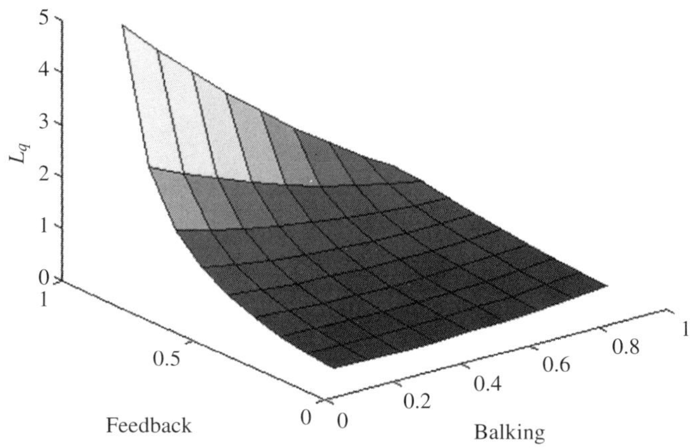

Figure 5.13 Increase of the number of vacations and decrease of the probability of idle, cause increase of the probability of idle during retrial time and increase of probability of that server on vacation; L_q versus p and b.

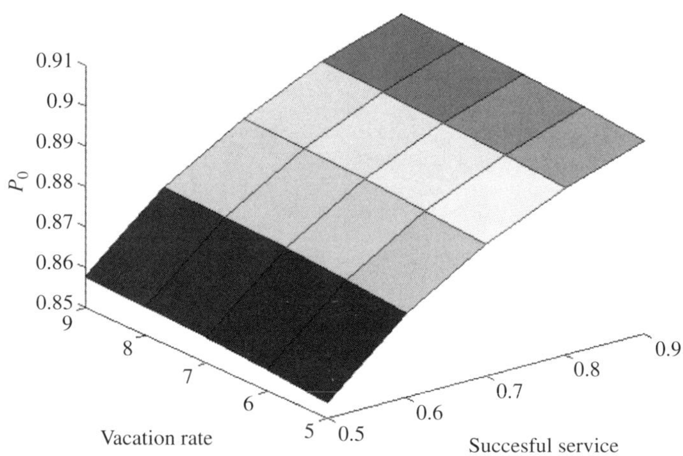

Figure 5.14 Increase of repair rate and the probability of server idle will cause decrease of the mean orbit size and the probability of server under repair; P_0 versus α and γ.

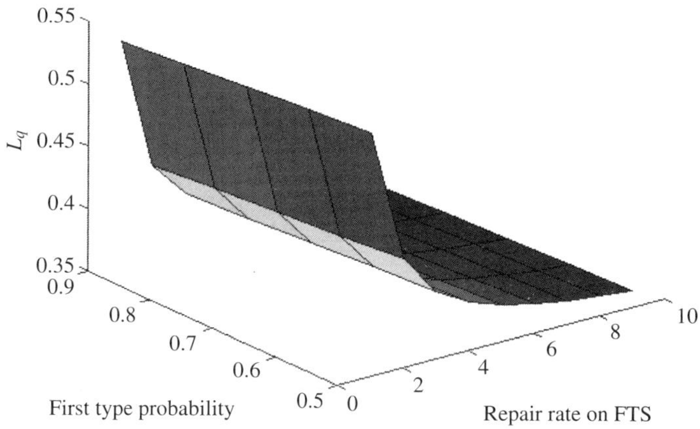

Figure 5.15 Decrease of mean orbit size when the first type probability increases; L_q versus p_1 and ξ_1.

5.10 MULTISERVER QUEUE WITH RETRIAL FEEDBACK QUEUING SYSTEM WITH TWO ORBITS

Feedback in the queuing literature represents customer dissatisfaction because of inappropriate quality of service. In case of feedback, after getting partial or incomplete service, customer retries for service. In computer communication, the transmission of protocol data unit is sometimes repeated due to occurrence of an error. This usually happens because of nonsatisfactory quality of service. Takács (1963) was the first to study feedback queuing model. He studied queue with feedback to determine the stationary process for the queue size, and the first two moments of the distribution function of the total time spent in the system by a customer. Montazer Haghighi (1981) considered a multiserver queuing system with feedback. Montazer Haghighi (1981) studied a multiserver Markovian queuing system with balking and reneging. Ayyapan et al. (2010) studied *M*/*M*/1 retrial queuing system with loss and feedback, under nonpreemptive priority service by matrix geometric method. Kumar and Sharma (2012) analyzed a single-server queuing system with retention of reneged customers. Arivudainambi and Godhandaraman (2012) considered a batch arrival queuing system with two phases of service, feedback, and *K* optional vacations under a classical retrial policy. Bouchentouf et al. (2014) analyzed a queuing model with two heterogeneous servers balking, reneging, and feedback.

Bouchentouf and Sakhi (2015) studied a queuing system presented in Figure 5.16.

The system is described as follows: The queuing model presented in Figure 5.16 consists of two reconnect orbits, two redial (retrial) orbits, s servers, and two independent Poisson streams of arrival tasks flow into a common infinite buffer. An arriving task of type i, $i = 1, 2$, is handled by an available server, if there is any; otherwise, it will wait in the buffer. The tasks are handled in the order of arrival, that is, FIFO. It is assumed that the required service time of each task is independent of its type. A waiting task of type i that did not get connected to a server will lose its patience

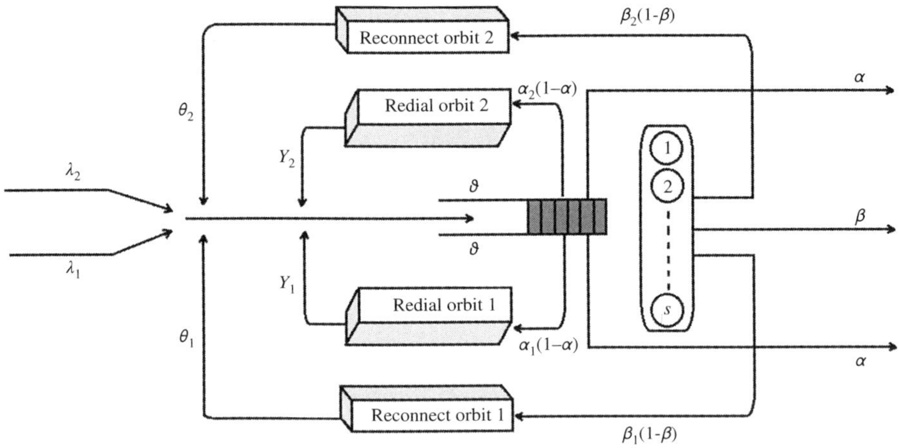

Figure 5.16 A queuing model with two redial and two reconnect orbits.

and abandon (***reneges***) after an exponentially distributed amount of time, ψ, with $E(\psi) = 1/\nu < \infty$, where ν is the abandonment rate. The abandoned task either leaves the system (a lost task) indefinitely with probability α or decides to stay requesting service with probability $1 - \alpha$. In other words, reneging occurs with probability α. If the task decides not to leave the system indefinitely, it will be routed to one of a separate type i redial queues that attempts to redispatch its jobs with probability $(1 - \alpha)\alpha_i, i = 1, 2, \alpha_1 + \alpha_2 = 1$. In other words, the task has to choose between two redial orbits. For instance, task of type 2 may be routed to redial either orbit 1 with probability $(1 - \alpha)\alpha_1$, or orbit 2 with probability $(1 - \alpha)\alpha_2$, the choice is random and does not depend on the threshold of the redial orbits or the type of the task.

We recall the reader that a redial orbit is an ordinary waiting room, where a task coming from the primary queue will be waiting there and then redial after an exponentially distributed amount of time $\omega_1, i = 1, 2$, with $E(\omega_i) = \gamma_i < \infty$. It is assumed that the service time of type i task has an exponential distribution with mean $1/\mu$. It is also assumed that service times are independent, and the required service time of each task is independent of its type. After a task has been served, it may leave the system with probability β or decide to return to the system for another service with probability $1 - \beta$. In other words, feedback probability for a task is $1 - \beta$. The type i task will enter one of the type i reconnect queues with probability $(1 - \beta)\beta_i$, where $\beta_1 + \beta_2 = 1$.

As it has been mentioned above (in the case of redial orbits), note that the task after being served, may return to the system, choosing between two reconnect orbits (for instance, task of type 2 may be routed to reconnect orbit 1 with probability $(1 - \beta)\beta_1$ or reconnect orbit 2 with probability $(1 - \beta)\beta_2$, the choice is random and does not depend on the threshold of the retrial orbits or the type of the task).

The reconnect orbit is an ordinary buffer, where a task coming back from the last service requiring for another service will be waiting there and then will reconnect after an exponentially distributed time ϕ_i, with $E(\phi_i) = \theta_i < \infty$. Therefore, an abandoned task of type i has to choose between one of the two retrial orbits and the one who decides to come back for another service has a choice between one of the two reconnect orbits; this means that a task of type 1 (respectively of type 2) is not directed automatically to the redial or reconnect orbits of type 1 (respectively of type 2).

It is assumed that α and β do not depend on tasks' experiences in the system. These experiences include holding time, waiting time, and the number of times that tasks have already tried to get a service. This creates a system with five dependent queues. Such a queuing system serves as a model for two competing job streams in a carrier sensing multiple access systems.

The goal is to present a type of call center, where the emphasis of reconnect and redial orbits is crucial in any telecommunication system.

The authors used a fluid model approximation to derive the first-order approximation for the number of tasks in the redials and reconnect orbits in the heavy traffic. The fluid limit of such a model is the unique solution to a system of three differential equations. Fluid models for call centers have been extensively studied. See Whitt (2006) and Mandelbaum et al. (2002).

Let us define the following random variables:

$Z_Q(t)$: the number of tasks in the system at time t.

$Z_{Ri}(t), i = 1, 2,$: the number of tasks in the redial queue i, for $i = 1$ and 2, at time t, with

$$Z_R(t) = Z_{R1}(t) + Z_{R2}(t). \tag{5.392}$$

$Z_{Oi}(t), i = 1, 2,$: the number of tasks in reconnect orbits i, for $i = 1$ and 2, at time t, with

$$Z_O(t) = Z_{O1}(t) + Z_{O2}(t). \tag{5.393}$$

Let us also define the following:

$\lambda_1(t) + \lambda_2(t)$: the external arrival rate at time t.

$(\gamma_1(t) + \gamma_2)E[Z_R(t)]$: the mean arrival rate due to redials at time t.

$(\theta_1(t) + \theta_2)E[Z_O(t)]$: the mean arrival rate due to reconnects at time t.

$E[\Delta(t)]$: the expected number of arrivals up to time t.

It can be shown that the stochastic process $\{Z(t), t \geq 0\}$, where

$$Z(t) = [Z_Q(t), Z_{R1}(t), Z_{R2}(t), Z_{O1}(t), Z_{O2}(t)]^T \tag{5.394}$$

is a five-dimensional Markov process that, considering (5.392) and (5.393), can be reduced to a three-dimensional one with

$$Z(t) = [Z_Q(t), Z_R(t), Z_O(t)]^T. \tag{5.395}$$

Now, let

$$E[\Delta(t)] = \lambda_1(t) + \lambda_2(t) + (\gamma_1(t) + \gamma_2)E[Z_R(t)] + (\theta_1(t) + \theta_2)E[Z_O(t)]. \quad (5.396)$$

As seen from (5.392), $E[Z_Q(t)]$ does not appear in (5.393). This is because both $E[Z_R(t)]$ and $E[Z_O(t)]$ depend on $E[Z_Q(t)]$. Thus, once $E[Z_Q(t)], E[Z_R(t)]$, and $E[Z_O(t)]$ are known, $E[\Delta(t)]$ will be calculated from (5.392).

Since analytic steady-state solution for distribution of $Z(t)$ does not seem to be imminent, the authors use the ***fluid limit approximation method***. For that, we consider an interval of nonnegative time with constant external arrival rates $\lambda_1 + \lambda_2$. Then, for the stochastic model under consideration, the following flow conversation equations hold:

$$Z_Q(t) = Z_Q(0) + \Delta_{\lambda 1}(t) + \Delta_{\lambda 2}(t) + D_R(t) + D_O(t) - D_s(t) - D_a(t), \quad (5.397)$$

with

$$D_s(t) = D_{s1}(t) + D_{s2}(t), \quad (5.398)$$

$$D_a(t) = D_{a1}(t) + D_{a2}(t), \quad (5.399)$$

$$Z_R(t) = Z_R(0) + \sum_{j=1}^{D_a(t)} B_j(1-\alpha) - D_R(t), \quad (5.400)$$

and

$$Z_O(t) = Z_O(0) + \sum_{j=1}^{D_s(t)} B_j(1-\beta) - D_O(t), \quad (5.401)$$

where

$\Delta_{\lambda_1}(t) + \Delta_{\lambda_2}(t)$ is the number of external arrivals of types 1 and 2 during the time interval $[0, t)$, (type 1 and 2 arrival processes are assumed Poisson with parameters λ_1 and λ_2, respectively),

$D_R(t) = D_{R1}(t) + D_{R1}(t)$ is the number of redials during the time interval $[0, t)$,

$D_O(t) = D_{O1}(t) + D_{O1}(t)$ is the number of reconnects during the time interval $[0, t)$,

$D_s(t) = D_{s1}(t) + D_{s1}(t)$ is the number of served tasks of types 1 and 2 during the time interval $[0, t)$,

$D_a(t) = D_{a1}(t) + D_{a1}(t)$ is the number of reneging (abandoned) tasks of types 1 and 2 during the time interval $[0, t)$,

$B_j(1-\alpha), \quad j = 1, 2, \dots, D_{ai}(t)$, is a Bernoulli random variable with success probability $1-\alpha$, such that

$$B_j(1-\alpha) = \begin{cases} 1, & \text{if the } j\text{th abandoned task decides to stay in the system} \\ & \text{and then enters one of the redial orbits,} \\ 0, & \text{otherwise,} \end{cases}$$

and

$$B_j(1-\beta) = \begin{cases} 1, & \text{if the } j\text{th abandoned task decides to stay in the system} \\ & \text{and then enters one of the reconnect orbits,} \\ 0, & \text{otherwise.} \end{cases}$$

Now, let us consider a sequence of models as represented by Figure (5.16). Choose the nth model. We will continue to use the index n for this chosen model. Denote the external arrival rates for the nth modes by λ_{1n} and λ_{2n} and the number served by s_n. Thus, Eqs (5.397), (5.400), and (5.401), using superscript n to denote variables for the nth model, can be rewritten as

$$Z_Q^{(n)}(t) = Z_Q^{(n)}(0) + \Delta_{\lambda 1n}^{(n)}(t) + \Delta_{\lambda 2n}^{(n)}(t) + D_R^{(n)}(t) + D_O^{(n)}(t) - D_s^{(n)}(t) - D_a^{(n)}(t), \quad (5.402)$$

$$Z_R^{(n)}(t) = Z_R^{(n)}(0) + \sum_{j=1}^{D_{a(t)}^{(n)}} B_j(1-\alpha) - D_R^{(n)}(t), \quad (5.403)$$

and

$$Z_O^{(n)}(t) = Z_O^{(n)}(0) + \sum_{j=1}^{D_{s(t)}^{(n)}} B_j(1-\beta) - D_O^{(n)}(t). \quad (5.404)$$

Definition 5.10.1 If there exists a limit in distribution for the scaled process $\{\overline{Z}^{(n)}(\cdot)\}_{n=1}^{\infty}$, that is, $\overline{Z}^{(n)}(\cdot) \xrightarrow{d} Z(\cdot)$, then $Z(\cdot)$ is called the ***fluid limit*** of the original stochastic model, where $\overline{Z}^{(n)}(\cdot)$ is given by

$$\overline{Z}^{(n)}(t) = [\overline{Z}_Q^{(n)}(t), \overline{Z}_R^{(N)}(t), \overline{Z}_O^{(n)}(t)]^T, \quad (5.405)$$

and by diving (5.402)–(5.404) by n, we have

$$\overline{Z}_Q^{(n)}(t) = \frac{Z_Q^{(n)}(t)}{n}, \overline{Z}_R^{(n)}(t) = \frac{Z_R^{(n)}(t),}{n}, \text{ and } \overline{Z}_O^{(n)}(t), = \frac{Z_O^{(n)}(t),}{n}. \quad (5.406)$$

As t approaches infinity, steady-state probabilities can be found as follows:

$$Z_Q = \lim_{t\to\infty} Z_Q(t) = \begin{cases} \dfrac{\lambda_1+\lambda_2}{\mu\beta}, & \text{if} \dfrac{\lambda_1+\lambda_2}{s\mu} < \beta, \\ s + \dfrac{\lambda_1+\lambda_2 - s\mu\beta}{\alpha\nu}, & \text{otherwise,} \end{cases} \quad (5.407)$$

$$Z_R = \lim_{t\to\infty} Z_R(t) = \begin{cases} 0, & \text{if} \dfrac{\lambda_1+\lambda_2}{s\mu} < \beta, \\ \dfrac{(1-\alpha)\nu(Z_Q - s)}{\gamma_1+\gamma_2}, & \text{otherwise,} \end{cases} \quad (5.408)$$

and

$$Z_O = \lim_{t\to\infty} Z_O(t) = \begin{cases} \dfrac{(1-\beta)\mu Z_Q}{\theta_a 1 + \theta_a 2}, & \text{if} \dfrac{\lambda_1+\lambda_2}{s\mu} < \beta, \\ \dfrac{s(1-\beta)\mu}{\theta_1+\theta_2}, & \text{otherwise}, \end{cases} \tag{5.409}$$

It should be noted that $\frac{\lambda_1+\lambda_2}{s\mu}$ is the load of the system due to the external arrivals and $\frac{\lambda_1+\lambda_2}{s\beta}$ is the total load when there is no redials. Hence, in (5.407)–(5.409), the value of $\frac{\lambda_1+\lambda_2}{\mu\beta}$ determines whether the fluid model is in heavy traffic or not. Therefore, now if $\frac{\lambda_1+\lambda_2}{\mu\beta} < 1$, since the fluid limit is deterministic, we have $Z_Q < s$ and there is no task in the two redial orbits, which means that $Z_R = 0$ would hold and that there is no abandonment at all in the fluid limit. In reality, due to the variability of the service duration and patience, abandonments would not be 0 though, but very small. Now, if $\frac{\lambda_1+\lambda_2}{\mu\beta} > 1$, by Equation (5.407), we have $Z_Q > s$. Consequently, the fluid model indicates that there will be $Z_Q - s$ amount of tasks waiting, each with rate ν and tasks will be routed to the redial orbits with rate $(1-\alpha)\nu(Z_Q - s)$.

5.11 STEADY-STATE STABILITY CONDITION OF A RETRIAL QUEUING SYSTEM WITH TWO ORBITS, RENEGING, AND FEEDBACK

We continue our discussion of retrial models from previous sections and, in particular, stability of the model studied by Bouchentouf et al. (2015). The feedback feature of the model is not in our focus, as other authors have recently discussed such models in detail. For instance, recently, Jain et al. (2012) have studied $M^{[x]}/G/1$ with Bernoulli feedback and Haghighi and Mishev (2013) have considered a more general model as a queue in tandem, Bernoulli feedback and splitting, that we will be discussed in the next chapter. However, our focus and reason to discuss it here will be on the stability part of the system Bouchentouf et al. (2015) have considered.

Figure 5.17, a special case of Figure 5.16, represents the Markovian retrial queuing system with two classes of tasks, constant retrial, abandonment, and feedbacks. Two independent Poisson streams of tasks, S_1 and S_2, flow into a single-server service system.

The service system can hold at most one task, that is, there is no buffer for the service station. The arrival rate of stream S_i, $i = 1, 2$, is α_i, $i = 1, 2$, with $\alpha_1 + \alpha_2 = \alpha$. The required service time of each task is independent of its type and is exponentially distributed with mean $1/\mu$. If an arriving type i, $i = 1, 2$, task finds the (main) server busy, it is routed to a dedicated retrial (orbit) queue from which tasks are retransmitted at an exponential rate. The rates of retransmissions may be different from the rates of the original input streams. Therefore, the blocked tasks of type i form a type i single-server orbit queue that attempts to retransmit tasks (if any) to the main service system at a Poisson rate of γ_i, $i = 1, 2$, that creates a system with three dependent queues.

The tasks in the orbit either attempts service again after a random time or gives up receiving service and leaves the system after a random time at rate δ_i, $i = 1, 2$. After completion of a service, a task will decide either to join the retrial group again

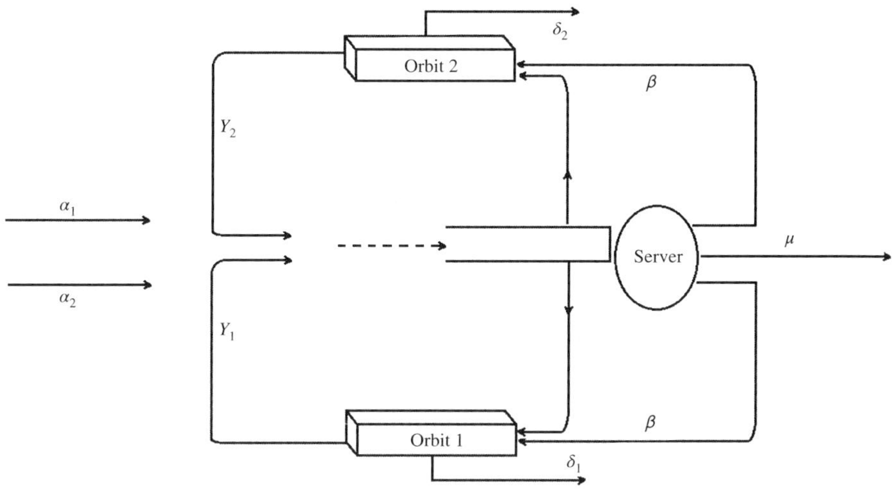

Figure 5.17 A queuing model with two orbits, abandonment, and feedback.

for another service with probability β or leave the system forever with probability $\overline{\beta} = 1 - \beta$.

Let us denote by $C(t)$ the number of tasks in the main queue and by $N_i(t)$, $i = 1, 2$, the number of tasks in orbit queue. Since there is no real buffer, the values of $C(t)$ will be 0 and 1, for all t. Thus, the Markov process $\{(N_1(t), N_2(t), C(t));\quad t \in [0, \infty]\}$ is irreducible on the state space $\{0, 1, 2, \infty\} \times \{0, 1, 2, \infty\} \times \{0, 1\}$.

An application of the model described is the local area computer network (LAN), which is an example of a model with two competing task streams in a carrier-sensing multiple access system (CSMA).

5.11.1 Necessary Stability Condition for the Steady-State System

Here, in this part, the authors are to give the necessary condition for stability of the model thus proposed. It is summarized in the following proposition:

Proposition 5.11.1 For a retrial queuing system with two orbits, constant retrials, abandoned, and feedback tasks, the following condition is necessary:

$$\frac{\alpha(\gamma_1 + \delta_1)(\gamma_2 + \delta_2)\left(1 + \dfrac{\alpha_i}{\gamma_i + \delta_i}\right)}{[\alpha + (1 + \beta)\mu](\gamma_1 + \delta_1)(\gamma_2 + \delta_2) - \alpha\gamma_1\gamma_2 - \alpha_1\delta_1\gamma_2 - \alpha_2\delta_2\gamma_1} < 1, \quad i = 1, 2, \tag{5.410}$$

and

$$[\alpha + (1 + \beta)\mu](\gamma_1 + \delta_1)(\gamma_2 + \delta_2) - \alpha\gamma_1\gamma_2 - \alpha_1\delta_1\gamma_2 - \alpha_2\delta_2\gamma_1 \neq 0.$$

Proof: We first show that

$$\frac{\alpha(\gamma_1 + \delta_1)(\gamma_2 + \delta_2)}{[\alpha + (1 + \beta)\mu](\gamma_1 + \delta_1)(\gamma_2 + \delta_2) - \alpha\gamma_1\gamma_2 - \alpha_1\delta_1\gamma_2 - \alpha_2\delta_2\gamma_1} < 1. \tag{5.411}$$

Denote the steady-state case of the Markov chain $\{(N_1(t), N_2(t), C(t)); t \in [0, \infty]\}$ by (N_1, N_2, C). Define the joint distribution $P_{n1,n2}(c)$ as

$$
\begin{aligned}
P_{n1,n2}(c) &= \lim_{t\to\infty} P\{N_1(t) = n_1, P\{N_2(t) = n_2, L(t) = c\} \\
&= P\{N_1 = n_1, N_2 = n_2, C = c\}, \quad n_1, n_2 = 0, 1, \dots, c = 0, 1, \qquad (5.412)
\end{aligned}
$$

if the limit exists. Let us define the following marginal probabilities:

$$
P_{n1} = \sum_{n_2=0}^{\infty} P_{n1,n2}(c) = P\{N_1 = n_1, N_2 = n_2, C = c\}, \quad n_1, n_2 = 0, 1, \dots, c = 0, 1 \qquad (5.413)
$$

and

$$
P_{n2} = \sum_{n_1=0}^{\infty} P_{n1,n2}(c) = P\{N_1 = n_1, N_2 = n_2, C = c\}, \quad n_1, n_2 = 0, 1, \dots, c = 0, 1. \qquad (5.414)
$$

Thus, the system of balance equations is as follows:

$$
\alpha P_{0,0}(0) = \overline{\beta}\mu P_{0,0}(1) + \delta_1 P_{1,0}(0) + \delta_2 P_{0,1}(0),
$$
$$
N_1 = n_1 = 0, N_2 = n_2 = 0, c = 0, \qquad (5.415)
$$

$$
(\alpha + \mu)P_{0,0}(1) = \alpha P_{0,0}(0) + \gamma_1 P_{1,0}(0) + \gamma_2 P_{0,1}(0) + \delta_1 P_{1,0}(1) + \delta_2 P_{0,1}(1), \quad N_1 = 0, c = 1, \qquad (5.416)
$$

$$
(\alpha + \gamma_1 + \delta_1)P_{n_1,0}(0) = \overline{\beta}\mu P_{n_1,0}(1) + \beta\mu P_{n_1-1,0}(1) + \delta_1 P_{n_1+1,0}(0) + \delta_2 P_{n_1,1}(0), \quad N_1 = n_1 \geq 1, c = 0, \qquad (5.417)
$$

$$
(\alpha + \mu + \delta_1)P_{n_1,0}(0) = \alpha P_{n_1,0}(0) + \alpha_1 P_{n_1-1,0}(1) + \gamma_1 P_{n_1+1,0}(0) + \gamma_2 P_{n_1,1}(0) + \delta_1 P_{n_1+1,0}(1) + \delta_2 P_{n_1,1}(1),
$$
$$
N_1 = n_1 \geq 1, c = 1, \qquad (5.418)
$$

$$
(\alpha + \gamma_2 + \delta_2)P_{0,n_2}(0) = \overline{\beta}\mu P_{0,n_2}(1) + \beta\mu P_{0,n_2-1}(1) + \delta_1 P_{1,n_2}(1) + \delta_2 P_{0,n_2+1}(0), \quad N_1 = 0, N_2 = n_2 \geq 1, c = 0, \qquad (5.419)
$$

$$
(\alpha + \mu + \delta_2)P_{0,n_2}(1) = \alpha P_{0,n_2}(0) + \gamma_1 P_{1,n_2}(0) + \gamma_2 P_{0,n_2+1}(0) + \delta_1 P_{1,n_2}(1) + \delta_2 P_{0,n_2+1}(1) + \alpha_2 P_{0,n_2-1}(1), \quad N_1 = 0, c = 1, \qquad (5.420)
$$

$$
(\alpha + \gamma_1 + \gamma_2 + \delta_1 + \delta_2)P_{n_1,n_2}(0) = \beta\mu P_{n_1-1,n_2}(1) + \beta\mu P_{n_1,n_2-1}(1) + \overline{\beta}\mu P_{n_1,n_2}(1) + \delta_1 P_{n_1+1,n_2}(0) + \delta_2 P_{n_1,n_2+1}(0),
$$
$$
N_1 = n_1 \geq 1c = 0, \qquad (5.421)
$$

$$(\alpha+\mu+\delta_1+\delta_2)P_{n_1,n_2}(1) = \alpha P_{n_1,n_2}(0) + \gamma_1 P_{n_1+1,n_2}(0) + \gamma_2 P_{n_1,n_2+1}(0) + \delta_1 P_{n_1+1,n_2}(1) + \delta_2 P_{n_1,n_2+1}(1) + \alpha_1 P_{n_1-1,n_2}(1) + \alpha_2 P_{n_1,n_2-1}(1), \quad N_1 = n_1 \geq 1 c = 1. \quad (5.422)$$

We now define the pgf of the steady-state Markov chain (N_1, N_2, C), denoted by $F(z_1, z_2, z)$, as follows:

$$F(z_1, z_2, z) = \sum_{n_1=0}^{\infty}\sum_{n_2=0}^{\infty}\sum_{c=0}^{1} P_{n1,n2}(c) z_1^{n_1} z_2^{n_2} z^c, \quad |z_1|, |z_2| \leq 1. \quad (5.423)$$

Let us also define the following partial pgfs:

$$R_{n_2}^{(c)}(z_1) = \sum_{n_1=0}^{\infty} P_{n1,n2}(c) z_1^{n_1}, \quad n_2 = 0, 1, 2, \ldots, c = 0, 1 \quad (5.424)$$

and

$$F^{(c)}(z_1, z_2) = \sum_{n_2=0}^{\infty}\sum_{n_1=0}^{\infty} P_{n1,n2}(c) z_1^{n_1} z_2^{n_2} = \sum_{n_2=0}^{\infty} P_{n2}^{(c)}(z_1) z_2^{n_2}, \quad c = 0, 1. \quad (5.425)$$

Thus, from (5.423) and (5.425), we have

$$F(z_1, z_2, z) = F^{(0)}(z_1, z_2) + zF^{(1)}(z_1, z_2), \quad |z_1|, |z_2| \leq 1. \quad (5.426)$$

Now, with standard generating function method for solving difference equations, we have

$$1 - F^{(0}(0, 1) = \left(1 + \frac{\alpha_1}{\gamma_1 + \delta_1}\right) F^{(1)}(1, 1) \quad (5.427)$$

and

$$1 - F^{(0}(1, 0) = \left(1 + \frac{\alpha_2}{\gamma_2 + \delta_2}\right) F^{(1)}(1, 1). \quad (5.428)$$

From (5.427) and (5.428), $F^{(1)}(1, 1)$ can be found as

$$F^{(1)}(1, 1) = \frac{\alpha(\gamma_1 + \delta_1)(\gamma_2 + \delta_2)}{[\alpha + (1+\beta)\mu](\gamma_1 + \delta_1)(\gamma_2 + \delta_2) - \alpha\gamma_1\gamma_2 - \alpha_1\delta_1\gamma_2 - \alpha_2\delta_2\gamma_1}, \quad (5.429)$$

with the denominator of (5.429) to be different from zero. Equation (5.410) can now be proved by contradiction. Let us assume that the left-hand side of (5.410) is zero so that from (5.426), we have $F^{(0)}(1, 0) = 0$. Since from (5.425),

$$F^{(0)}(1, 0) = \sum_{n_1=0}^{\infty} P_{n1,0}(0), \quad (5.430)$$

$F^{(0)}(1,0) = 0$ implies that

$$P_{n1,0}(0) = 0, \quad n_1 = 0, 1, 2, \ldots . \tag{5.431}$$

Thus, from (5.415) and (5.416), we have

$$P_{n1,0}(1) = 0, \quad n_1 = 0, 1, 2, \ldots . \tag{5.432}$$

Applying mathematical induction, as shown below, yields

$$P_{n1,n2}(0) = 0, \quad n_1, n_2 = 0, 1, 2, \ldots . \tag{5.433}$$

It can be seen from (5.431) that (5.433) is true for $n_2 = 0$. Now, let us assume that (5.433) is true for $n_2 = k$. In order to show that (5.433) is true for $k + 1$, we see from the induction hypothesis that

$$P_{n1,k}(0) = P_{n1,k}(1) = 0. \tag{5.434}$$

Hence, for $n_2 = 0$ and $c = 0$, by multiplying (5.415) and (5.417) by $z_1^{n_1}$, we have

$$\begin{aligned} &[(\alpha + \gamma_1 + \delta_1)z_1 - \delta_1]R_0^{(0)}(z_1) - [(\gamma_1 + \delta_1)z_1 - \delta_1]P_{0,0}(0) = (\overline{\beta}\mu z_1 + \beta\mu z_1^2) \\ &R_0^{(1)}(z_1) + \delta_2 z_1 R_1^{(0)}(z_1). \end{aligned} \tag{5.435}$$

From (5.434) and (4.435), we obtain

$$P_{n1,k+1}(1) = 0. \tag{5.436}$$

We now prove that

$$P_{n1,n2}(1) = 0, \quad n_1, n_2 = 0, 1, 2, \ldots . \tag{5.437}$$

From (5.422) and (5.432), (5.437) is true. Thus, only the case of $n_1 = 0$ and $n_2 = 0, 1, 2, \ldots$ remain to be proven. In order to prove this part, assuming the left-hand side of (5.410) equals 1 along with (5.421), we have

$$P_{0,n2}(1) = 0, \quad n_2 = 0, 1, 2, \ldots . \tag{5.438}$$

We have already noted that $P_{0,0}(1) = 0$. Thus,

$$P_{n1,n2}(0) = P_{n1,n2}(1) = 0, \quad n_1, n_2 = 0, 1, 2, \ldots , \tag{5.439}$$

which completes the proof of the proposition.

We note that the sufficient condition of stability of such a model is yet an open problem.

5.12 BATCH ARRIVAL QUEUE WITH GENERAL SERVICE IN TWO FLUCTUATING MODES AND RENEGING DURING VACATION AND BREAKDOWNS

We conclude the chapter by presenting one more single-server queue with breakdown with different features than previous ones. This model is published by Baruah et al. (2014). The authors study the behavior of a single-server batch arrival queuing system with an arbitrary general service with different service rates in two modes 1 and 2 of service, full efficiency and reduced efficiency with probabilities p_1 and p_2, respectively. The server is subject to random breakdown that will stop the service until the server is repaired. As soon as the server breaks down, the task whose service is interrupted returns to the head of the queue. The server immediately starts the service in mode 1, as soon as it is repaired. The server, while providing service, may also take a vacation. During the breakdown or vacation, the tasks waiting for service may renege. It is assumed that failure or breakdown can occur even when the server is in idle state.

The authors derive the steady-state queue size distribution and with the help of a numerical example show the validity of their results.

5.12.1 The Model

The model is defined under the following assumptions and terms:

1. Tasks arrive in batches of variable size according to a compound Poisson process.
2. Let the probability of arrival of batch "k" tasks in the system at a short interval of time $(t, t + \Delta t)$ be denoted by $\lambda c_k \Delta t, \quad k = 1, 2, \ldots, 0 \leq c_k \leq 1$, $\sum_{k=1}^{\infty} c_k = 1$, where $\lambda > 0$ is the arrival rate of batches.
3. There is a single server providing service in two fluctuating modes.
4. The tasks are served one by one based on FCFS discipline.
5. The probabilities of providing service in modes 1 and 2 are p_1 and p_2, respectively, $p_1 + p_2 = 1$.
6. The service time follows general distributions G_1 and G_2 at modes 1 and 2 with rates of service μ_1 and μ_2, respectively.
7. Let $S_i(x)$ and $s_i(x)$, $i = 1, 2$, be the distribution and density function of the service time, respectively.
8. The conditional probability of the service time during the interval $(t, t + \Delta t)$, given that elapsed service time is x is

$$\mu_i(x) = \frac{s_i(x)}{1 - S_i(x)}. \tag{5.440}$$

Hence,

$$s(t) = \mu_i(t) e^{-\int_0^t \mu_i(x)dx}, \quad i = 1, 2. \tag{5.441}$$

9. The system may be subjected to breakdown at random. The breakdowns are time-homogeneous in the sense that the server can fail even while it is idle.
10. The customer receiving service during breakdown returns to the head of the queue.
11. Once repair is complete, the server immediately provides service to the task in mode 1.
12. It is assumed that time between breakdowns occur according to a Poisson process with mean rate of breakdown as $\alpha > 0$.
13. The repair times follow a general distribution with distribution function (x) and density function (x).
14. Let the conditional probability of completion of the repair be $\beta(x)dx$ such that

$$\beta(x) = \frac{B(x)}{1 - B(x)}. \tag{5.442}$$

Hence,

$$b(u) = \beta(u)e^{-\int_0^u \beta(x)dx}. \tag{5.443}$$

15. After each service completion, the server may take a vacation of random length with probability ϕ or remain in the system with probability $1 - \phi$.
16. The server's vacation time follows an arbitrary distribution with distribution and density functions $F(v)$ and $f(v)$, respectively.
17. Let the conditional probability of completion of vacation during the interval $(x, x + \Delta x)$ during the elapsed vacation time x be denoted by $\eta(x)$. Thus,

$$\eta(x) = \frac{f(v)}{1 - F(v)} \tag{5.444}$$

and

$$f(v) = \eta(v)e^{-\int_0^v \eta(x)dx}. \tag{5.445}$$

18. It is assumed that tasks may renege when the server is on vacation.
19. Reneging is assumed to follow exponential distribution with parameter γ. Thus,

$$f(t) = \gamma e^{-\gamma t}dt, \gamma > 0. \tag{5.446}$$

20. Let the probability that a task may renege during a short interval of time $(t, t + \Delta t)$ be denoted by $\gamma \Delta t$.
21. If there are $n (n \geq 0)$ tasks waiting in the waiting line and elapsed time is x, then the probability that at time t the server is providing service in mode i, $(i = 1, 2)$ is denoted by $P_n^{(i)}(t, x)$, $i = 1, 2$. Thus, the probability that at time t there are $n (n \geq 0)$ tasks in the waiting line, irrespective of the value x, is

$$P_n^{(i)}(x) = \int_0^\infty P_n^{(i)}(t, x)dx. \tag{5.447}$$

22. If there are $n(n \geq 0)$ tasks waiting in the waiting line and elapsed time is x, then the probability that the server is under repair is denoted by $R_n(t,x)$. Thus, the probability that the server is under repairs when there are $n(n \geq 0)$ tasks in the waiting line, irrespective of the value x, is

$$R_n(t) = \int_0^\infty R_n(t,x)dx. \tag{5.448}$$

23. If there are $n(n \geq 0)$ tasks waiting in the waiting line and elapsed time is x, then the probability that at time t the server is on vacation is denoted by $V_n(t,x)$. Thus, the probability that at time t there are $n(n \geq 0)$ tasks in the waiting line, irrespective of the value x, the server is on vacation is

$$V_n(t) = \int_0^\infty V_n(t,x)dx. \tag{5.449}$$

24. The steady-state probability that the server is idle is denoted by Q.

5.12.2 Analysis

We leave it as an exercise for the reader to write the system of steady-state difference equations for the queue length. Assume the boundary conditions as follows:

$$\begin{aligned} P_n^{(1)}(0) &= \lambda c_{n+1} p_1 Q + (1-Q) \\ &\times \left[p_1 \int_0^\infty P_{n+1}^{(1)}(x)\mu_1(x)dx + p_1 \int_0^\infty P_{n+1}^{(2)}(x)\mu_2(x)dx \right] \\ &+ p_1 \int_0^\infty R_{n+1}(x)\beta(x)dx + p_1 \int_0^\infty V_{n+1}(x)\eta(x)dx, \quad n \geq 0. \end{aligned} \tag{5.450}$$

Note that the left-hand side of the above equation indicates the probability that there is one task in service and n tasks in the queue. Zero in the parentheses of the left-hand side of the boundary condition (5.450) indicates the moment when the service starts in mode 1. The right-hand side of (5.450) shows five mutually exclusive cases, each contributing to the immediate start of service in mode 1. The first term on the right-hand side means that as soon as a batch of size $n+1$ arrives when the system is empty, the service in mode 1 starts immediately. Similarly, other terms on the right-hand side of (5.450) indicate that just after the completion of a service in mode 1 or mode 2, or completion of a vacation or completion of repairs, the service in mode 1 immediately starts. Utilizing similar reasoning, we get the other boundary conditions as follows:

$$\begin{aligned} P_n^{(2)}(0) &= \lambda c_{n+1} p_2 Q + (1-\phi) \\ &\times \left[p_2 \int_0^\infty P_{n+1}^{(1)}(x)\mu_1(x)dx + p_2 \int_0^\infty P_{n+1}^{(2)}(x)\mu_2(x)dx \right] \\ &+ p_2 \int_0^\infty R_{n+1}(x)\beta(x)dx + p_2 \int_0^\infty V_{n+1}(x)\eta(x)dx, \quad n \geq 0. \end{aligned} \tag{5.451}$$

$$R_{n+1}(0) = \alpha \int_0^\infty P_n^{(1)}(x)dx + \alpha \int_0^\infty P_n^{(2)}(x)dx = \alpha P_n^{(1)} + \alpha P_n^{(2)}, \tag{5.452}$$

$$R_0(0) = \alpha Q, \tag{5.453}$$

$$V_n(0) = \phi \left[\int_0^\infty P_n^{(1)}(x)\mu_1(x)dx + \int_0^\infty P_n^{(2)}(x)\mu_2(x)dx \right]. \tag{5.454}$$

In order to solve the system of difference equations for the model, define the following generating functions:

$$P^{(i)}(x, z) = \sum_{n=0}^{\infty} z^n P_n^{(i)}(x), P(z) = \sum_{n=0}^{\infty} z^n P_n^{(i)}, \tag{5.455}$$

$$R(x, z) = \sum_{n=0}^{\infty} z^n R_n(x), R(z) = \sum_{n=0}^{\infty} z^n R_n, \tag{5.456}$$

$$V(x, z) = \sum_{n=0}^{\infty} z^n V_n(x), V(z) = \sum_{n=0}^{\infty} z^n V_n, C(z) = \sum_{n=0}^{\infty} z^n C_n. \tag{5.457}$$

We leave it as an exercise for the reader to solve the system of difference equations using the above generating functions to obtain the queue length of the system.

EXERCISES

5.1 Prove that denominator of (5.41) has a unique root in the interval (0, 1).

5.2 Find a simplified expression for L_b used in (5.81).

5.3 Prove the relation (5.113).

5.4 Prove the relation (5.118).

5.5 In Section 5.6.1, show that the resulting new path is in one-to-one correspondence to the original skeleton path.

5.6 In Section 5.8, show that the LSTs and moment generating functions of the state probabilities are given by (5.254)–(5.257).

5.7 Write the system of steady-state difference equations for the queue length of model described in Section 5.12.

5.8 Solve the system of difference equations found in Exercise 5.8, using the above generating functions (5.455)–(5.457) to obtain the queue length of the system.

6

NETWORKS OF QUEUES WITH DELAY

6.1 INTRODUCTION TO NETWORKS OF QUEUES

As the modern world has become highly technological with the advent of the Internet and wireless communication, as well as advancements in manufacturing and production systems, public transportation, logistics, and health care, performance evaluation of such systems is best possible by modeling them as a network of queues, in which many theoretical ideas have been developed during the past six decades. A network of queues is applied to model a real-world system when a set of resources is shared by some components of the system. In other words, a network of queues is a ***flow system***, which indicates that a task flows (transfers or moves) through two or more finite- or infinite-capacity stations to receive a complete service. Several examples of networks of queues appear in our day-to-day lives and perhaps we do not notice them. Assembly of an automobile in a car manufactory is a simple example. The digital computer developed in the early 1960s highlighted more applications of queueing theory. Examples are in a sequence of application in the field of information processing and queueing networks of various types. Applications demanded expansion of analytical and numerical methodologies of various queueing models. The solution of Riemann–Hilbert problems is an example. In many cases, however, exact analysis of networks is impossible and, thus, it is necessary to introduce assumptions to offer solution for a model.

A ***network of queues*** or a ***queueing network*** or ***stochastic network*** is defined as a queueing system with two or more ***service stations*** or ***nodes*** (each with one or more severs) arranged in tandem. These service stations could serve as places for machine repairs, computing servers, or loading and unloading docs. The system may consist

Delayed and Network Queues, First Edition. Aliakbar Montazer Haghighi and Dimitar P. Mishev.

of one or multiple lines of queues in tandem. In other words, there will be queues in parallel and tandem. In an ***open (one tandem queue line) system*** (vs a ***closed*** or ***cyclic*** system), arriving tasks to the first station are served and moved to the next station until all stations have been visited. In such a system, arrivals to each station will be from departure of another station as well as arrivals from outside. In the basic scenario, a task moves in order, from one station to another, until it exits the system. In such a case, arrivals to the first station are from outside. However, for every other one will be departures from the previous station. This leads to two types of queueing networks, namely ***open*** and ***closed***. A stochastic network is open if tasks enter from a source outside of it and after being completely served at service stations, it leaves the network with a positive probability that at least one exit gate is present.

The relationship between open and closed networks of queueing models is that in the former, tasks arrive from outside with a constant rate, receive service at one or more stations, and then depart from the system, whereas in the latter, tasks neither arrive into nor depart from the system, but there would be a finite number of tasks that circulate within the system . In an open network system, the total external arrivals or the ***throughput*** is an independent variable and the number of tasks in the system is a dependent variable. Distribution of this random variable is usually the solution of the problem that arises for a network queueing system. On the contrary, in a closed system, the fixed number of tasks in the system is an independent variable. The throughput in this case could be defined as the departure rate whose value is desirable. The individual service rate is part of the system data. Hence, knowing the throughput is the same as knowing the utilization factor for the system that is expected proportion of the servers at the designated station that are busy in the stationary case.

There is a possibility of ***mixed queueing networks*** and that is a system which is open for a ***class*** of tasks and closed for others. For instance, consider breakdown of computers in a computer laboratory. An arrival in that case is a computer that breaks, and because there is no external possibility, this system is closed. The system will be cyclic, unless, if a broken computer is taken out from the inventory. In the latter case, the system will become an open system in the sense that there is an exit gate in the system.

In a simple mixed parallel and tandem system, tasks arrive from one end, distribute among parallel tandem queues, receive service, and finally leave from only one exit. This type of system is referred to as ***Fork–Join*** queueing system. In the simplest case, for every task entering the Fork–Join queue, only one task leaves the system. This is where the name fork (distribute) – join (back together) comes from. In more complicated systems, tasks may arrive at different stages and leave from different exits.

It is important to note that characterization of the open and closed systems is not always valid. For instance, in the Jackson model, the external arrival rate may depend on the total number of tasks in the system. Thus, neither the external rate nor the number of tasks in the system is fixed.

There is a possibility that a closed system be approximated by open networks. Whitt (1984) studied this ***approximation*** and suggested using open networks with specified expected stationary populations. He called this approach the ***fixed-population mean*** (***FPM***) method. In some cases, the open system methods also yield bounds for the performance measures in the closed networks. The FPM method is also used to determine approximate congestion measures for closed non-Markovian networks that include multiserver stations with nonexponential service time distributions and first-come, first-served (FCFS) discipline. In the basic Jackson model, Whitt considered a closed model with a designated exit–entry station.

Ward Whitt

In this model, an arriving task to this node from other entrance in the network completes its service and its departure indicates arrival of a new task. In order to obtain the associated open system, Whitt proposed cutting the flow into this exit–entry node. He, then, assumed all internal arrivals into this station depart from the system and insert an external passion arrival process. The FPM throughput in this scenario is the rate or intensity of the external Poisson arrival process for which the expected stationary queue length has a specific value. The approximate expected number of tasks at each node is the expected number in the open system with the FPM arrival rate. In this system, any node can serve as the exit–entry node.

Example 6.1.1 [Whitt (1984, Example 1)] Let us consider a cyclic Markovian network with a single-server, K tasks (of the same type), and FCFS discipline. Let us also assume that there are n_1 nodes having mean service times 1 and n_2 nodes with mean service times τ. Nodes are arranged in an arbitrary order. As the cyclic indicates, all departures from node j move to node $j + 1$, $1 \leq j \leq n_1 + n_2 - 1$, and all departures from node $n_1 + n_2$ move to node 1. Thus, we apply FPM method.

We cut the flow into one node, let all original arrivals on that arc depart the system and insert an external Poisson arrival process with rate λ. We also assume that for the stationary process, the mean total number of tasks in the system is K. Considering a cyclic system implies that the arrival rate at each node is the external arrival rate. The stationary distribution in the open network system is the same as independent $M/M/1$ queue. Thus, we obtain λ from the following quadratic equation:

$$\frac{n_1 \lambda}{1-\lambda} + \frac{n_2 \lambda \tau}{1-\lambda \tau} = K. \tag{6.1}$$

It is important to note that if we were not to use the property of cyclic network, as earlier, we should have used the traffic rate equations to find λ. We also note that if there were m different service rates, then Equation (6.1) would have been a polynomial of mth degree.

Thus, if $K = 20$, $n_1 = n_2 = 10$, and $\tau = 1.2$, then the approximate throughput is 0.45. Whitt proved that this number is the lower bound for the throughput in the original closed network. He suggested a possible improvement and, hence, obtained $\lambda = 0.46$. He also obtained the actual throughput in the original closed network as 0.46. In order to obtain this number, he used the software package PANACEA for a closed Markovian network of queues. Using $\lambda = 0.45$ as an approximation for the throughput by FPM method, the expected number of tasks at each station with mean service times 1 (1.2) is found to be 0.82 (1.18) by applying $M/M/1$'s formula. For a closed node, this value becomes 1.79 (2.53).

It should be noted that for practical purposes, the standard congestion measures computed by the FPM method agreed with those for the closed system in this example. In addition, the FPM solution remains unchanged if the number of tasks and nodes of each type are multiplied by a constant. Further, the quality of approximation improves as the network grows in this manner. In other words, FPM method does not perform well when few nodes are present, such as $n_1 + n_2 = 1$.

6.2 HISTORICAL NOTES ON NETWORKS OF QUEUES

In the 1960s, the computer industry was not aware of the need of queueing network analysis; instead, the time-sharing idea was used in the data processing industry. The ***networks*** became modern by the end of the decade. New challenges developed in the 1970s when the wireless communication system (such as broadcast satellite communications and packet radio) was introduced. The local area network (LAN) was first studied in the 1980s, and is still under study.

Networks of queues or tandem queueing systems have attracted much interest in the literature because of their real-world applicability. See, for example, Haghighi and Mishev (2013) and references therein. We have already addressed systems of networks of queues in Haghighi and Mishev (2014). It is worth noting that in modern computer communication networks, queueing theory is a quite useful tool to analyze node-to-node communication parameters. This is particularly true in Packet-Switched Computer Communication Systems (Muh, 1993). Packet switching is a digital networking communications method. All the transmitted data are broken into suitably smaller units (or sized blocks), regardless of their contents, types, or structures. These grouped data are called packets.

In this chapter, we discuss such systems with an additional feature "delay." However, a brief presentation of the history of queueing network is instructional and we present it from the words of the person who viewed network through queues and finally developed the Internet, without which we cannot live. He is Leonard Kleinrock, who explained the history in his Kleinrock (1964). We also add comments of Jackson (2002), who acknowledged the work of Kleinrock.

Leonard Kleinrock

While in his graduate program at the Massachusetts Institute of Technology (MIT) in the Electrical Engineering Department, studying data networks, Leonard Kleinrock was surrounded by computers, and it was clear to him that "some technological breakthroughs were necessary to allow them to communicate with each other efficiently." Hence, in 1961, he submitted a doctorate proposal (Kleinrock 1961a) to study data networks, thus launching the technology that eventually led to the invention of the Internet. In his paper (Kleinrock 1961b), he laid the fundamental of the mathematical theory of data networking. He introduced the ideas of segmenting messages (jobs) into smaller pieces (later called "***packets***," as defined above) in the early 1962 (Kleinrock 1961a). Kleinrock completed his PhD in 1962 (Kleinrock 1961b), which was later published by McGraw-Hill as an MIT book entitled *Communication Nets* (Kleinrock 1964). In these works, Kleinrock developed the theory of stochastic flow of message traffic in connected networks of communication centers. He also developed the basic principles of packet switching. With these developments, he provided the fundamental underpinnings for the Internet technology. He used the phrase "***Internet technology***" to mean the fundamental analytic and design principles and algorithms, and "not to a wider use of the term, which might include, for example, the ***World Wide Web***, ***HTML***, and ***Java***.

Kleinrock set up the mathematical model, using queueing theory, to evaluate network performance and developed optimal design procedures to determine the capacity assignment, topology, routing procedure, and message size. He introduced and evaluated distributed adaptive routing control procedures. He also evaluated different queueing disciplines for handling traffic in the nodes (particularly, chopping messages into smaller segments, known as packets). The principles he uncovered (along with his subsequent research) continue to provide a basis for the present Internet technology. It is for this work that he is known as the inventor of the Internet technology and one of the fathers of the Internet.

It was clear to Kleinrock that message traffic was stochastic and hence the tools of stochastic processes would be needed for analysis. He had already developed a mathematical model that reflected this concept. The basic structure chosen by him was that of a queue (Kleinrock, 1975). "A queue is a perfect resource sharing mechanism," because it "is dynamic, adaptive, and efficient," he said in Kleinrock (1975).

The quantities of interests in queueing theory such as throughput, response time, efficiency, loss, and priorities are also of interest in data networks. Indeed, it was clear to Kleinrock , as he stated in his paper that "a queue was just the right structure for implementing demand access and that queueing theory was perfect for describing and analyzing data networks."

Networks of queues are extensively used in computer system modeling, signal processing, and electrical engineering. The flow of computer programs through a time-sharing computer system is another example. Other examples are airline routing system and hiring processes in large companies. For the latter, consider personnel hiring policy in a large company such as IBM, AT&T, Dell, or Exxon. Because of turnover and retirement of employees, it may be assumed that hiring service is a continuous task in any such company and applications continuously arrive, electronically, at any time through the companies' website. In general, such a website would be capable of receiving a group (bulk) of applications simultaneously.

6.3 JACKSON'S NETWORK OF QUEUES

Research papers on networks of queues started to appear in the literature in the second half of the 1950s, of which Hunt (1957) is the first paper discussing the networks of queues to some extent.

James R. Jackson (1924–2011)

However, the classic work of James R. Jackson (1957), modeling a "job shop," where the nodes were workstations and the customers were the jobs, seems to be the first mathematically relatively rich paper published on open networks of queues. "***Jobs***" in Jackson's network may be viewed as ***messages*** in an information network or as phases of the life of individuals, for example. In brief, ***Jackson's queueing network*** is modeling operation of a job shop. In particular, it is a special case of an open stochastic network, which consists of arbitrary but finite number of interconnected queues, each with an infinite-sized buffer before its station (or node). A task that arrives at a station waits there until it is served at that station and after some random service time, it leaves the station and randomly chooses where to go next. Service rate at each station can be both node-dependent and state-dependent.

We now give a detailed description of an open stochastic network process by Jackson (1957).

6.3.1 Jackson's Model

An open Jackson network with N nodes (or stations), numbered as $1, 2, \dots, N$, is an open stochastic network such that (indistinguishable) jobs (or tasks or customers) arrive from outside and enter the "shop" following a Poisson process with rate λ_0. An arriving job is independently routed to a node with some probability. After completion of service on a job at a node i, it may be routed randomly (independent of its past history) to another node, say j, with transition probability $p_{i,j}$, $i = 0, 1, 2, \dots, N$, where 0 indicates the outside (an artificial station) and $j = 1, 2, \dots, N$. The job with completed service at node i may leave the system with a nonnegative probability $p_{i,0} \geq 0$, where

$$p_{i,0} = 1 - \sum_{j=1}^{N} p_{i,j}, \quad i = 0, 1, 2, \dots, N, \tag{6.2}$$

with $p_{0,0} = 0$.

The state of the system at any time is indicated by the vector $(m_i, \ n_j)$, $i, j = 1, 2, \dots, N$. Arrival rate at node j is denoted by λ_j, $j = 1, 2, \dots, N$, where λ_0 denotes the external arrival rate. Thus, for an open Jackson network, the mean arrival rate at a node j, sometimes referred to as the ***traffic equation***, is the sum of external and internal arrivals. That is,

$$\lambda_j = \lambda_0 p_{0,j} + \sum_{i=1}^{N} \lambda_i p_{i,\ j}, \quad j = 1, 2, \dots, N. \tag{6.3}$$

Each station i, $i = 1, 2, \dots, N$, consists of an infinite-sized buffer and single server with independent exponential service times with parameter μ_i. In addition, all inter-arrival times constitute a set of independent random variables. Service at each station is based on a FCFS discipline. The system is with Bernoulli switches or switching mechanism for jobs among stations.

This network system of queues has been extensively reviewed in the literature. The equilibrium distribution for this system is simple to compute as the network has a product-form solution, which is the essence of Jackson (1957). Jackson (1963) is a more rigorous, slightly generalized, and "neatened presentation," as he described it in Jackson (2002). It was found that in stationary case, the traffic processes on the so-called exit arcs with single-server nodes constitute Poisson processes. This was a result of Burke's theorem from single-server queues to multiserver queues and the networks of queues. Burke (1956) and others conjectured that the traffic processes on nonexit arcs cannot be Poisson in the steady-state case. Melamed (1979) proved the conjecture true.

It turns out that networks exhibiting these three properties (1) Poisson arrivals, (2) exponential services, and (3) state-independent routing probabilities are the class of queueing networks called ***separable networks*** or ***product-form networks***. Jackson showed that this class of networks exhibits the following unique property: the joint probability distribution for the number of tasks at each node can be written as a product of the marginal probability distributions at each of the nodes, as we noted earlier. In other words, in terms of queueing length distribution, the network *acts as if* each node can be viewed as an independent $M/M/1$ queue with arrival and service rates λ_i and μ_i, respectively, although the actual internal flow in this type of networks is not always Poisson. Jackson assumed Poisson job arrivals, independent exponential job service times at each station, and paths through the network that were governed by independent transition probabilities among the workstations. He solved for the steady-state joint distribution of the number of jobs in each of the stations and showed some remarkable properties of the solution. Hence, suppose the network system contains N independent nodes with arrival rates λ_i, $i = 0, 1, 2, \ldots, N$, and service rates μ_i, $i = 1, 2, \ldots, N$. We have

$$\rho_i = \frac{\lambda_i}{\mu_i}, \quad i = 1, 2, \ldots, N. \tag{6.4}$$

Let the number of jobs at the ith node, including the ones being served, at time t be represented by the random variable $\xi_i(t)$, $i = 1, 2, \ldots, N$. We also let the joint probability of m_i jobs at the ith node, including the ones being processed, at time t be denoted by $P_{m_i}(t)$, that is,

$$P_{m_i}(t) = P\{\xi_i(t) = m_i\}, \quad i = 1, 2, \ldots, N. \tag{6.5}$$

We further let

$$P_{m_i} = \lim_{t \to \infty} P\{\xi_i(t) = m_i\}, \quad i = 1, 2, \ldots, N. \tag{6.6}$$

Then, Jackson's theorem (Jackson, 1957) states that the joint probability distribution of number of jobs in the system is given by

$$P_{m_1, m_2, \ldots, n_N} = \prod_{i=1}^{N} P_{m_i}, \tag{6.7}$$

which is the probabilities of m_i tasks in the ith node.

Example 6.3.1 An Economic Application Bellman (1957) presented the theory of a new numerical method for the solution of sequential decision problems. His idea was to observe the system under consideration over a finite or infinite time horizon split up into periods or stages. At each stage, the state of the system was to be observed and a decision had to be made. The decision influences (deterministically or stochastically) the state to be observed at the next stage, and depending on the state and the decision made, an immediate reward was to be gained. The expected total rewards from the present stage until the end of the planning horizon was expressed by a ***value function***. The relationship between the value function at the present stage and the next stage was expressed by the ***functional equation***. Optimal decisions depending on stage and state were determined backward step by step as those maximizing the right-hand side of the functional equation.

Howard (1960) combined the dynamic programming technique with Markov chain. His contribution was to the solution of infinite-stage problems, where the policy iteration method was created as an alternative to the stepwise backward contraction method, which he called ***value iteration***. Hence, the policy iteration was a result of the application of the Markov chain environment, and it was an important contribution to the development of optimization techniques. Thus, Markov decision programming is directly able to take the variability in traits and the cyclic production into account without any modification.

We pause here to mention some interesting comments on the origin and application of Markov decision processes by Ronald A. Howard (2002). He stated that his one successful application was the original application that sparked his interest in this whole research area. His research interest began when he was a graduate student in electrical engineering at MIT, working part time with the Operations Research Group of Arthur D. Little, Inc. The group had an important relationship with Sears, Roebuck and Company and helped that firm with a number of its problems.

Ronald A. Howard

Management of Sears was becoming quite concerned with the effectiveness of the operation by which they sent catalogs to present and prospective customers. There was variety of options at the time. A customer could receive up to 14 individual mailings a year, ranging from the general catalog to individual sales fliers. Or, he/she might receive any part of these mailings. The cost of any particular mailing was easily determined, but the question for the management was the benefit?

In order to answer the question, Howard presented his experience at Sears' catalog information system, 20 years back from then. Therefore, let us go back three decades in the city of Chicago and see what Howard has to say. We should note that in those days, copy machines and printers were using stencil sheets. Now suppose we had two or three acres of green steel filing cabinets. Each cabinet contained steel Addressograph plates, about four inches square. Each plate had a stencil for printing the customer's name and address, and, as inserts, three small cards with several punched holes. These holes provided a limited summary over three seasons of the customer's purchasing history as a Sears' mail order client. About 100 young women continually circulated among the filing cabinets. They were supplied with one copy of the customer's latest mail order, and their jobs were to update punches on the cards to incorporate the effect of the order. The information recorded was highly quantized in terms of the number and amount of the orders. The key to the system was the machine that used steel plates to print labels for the catalog to be mailed. A drawer of plates was stacked into the machine. The machine examined the punched holes in the cards on each plate and determined, according to wired-board logic, whether this pattern of holes qualified the customer to receive the particular catalog being distributed at that time. If the pattern was favorable, a label was printed; otherwise, not. Thus, for example, the management could decide to send the general catalog only to customers who purchased for more than $20 during the present season. By wiring the machine appropriately, this decision was implemented simply by passing every drawer of cards through the labeling machine. In making this decision, the management had traditionally looked at the direct profit to be expected from a customer as measured by the difference between the marginal profit on his/her purchases and the cost of sending him the catalog.

The mentioned operation was examined to see if it might be profitable to send catalogs, not just based on the profit they might produce during the present season, but also for the impact they might have in moving the customer into more profitable categories in the future.

The team decided to model this system by what is known currently as a ***Markovian decision process***. The process was as follows: Each customer's state was described by his/her purchase history; there were a total of about 50 states. Transitions occurred in each season. The reward for a transition, under a given catalog mailing policy, was simply the marginal profit from the transition less the cost of the catalogs mailed. Finally, the transition probabilities were computed by special runs from the Addressograph system. In fact, it was the existence of this system that made the entire approach feasible.

Howard continued to explain the optimization part of the operation. He worked on the problem for about 6 months and was able to develop a policy iteration method discussed in Howard (1960).

On the basis of Howard (1960), Matalytski, in a series of similar and overlapping papers (Matalytski and Pankov 2003a,b,c, 2004a,b,c, 2007, 2008, 2009, 2015) considered a banking network described as follows: For a banking network system, there is a central bank (CB) on the top level and large peripheral banks with their branches at levels below. All interbank transactions are performed by means of bank computer networks on payment transactions of banks and their clients. All branches operate according to the CB's rules, which are controlled electronically. Inquiries are messages, exchanges, or claims (the latter is used in this study hereafter), within the branches from the CB that bring payments to the CB.

Because finding characteristics of different objects depending on time is important for a particular system, transient analysis of such a system becomes necessary. The purpose of this chapter is to find the expected income from the passages between the states of banking network, which correspond to the entering of requests from the commercial banks in central bank and obtaining answers on these demands. They can serve as requests for transaction for the transfer of money into another bank.

The network system studied is a Markov network with incomes, where durations of claims residing in the network queueing system are represented by stochastic values distributed according to the exponential distribution function. Thus, suppose that the duration of claims waiting at the queue of the ith branch, referred as "station" hereafter, is a random variable distributed according to the exponential law with parameter θ_i and it does not depend on other factors such as the number of claims in the queue and waiting time of other claims in the queue. An arriving claim may have to wait to gain access to some computational resource and then, if it has not been serviced over a period of random time, it may go on to another station, that is, jugging of impatient claims is possible.

Hence, the system to be considered is assumed as (1) an open Jackson exponential queueing network, with two cases: Case 1 (single-server stations) and Case 2 (multiserver stations) or (2) a closed network with arbitrary service distributions as discussed in Case 3.

Case 1. Single-Server Stations For this case, the following terminologies and assumptions are used:

1. In this Markov chain with continuous time parameter, claims are considered to be the arrival jobs.
2. Claims are assumed to be of the same kind.
3. The system has n service stations, S_i, $i = 1, 2, \ldots, n$, with S_0 representing the outside station, from which external claims arrive with intensity λ claims per unit time.
4. The parameter $p_{0,j}$ denotes the probability of a claim arriving from S_0 and attending the station S_j, $\sum_{j=1}^{n} p_{i,j}0 = 1$.
5. Before service of a claim starts, it may be transferred from one station to another, say from S_i to S_j, with transition probability of $p_{i,j}$, $\sum_{j=1}^{n} p_{i,j} = 1$, $i = 1, 2, \ldots, n$. Thus, the transition movement of claims before completion of services may be described by an $(n+1) \times (n+1)$ matrix $\mathbf{P} = \|p_{i,j}\|$.
6. A state of the system is denoted by $k(t)$ and is defined as

$$k(t) \triangleq (k_1, k_2, \ldots, k_n; t), \tag{6.8}$$

where k_i is the number of claims in the station S_i, $i = 1, 2, \ldots, n$, at time t, $t \in [0, \ +\infty)$.
7. Claims are served according to the FCFS discipline.

8. Service times of claims at the queue S_i are random variables distributed exponentially with parameter $\mu_i(k_i)$, $i = 1, 2, \ldots, n$.
9. The intensity of claims leaving station S_i after waiting for a very long time in this station is $\theta_i(k_i)$, $i = 1, 2, \ldots, n$, which depends on the number of claims in the queue. This is the reneging rate.
10. Initial income of the company at station S_i is assumed to be $v_{i,0}$ and its income at time t is denoted by $V_i(t)$.

For the **purpose** of finding the expected income of the bank, the time segment $[0,\ t]$, for a large t, is divided into m equal subintervals of length

$$\Delta t \triangleq \frac{t}{m}, \tag{6.9}$$

and after completion of analysis, Δt will be set to zero, to get back to the time continuous case.

In order to find the system's income, the lth time subinterval; $l = 1, 2, \ldots, m$, is considered. Let us consider the possible events that may occur over this subinterval.

(i) The station S_i may receive a claim from the external source, which increases the station's income by $r_{0,i}$, where $r_{0,i}$ is a random variable with cumulative distribution function (cdf) denoted by $F_{0,i}(x)$, $i = 1, 2, \ldots, n$. The probability of this event is $\lambda p_{0,i}\Delta t + o(\Delta t)$, where $o(\Delta t)$ is the little o of Δt.

(ii) After completion of service at station S_i, the served claim may leave the station S_i and the network system permanently, which reduces the $t \in [0,\ +\infty)$. station's income by $R_{i,0}$, where $R_{i,0}$ is a random variable with cdf denoted by $F_{i,0}(x)$, $i = 1, 2, \ldots, n$. The probability of this event is $\mu_i(k_i(l))p_{i,\ 0}\Delta t + o(\Delta t)$.

(iii) After completion of service at station S_j, the served claim may leave the station S_j, and moves to station $S_{i,}$ which increases the station $S_{i,}$'s income by $r_{j,i}$, while the income of station S_j reduces by the same amount, $i, j = 1, 2, \ldots, n,\ \ j \neq i$, where $r_{j,i}$ is a random variable with cdf denoted by $F_{1j,i}(x)$. The probability of this event is $\mu_j(k_j(l))p_{j,i}\Delta t + o(\Delta t)$.

(iv) After completion of service at station S_i, the served claim may leave the station S_i and moves to station S_j, which reduces the station S_i's income by $R_{i,j}$, while the income of station S_j increases by the same amount, where $R_{i,j}$ is a random variable with cdf denoted by $F_{2i,j}(x)$, $i, j = 1, 2, \ldots, n, j \neq i$. The probability of this event is $\mu_i(k_i(l))p_{i,j}\Delta t + o(\Delta t)$.

   It is important to note that $r_{j,i} = R_{i,j}$ with probability 1, that is,

$$F_{1j,i}(x) = F_{2i,j}(x),\ \ i, j = 1, 2, \ldots, n. \tag{6.10}$$

(v) A claim may renege, that is, after waiting for a long time for service at station S_i, a claim may leave the service at station S_i forever, which reduces the station S_i's income by $H_{i,0}$, where $H_{i,0}$ is a random variable with cdf denoted by $Q_{i,0}(x)$, $i = 1, 2, \ldots, n$. The probability of this event is $\theta_i(k_i(l))q_{i,0}\Delta t + o(\Delta t)$.

(vi) A claim may jug, that is, after waiting for a long time for service at station S_j, a claim may leave the service station and move to another, say S_i, which increases the station S_i's income by $h_{j,i}$, while reduces the station S_j's income by the same amount, $i, j = 1, 2, \ldots, n, j \neq i$, where $h_{j,i}$ is a random variable with cdf denoted by $Q_{1j,i}(x)$. The probability of this event is $\theta_j(k_j(l))q_{j,i}\Delta t + o(\Delta t)$.

(vii) A claim may jug, that is, after waiting for a long time for service at station S_i, a claim may leave the service station and move to another, say S_j, which increases the station S_i's income by $H_{i,j}$, while reduces the station S_j's income by the same amount, $i, j = 1, 2, \ldots, n,\ \ j \neq i$, where $H_{i,j}$ is a random variable with cdf denoted by $Q_{2i,j}(x)$. The probability of this event is $\theta_i(k_i(l))q_{i,j}\Delta t + o(\Delta t)$.

It is important to note that $h_{j,i} = H_{i,j}$ with probability 1, that is,

$$Q_{1j,i}(x) = Q_{2i,j}(x), i, j = 1, 2, \ldots, n. \tag{6.11}$$

(viii) Over the time interval of size Δt, there may not be a change in the state of station S_i, which increases the station S_i's income by $r_i\Delta t$ through interest for the monetary resources circulating in the station, where r_i is a random variable with cdf denoted by $F_i(x)$. The probability of this event is $1 - (\lambda p_{0,i} + \mu_i(k_i(l)) + \theta_i(k_i(l)))\Delta t + o(\Delta t)$.

It is assumed that random variables $r_{j,i}$, $r_{o,i}$, $R_{0,j}$, r_i, and $h_{j,i}$, $i, j = 1, 2, \ldots, n$, are pairwise independent.

We now denote the change in the income of station S_i on the lth time interval of size Δt by $\Delta V_{i,l}(t)$, defined by

$$\Delta V_{i,l}(t) = V_{i,l}(t + \Delta t) \ - \ V_{i,l}(t). \tag{6.12}$$

From the possible events and their probabilities discussed in items (i)–(viii) and Equation (6.12), we obtain

$$\Delta V_{i,l}(t) = \begin{cases} r_{0,i} + r_i\Delta t, & \text{with probability, } \lambda p_{0,i}\Delta t + o(\Delta t), \\ -R_{i,0} + r_i\Delta t, & \text{with probability, } \mu_i(k_i(l))u(k_i(l))p_{i,0}\Delta t + o(\Delta t), \\ r_{j,i} + r_i\Delta t, & \text{with probability, } \mu_j(k_j(l))u(k_j(l))p_{j,i}\Delta t + o(\Delta t), \\ & \qquad j = 1, 2, \ldots, n, \ \ j \neq i, \\ -R_{i,j} + r_i\Delta t, & \text{with probability, } \mu_i(k_i(l))u(k_i(l))p_{i,j}\Delta t + o(\Delta t), \\ & \qquad j = 1, 2, \ldots, n, \ \ j \neq i, \\ -H_{i,0} + r_i\Delta t, & \text{with probability, } \theta_i(k_i(l))u(k_i(l))q_{i,0}\Delta t + o(\Delta t), \\ h_{j,i} + r_i\Delta t, & \text{with probability, } \theta_j(k_j(l))u(k_j(l))q_{j,i}\Delta t + o(\Delta t), \\ & \qquad j = 1, 2, \ldots, n, \ \ j \neq i, \\ -H_{i,j} + r_i\Delta t, & \text{with probability, } \theta_i(k_j(l))u(k_i(l))q_{i,j}\Delta t + o(\Delta t), \\ & \qquad j = 1, 2, \ldots, n, \ \ j \neq i, \\ r_i\Delta t, & \text{with probability, } 1 - (\lambda p_{0,i} + \mu_i(k_i(l))u(k_i(l)) \\ & \qquad +\theta_i(k_i(l))u(k_i(l))\Delta t + o(\Delta t), \end{cases} \tag{6.13}$$

where $u(x)$ is the Heaviside function, that is,

$$u(x) = \begin{cases} 1, & x > 0, \\ 0, & x \leq 0. \end{cases} \tag{6.14}$$

The income of the station S_i can be written as

$$V_i(t) = v_{i,0} + \sum_{l=1}^{m} \Delta V_{i,l}(t). \tag{6.15}$$

For $i, j = 1, 2, \ldots, n$, denoting by $E(\cdot)$ the expected value, mean, or average of a random variable, we define the following:

$$\left.\begin{array}{ll} E\{r_{j,i}\} = \int_0^\infty x\ dF_{1i,j}(x) = a_{j,i}, & E\{R_{i,j}\} = \int_0^\infty x\ dF_{2i,j}(x) = b_{i,j}, \\ E\{r_i\} = \int_0^\infty x\ dF_i(x) = c_i, & E\{r_{0,i}\} = \int_0^\infty x\ dF_{0,i}(x) = a_{0,i}, \\ E\{R_{i,0}\} = \int_0^\infty x\ dF_{i,0}(x) = b_{i,0}, & E\{h_{j,i}\} = \int_0^\infty x\ dQ_{1j,i}(x) = \overline{h}_{j,i}, \\ E\{H_{i,j}\} = \int_0^\infty x\ dQ_{2i,j}(x) = \overline{H}_{i,j}, & E\{H_{i,0}\} = \int_0^\infty x\ dQ_{i,0}(x) = \overline{H}_{i,0}. \end{array}\right\} \tag{6.16}$$

From Eqs. (6.10), (6.11), and (6.15), we have

$$a_{j,i} = b_{j,i},\ \ \overline{h}_{j,i} = \overline{H}_{j,i},\ \ i, j = 1, 2, \ldots, n. \tag{6.17}$$

Since exact value of expected income is difficult to find, if not formidable, an approximation for station S_i at time t was obtained. Thus, for a fixed realization $k(l)$, using Eqs. (6.13) and (6.16), approximation can be found, up to $o(\Delta t)$, for the expected value, denoted by $E[V_{i,l}(t)/K(l)],\ \ i = 1, 2, \ldots, n$. Thus, from Eqs. (6.9) and (6.17), we have

$$\begin{aligned} E\left[\frac{V_i(t)}{k(t)}\right] &= \sum_{l=1}^{m} E\left[\frac{V_{i,l}(t)}{k(t)}\right] \\ &= v_{i,0} + (\lambda a_{0,i} p_{0,i} + c_i)t + \sum_{j=1}^{n} a_{j,i} p_{j,i} \sum_{l=1}^{m} \mu_j(k_j(l)) u(k_j(l)) \Delta t \\ &\quad + \sum_{j=1}^{n} \overline{h}_{j,i} q_{j,i} \sum_{l=1}^{m} \theta_j(k_j(l)) u(k_j(l)) \Delta t - \sum_{j=0}^{n} b_{i,j} p_{i,j} \sum_{l=1}^{m} \mu_i(k_i(l)) u(k_i(l)) \Delta t \\ &\quad - \sum_{j=0}^{n} \overline{H}_{i,j} q_{i,j} \sum_{l=1}^{m} \theta_i(k_i(l)) u(k_i(l)) \Delta t + o(\Delta t), i = 1, 2, \ldots, n. \end{aligned} \tag{6.18}$$

As m approaches infinity, Δt will approach zero, and hence

$$\sum_{l=1}^{m} \mu_j(k_j(l))u(k_j(l))\Delta t \underset{\Delta t \to 0}{\longrightarrow} \int_0^{\infty} \mu_j(k_j(s))u(k_j(s))ds, \quad j = 1, 2, \ldots, n. \tag{6.19}$$

Therefore,

$$\begin{aligned} E\left[\frac{V_i(t)}{k(t)}\right] &= v_{i,0} + (\lambda a_{0,i} p_{0,i} + c_i)t + \sum_{j=1}^{n} a_{j,i} p_{j,i} \int_0^t \mu_j(k_j(s))u(k_j(s))ds \\ &\quad + \sum_{j=1}^{n} \overline{h}_{j,i} q_{j,i} \int_0^t \theta_j(k_j(s))u(k_j(s))ds \\ &\quad - \sum_{j=0}^{n} b_{i,j} p_{i,j} \int_0^t \mu_i(k_i(s))u(k_i(s))ds \\ &\quad - \sum_{j=0}^{n} \overline{H}_{i,j} q_{i,j} \int_0^t \theta_i(k_i(s))u(k_i(s))ds, \quad i = 1, 2, \ldots, n. \end{aligned} \tag{6.20}$$

Averaging over $k(t)$, applying the normalizing condition $\sum_k P\{k(t) = k\} = 1$, the expected income for station S_i, $i = 1, 2, \ldots, n$, is

$$\begin{aligned} E\{V_i(t)\} \equiv v_i(t) &= v_{i,0} + \sum_k P\{k(t) = k\} E\left\{\frac{V_i(t)}{k(t)}\right\} \\ &= v_{i,0} + (\lambda a_{0,i} p_{0,i} + c_i)t + \sum_k P\{k(t) = k\} \\ &\quad \times \left\{ \sum_{j=1}^{n} \left[a_{j,i} p_{j,i} \int_0^t \mu_j(k_j(s))u(k_j(s))ds + \overline{h}_{j,i} q_{j,i} \int_0^t \theta_j(k_j(s))u(k_j(s))ds \right] \right. \\ &\quad \left. - \sum_{j=0}^{n} \left[b_{i,j} p_{i,j} \int_0^t \mu_i(k_i(s))u(k_i(s))ds + \overline{H}_{i,j} q_{i,j} \int_0^t \theta_i(k_i(s))u(k_i(s))ds \right] \right\}, \\ &\quad i = 1, 2, \ldots, n. \end{aligned} \tag{6.21}$$

Case 2 Multiserver Stations For a multiserver station, let S_i, $i = 1, 2, \ldots, n$, contain m_i identical service lines, where each line's intensity is μ_i, $i = 1, 2, \ldots, n$. Thus, for this case, we have

$$\begin{aligned} \mu_i(k_i(s))u(k_i(s)) &= \begin{cases} \mu_i k_i(s), & k_i(s) \le m_i, \\ \mu_i m_i, & k_i(s) > m_i, \end{cases} \\ &= \mu_i \min(k_i(s), m_i), i = 1, 2, \ldots, n. \end{aligned} \tag{6.22}$$

Suppose that

$$\theta_i(k_i(s))u(k_i(s)) = \begin{cases} 0, & k_i(s) \le m_i, \\ \theta_i(k_i(s) - m_i), & k_i(s) > m_i, \end{cases}$$
$$= \theta_i(k_i(s) - m_i)u(k_i(s) - m_i), i = 1, 2, \dots, n. \quad (6.23)$$

Then, from Equation (6.21), we have

$$v_i(t) = v_{i,0} + (\lambda a_{0,i} p_{0,i} + c_i)t + \sum_k P\{k(t) = k\}$$
$$\times \int_0^t \left\{ \sum_{j=1}^n [\mu_j a_{j,i} p_{j,i} \min(k_j(s), m_j) + \theta_j \overline{h}_{j,i} q_{j,i}(k_j(s) - m_j)u(k_j(s) - m_j)] \right.$$
$$\left. - \sum_{j=0}^n [\mu_i b_{i,j} p_{i,j} \min(k_i(s), m_i) + \theta_i \overline{H}_{i,j} q_{i,j}(k_i(s) - m_i)u(k_i(s) - m_i)] \right\} ds,$$
$$i = 1, 2, \dots, n. \quad (6.24)$$

Let us assume the following:

$$E[\min(k_i(s), m_i)] = \min(N_j(s), m_i), \quad j = 1, \dots, n, \quad (6.25)$$

$$E[(k_j(s) - m_j)u(k_j(s) - m_j)] = (N_j(s) - m_j)u(N_j(s) - m_j), \quad j = 1, \dots, n, \quad (6.26)$$

$$E\left[\int_0^t \min(k_j(s), m_j) ds\right] = \int_0^t \min(N_j(s), m_j) ds, \quad j = 1, \dots, n, \quad (6.27)$$

where $N_j(s)$ is the expected number of claims (waiting and being served) in the station S_j over the time interval $[0, s]$, $j = 1, \dots, n$.

It is important to note that relations (6.25)–(6.27) hold if the system network operates under high load, that is, when $k_j(t) > m_j$, $\forall t$, since in this case $\min(k_j(s), m_j) = m_j, j = 1, \dots, n$.

On the basis of assumptions (6.25)–(6.27), from Equation (6.23), we have

$$v_i(t) = v_{i,0} + (\lambda a_{0,i} p_{0,i} + c_i)t$$
$$+ \int_0^t \left\{ \sum_{j=1}^n [\mu_j a_{j,i} p_{j,i} \min(N_j(s), m_j) + \theta_j \overline{h}_{j,i} q_{j,i}(N_j(s) - m_j)u(N_j(s) - m_j)] \right.$$
$$\left. - \sum_{j=0}^n [\mu_i b_{i,j} p_{i,j} \min(N_i(s), m_i) + \theta_i \overline{H}_{i,j} q_{i,j}(N_i(s) - m_i)u(N_i(s) - m_i)] \right\} ds,$$
$$i = 1, 2, \dots, n. \quad (6.28)$$

From Eqs. (6.17) and (6.28), the expected income of the network system can be written as

$$\sum_{i=1}^{n} v_i(t) = \sum_{i=1}^{n} \{v_{i,0} + (\lambda a_{0,i} p_{0,i} + c_i)t$$
$$- \int_0^t [\mu_i b_{i,0} p_{i,0} \min(N_i(s), m_i) + \theta_i \overline{H}_{i,0} q_{i,0} (N_i(s) - m_i) u(N_i(s) - m_i)] \Big\} ds. \tag{6.29}$$

Case 3 Closed Network with Multiserver, Arbitrary Service Distributions and Bounded Waiting Times We now consider the system as a closed network with arbitrary claim service time distributions with bonded waiting times in the queue. Let K be the number of claims in the closed network system. Suppose station S_i contains m_i identical servers, $i = 1, 2, \dots, n$, with μ_i as the servicing intensity for claims in each line of S_i. Suppose also that τ_i and ρ_i are, respectively, the average time a claim resides in the station (both waiting and servicing) and the average number of occupied lines in the station S_i, $i = 1, 2, \dots, n$, over the time interval $[0, t]$, $t < T$, where T is the maximum time limit for waiting for a claim in a station.

Statkevich and Matalytski (2012) showed the following relations, recurrent in t,

$$\sum_{i=1}^{n} N_i(0) = K, \tag{6.30}$$

$$\rho_i = \min\{N_i(t), m_i\}, \quad i = 1, 2, \dots, n, \tag{6.31}$$

$$N_i(t+1) = \tau_i(t) \sum_{i=1}^{n} \{\mu_l \rho_l(t) p_{l,i} + [N_i(t) - m_i]\theta_i u(N_l(t) - m_l) q_{l,i}\}, \quad i = 1, 2, \dots, n, \tag{6.32}$$

where θ_i denotes the intensity of claims leaving the queue of the station S_i, $i = 1, 2, \dots, n$. Using Eqs. (6.30)–(6.32), the average characteristics of the network at any time moment in the transition mode can be estimated. For instance, let T_i denote the time instance when $N_i(t)$ reaches its stationary value. Assume that this value has been reached on the l_ith iteration. Then, as an estimate for $N_i(t)$ at time $t = (T_i / l_i)u_i$, using this method, the value of $N_i(u_i)$ can be obtained on the u_ith iteration, $u_i = 1, 2, \dots, l_i$. Similarly, other estimated values of $N_i(t)$ at other time instances in the interval $[0, T_i]$, $i = 1, \dots, n$, can be obtained by interpolation.

Example 6.3.2 Closed Exponential Network System Consider Example 6.3.1 in Case 3. Let us consider a closed exponential network system. Claims after servicing in a peripheral station S_i arrive at the central branch S_n and will return to the station S_i with probability $p_{n,i}$, $i = 1, 2, \dots, n-1$. We choose $n = 5$, $K = 45$, and the following data:

I	1	2	3	4	5
$m_i =$	2	2	6	6	7
$\mu_i =$	5	3	4	2	5.3
$\theta_i^{-1} =$	1.400	0.667	0.850	2.600	1.760
$p_{i,5} = q_{5,i} =$	1	1	1	1	0
$p_{5,i} = q_{5,i} =$	1/4	1/4	1/4	1/4	0
$p_{i,i} = q_{i,i} =$	0	0	0	0	0
$v_i(0) =$	20	20	20	20	80
$N_i(0) =$	5	5	5	5	25

It is important to note that for $i = 1, 2, 3, 4, 5$, μ_i are service rates at stations and θ_i^{-1} are the average waiting time of claims at stations. Then, the expected incomes from transitions between the network states are calculated as follows:

$$(a_{5,1};\ a_{5,2};\ a_{5,3};\ a_{5,4}) = (a_{1,5};\ a_{2,5};\ a_{3,5};\ a_{4,5}) = (3.5;\ 2.0;\ 3.5;\ 4.5),$$

$$(b_{5,1};\ b_{5,2};\ b_{5,3};\ b_{5,4}) = (b_{1,5};\ b_{2,5};\ b_{3,5};\ b_{4,5}) = (4.8;\ 2.6;\ 1.1;\ 3.4),$$

$$(\overline{h}_{5,1};\ \overline{h}_{5,2};\ \overline{h}_{5,3};\ \overline{h}_{5,4}) = (\overline{h}_{1,5};\ \overline{h}_{2,5};\ \overline{h}_{3,5};\ \overline{h}_{4,5}) = (6.0;\ 9.0;\ 4.0;\ 3.0),$$

$$(\overline{H}_{5,1};\ \overline{H}_{5,2};\ \overline{H}_{5,3};\ \overline{H}_{5,4}) = (\overline{H}_{1,5};\ \overline{H}_{2,5};\ \overline{H}_{3,5};\ \overline{H}_{4,5}) = (1.5;\ 1.9;\ 3.0;\ 3.6).$$

We leave it as an exercise for the reader to find $N_i(s)$, $i = 1, 2, 3, 4, 5$, using Eqs. (6.30)–(6.32) and to find expressions for expected incomes of the network's systems using Equation (6.26). For instance, the expression for the expected income of the peripheral station S_3 has the form

$$u_3(t) = 12.03e^{-5.51t} + 5.902e^{-4.023t} + 0.696e^{-3.45t} + 14.32e^{-2t} + 15.006t - 12.237.$$

Similar expressions for expected incomes of other stations in the network, that is, expected incomes, depend approximately linearly on time.

Example 6.3.3 Busy Period of a Cyclic Queueing Network As we have already alluded to, queueing networks are widely used for modeling computer and communications systems. We also have already discussed the product-form solution for the stationary probabilities, from which performance values as throughput and average queue size can be derived. In some applications, however, the interest is in the delay or response time of an individual task. In particular, the interest may be in finding the distributions of time periods during which a service station is operating continuously (the busy period), or during which several stations are operating simultaneously. Such characteristics cannot be obtained from the steady-state analysis and require a time-dependent analysis of processes of finite lengths.

Recently, Boxma and Daduna (2014) have considered a model that is an expansion of the basic cyclic tandem queue by Kuehn (1983). They determined the joint sojourn time distribution of a tagged task in, first, the general queue and, then, the exponential queue. Subsequently, they indicated how the approach toward this closed system also allowed to study the joint sojourn time distribution of a tagged task in the equivalent open two-queue system, consisting of FCFS single-server queues with general and exponential service times, respectively, in the case that the input process to the first queue is a Poisson process. In this study, we discuss the Kuehn's model.

Efficient algorithms were developed by Kuehn (1983) for the stationary state analysis, from which the performance can be evaluated in terms of throughput and ***average delays***. He used the ***method of first passage times*** (Chapter 2) to analyze time-dependent processes within Markovian queueing networks as busy periods and response (cycle) times. The first passage time method has been used as a powerful means to study waiting time problems for queueing systems with different service disciplines (Cohen, 1982; Syski, 1964; Kuehn, 1972, 1973). One of the advantages of the method of first passage times is for cases with higher degrees of dependence as, for example, networks with state-dependent service rates, queue disciplines other than FIFO, or cycle paths with overtaking.

Kuehn (Kühn) derived closed-form expressions for the cyclic queueing system with two service stations. He also discussed generalizations to more complex networks. The analysis of lifetime processes leads to a much deeper insight into the behavior of queueing networks than that with the stationary state analysis allowing for individual task delay distributions or distributions of periods of continuous or simultaneous service. In fact, he wanted to show that the method of first passage times is an adequate tool to solve ***lifetime*** processes in Markovian queueing networks.

Consider a class of Markovian queueing systems, which can be described by a Markov chain with a countable set of states and a continuous time parameter. The waiting time or the ***response time*** of a task and the busy period of a service station can be considered as "***lifetimes***," T, of one or several tagged (test) tasks within a properly defined set S of states. The lifetime terminates when the test task leaves S for the first time entering a "taboo" set $H = \overline{S}$; his lifetime is equal to the "first passage time" to H. The lifetime process can be considered as a special system-state process with "absorbing" states in H from which a test task, once having entered, does not return. This modified process can be constructed from the system-state transition probabilities under the condition that states in H are excluded. The state of the modified process must be specified such that all effects, which may influence the lifetime T of the test task directly or indirectly, are reflected properly.

We now present the model studied by Kuehn (Kühn), a basic cyclic queueing network, depicted in Figure 6.1, consisting of two stations, each with a single server.

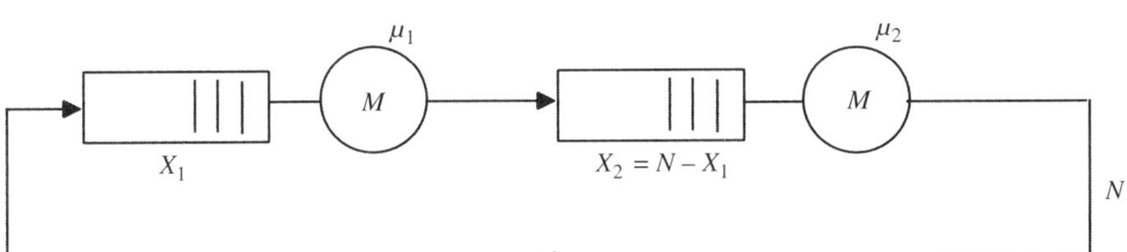

Figure 6.1 Cyclic queueing system.

Service distribution at each station is exponential with average service times at stations 1 and 2 as $h_1 = 1/\mu_1$ and $h_2 = 1/\mu_2$, respectively. Each station has a buffer in front of it with capacity of stations. Let the total capacity (buffer plus service) be represented by X_1 and X_2, respectively, with total capacity of the system to be N. Thus, $X_2 = N - X_1$, where each X_1 and X_2 may take values as $0, 1, 2, \dots, N$.

Let $T_{B\nu}$, $\nu = 1, 2$, denote the busy period of station ν, $\nu = 1, 2$. Let also $T_{BS\nu}$, $\nu = 1, 2$, denote the simultaneous busy period of both stations initiated by a busy period of station ν, $\nu = 1, 2$. Further, let T_C denote the cycle time or response time of a test task between two successive arrivals at either station, for instance, station 1. Finally, let the steady-state distribution of the number in the system be denoted by P_X, which is expressed as

$$P_X = P\{X_1 = x, X_2 = N - x\}, \quad x = 0, 1, 2, \dots, N. \tag{6.33}$$

It is known that (see, for example, Haghighi and Mishev, 2013, Chapter 5, Section 6), if $\rho = \frac{\mu_2}{\mu_1}$, then

$$P_X = \begin{cases} \dfrac{(1-\rho)\rho^X}{1-\rho^{N+1}}, & \text{if } \rho \neq 1, \\ \dfrac{1}{N+1}, & \text{if } \rho = 1. \end{cases} \tag{6.34}$$

The objective of the model is to find the probability distribution function of the variables $T_{B\nu}$, $\nu = 1, 2$, $T_{BS\nu}$, $\nu = 1, 2$, and T_C using the mean of the first passage time method.

In order to achieve the purpose, let $\{\xi(t), t \geq 0\}$ denote the lifetime process with state space S. We also let $\xi(t) = i$ define the state of the busy period lifetime process in station 1, where i represents the number of tasks in station 1, $i \in S = \{1, 2, \dots, N\}$. It is important to note that H in this case is $H = \overline{S} = \{0\}$. The busy period in station 1 starts with the first arrival task from station 2, while the station 1 is empty and the server is idle. This busy period will terminate when the last task leaves station 1, that is, when i transits from 1 to 0. Thus, a busy period starts with initial state 1. From that state, either it terminates or the process changes into state 2 by another arrival from station 2. Further arrivals may change the state of the process up to state N. However, the process can only terminate from state 1.

Let us denote by $w(t \mid i)$, the conditional complementary lifetime distribution function, defined by

$$w(t \mid i) = P\{T > t \mid \xi(0) = i\}, \quad i \in S, \tag{6.35}$$

with its Laplace transform denoted by $W(s \mid i)$. It is clear that $w(0 \mid i) = 1$. We also denote the probability density function (pdf) of T and its Laplace transform, respectively, by $d(t \mid i)$ and $D(s \mid i)$. We further denote by $q_{i,j}$, the instantaneous transition rate for the transition of the $\xi(t)$ process from i to j. Furthermore, q_i represents the transition rate for leaving state i to any other state in S or H, and ε_i represents the transition rate for leaving state i into H, where

$$q_i = \sum_{j \neq i} q_{i,j} + \epsilon_i, \quad i, j \in S, \tag{6.36}$$

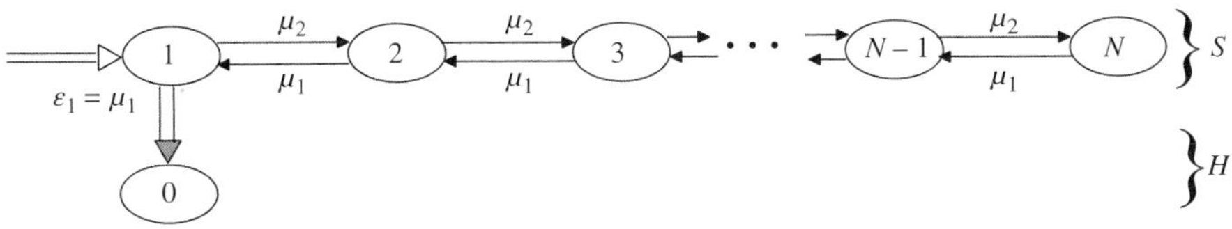

Figure 6.2 Busy period process state transition for station 1.

and

$$\epsilon_i = \lim_{t\to 0}\left(-\frac{d}{dt}w(t\,|\,i)\right),\quad i,j\in S. \tag{6.37}$$

It should be noted that the lifetime starts and terminates from any state in S. However, in many applications, it starts or terminates from a particular subset of S.

The busy period process of station 1 is summarized and represented by the state transition diagram of Figure 6.2.

We leave it as an exercise for the reader to show that the busy period process is described by

$$(q_i + s)D(s\,|\,i) - \sum_{j\neq i} q_{i,j}D(s\,|\,j) = \epsilon_i,\quad i,j\in S, \tag{6.38}$$

which results in

$$\begin{pmatrix} s+\mu_1+\mu_2 & -\mu_2 & 0 & \cdots & 0 \\ \mu_1 & s+\mu_1+\mu_2 & -\mu_2 & \cdots & 0 \\ 0 & -\mu_1 & s+\mu_1+\mu_2 & \ddots & 0 \\ 0 & 0 & -\mu_1 & \ddots & 0 \\ \vdots & 0 & \ddots & \ddots & 0 \\ \vdots & \vdots & \vdots & s+\mu_1+\mu_2 & -\mu_2 \\ \cdots & \cdots & \cdots & -\mu_1 & s+\mu_1 \end{pmatrix}\cdot\begin{pmatrix} D(s|1) \\ D(s\,|\,2) \\ D(s\,|\,3) \\ \vdots \\ D(s\,|\,N-2) \\ D(s\,|\,N-1) \\ D(s\,|\,N) \end{pmatrix} = \begin{pmatrix} \mu_1 \\ 0 \\ 0 \\ \vdots \\ 0 \\ 0 \\ 0 \end{pmatrix}. \tag{6.39}$$

It is important to note that the eigenvalues of Equation (6.39) are negative-real, distinct, and bounded by the interval $[-2(\mu_1+\mu_2), 0)$.

The matrix equation (6.39) can be solved by standard methods. Thus, the first two moments of the busy period for station 1, T_{B1}, are, respectively as follows:

$$E[T_{B1}] = \frac{1}{\mu_1}\frac{1-\rho^N}{1-\rho} = \frac{1}{\mu_1}\sum_{k=0}^{N-1}\rho^k, \tag{6.40}$$

and

$$E[T_{B1}^2] = \frac{2}{\mu_1^2}\frac{1}{(1-\rho)^3}[1-\rho^{2N+1}-(2N+1)(1-\rho)\rho^N]. \tag{6.41}$$

Remark Let the average number of tasks cycling around per time unit (the throughput rate) be denoted by λ. Also, let the idle period of station 1 be denoted by T_{I1}.

We assume that the idle busy periods are independent, identically distributed (iid) random variables with exponential distribution as

$$P\{T_{I1} \leq t\} = 1 - e^{-\mu_2 t}, \quad t \geq 0. \tag{6.42}$$

Then,

$$\lambda = \frac{E[T_{B1}]}{E[T_{B1}] + E[T_{I1}]}\mu_1 = \frac{1 - \rho^N}{1 - \rho N + 1}\mu_2. \tag{6.43}$$

We, finally note that the results for T_{B2} can be obtained by likewise interchanging of indices.

Example 6.3.4 Healthcare Case Studies as Network of Queues Some studies on case studies of hospital systems are found in the literature (Bruin et al., 2007; Sobolov et al., 2006). The questions to address are as follows: (1) how many servers (nurses) are required (staffing) and (2) how many fixed resources (beds) are needed (allocation) to minimize costs while sustaining a certain service level? Yom-Tov (2008) answered these questions in her PhD dissertation "Queues in Hospitals, Semi-Open Queueing Networks in the QED Regime (QED = Quality- and Efficiency-Driven)" at Industrial Engineering and Management Technion, Israel Institute of Technology, studies queueing in healthcare systems. We take part of his work as a case study in this section.

In general, hospitals consist of several medical units (MU) specializing in different medical areas such as internal (MU_1), surgery (MU_2), cancer (MU_3), women (MU_4), and children (MU_5), … (MU_n)They also include service units providing services for the patients such as imaging units (IU), laboratory units (LU), and information technology (IT) units. Furthermore, inpatients arrive randomly via emergency room (ER). These patients are dismissed after being treated or admitted to stay and discharged after treatment. These operations are graphically represented by Yom-Tov as a complex stochastic network in Figure 6.3.

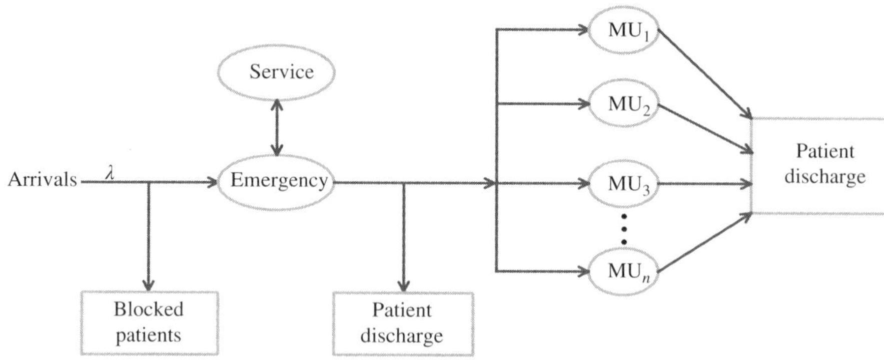

Figure 6.3 Basic operational model of a hospital system.

Capacity restrictions of a hospital can lead to a blocking system. Thus, the ER and MUs can be blocked and a situation where ambulances are turned away. In large medical centers, there are several MUs of the same type, presenting in a parallel setting as shown in Figure 6.3. Regardless of how many similar medical units are put in place, blocking yet could occur. In surgery and organ transplant, sophisticated tests such as computed tomography (CT) scans and magnetic resonance imaging (MRI) become necessary and, thus, cause weeks and months of delays in treatments of patients in queues.

When a patient in the unit requires the assistance of a nurse, the patient's state is referred to as ***needy***. When no such assistance is required, the patient's state is referred to as ***dormant***. It is assumed that when patients arrive, they start in a needy state and then alternate between needy and dormant states. When patients are discharged from the hospital, they leave from the needy state. The last treatment will be the discharge process. A vacant bed after a patient leaves that medical unit is to be made available for a new patient. Bed cleaning is usually the job of cleaning crew and, thus, that state of a bed is referred to as ***cleaning***. Responses of a needy if there is an idle nurse is immediate. Otherwise, needy is treated based on an FCFS policy.

Jennings and Véricourt (2007) considered the hospital queues in QED regime based on a special Jackson closed-network (machine-repairmen) model of the MU, where the circulating customers are patients' requests for nursing assistance. Their model is, in fact, an asymptotic $M/M/s/\infty/n$, as the number of servers increase indefinitely (Fig. 6.4). Their model consists of n beds, all occupied by patients, and s nurses. They assumed that treatment and dormant times are iid exponential with rates μ and λ, respectively. It is also assumed that the needy and dormant times are independent of each other.

Jennings and Véricourt found that the ***approximate probability of delay*** has a nondegenerate limit, say α, $\alpha \in (0, 1)$, if and only if $\beta_n = \left(\frac{s_n}{n} - rg\right)\sqrt{n} \to \beta$, as $n \to \infty$, for some $\beta \in (-\infty, \infty)$, where

$$\alpha = \left[1 + e^{\frac{-\beta^2}{r^2}} \sqrt{r} \frac{\Phi\left(\frac{\beta}{\sqrt{r(1-r)}}\right)}{\Phi\left(\frac{-\beta}{\sqrt{r(1-r)}}\right)}\right]^{-1}, \tag{6.44}$$

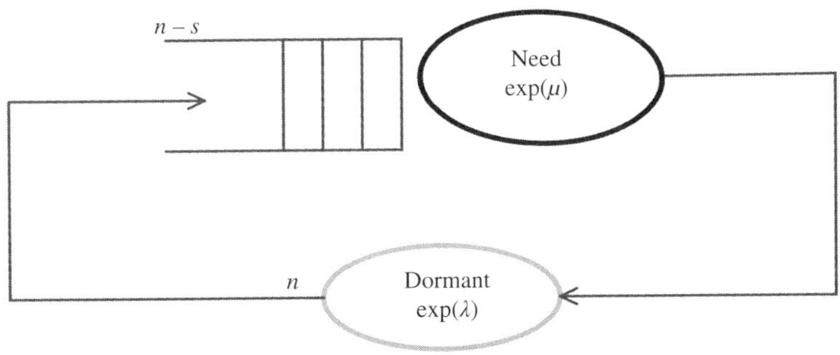

Figure 6.4 Jennings and Véricourt's model.

where s_n, $s_n = [rn + \beta\sqrt{n}]$, is the number of nurses in the nth system (the staffing rule) and $r = \lambda/(\lambda + \mu)$.

Galit Yom-Tov's medical unit (MU) model is described as follows: There is a maximum of n number of beds available and s number of nurses serving patients in the unit. It is a standard practice and is assumed that the number of nurses does not exceed the number of beds, that is, $s \leq n$, that is, beds are either partly or fully occupied by patients. All other aforementioned standard terms in healthcare system apply here. The queueing policy is FCFS. Because of a possible complication, after completing treatment, a patient is discharged from the hospital with a probability of $1 - p$ or returns to a dormant state with a probability of p for further treatment.

Treatment, dormant, and cleaning times are assumed iid random variables with exponential distributions represented by μ, δ, and σ, respectively. It is assumed that the needy, dormant, and cleaning times are independent of each other and the arrival process. The arrival process is assumed to be a Poisson process at rate λ. It is further assumed that if patients arrive at the MU to be hospitalized but the unit is full, they are diverted elsewhere, say, back to the ER. This assumption creates a blocking system, that is, the MU is blocked and the request is lost. In a call center correspondence, this situation is of receiving a busy signal. Thus, a service system is considered in which the patients need medical care provided by the nurses, if a bed is available in the medical unit for hospitalization. This model is a semi-open queueing network with multiple statistically identical customers and servers. For convenience, this model is referred to as ***IW*** (internal ward, Fig. 6.5).

The aforementioned semi-open queueing network, represented by Figure 6.5, can be analyzed as a closed Jackson network and, hence, it yields a product-form steady-state solution (Fig. 6.6). This is done by representing the model as a system with four nodes as follows:

Node 1 represents beds with patients in a needy state.

Node 2 represents beds with patients in a dormant state. For convenience, one sometimes refers to a bed with a patient as simply a patient.

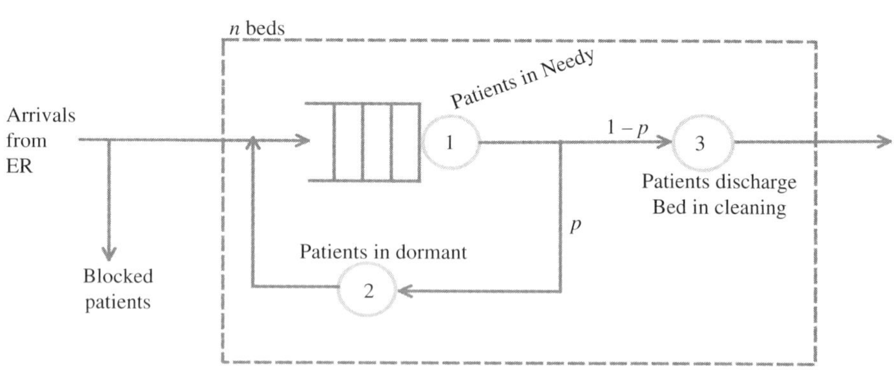

Figure 6.5 The IW model as a semi-open queueing network.

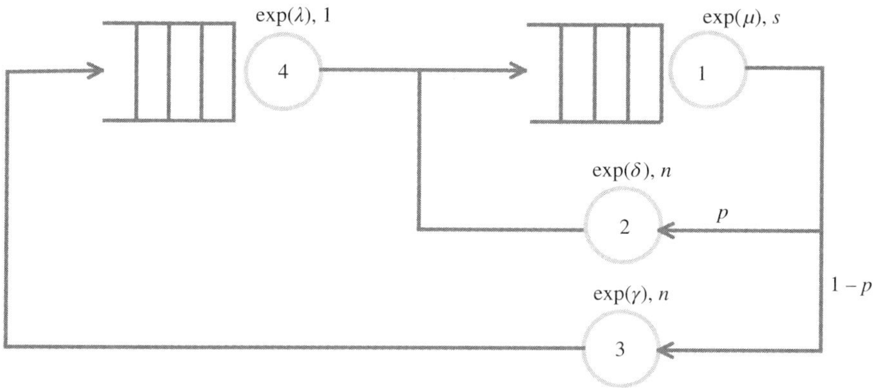

Figure 6.6 IW model as a closed Jackson network.

Node 3 represents beds in preparation, that is, in a cleaning state.

Node 4 represents prepared beds, awaiting a patient.

Nodes 1–3 are multiserver queues. Node 1 can handle, at most, s patients at one time. Nodes 2 and 3 can "handle" or contain at most n patients at a time. The fourth node has been added to formulate the model, which is an exponential server system with rate λ.

The following three parameters were found: (1) probability of blocking, (2) probability of waiting, and (3) expected waiting time. In addition, fluid limits for a closely related open network was found, in which the restriction on n was relaxed, and thus the blocking phenomenon cannot occur. Then, these findings were generalized to any semi-open network which has the following structure: one arrival stream ($M/M/1$), one service node with multiple identical servers ($M/M/s$), and any arbitrary finite number of delay nodes ($M/M/\infty$).

Let $Q(t) = (N(t), D(t), C(t))$ represent the number of beds in needy, dormant, or cleaning states, respectively, with $N(t) + D(t) + C(t) \leq n$ for all $t \geq 0$. The process $\{Q(t), t \geq 0\}$ is a continuous Markov chain with the following properties: The state (i, j, k) represents a case when i needy patients are being served or wait for service, j dormant patients are in the unit, but need no service at the time, and k beds are being prepared for future patients, $i + j + k \leq n$.

In comparison with the closed Jackson model, for Yom-Tov's model, we have the following situations: The needy node (or the first node) can be thought of as a system with s servers, each with a buffer in form of its stations and exponential service time with parameter μ. The dormant node (or the second node) is an infinite-server node with exponential service time distribution with parameter δ. The cleaning node (or the third node) is also an infinite-server node with exponential service time distribution with parameter γ. The capacity being finite, a new patient can enter the unit only if $N(t) + D(t) + C(t) < n$. Hence, the rate of admitting new patients into the unit (arrival) is $\lambda(i, j, k)$,, where

$$\lambda(i, j, k) = \begin{cases} \lambda, & \text{if } i+j+k < n, \\ 0, & \text{otherwise.} \end{cases} \tag{6.45}$$

Denoting by $\pi^m(i)$ the steady-state probability for node m, $m = 1, 2, 3, 4$, ($M/M/s$, $M/M/\infty$, $M/M/\infty$, $M/M/1$), based on Gordon and Newell (1967), the joint steady-state distribution for the four closed-node Jackson network model, denoted by $\pi(i, j, k, l)$, has the following product form:

$$\pi(i, j, k, l) = \begin{cases} \dfrac{\pi^1(i)\pi^2(j)\pi^3(k)\pi^4(l)}{\sum_{a+b+c+d=n}\pi^1(a)\pi^2(b)\pi^3(c)\pi^4(d)}, & i+j+k+l = n, \\ 0, & \text{otherwise.} \end{cases} \tag{6.46}$$

Thus, the steady-state probability $\pi(i, j, k)$ of having *i needy patients*, *j dormant patients*, and *k beds in cleaning* is given in a product form as follows:

$$\pi(i, j, k) = \begin{cases} \pi_0 \frac{1}{\nu(i)} \left(\frac{\lambda}{(1-p)\mu}\right)^i \frac{1}{j!}\left(\frac{p\lambda}{(1-p)\delta}\right)^j \frac{1}{k!}\left(\frac{\lambda}{(\gamma}\right)^k, & 0 \le i+j+k \le n, \\ 0, & \text{otherwise,} \end{cases} \tag{6.47}$$

where

$$\nu(i) = \begin{cases} i!, & i \le s, \\ s!s^{i-s}, & i \ge s, \end{cases} \tag{6.48}$$

and π_0 is given by

$$\begin{aligned} \pi_0^{-1} &= \sum_{0\le i+j+k\le n} \frac{1}{\nu(i)}\left(\frac{\lambda}{(1-p)\mu}\right)^i \frac{1}{j!}\left(\frac{p\lambda}{(1-p)\delta}\right)^j \frac{1}{k!}\left(\frac{\lambda}{\gamma}\right)^k \\ &= \sum_{l=0} \frac{1}{l!}\left(\frac{\lambda}{(1-p)\mu} + \frac{p\lambda}{(1-p)\delta}\right)^l \\ &\quad + \sum_{l=s+1}^{n}\sum_{m=s+1}^{l}\sum_{i=s+1}^{m}\left(\frac{1}{s!s^{i-s}} - \frac{1}{i!}\right)\frac{1}{(m-i)(l-m)}\left(\frac{\lambda}{(1-p)\mu}\right)^i \frac{1}{k!} \\ &\quad \times \left(\frac{p\lambda}{(1-p)\delta}\right)^{m-i}\left(\frac{\lambda}{\gamma}\right)^{l-m}. \end{aligned} \tag{6.49}$$

For a detailed derivation of Equation (6.49), see Yom-Tov (2008). It should be noted that π in Eqs.(6.45) and (6.49) is also a function of n and s. Because of this property, the author also uses notations such as $\pi_n(\cdot)$ and $\pi_{n,x}(\cdot)$.

The focus of Yom-Tov in her work is on the following managerial questions, and her attempt to have answers to both questions to be synchronized.

Question 1. How many beds should be planned for in the unit?

Answer to this question could be provided by the viewpoint of service quality, that is, providing reasonable availability, or from the aspect of cost optimization. Hence, one may ask the following subquestion to answer question 2:

What is the probability of blocking?

In other words, what is the duration of a full capacity ($i + j + k = n$), which translates into the percentage of patients not admitted into the MU?

Question 2. How many nurses should be planned for the unit?

This question is in the context of providing reasonable service levels and from the viewpoint of cost/profit optimization. Hence, one may ask the following subquestions to answer question 2:

(a) What is the probability of waiting for a nurse?
(b) What is the probability of waiting for more than T units of time?

To answer question 1, let us denote the steady-state probability that there are l beds occupied, $0 \leq l \leq n$, by P_l. The beds may be occupied by patients in needy, dormant states or in the cleaning state. Then, the author states that P_l is given by the following relation:

$$P_l = \sum_{\substack{i,j,k>0 \\ i+j+k=l}} \pi(i, j, k) = \sum_{i=0}^{l} \sum_{j=0}^{l-i} \pi(i, j, l-i-j). \tag{6.50}$$

We can recognize two cases: (a) $l \leq s$ and (b) $l > s$. Hence, from Equation (6.45), for these cases, we have

(a) $l \leq s$:

$$\begin{aligned} P_l &= \pi_0 \sum_{i=0}^{l} \sum_{j=0}^{l-i} \frac{1}{i!}\left(\frac{\lambda}{(1-p)\mu}\right)^i \frac{1}{j!}\left(\frac{p\lambda}{(1-p)\delta}\right)^j \frac{1}{(1-i-j)!}\left(\frac{\lambda}{\gamma}\right)^{l-i-j} \\ &= \pi_0 \frac{1}{l!}\left(\frac{\lambda}{(1-p)\mu} + \frac{p\lambda}{(1-p)\delta} + \frac{\lambda}{\gamma}\right)^l \end{aligned} \tag{6.51}$$

(b) $l > s$:

$$\begin{aligned}
P_l &= \pi_0 \sum_{i=0}^{l} \sum_{j=0}^{l-i} \frac{1}{v(i)} \left(\frac{\lambda}{(1-p)\mu} \right)^i \frac{1}{j!} \left(\frac{p\lambda}{(1-p)\delta} \right)^j \frac{1}{(l-i-j)!} \left(\frac{\lambda}{\gamma} \right)^{l-i-j} \\
&= \pi_0 \left[\sum_{i=0}^{s} \sum_{j=0}^{l-i} \frac{1}{i!} \left(\frac{\lambda}{(1-p)\mu} \right)^i \frac{1}{j!} \left(\frac{p\lambda}{(1-p)\delta} \right)^j \frac{1}{(l-i-j)!} \left(\frac{\lambda}{\gamma} \right)^{l-i-j} \right. \\
&\quad \left. + \sum_{i=s+1}^{l} \sum_{j=0}^{l-i} \frac{1}{s!s^{i-s}} \left(\frac{\lambda}{(1-p)\mu} \right)^i \frac{1}{j!} \left(\frac{p\lambda}{(1-p)\delta} \right)^j \frac{1}{(l-i-j)!} \left(\frac{\lambda}{\gamma} \right)^{l-i-j} \right] \\
&= \pi_0 \left[\frac{1}{l!} \left(\frac{\lambda}{(1-p)\mu} + \frac{p\lambda}{(1-p)\delta} + \frac{\lambda}{\gamma} \right)^l + \sum_{i=s+1}^{l} \sum_{j=0}^{l-i} \left(\frac{1}{s!s^{i-s}} - \frac{1}{i!} \right) \right. \\
&\quad \left. \times \left(\frac{\lambda}{(1-p)\mu} \right)^i \frac{1}{j!} \left(\frac{p\lambda}{(1-p)\delta} \right)^j \frac{1}{(l-i-j)!} \left(\frac{\lambda}{\gamma} \right)^{l-i-j} \right]. \qquad (6.52)
\end{aligned}$$

Therefore,

$$\begin{aligned}
P_l &= \pi_0 \left[\frac{1}{l!} \left(\frac{\lambda}{(1-p)\mu} + \frac{p\lambda}{(1-p)\delta} + \frac{\lambda}{\gamma} \right)^l + I_{\{l>s\}}, \sum_{i=s+1}^{l} \sum_{j=0}^{l-i} \left(\frac{1}{s!s^{i-s}} - \frac{1}{i!} \right) \right. \\
&\quad \left. \times \left(\frac{\lambda}{(1-p)\mu} \right)^i \frac{1}{j!} \left(\frac{p\lambda}{(1-p)\delta} \right)^j \frac{1}{(l-i-j)!} \left(\frac{\lambda}{\gamma} \right)^{l-i-j} \right], \qquad (6.53)
\end{aligned}$$

where $I_{\{l>s\}}$ is the indicator function. Substituting n for l in Equation (6.52), P_n, the probability of blocking of the medical unit, can be found.

Let us denote by W the steady-state random variable representing ***in-queue waiting time*** (the time that a patient may have to wait to be treated) for a virtual patient who just became needy. If a patient becomes needy when there are already i other needy patients in the unit, he/she will need to wait. Obviously, the patient only waits if $i \geq s$. The random variable representing this in-queue random waiting time follows an Erlang distribution with $(i - s + 1)^+$ stages, each with rate $s\mu$. The probability that this Erlangian distributed random variable is greater than t is $e^{-s\mu t} \sum_{j=0}^{i-s} (s\mu t)^j / (j!)$.

Given n beds and s nurses, we also denote by $p_{n,s}(t)$ the tail steady-state distribution of W, that is, $p_{n,s}(t) = P(W > t)$. We further observe that for a closed system, the rate at which the collective stable patient population produces needy patients is modulated by the number of needy, dormant, and cleaning beds. In order to find the distribution of W, we need the following theorem (Chen and Yao, 2001):

Theorem 6.3.1 Arrival Theorem In a closed Jackson network, the arrival at (or departure from) any node obscures time averages, with the job itself excluded. In particular, the probability that the network is in state $x - c_i$ immediately before an arrival (or immediately after a departure) epoch at node i is equal to the ergotic distribution of a closed network with one less job, in state $x - c_i$.

We now denote by $\pi^A(x - c_i)$ the probability that the system is in state $x - c_i$ at the arrival epoch of a patient to node i. Hence, at such an epoch, the state is x. Thus, the arriving patient sees in front of him/her the state $x - c_i$, which corresponds to a network with one less job. Therefore, by Theorem 6.3.1, we have

$$\pi^A(x - c_i) = \pi_{n-1}(x - c_i). \tag{6.54}$$

In particular, for the needy state (node 1)

$$\pi^A(x - c_1) = \pi_{n-1}(x - c_1) = \pi_{n-1}(i - 1, j, k). \tag{6.55}$$

The probability of $W = 0$, that is, when a patient will get service immediately as he/she becomes needy is the sum of probabilities that the patient arriving at the needy state will see fewer than s needy patients. In other words,

$$P(W = 0) = \sum_{l=0}^{n-1} \sum_{m=0}^{l} \sum_{i=0}^{\min(m,s-1)} \pi^A(i, m - i, l - m). \tag{6.56}$$

Yom–Tov found the distribution function of waiting time of a patient as follows:

$$\begin{aligned}
P(W \le t) &= P(W = 0) \\
&\quad + \sum_{i=s}^{n-1} P(\text{there are } (i - s + 1) \text{ patients who ended their services} \\
&\quad \text{on time} \le t \mid \text{ arrival at the needy state found } i \text{ needy patients}) \\
&\quad \times \ \pi^A(i, m - i, l - m) \\
&= P(W = 0) + \sum_{l=s}^{n-1} \sum_{m=s}^{l} \sum_{i=s}^{m} \pi^A(i, m - i, l - m) \int_0^t \frac{s\mu(s\mu x)^{i-s}}{(i - s)!} e^{-s\mu t}\, dx \\
&= P(W = 0) + \sum_{l=s}^{n-1} \sum_{m=s}^{l} \sum_{i=s}^{m} \pi^A(i, m - i, l - m) \left[1 - \sum_{h=0}^{i-s} \frac{(s\mu x)^h}{h!} e^{-s\mu t} \right] \\
&= \sum_{l=0}^{n-1} \sum_{m=0}^{l} \sum_{i=0}^{\min(m,s-1)} \pi^A(i, m - i, l - m) + \sum_{l=s}^{n-1} \sum_{m=s}^{l} \sum_{i=s}^{m} \pi^A(i, m - i, l - m) \\
&\quad - \sum_{l=s}^{n-1} \sum_{m=s}^{l} \sum_{i=s}^{\min(m,s-1)} \pi^A(i, m - i, l - m) \sum_{h=0}^{i-s} \frac{(s\mu x)^h}{h!} e^{-s\mu t} \\
&= 1 - \sum_{l=s}^{n-1} \sum_{m=s}^{l} \sum_{i=s}^{m} \pi_{n-1}(i, m - i, l - m) \sum_{h=0}^{i-s} \frac{(s\mu x)^h}{h!} e^{-s\mu t}.
\end{aligned} \tag{6.57}$$

From Equation (6.57), the tail steady-state distribution of W is

$$P(W > t) == \sum_{l=s}^{n-1} \sum_{m=s}^{l} \sum_{i=s}^{m} \pi_{n-1}(i, m-i, l-m) \sum_{h=0}^{i-s} \frac{(s\mu x)^h}{h!} e^{-s\mu t} \tag{6.58}$$

and from Equation (6.58), the expected waiting time of a patient, denoted by $E(W)$, is

$$\begin{aligned}
E(W) &= \int_0^\infty P(W > t)dt \\
&= \int_0^\infty \sum_{l=s}^{n-1} \sum_{m=s}^{l} \sum_{i=s}^{m} \pi_{n-1}(i, m-i, l-m) \sum_{h=0}^{i-s} \frac{(s\mu x)^h}{h!} e^{-s\mu t}\ dt \\
&= \sum_{l=s}^{n-1} \sum_{m=s}^{l} \sum_{i=s}^{m} \pi_{n-1}(i, m-i, l-m) \sum_{h=0}^{i-s} \int_0^\infty \frac{(s\mu x)^h}{h!} e^{-s\mu t}\ dt \\
&= \sum_{l=s}^{n-1} \sum_{m=s}^{l} \sum_{i=s}^{m} \pi_{n-1}(i, m-i, l-m) \sum_{h=0}^{i-s} \frac{1}{s\mu} \\
&= \frac{1}{s\mu} \sum_{l=s}^{n-1} \sum_{m=s}^{l} \sum_{i=s}^{m} \pi_{n-1}(i, m-i, l-m)(i-s+1).
\end{aligned} \tag{6.59}$$

Relation (6.59) is the same as the mean waiting time, $E(W_j(n))$, at node j in a closed Jackson network with $M/M/c_j$ nodes containing n jobs, which is

$$E(W_j(n)) = \frac{1}{\mu_j c_j} \sum_{i=c_j}^{n-1} (i - c_i + 1) p_j(i, n-1), \tag{6.60}$$

where $p_j(i, n-1)$ is the marginal probability of i in an $(n-1)$ job system at node j (Gross and Harris, 1998, p. 193). Thus, from Equation (6.60) for the Yom-Tov's model and Equation (6.59), we have

$$\begin{aligned}
E(W) &= \frac{1}{s\mu} \sum_{i=s}^{n-1} (i-s+1)\pi_1(i, n-i) \\
&= \frac{1}{s\mu} \sum_{i=s}^{n-1} (i-s+1) \sum_{l=i}^{n-1} \sum_{m=i}^{l} \pi_{n-1}(i, m-i, l-m) \\
&= \frac{1}{s\mu} \sum_{l=i}^{n-1} \sum_{m=i}^{l} \sum_{i=s}^{m} (i-s+1)\pi_{n-1}(i, m-i, l-m).
\end{aligned} \tag{6.61}$$

It is important to note that $P(W > 0)$ is the probability of delay, that is, waiting time. In fact, we should find the performance measures of an MU with n beds and s nurses by the steady-state distribution of the same system with $n-1$ beds and s nurses. The probability that a patient who becomes needy has to wait is the probability that there are more than s needy patients in a system with n beds, and this is exactly the steady-state probability of having more than s needy patients in a system with $n-1$ beds. From Theorem 6.3.1, a patient entering node 1 at his/her arrival, sees the system in state $x - e_i$ with probability $\pi_{n-1}(x - e_i)$, regardless of where he/she came from. After his/her arrival, the system's state will be x. Thus, for the arriving patient to see the system full, the probability will be the sum of probabilities of all x arriving patients such that $x_1 - 1 \geq s$ and $|x| = \sum_i x_i \leq n$. Therefore, we have

$$P(W > 0) = \sum_{x||x|\leq n;\ x_1-1\geq s} \pi^A(x - e_1) = \sum_{i,j,k\,|\,i+j+k\leq n-1;\ i\geq s} \pi^A(i,j,k)$$

$$= \sum_{i\geq s} \pi_{n-1}(i,j,k)$$

$$= \sum_{l=s}^{n-1} \sum_{m=s}^{l} \sum_{i=s}^{m} \pi_{n-1}(i, m-i, l-m). \tag{6.62}$$

From Equation (6.62), we see that for the system with n beds and s nurses, the percentage of patients who are required to wait before being served coincides with the probability, that is, all nurses would be busy, in a system with $n-1$ beds and s nurses. In other words:

$$P(W > 0) = P_{n-1}(N(\infty) \geq s). \tag{6.63}$$

Yom-Tov developed QED regime policies that are asymptotically optimal at the limit, as the number of patients entering the system (λ), the number of beds (n), and the number of servers (s) grows to infinity together. These approximations lead to accurate values of parameters that are realistic in a hospital setting. ***QED regime is essentially the mathematical framework for the patient flow from ER to MUs.***

A QED may be characterized as follows: (1) many server queues; (2) high service quality (i.e., waiting time less than service time); (3) high server utilization (percentage of occupied beds); and (4) adhere to some version of the so-called ***square root staffing rule*** (defined by Erlang in 1924, see Erlang 1948 and for formal discussion see Halfin and Whitt 1981).

For the last characteristic of QED, Erlang offers the following example:

Consider a sequence of s-server queues, indexed by n. Assume that $\lambda_n \to \infty$ as $n \to \infty$ and the service rate μ is fixed. Define the system's ***offered load*** as $R_n = \frac{\lambda_n}{(1-p)\mu}$. Let also $\rho_n = \lambda_n / [(1-p)s\mu]$. Then, the QED regime is achieved by selecting λ_n and s_n such that $\sqrt{s_n}(1-\rho_n) \to \beta$, as $n \to \infty$, for some β. If a patient is extremely patient, then ρ_n may be interpreted as the offer load per server. In that case, $-\infty < \beta < \infty$ will be allowed. In this situation, the staffing level may be approximated as

$$s_n \approx R_n + \beta\sqrt{R_n}, \quad -\infty < \beta < \infty.$$

Let λ, s, and n approach infinity simultaneously so that

$$s = \frac{\lambda}{(1-p)\mu} + \beta\sqrt{\frac{\lambda}{(1-p)\mu}} + o(\sqrt{\lambda}), -\infty < \beta < \infty, \tag{6.64}$$

and

$$\begin{aligned} n - s = \eta_1\sqrt{\frac{\lambda}{(1-p)\mu}} + \frac{p\lambda}{(1-p)\delta} + \eta_2\sqrt{\frac{p\lambda}{(1-p)\delta}} + \frac{\lambda}{\gamma} \\ + \eta_3\sqrt{\frac{\lambda}{\gamma}} + o(\sqrt{\lambda}), -\infty < \eta_1, \eta_2, \eta_3 < \infty. \end{aligned} \tag{6.65}$$

Yom-Tov proved that Equation (6.65) is equivalent to the following:

$$n - s = \frac{p\lambda}{(1-p)\delta} + \frac{\lambda}{\gamma} + \eta\sqrt{\frac{p\lambda}{(1-p)\delta} + \frac{\lambda}{\gamma}} + o(\sqrt{\lambda}), \quad -\infty < \eta < \infty,$$

where

$$\eta = \eta_1\sqrt{\frac{\delta\gamma}{\mu(\gamma p + (1-p)\delta)}} + \eta_2\sqrt{\frac{\gamma p}{\gamma p + (1-p)\delta}} + \eta_3\sqrt{\frac{(1-p)\delta}{\gamma p + (1-p)\delta}}. \tag{6.66}$$

6.4 ROBUSTNESS OF NETWORKS OF QUEUES

Mehmet Ali and Gu (2006) showed that the product-form results turned out to be extremely ***robust*** and insensitive to violations of the assumptions, at least in the steady-state situation. They came to this conclusion while using the steady-state results of Sauer and Daduna (2003) to analyze the performance of a wireless sensor network model for globally unstable networks.

The most common choice for telecommunication network design is based on the exponential assumption for interarrival and service times. Although general arrival and service distributions have been considered extensively, the current networks and applications generate a traffic that is bursty over a wide range of timescales. A number of empirical studies have shown that the network traffic is ***self-similar*** (the whole has the same shape as one or more of the parts) or ***fractal*** (a repeating pattern that displays at every scale) in nature. (Note that heavy-tailed distributions have been suggested as a cause of self-similarity in network traffic.) Using Poisson processes for packet arrivals in a computer network may lead to a failure since there is far too much correlation among packet arrivals to have an assumption of independent arrivals. A unique characteristic of Poisson or Markov processes, however, is that the burst lengths tend to be smoothed by averaging over a sufficiently long timescale. However, the traffic measures have shown that there is a significant traffic variance over a wide range of timescales. A range of characteristics of "modern" traffic could be heavy-tailed distributions.

Recently, robustness of queueing network has been examined by Bandi et al. (2015). These authors study queues based on robust optimization, which are essentially deterministic queues. They are to avoid probabilistic assumptions for network queues. In order to justify their moving away from such assumptions and the fact that the key performance analysis of queueing networks is yet an open research topic, they quoted a message by one of the frontiers of queueing theory, J.F.C. Kingman.

Sir John F.C. Kingman

Kingman was the first who offered an approximation for the mean waiting time in a *G*/*G*/1 queue with arrival and service rates λ and μ, respectively, denoted by $E(W_q)$, as

$$E(W_q) \approx \frac{1}{\mu}\left(\frac{\rho}{1-\rho}\right)\left(\frac{c_a^2 + c_s^2}{2}\right), \tag{6.67}$$

where $\rho = \lambda / \mu$, c_a and c_s are the coefficient of variations for arrivals and service times, respectively (Kingman, 1961). (Coefficient of variation is the ratio of the standard deviation and the mean.) In his remarks, that appeared in Kingman (2009, p. 9), during his opening lecture at the conference "100 Years of Queueing – The Erlang Centennial," Kingman said:

> *If a queue has an arrival process which cannot be well modeled by a Poisson process or one of its near relatives, it is likely to be difficult to fit any simple model, still less to analyze it efficiently. So why do we insist on regarding the arrival times as random variables, quantities about which we can make sensible probabilistic statements? Would it not be better to accept that the arrivals form an irregular sequence, and carry out calculations without positing a joint probability distribution over which that sequence can be averaged?*

There are basically two popular approaches to optimization under uncertainty. One is the ***stochastic optimization*** **(*SO*) or *stochastic programming approach*** that models uncertain parameters as random variables with a specified probability distribution. In this case, optimization is applied to the expected value of the objective function (Prékopa, 1995; Birge and Louveaux, 2011). The other is the ***robust optimization*** **(*RO*) *approach***, a more recent approach to optimization under uncertainty, in which the uncertainty model is not stochastic, but rather deterministic and set-based. In other words, ***RO approach*** optimizes a worst possible case of a considered problem. Basic versions of RO assume "hard" constraints, that is, constraint violation cannot be allowed for any realization of the data in the uncertainty set. The first published study regarding RO was conducted by Soyster (1973). Nonetheless, the field is yet relatively young and under active research, and has been mainly developed in the last 15 years.

It should be noted that that RO is distinctly different from sensitivity analysis, which is typically applied as a postoptimization tool for quantifying the change in cost for small perturbations in the underlying problem data.

Optimization methods have been in use for the past several decades, starting with Dantzig (1949). Its objective is to solve multidimensional problems efficiently from a practical point of view. Robust optimization is the fastest growing area in optimization theory during the past decade (Ben-Tal et al., 2009; Bertsimas et al., 2011). As probability distributions are not inherent to the queueing system, they rather represent a modeling choice that attempts to approximate the actual underlying behavior of the arrival and service processes.

As mentioned earlier, a key result that allows generalizations to networks of queues is Burke theorem (Burke, 1956) that essentially leads to a product-form solution. However, this property would not hold if the queue is not Markovian in arrival and service, that is, the departure process is not renewal and interdeparture times are no longer independent. Thus, it would be difficult to exactly determine the performance measures, even in the case of a simple tandem queueing system. In such a case, two avenues are open, ***simulation*** and ***approximation***. Simulation provides an accurate depiction of the system's performance measures. However, the amount of time to run the simulation to obtain a result that is statistically significant, particularly for heavy-tailed system in heavy traffic is huge. In addition, simulation models are complex and hence make it too difficult to isolate and comprehend the key quantitative insights. On the contrary, approximation methods such as queueing analyzer (QNA), developed by Whitt (1983), and QNET, developed by Dai and Harrison (1992), provide a fair estimation of performance measures. However, these methods have one problem in common, that is, lack of generalizability to model heavy-tailed behavior.

Bandi et al. (2015) considered the uncertainties of arrivals and services via ***polyhedral uncertainty sets*** (that is, uncertainty sets are polytopes) and used the generalized central limit theorem to model ***heavy-tailed*** behavior characterized by bursts of rapidly occurring arrivals and long service times. The assumption of polyhedral uncertainty leads to compact and efficiently solvable linear formulations. The authors chose the worst case approach and obtained closed-form upper bounds on the system time in a multiserver queue. Interestingly, the results are the same as those obtained in the probability setting for light-tailed arrival and services and generalize them to the case of heavy-tailed behavior. In addition, the authors developed a calculus for analyzing a network of queues based on three main principles: departure, superposition, and thinning of arrival processes.

The classical product-form theory of Jackson for stochastic networks is concerned with stations (nodes) which have completely reliable servers. If this assumption is violated, that is, unreliable nodes were to be considered, the method of adjusted service rates, discussed later in this chapter, is to be used.

Kleinrock built up on Jackson's model for modeling data networks. As mentioned earlier, he modeled a data network as a network of communication channels whose purpose was to move data messages from their origin to their destination. Each channel was modeled as a server serving a queue of data messages awaiting transmission. He used the average time it took for messages to move across the net. He refers to this mean time as "***average network delay.***"

It is interesting to note Jackson's words (2002): "In every case that I've heard of, the networks-of-queues model provides an abstract and grossly simplified model of messy phenomena, which one hopes has provided a foothold from which more realistic views can be developed. Perhaps the most interesting such use has been that of Leonard Kleinrock and his associates as a first formal conceptualization of the systems from which the Internet has descended."

It should be noted that in Jackson's model it was assumed that the service time at each station in the network was an independent random variable. However, this is not the case in data networks, because messages and packets (mentioned earlier) basically maintain a constant length as they move from one station to another, and the service time (that is, transmission time) at each station is directly proportional to this length. In addition, there is a definite correlation between message lengths and interarrival times for the message stream. This dependence invalidates the Jackson's model. After over 15 years, Kleinrock introduced the so-called ***independence assumption*** to resolve the problem. The independence assumption indicates, as Kleinrock put it, "each time that a message is received at a node within the net, a new length is chosen for this message independently from an exponential distribution."

Therefore, the network went through development. By December 1969, the following four sites were connected: University of California-Los Angeles (UCLA), Stanford Research Institute, University of California-Santa Barbara, and the University of Utah. By mid-1970, 10 nodes were connected, spanning the United States. In 1972, electronic mail (e-mail) was an ad-hoc add-on to the network, and it immediately began to dominate network traffic; indeed, the network was already showing its most attractive characteristic, the ability to promote "people-to-people" interaction. At present, the Internet has millions of users!

It should also be noted that usually analysis of transient (time-dependent) network of queues is not an easy task. Often, a closed-form solution for such a case is almost an impossible expectation, except in very simple cases. Because of the limitations of closed-form solutions and exact algorithms to analyze general queueing networks, approximations are the tools researchers are focusing on. In most such queueing systems, the approximations are used, generally, on a few parameters such as the interarrival and service time distributions and the first two moments. Hence, large errors in many conditions are possible.

It can be predicted that queueing networks start with a very basic model and can climb up to be a very complex situation. For instance, after a job is served in a service station, on its way out, it may reenter the same station for more service, that is, feedback, or move to another service station, or leave the system forever. We will give more examples as we move ahead. Even within a basic network, one may experience difficulty, for instance, when any one of the station's buffer is finite in capacity. Then, if the work of a server takes less time than the next station, then the task needs to go, the station that has not completed its service will block the task where service has ended. This is a ***blocking*** network system. Bulk arriving and batch service may, even, complicate the model much more so that the system may have to be decomposed to be solved.

We provide an example considered by the authors of this book. Hiring in a large company is normally set so that an application must go through a chain of channels before the last decision is made. One such company may set three stages for its hiring services as follows: (1) the initial screening of arrived applications, (2) the stage of series of interviews of applicants whose applications met the minimum qualifications in the first stage, and (3) completing the hiring service of the selected candidates after interviews. As it can be seen, since applications and their applicants have to wait at each stage to be served, hiring service in a large company is a network of queues. Haghighi and Mishev (2013) considered such a network of queues with bulk arrivals, that is, $M^X / M^{(k,K)} / 1 - M^Y / E_r / 1 - \infty$. Assuming that there are continuous arrivals of applications, vacant positions are continuously available, infinite sizes of buffers, and using decomposition of the system, they found generation functions and mean of the number of applications in each stage. In some cases, they found explicit distributions of the number of applications, as well. We will discuss this in detail later.

6.5 A *MAP* SINGLE-SERVER QUEUEING SYSTEM WITH DELAYED FEEDBACK AS A NETWORK OF QUEUES

In this section, we consider a model that not only presents another practical application but also shows a queue with delayed feedback that is presented as a network of queues.

In reality, in many manufacturing processes tasks (jobs, customers, or orders) arrive from various sources, such as vendors, shifts, and assembly lines, to a common processing area. In such cases, the arrival process can no longer be assumed to form a renewal process. However, in most such cases, ***Markovian arrival process* (*MAP*)** seems to be a good choice.

MAP was introduced by Neuts (1989) and its queueing systems have been extensively analyzed through the matrix analytic method in the literature. Thus, it is a fairly general arrival process. By appropriately choosing the parameters, the underlying *MAP* can be made a renewal process. The *MAP* and its extension allowing for bulk arrivals were shown by Chakravarthy (1993) to be equivalent to Neuts' versatile point process. Chakravarthy and Alfa (1994) discussed a finite-capacity queue with Markovian arrivals and two servers with batch services. He discussed efficient algorithmic procedures for computing the steady-state queue length densities and other system performance measures. He showed that the stationary waiting time distribution is to be of phase type when the interarrival times are of phase type. He also obtained numerically stable expressions for the steady-state queue length densities at arrivals and arbitrary time points.

Not many studies conducted on delayed-feedback queues, in particular, delayed-batch feedback can be found in the literature. Studies of queueing systems with feedback originally started with the consideration of a single queue with instantaneous Bernoulli feedback by Takács (1963) and was extended by Montazer-Haghighi (1976) to a multichannel system, while briefly mentioning the delayed-feedback single server as a particular case of a tandem queueing model, but without offering a solution. A delayed-feedback queue, however, appears to have been first discussed in Nakamura (1971), followed by Hannibalsson and Disney (1977), who looked at a queue length with exponential service times and subsequently by Foley and Disney (1983), and Fayolle's (1979) finite-buffer $M/G/1$ queue with delayed feedback as a multiserver queue. The system was thus viewed as a tandem multiserver queue. Haghighi et al. (2011) considered an $M/M/1$ queue with random delayed feedback, including an additional feature, a splitting device set after the service station that a serviced task might choose to go through. However, the model they discussed involved only a single task feedback. A detailed discussion of this model has been provided in Haghighi and Mishev (2013).

In a machine shop, whenever a machine fails, it not only causes delay in the expected production but also reduction in expected profit. A delayed-service model treated in Haghighi et al. (2008) was expanded in Haghighi and Mishev (2014) as an $M/G/1$ queueing system with processing time consisting of two independent parts, delay time and service time. We outlined this study in Chapter 5. A task was to undergo service with delay. In a special case, the authors obtained the Laplace transform of the generating function of the transient distribution and the generating function of the steady-state queue length. Several queue measures were found from these transforms. Recently, few research papers on delayed-batch-feedback have been published.

Aliakbar Montazer Haghighi

Dimitar P. Mishev (Michev)

In this section, we consider a *MAP* model recently studied in Haghighi and Mishev (2016a,b), which extends Haghighi et al. (2011) and generalizes Haghighi and Mishev (2013) by recognizing that delayed-feedbacks may occur in batches of varying size between a minimum and maximum number.

The *MAP* model that we are to discuss in this section exhibits several features, including the following:

1. Two sources of arrivals by batch internally and one at a time externally
2. Varying batch sizes
3. Feedback with delay
4. Splitting
5. Infinite buffers
6. A two-station tandem queue equivalent system and
7. Approximation of the explicit joint probability distribution of the system's queue length through ***augmentation*** and ***triangularization*** method.

In order to understand the process of splitting (4) in the sense we use the term not only in our previous research papers, but also in this book, we offer an example. In the United States, the creation of the Health Protection Agency (HPA), the Departments of Health (DH) and the Home Office (HO) agreed that there should be a review of port health. A joint DH, HO, and HPA Steering Group was established to oversee the work, and a joint HPA and HO Project Group was asked to undertake the review. Let us consider the service station as the port of inspection of regulations of both DH and HO. Suppose a person subject to immigration control (an arrival) attends the service station. The inspection of all documents may lead to the release of the person with no problem. However, it may be necessary to reinspect the documents and hence the person is sent back to the service station. On the contrary, it might be necessary for the documents to be reviewed by the HPA and HO Joint Project. At this review point, the individual with all documents is sent back to the service-station for further inspection, where the documents will be reviewed. This may lead to releasing the documents with no problem or sending them (as a new individual) to the service station for yet another inspection. The rest of the model follows as stated above.

The feature (6) will be discussed in the busy period of the services station below.

6.5.1 Description of the Model

As mentioned in feature (1), external tasks arrive singly according to a Poisson distribution with parameter λ to an infinite-sized buffer in front of a single-server station that we henceforth refer to as the ***service station***. The service distribution at the service station is assumed to be exponential with parameter μ. Internal arrivals come from the ***delay station*** in batches that we will explain below: once a task leaves the service station, one of the following three events may occur: it may

(i) leave the system forever with probability q_o, $0 \leq q_o \leq 1$;

(ii) ***feedback*** the service station for further service with probability p_f, $0 \leq p_f \leq 1$; or

(iii) go to a unit called ***splitter*** with probability p_s, $0 \leq p_s \leq 1$; the splitting is immediate, that is, the time required to split is negligible;

such that $q_o + p_f + p_s = 1$.

The splitter unit receives a task and splits it into two subtasks, one returns to the service station with probability 1 and the other either leaves the system forever with probability q_{so}, $0 \leq q_{so} \leq 1$, or moves to the service station with probability p_{sf}, $0 \leq p_{sf} \leq 1$, such that $q_{so} + p_{sf} = 1$. However, all feedback tasks must go through the delay station that contains a ***mover*** and an infinite-size buffer in front of it. After processing at the delay station is complete, the mover carries these tasks from the delay station to the service station in batches of varying sizes, say k, where $1 \leq k \leq K$. The natural numbers k and K are the minimum and maximum batch sizes. Hence, the batch sizes may be represented by a random variable, denoted by X, with values between k and K, inclusive, and probability distribution function

$$P\{X = x\} = \alpha_x, \quad k \leq x \leq K, \tag{6.68}$$

with the mean random size of a batch denoted by $\overline{\alpha}$. It is assumed that both the mean and variance of X are positive and finite. It is also assumed that the delay times have exponential distribution with parameter ν. Thus, tasks arriving from the internal source to the service station follow a Poisson distribution with mean ν batches per unit time or $\overline{\alpha}\nu$ tasks per unit time. Although the number of departures from the delay station depends on the number of tasks attend by it, by assigning probability to each arrival to the delay station, we cut off this dependence and thus assume the departure distribution from the delay station independent of any other event in the system. This implies that the arrival stream into the service station is the sum of two independent Poisson processes with rate $\lambda + \overline{\alpha}\nu$.

Once a batch arrives at the service station's buffer from the delay station, tasks in it will be served one at a time according to their order of arrivals into the delay station's buffer. It is, of course, assumed that the return of a task is an event independent of any other event and, in particular, independent of the number of its returns.

The model described may be interpreted as a two-station (service and delay) tandem queue with Poisson external arrival, an exponential single server at the service station, with the possibility of immediate splitting after exiting the service station, and an exponential batch-processing distribution at the delay station. It may also be described that after departure from the service station, a task may (1) leave the system forever with probability $q_o + p_s q_{so}$ or (2) join the delay station's buffer with probability $p_f + p_s(1 + p_{sf})$. Thus, provided that the stability conditions are met, the system that is described in Figure 6.7 may be considered as a network of two queues.

It is important to note that one should not expect explicit solution for any type of network queue as in the case of nonnetwork ones. Thus, algorithmic method is used is such cases. We develop an algorithm to find not only the mentioned characteristics, but also the joint distribution of the number of tasks in the system, that is, in both stations. To the best of the authors' knowledge, this is the first study of its kind to

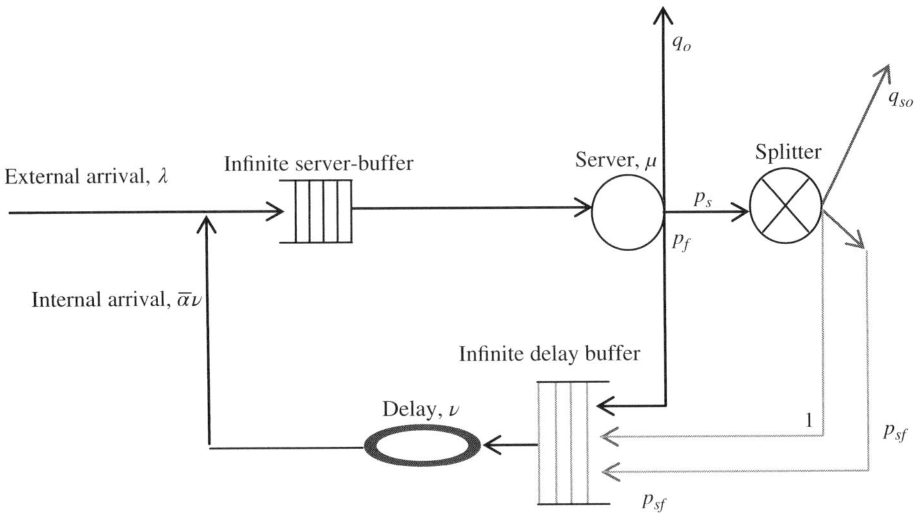

Figure 6.7 Single processor with infinite server-buffer, splitting and delay-general batch feedback.

describe a model with stepwise solution for joint distribution of the queue length of the system and enrichment of the field.

Let the random variables $\xi_1(t)$ and $\xi_2(t)$ denote the number of tasks at time t, $t \geq 0$, at the service station, including the one being served, and at the delay station, including those being processed (i.e., the batch ready to be moved to the service station), respectively, we let the joint probability of these random variables be denoted by $\Psi_{m,n}(t)$, that is,

$$\Psi_{m,n}(t) = P\{\xi_1(t) = m, \xi_2(t) = n\}. \tag{6.69}$$

Thus, $\{\xi_1(t) = m, \xi_2(t) = n\}$ is an irreducible continuous-time Markov chain.

Considering the stationary process, we define the state of the system as (m, n), where m denotes the number of tasks in the service station and n denotes the number of tasks in the delay station. Hence,

$$\Psi_{m,n} = \lim_{t\to\infty} \Psi_{m,n}(t). \tag{6.70}$$

Of course, tasks in service and being processed are included in m and n, respectively, and since splitting is immediate, only those on their way to the delay station are to be considered as part of the delay station. That is, the state of the system is the event $\{\xi_1 = m, \xi_2 = n\}$. Thus,

$$\begin{aligned}\Psi_{m,n} &= \lim_{t\to\infty} \Psi_{m,n}(t)\\ &= \lim_{t\to\infty} P\{\xi_1(t) = m, \xi_2(t) = n \,|\, \xi_1(0) = m_0, \xi_2(0) = n_0\},\end{aligned} \tag{6.71}$$

that is, the stationary probability of having m tasks in the service station and n tasks in the delay station, given the initial states in each station, exits and is independent of the initial state.

Let us denote by λ_1 and λ_2 the expected attendance rates to the service station and delay station, respectively. Then, from the description of the model, it can be seen that

$$\begin{cases} \lambda_1 = \lambda + \overline{\alpha}\nu, \\ \lambda_2 = \lambda_1[p_f + p_s(1 + p_{sf})], \end{cases}$$

or

$$\begin{cases} \lambda_1 = \lambda + \overline{\alpha}\nu, \\ \lambda_2 = (\lambda + \overline{\alpha}\nu)[p_f + p_s(1 + p_{sf})]. \end{cases} \tag{6.72}$$

Then, we may write the traffic intensities in the system, for the service station, ρ_1, delay station, ρ_2, and system, ρ_{sys}. We note that in our case, the traffic intensity of the system is the ratio of the total arrivals into the system, that is, the external plus internal, that is, λ_1, and the total processing rate, that is, the service and moving rates $\mu + \overline{\alpha}\nu$. Thus, we have

$$\rho_1 = \frac{\lambda_1}{\mu}, \ \rho_2 = \frac{\lambda_2}{\overline{\alpha}\nu}, \text{ and } \rho_{\text{sys}} = \frac{\lambda_1 + \lambda_2}{\mu + \overline{\alpha}\nu} = \frac{(\lambda + \overline{\alpha}\nu)[1 + p_f + p_s(1 + p_{sf})]}{\mu + \overline{\alpha}\nu}, \tag{6.73}$$

each restricted to less than one.

Among our goals is finding the steady-state joint probabilities of the number of tasks in both stations, denoted by $\Psi_{m,n}$, for $m, n \geq 0$. We first analyze the service station.

6.5.2 Service Station

We consider two characteristic measures for the service station: queue length and busy period.

Number of Tasks in the Service Station Since all feedback items have to go through the delay station with some probability and that a feedback event is assumed to be independent of any other event in the system, we can consider the service station as an *MAP*/*M*/1 with two arrival sources with rates λ and $\overline{\alpha}\nu$ from the external and internal sources, respectively, and service rate μ.

It is well known that *G*/*M*/1 may be considered the dual of the *M*/*G*/1 queue (Niu and Cooper, 1989; Greenberg, 1969). The *G*/*M*/1 queue may also be analyzed in a variety of ways such as an alternative approach using the ***method of supplementary variables***. For example, Choi et al. (1998) considered the *MAP*/*G*/1 queueing system with infinite capacity, using the supplementary variable method and derived the double transform of the queue length. Their approach is based on an appropriately imbedded Markov chain at the arrival epoch of a task. It seems that the supplementary variables approach is easier to follow. It may also be easily extended to the analysis of a general *G*/*M*/*c* queue (see Kleinrock, 1975).

The service discipline is assumed to be FCFS. Hence, on the basis of the assumptions mentioned above,

$$\text{the mean arrival rate} = \frac{1}{\lambda + \overline{\alpha}\nu}, \tag{6.74}$$

the service times are independent and identically exponentially distributed with the mean service times as $1/\mu$, regardless of the source of arrivals and traffic load ρ_1 given in Equation (6.73).

It should be noted that, although the arrivals at the delay station are dependent on departures from the service station, as long as the service station is not empty, nonexiting departures from the service station arrive to the delay station with assigned probability and, thus, there is virtually no dependence between the service station and delay station. Therefore, at the arrival to the service station, the two Poisson streams are independent and hence the mean arrival would be as mentioned. When the service station is empty, the delay station is either empty, as we will have the state $(0, 0)$, or it continues its processing without new arrival, which does not constitute any dependence.

Recognizing the service station as an *MAP/M/*1 queueing system with the mean interarrival rate as in Equation (6.74) and mean service times $1/\mu$, many measures will be known and are available. Some of those measures are described below.

In this subsection, we focus only on the number of tasks in the service station. The busy period of the server in the service station will be discussed later. Thus, we assume that the service at the service station starts with i, $i \geq 0$, tasks in its buffer. Denote by $\phi_m(t)$ the probability of having m tasks in the service station at time t, that is,

$$\phi_m(t) = P\{\xi_1(t) = m\}. \tag{6.75}$$

For the stationary case, we let

$$\phi_m = \lim_{t\to\infty} \phi_m(t) = \lim_{t\to\infty} P\{\xi_1(t) = m \mid \xi_1(0) = i\}. \tag{6.76}$$

As one of the known properties of the *MAP/M/*1 queue, the transient probability distribution of the number of tasks in the service station at time t is

$$\begin{aligned}\phi_m(t) = e^{-(\lambda_1+\mu)t} & \left[\rho_1^{\frac{m-i}{2}} I_{m-i}(2\mu\sqrt{\rho_1}t) + \rho_1^{\frac{m-i-1}{2}} I_{m+i+1}(2\mu\sqrt{\rho_1}t)\right. \\ & \left. + (1-\rho_1)\rho_1^m \sum_{j=m+i+2}^{\infty} \rho_1^{-\frac{i}{2}} I_j(2\mu\sqrt{\rho_1}t)\right],\end{aligned} \tag{6.77}$$

where ρ_1 is given in Equation (6.76), I_r is the modified Bessel function of the first kind given by

$$I_r(x) = \sum_{m=0}^{\infty} \frac{1}{m!\Gamma(m+r+1)} \left(\frac{x}{2}\right)^{2m+r}, \tag{6.78}$$

$\Gamma(u)$ is the gamma function given by

$$\Gamma(u) = \int_0^{\infty} t^{u-1} e^{-t} \, dt, \tag{6.79}$$

with $\Gamma(n) = (n-1)!$, where n is a positive integer and i is the initial number of tasks in the station.

The stationary queue length distribution at the service station, including the one being serviced, is well known and is geometric as follows:

$$\phi_m = \rho_1(1-\rho_1)^m, m = 0, 1, 2, \ldots. \tag{6.80}$$

Thus, the transient and stationary distributions of the number of tasks in the service station are given by Eqs. (6.77) and (6.80), respectively. Hence, denoting by ξ_1 the stationary queue length of the service station, the mean and variance of the queue length are expressed, respectively, as

$$E(\xi_1) = \frac{\rho_1}{1-\rho_1} \quad \text{and} \quad \text{Var}(\xi_1) = \frac{1+\rho_1}{\mu^2(1-\rho_1)^3}, \tag{6.81}$$

where ρ_1 is given in Equation (6.73).

Busy Period of the Service Station We now discuss the busy period of the service station that has been discussed in Haghighi and Mishev (2016). We use theoretically proven equivalence of two queueing systems to obtain the transient and steady-state distributions of the number of tasks in the service station. More importantly, we theoretically find the distribution and mean of a busy period, as well as the mean number of busy periods within an interval of time. With a singularity at 0, we present a method to circumvent this difficulty by an example. Hence, we can calculate the density and distribution functions of a busy period. In the method of approximation, we use an elegantly simple method of dealing with the singularity. In order to calculate the mean number of busy periods given by Takács's renewal equation (including the modified Bessel function of the first kind), we introduce an algorithmic method, a new technique that we initiate.

Recall that there are two types of arrivals, singly with Poisson and in bulk, also with Poisson, but with a different parameter. Thus, the distribution of arrival is not immediately known, and the service station may be looked at as a *G*/*M*/1 with two arrival types: external (tasks arrive singly) of rates λ and internal (tasks arrive in bulks) of rates ν, and service (in batches) of rate μ. It is well known that *G*/*M*/1 could be considered the dual of the *M*/*G*/1 queue. However, the *G*/*M*/1 queue may be analyzed in a variety of ways or different approaches, some using the method of supplementary variables.

We use the imbedded Markov chain approach at the arrival epoch of a task. This approach is seemingly easier to follow and may also be easily extended to the analysis of a general *G*/*M*/*c* queue (see Kleinrock, 1975). The service station, therefore, becomes an *M*/*M*/1 queueing system with two sources of arrivals. Hence, on the basis of the assumptions mentioned above, the mean arrival rate is given in Equation (6.73), the service times are independent and identically exponentially distributed with mean service times $1/\mu$, regardless of the source of arrivals, and the traffic load ρ_1 is given in Equation (6.73).

Recognizing the service station as an *M/M/*1 queueing system, many measures will be known and available, some of which are listed below. Let us assume that the service station starts with i, $i \geq 0$, tasks. As long as the server is serving a task, we say the *server is busy*. By the ***server's busy period,*** we mean the period starting from the epoch when there are $i + 1$ tasks in the service station and server starts serving them until the epoch at which the number of tasks is reduced to i for the first time, when the server stops working. If $i = 0$, then the busy period starts when the first task arrives. By an *idle period*, we mean the period during which the server is ready to work, but there is no work to be done, that is, the number of tasks in the service station is less than $i + 1$. By a *busy cycle* we mean a busy period followed by an idle period. The literature is quite rich in this subject. We refer the reader to Haghighi and Mishev (2006), Haghighi et al. (2011), and Haghighi and Mishev (2014).

It is important to note that if, during a busy period, a total of l tasks join the service station (internally and externally), where l is a nonnegative integer, then the server's busy period will go through the state transitions: $i + 1 \to i + 1 + l \to i$, $l = 0, 1, 2, \ldots$. Furthermore, during a server's busy period, when an individual task's service is completed, another task will take its place. That is, tasks are born and die during this period. Hence, a busy period contains a renewal process.

Now, we denote the length of a busy period by B. Let us also assume that the service station is empty at time $t = 0$, that is, $i = 0$ in Equation (6.76), and service starts when the first task arrives. Then, as is well known, the distribution of B can be determined through the adjusted transition system of difference differential equations. Thus, the pdf of a busy period, B, denoted by $f(t)$, will be

$$f(t) \triangleq \phi_0'(t), \tag{6.82}$$

where $\phi_0(t)$ is the probability of the service station being empty at time t, which may be interpreted as the probability of $B < t$. Hence,

$$f(t) = \frac{1}{t\sqrt{\rho_1}} e^{-(\lambda_1+\mu)t} I_1(2t\sqrt{\lambda_1 \mu}), t > 0, \tag{6.83}$$

where I_1 is the modified Bessel function of the first kind given by Equation (6.78). Distribution of the length of a busy period may be obtained by integrating Equation (6.83) over the interval $(0, t]$. The mean and variance of a busy period are

$$E\{B\} = \frac{1}{\mu - \lambda_1} \quad \text{and} \quad \text{Var}\{B\} = \frac{1 + \rho_1}{\mu^2 - (1 - \rho_1)^3}, \tag{6.84}$$

respectively. It is clear from Equation (6.84) that if $\mu \leq \lambda_1$, B can be infinity with some positive probability.

Denoting the length of an idle period by Δ, since the interarrival times to the service-station are exponentially distributed with parameter λ_1, the pdf of Δ, denoted by $\psi_\Delta(t)$, is

$$\psi_\Delta(t) = \lambda_1 \ e^{-\lambda_1 t}, t > 0. \tag{6.85}$$

Thus, we have found the distribution of the busy and idle periods through their pdfs in Eqs. (6.87) and (6.88), respectively.

Example 6.5.1 Suppose that the distribution of the random variable representing the batch sizes, α_x, is a uniform distribution:

$$P\{X = x\} = \alpha_x = \begin{cases} \dfrac{x-k}{K-k}, & k \le x \le K, \\ 0, & \text{otherwise.} \end{cases} \tag{6.86}$$

Hence, from Equation (6.86), we have

$$\overline{\alpha} = \frac{k+K}{2}. \tag{6.87}$$

The system under consideration may be thought of as a discrete-time single-server tandem queueing system with two queues each having an infinite space buffer. Two types of tasks are arriving in the system: single tasks arriving from outside and batches from the second queue. The first queue feeds the service station and some of the departures from the service station feed the second queue and there is no external arrival to the second queue. This tandem queueing system has multiple exits.

In order to calculate the density and distribution of a busy period, we split the time interval into subintervals. As an illustration, we choose a finite time interval of length two units, that is, $[0, T] = [0, 2]$. The left endpoint of this interval is a singularity of the pdf defined in Equation (6.83). Problems that occur because of this singularity such as biasness may be addressed using the method proposed by Kim and Whitt (2013) in the estimation of Little's law parameters when the interval under consideration is finite and initial busy period is not zero. However, we choose a more basic method as follows:

Choose subintervals with equal lengths, denoted by h, and in this case, $h = 0.0001$. Thus, for the interval $[0, T] = [0, 2]$, the number of subintervals is $n = \frac{T}{h} = \frac{2}{.0001}$, with the beginning and end partition points as $t_0 = 0$, $t_n = 2$. We also choose the following arbitrary data satisfying conditions imposed previously: $p_f = 0.1$, $q_o = 0.3$, $p_s = 0.6$, $q_{so} = 0.1$, $p_{sf} = 0.9$, $k = 2$, and $K = 5$.

It is important to note that the exact values of the mean and variance of a busy period are given by Equation (6.84). However, we can find their approximated values using discretized values of $f(t)$. Thus, we verify our approximation of $f(t)$ using a partitioning of the time interval. Table 6.1 shows these values.

On the basis of the data and corresponding parameter values, the density function (pdf) and distribution function (cdf) of the length of a busy period are given in Figure 6.8a–c for three given values of the data set (λ, μ, υ).

TABLE 6.1 Expected Value and Variance of Length of a Busy Period for Three Sets of Data Points

Data Set	Λ	μ	υ	$E(B)$ Using Formula	$E(B)$ Using Approximation	Var(B) Using Formula	Var(B) Using Approximation
1	15	180	30	0.0167	0.0167	0.0001	0.0014
2	10	60	10	0.0667	0.0637	0.0005	0.0233
3	3	30	3	0.6667	0.1818	0.0084	0.1059

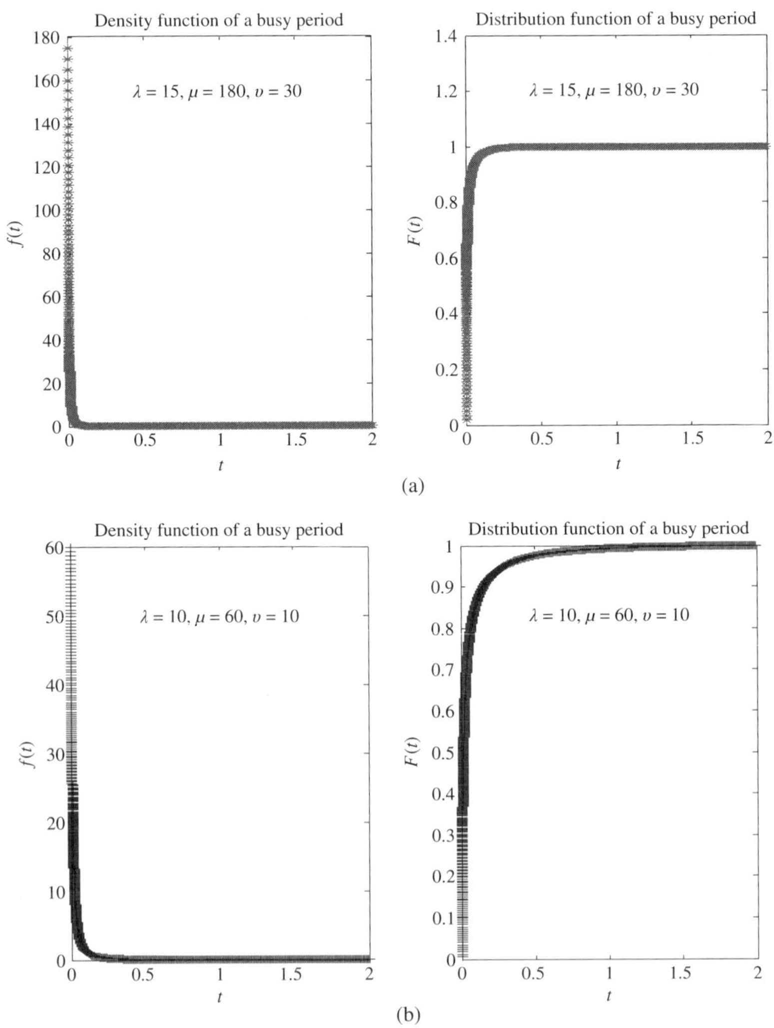

Figure 6.8 Density and distribution functions of a busy period with small values of h.

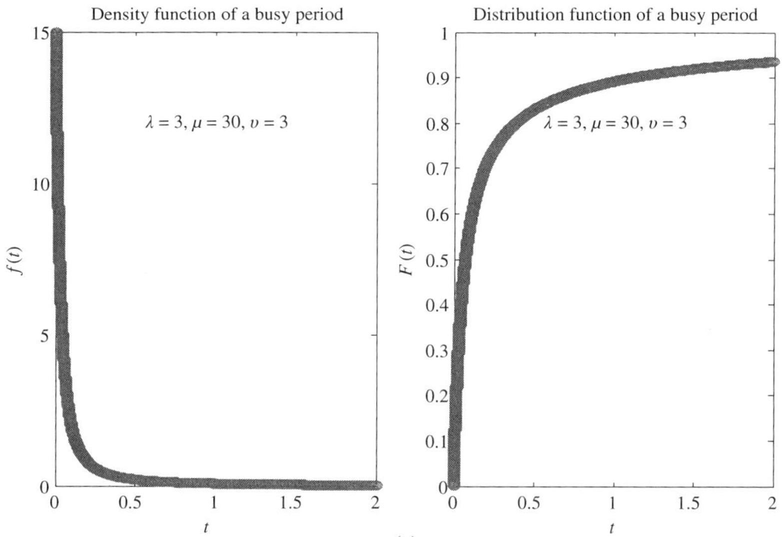

(c)

Figure 6.8 *(continued)*.

Number of Busy Periods In general, the elapsed time from $t = 0$ to the epoch of observing $i + 1$ tasks in the service station is, sometimes, referred to as the *age of the busy period at* $t = 0$. (This is actually the initial idle period.) Using this terminology,

(a) we denote by ψ_0 the age of the busy period at $t = 0$, which from Equation (6.85) is $\psi_0(0) = \lambda_1$, and by X_1 the remaining time of the first period, that is, the duration of time from arrival of the $(i + 1)$st task until the epoch the station returns to containing j tasks for the first time thereafter; and

(b) keeping the age in mind, we denote by $X_1, X_2, \ldots,$ the random variables representing the length of successive busy periods, after the age before beginning the first one.

We assume that the random variables $X_2, X_3, \ldots,$ are independent and identically distributed with cdf as $F(t)$ and are independent of X_1, where

$$F_1(t) = P\{X_1 \leq t\} \tag{6.88}$$

and

$$F(t) = P\{X_n \leq t, n = 2, 3, \ldots\}. \tag{6.89}$$

It is important to note that derivative of $F(t)$, given in Equation (6.89), is the same as $f(t) \equiv \phi_0'(t)$, given in Equation (6.80), for $t = 2, 3, \ldots$. In general, $F_1(t) \neq F(t)$, except when $\psi_0 = 0$. Thus, clearly, we have

$$F_1(t) = P\{X - \psi_0 \le t \,|\, X > \psi_0\} = \frac{P\{\psi_0 < X < t + \psi_0\}}{P\{X > \psi_0\}}$$
$$= \frac{F(t+\psi_0) - F(\psi_0)}{1 - F(\psi_0)}, t \ge 0. \tag{6.90}$$

Suppose we initially choose a busy period with age distribution $\alpha(\psi_0)$ with its pdf $d\alpha(\psi_0)$. Then, Equation (6.90) can be rewritten as

$$F_1(t) = \int_0^\infty \frac{F(t+\psi_0) - F(\psi_0)}{1 - F(\psi_0)} d\alpha(\psi_0). \tag{6.91}$$

Concentrating on only the busy periods of the service station and ignoring possible idle periods, it is clear that successive busy periods of the server constitute renewal cycles. Hence, we can easily answer the following two questions:

(Q1) How old is the current busy period?
(Q2) On average, how many busy periods will be observed?

In order to answer these questions, first note that when we ignore the idle periods, the epoch of ending each busy period can be considered as an arrival epoch. Then, we have a sequence of arrivals, as each busy period ends, that is, $\{X_n\}$; in other words, ending busy periods form a renewal or recurrent process.

Now, let the partial sums of sequence $\{X_n\}$ of the busy periods be denoted by

$$S_n = \begin{cases} 0, & n = 0, \\ \sum_{i=1}^n X_i, & n = 1, 2, \ldots. \end{cases} \tag{6.92}$$

That is, S_n counts the number of busy periods. Let us denote the distribution of S_n by $F_n(t)$, that is,

$$F_n(t) = P\{S_n \le t\}, \quad t \ge 0, n \ge 0. \tag{6.93}$$

It is important to note that $F_n(t)$ denotes the n-fold convolution of F with itself. Furthermore, $F_0(t) = 0$ for $t \ge 0$, and $F_1(t)$ has already been defined above through (6.90) and (6.91). Thus, we have

$$F_{n+1}(t) = \int_0^t F_n(t-s)dF(s), \quad n = 1, 2, \ldots. \tag{6.94}$$

We now define the random variable $N(t)$ to count the number of busy periods within the interval $(0, t]$.Then, values of $N(t)$ in the sequence $\{N(t), t > 0\}$ are positive integers. The sequence $\{N(t)\}$ is increasing, that is, $N(s) < N(t)$, if $s < t$. It is important to note that if $s < t$, then the difference $N(t) - N(s)$ is the number of busy periods that occurred during the interval $[s, t]$. Accordingly, $S_n = \tau$ will be the time the nth busy period has concluded. Counting from this point, the ith subsequent busy period will occur at $S_{n+i} - S_n = X_{n+1} + \cdots + X_{n+i}$. Thus, given $S_n = \tau$, $\{N(\tau + t) - N(\tau), 0 \le t < \infty\}$ is a renewal counting or recurrent process.

An extensive study was conducted on the behavior of the ratio $N(t)/t$, which is a random variable for each value of t and the time average renewal rate over the interval $(0, t]$, that is, the number of busy periods per unit time. It is well known that

$$\lim_{t\to\infty} \frac{N(t)}{t} = \frac{1}{E(X)}, \tag{6.95}$$

where $E(X)$ is given in Equation (6.84). See, for example, Täcklind (1944). Hence, in our case, we have

$$\lim_{t\to\infty} \frac{N(t)}{t} = \mu - \lambda_1. \tag{6.96}$$

We note that the process $\{S_n, n = 0, 1, 2, \dots, S_0 = 0\}$ can be considered as a random walk in the sense that when the system has completed its $n - 1$ busy periods, that is, the system is in state S_{n-1}, then X_n units of time later, it would be in state S_n. Thus, the following equivalence holds:

$$\{N(t) \ge n\} \equiv \{S_n \le t\}. \tag{6.97}$$

Relation (6.97) implies that

$$P\{N(t) \ge n\} = P\{S_n \le t\}. \tag{6.98}$$

Therefore, from Eqs. (6.93) and (6.98) we have

$$\begin{aligned} P\{N(t) = n\} &= P\{S_n(t) \le t\} - P\{S_{n+1}(t) \le t\} \\ &= F_n(t) - F_{n+1}(t), n = 0, 1, 2, \dots. \end{aligned} \tag{6.99}$$

If we denote the expected number of busy periods in the time interval $(0, t)$ by $H(t)$, then from Equation (6.99), we have

$$H(t) \equiv E\{N(t)\} = \sum_{n=1}^{\infty} n[F_n(t) - F_{n+1}(t)]. \tag{6.100}$$

As

$$n[F_n(t) - F_{n+1}(t)] = F_1(t) - F_2(t) + 2[F_2(t) - F_3(t)] + 2[F_3(t) - F_4(t)] + \cdots,$$

from Equation (6.100) we have

$$H(t) \equiv E\{N(t)\} = \sum_{n=1}^{\infty} F_n(t). \tag{6.101}$$

Thus, if we assume $X_1 = \eta, 0 \le \eta \le t$, then we have the ***Takács's renewal equation***

$$H(t) = F(t) + \int_0^t H(t - \eta)dF(\eta). \tag{6.102}$$

See Takács (1958). It can also be shown that

$$E\{N(t)^2\} = E\{N(t)\} + 2\int_0^t H(t - \eta)dH(\eta), \tag{6.103}$$

See Prabhu (2007, p. 111).

From Eqs. (6.83) and (6.102), we can find the average duration that the server will be busy within any time interval $(0, t]$. We will later offer steps to solve the recursive equation (6.102). However, if the duration of a busy period rather than Equation (6.83) had an exponential distribution with parameter β, that is, $F(t) = 1 - e^{\beta t}$, $t \ge 0$, then from Equation (6.89), we would have had

$$F_1(t) = \frac{(1 - e^{-\beta(t+\psi_0)}) - (1 - e^{\beta\psi_0})}{1 - e^{\beta\psi_0}} = F(t), \tag{6.104}$$

and from Equation (6.93) we would have

$$F_n(t) = 1 - \sum_{i=0}^{n-1} e^{-\beta t}\frac{(\beta t)^i}{i!}, n = 1, 2, \ldots. \tag{6.105}$$

Hence, in this hypothetical case,

$$P\{N(t) = n\} = F_n(t) - F_{n+1}(t) = e^{-\beta t}\frac{(\beta t)^n}{n!}, \quad n = 0, 1, 2, \ldots, \tag{6.106}$$

which is a Poisson distribution with parameter β. Therefore, the average number of busy periods in this case can be expressed as

$$E\{N(t)\} = \beta. \tag{6.107}$$

Thus, Eqs. (6.100) and (6.102) in general case and Eqs. (6.106) and (6.107) for the special case answer questions (Q1) and (Q2), respectively.

Computation of Takács's Renewal Equation (6.102), Mean Number of Busy Periods Referring to the density function given by Equation (6.83), it is known that Takács's renewal equation (6.102) is a special case of the so-called Volterra integral equation of the second kind. It is also well known that this has a unique solution that is bounded on a finite interval. See Tijms (2003, p. 309, Theorem 8.1.1). For practical purposes, solving Equation (6.102) could be complicated since it involves the integral of $f(t)$ given in Equation (6.83), which contains the modified Bessel function of the first kind. Coxian-2 distribution usually leads to an explicit expression of the renewal function. However, as in our case, we need to numerically approximate the solution, we avoid excessive round off error by applying a modified version of the Riemann–Stieltjes method given in Xie (1989). This method discretizes t and recursively computes $H(t)$ on partition points of the given interval with good convergence properties; t, however cannot be large.

We approach our numerical approximation as follows. For fixed $t > 0$, we let $[0, t]$ be partitioned and equally spaced as follows: $0 = t_0 < t_1 < \cdots < t_n = t$, where $t_i = ih$, for a given $h > 0$, and n is the number of partitions. In other words, $t_{i+1} - t_i = h = t/n$, $i = 1, 2, \ldots, n$. Then, the following steps form the recursion scheme for computing H_i as an approximation to Equation (6.102).

Step 1 Start with an h, where $h = t_{i+1} - t_i$, $i = 1, 2, \ldots, n$, and $n = t/h$. Using an approximation method or Matlab built-in function "integral," find

$$A_i = F\left(\left(i - \frac{1}{2}\right)h\right) = \int_0^{\left(i-\frac{1}{2}\right)h} f(\tau)d\tau, \quad i = 1, 2, 3, \ldots, n, \tag{6.108}$$

and

$$F_i = F(ih) = \int_0^{ih} f(\tau)d\tau, i = 1, 2, 3, \ldots, n, \tag{6.109}$$

where $f(t)$ is given in Equation (6.83).

We note that the discretization method requires one be mindful of the amount of errors accumulated and so must exercise common sense to accept or reject the results. We also note that $f(t)$ given in Equation (6.83) has a singularity at $t = 0$. However, for computational purposes, this singularity may be removed as follows. The integral formula for the modified Bessel function of the first kind given in Equation (6.78) is given by

$$I_r(x) = \frac{1}{\pi}\int_0^{\pi} e^{x\cos\theta}\cos(r\theta)d\theta - \frac{\sin(r\pi)}{\pi}\int_0^{\infty} e^{-x\cosh y - ry}\,dy. \tag{6.110}$$

See Abramowitz and Stegum (1972, p. 376). When r is an integer, say n, as in our case that it is 1, Equation (6.110) becomes

$$I_n(x) = \frac{1}{\pi}\int_0^{\pi} e^{x\cos\theta}\cos(n\theta)d\theta. \tag{6.111}$$

See Andrews (1992, p. 298). Hence, using L'Hôpital's rule and evaluating the result at $t = 0$, we obtain

$$f(0) = \frac{\lambda_1\mu}{\rho_1}. \tag{6.112}$$

Step 2 Set H_0, H_1, and H_i, $i = 2, 3, \ldots, n$, respectively, as follows:

$$H_0 = 0, \tag{6.113}$$

$$H_1 = \frac{F_1}{1 - A_1}, \tag{6.114}$$

and

$$H_i = \frac{1}{1 - A_1}\left[F_i + \sum_{j=1}^{i-1}(H_j - H_{j-1})A_{i-j+1} - H_{i-1}A_1\right], \quad \text{for } i = 2, 3, \ldots, n. \tag{6.115}$$

Step 3 From Step 2, set the last term (the value of H_i at n) as $H(t)$, that is, $H_n = H(t)$.

Step 4 Repeat Step 1 through Step 3 starting with $h/2$ and ending with $H_{2n} = H(t)$. It is suggested that the computation be done at least for both h and $h/2$.

Step 5 Choose a small preassigned ε, $\varepsilon > 0$.

Step 6 Check $|H_{2n} - H_n|$. If $|H_{2n} - H_n| < \varepsilon$, then write $H(t) = H_{2n}$. However, if $|H_{2n} - H_n| \geq \varepsilon$, then repeat the steps as many times as necessary, starting with $h/4$, etc., and ending with $H(t) = H_{4n}$, H_{8n}, etc., respectively, until the inequality $|H_{2n} - H_n| < \varepsilon$ holds true.

Example 6.5.2 For Takács's Renewal Equation (6.102). According to Tijms (2003, p. 311), choosing h, in Step 1 above, depends on

1. the desired accuracy,
2. the shape of the distribution function $F(t)$, and
3. the length of the interval $[0, t]$.

In general, in many cases to obtain a four-digit accurate result, h must be between 0.05 and 0.01 (in descending order). A choice of h in this range, however, may lead to a length of interval that is asymmetrically skewed to one end of $[0, t]$ in item (3), that is, t may be very small or very large and because $f(t)$ has a singularity at $t = 0$ and involves Bessel function, such values may pose some difficulties. In consequence, it is important that the set of three items above be contingent on the practicality of the situation they must be confined to "what works." The following is an illustration of workable case of the aforementioned steps to determine the average number of busy periods within a finite time interval [0, 2]. We present the following data:

$$p_f = 0.1, q_o = 0.3, p_s = 0.6, p_{sf} = 0.1,$$
$$k = 2, K = 5, \lambda = 3, \mu = 15, \nu = 3, h = 0.02, \epsilon = 0.002,$$

starting with 0.0002 and then 0.0001.

Following the aforementioned steps, the results obtained are presented in Table 6.2. Then, from Equation (6.106), $H(t)$, the expected number of busy periods

TABLE 6.2 Number of Busy Periods for Example 6.5.2

r	$\frac{h}{2^r}$	$n = \left[\frac{t}{h/2^r}\right]$	H_n	$\|H_n - H_{n-1}\|$
0	0.0200000000000	100	7.4050	
1	0.0100000000000	200	8.8900	1.4850
2	0.0050000000000	400	9.9329	1.0429
3	0.0025000000000	800	10.5526	0.6197
4	0.0012500000000	1,600	10.8908	0.3382
5	0.0006250000000	3,200	11.0674	0.1766
6	0.0003125000000	6,400	11.1577	0.0903
7	0.0001562500000	12,800	11.2034	0.0457
8	0.0000781250000	25,600	11.2264	0.0230
9	0.0000390625000	51,200	11.2379	0.0115
10	0.0000195312500	102,400	11.2436	0.0057
11	0.0000097656250	204,800	11.2465	0.0029
12	0.0000048828125	409,600	11.2480	0.0015

in the time interval [0, 2], is 11.2480. Hence, with this set of values, the average busy period is 2/11.2480 = 0.17781.

Using the results in Table 6.2 and formula (6.84), we see from Table 6.2 that the average lengths of the time interval are the ones listed in the last column.

6.5.3 Stepwise Explicit Joint Distribution of the Number of Tasks in the System: General Case When Batch Sizes Vary Between a Minimum *k* and a Maximum *K*

When $k = K = 1$, the joint distribution function of the system's queue length, using the generation function equation, is given in Haghighi et al. (2011) by the following Theorem:

Theorem 6.5.1 Let $a = q_o\mu$, $b = (p_f + p_s q_{so})\mu$, $c = p_s p_{sf}\mu$ with $a + b + c = \mu$, and $d = q_0 - p_s p_{sf}$. If $d > 0$, $\frac{\lambda_1}{\mu} < 1$,, and $\frac{\lambda_2}{\nu} < 1$, then for $k = K = 1$, the joint distribution function of the number of tasks in each station exits and can be obtained by the coefficient of the Maclaurin expansion of solution of the functional equation for the generating function of the distribution function

$$A(w, z)G_1(w, z) = B_1(w, z)G_1(0, z) + B_2(w, z)G_1(w, 0), \tag{6.116}$$

where A, B_1, B_2, and $G_1(w, z)$ are defined as

$$A(w, z) = cz^3 + bz^2 + [a + \lambda w^2 - (\lambda + \mu + \nu)w]z + \nu w^2, \tag{6.117}$$

$$B_1(w, z) = cz^3 + bz^2 + (a - \mu w)z, \tag{6.118}$$

$$B_2(w, z) = \nu w^2 - \nu wz, \tag{6.119}$$

and

$$G_1(w,z) = \sum_{m=0}^{\infty}\sum_{n=0}^{\infty} \Psi_{m,n} w^m z^n,\ |z| < 1,\ |w| < 1, \tag{6.120}$$

which is the generating function of the joint distribution function of the number of tasks in each station. The index of G indicates the value of k, which, in this case, is equal to 1.

It is important to note that because the batch sizes are the same and equal to 1, we have $\overline{\alpha} = 1$.Now, for the general case, when $1 \le k \le K$ and $K \ge 2$, the matrix for the system of balance equations for the queue length is as follows (Fig. 6.7):

$$\widehat{\mathbf{Q}} = \begin{bmatrix}
\widehat{\mathbf{B}}_1 & \widehat{\mathbf{A}}_0 & 0 & 0 & 0 & \cdots & 0 & 0 & 0 & 0 & 0 & 0 & \cdots \\
\widehat{\mathbf{B}}_2 & \widehat{\mathbf{A}}_1 & \widehat{\mathbf{A}}_0 & 0 & 0 & \cdots & 0 & 0 & 0 & 0 & 0 & 0 & \cdots \\
\widehat{\mathbf{B}}_3 & \widehat{\mathbf{B}}_2 & \widehat{\mathbf{A}}_1 & \widehat{\mathbf{A}}_0 & 0 & \cdots & 0 & 0 & 0 & 0 & 0 & 0 & \cdots \\
\widehat{\mathbf{B}}_4 & \widehat{\mathbf{B}}_3 & \widehat{\mathbf{B}}_2 & \widehat{\mathbf{A}}_1 & \widehat{\mathbf{A}}_0 & \cdots & 0 & 0 & 0 & 0 & 0 & 0 & \cdots \\
\ddots & \ddots & \ddots & \ddots & \ddots & \ddots & \ddots & \cdots & \cdots & \cdots & 0 & 0 & \cdots \\
\vdots & \vdots & \ddots & \ddots & \ddots & \ddots & \ddots & \ddots & \ddots & \vdots & 0 & 0 & \cdots \\
\widehat{\mathbf{B}}_K & \widehat{\mathbf{B}}_{K-1} & \widehat{\mathbf{B}}_{K-2} & \cdots & 0 & \widehat{\mathbf{B}}_2 & \widehat{\mathbf{A}}_1 & \widehat{\mathbf{A}}_0 & 0 & \cdots & 0 & 0 & \cdots \\
\widehat{\mathbf{B}}_{K+1} & \widehat{\mathbf{B}}_K & \widehat{\mathbf{B}}_{K-1} & \cdots & 0 & 0 & \widehat{\mathbf{B}}_2 & \widehat{\mathbf{A}}_1 & \widehat{\mathbf{A}}_0 & 0 & 0 & 0 & \cdots \\
0 & \widehat{\mathbf{B}}_{K+1} & \widehat{\mathbf{B}}_K & \cdots & 0 & 0 & 0 & \widehat{\mathbf{B}}_2 & \widehat{\mathbf{A}}_1 & \widehat{\mathbf{A}}_0 & 0 & 0 & \cdots \\
0 & 0 & \widehat{\mathbf{B}}_{K+1} & \cdots & 0 & 0 & 0 & 0 & \widehat{\mathbf{B}}_2 & \widehat{\mathbf{A}}_1 & \widehat{\mathbf{A}}_0 & 0 & \cdots \\
\vdots & \vdots & \vdots & \vdots & \vdots & \vdots & \vdots & \vdots & \vdots & \vdots & \vdots & \vdots & \vdots
\end{bmatrix}, \tag{6.121}$$

where block matrices $\widehat{\mathbf{A}}_0, \widehat{\mathbf{A}}_1, \widehat{\mathbf{B}}_1, \widehat{\mathbf{B}}_2, \widehat{\mathbf{B}}_3, \ldots, \widehat{\mathbf{B}}_K$, and $\widehat{\mathbf{B}}_{K+1}$ are square infinite-size matrices with the elements of each described, respectively, through Eqs. (6.122)–(6.127), below (note that rather than using $\overline{\alpha}$, we write all possible discrete values of k):

$$\widehat{a}_0(i,j) = \begin{cases} q_o\mu, & i = j, \quad j = 1,2,3,\ldots, \\ (p_f + p_s q_{so})\mu, & i = j+1, \quad j = 1,2,3,\ldots, \\ p_s p_{sf}\mu, & i = j+2, \quad j = 1,2,3,\ldots, \\ 0, & \text{otherwise.} \end{cases} \tag{6.122}$$

$$\widehat{a}_1(i,j) = \begin{cases} -(\lambda+\mu), & i = j = 1, \\ -(\lambda+\mu+\nu), & i = j = 2, \\ -(\lambda+\mu+2\nu), & i = j = 3, \\ \vdots & \vdots, \\ -[\lambda+\mu+(K-1)\nu], & i = j = K, \\ -(\lambda+\mu+K\nu), & i = j = K+1, K+2, K+3, \ldots, \\ 0, & \text{otherwise.} \end{cases} \tag{6.123}$$

$$\widehat{b}_1(i,j) = \begin{cases} -\lambda, & i = j = 1, \\ -(\lambda + \nu), & i = j = 2, \\ -(\lambda + 2\nu), & i = j = 3, \\ \vdots & \vdots, \\ -[\lambda + (K-1)\nu], & i = j = K, \\ -(\lambda + K\nu), & i = j = K+1, K+2, K+3, \ldots, \\ 0, & \text{otherwise.} \end{cases} \tag{6.124}$$

$$\widehat{b}_2(i,j) = \begin{cases} \lambda & i = j, i = 1, 2, 3, \ldots, \\ \nu, & i = 1, j = 2, \\ 0, & \text{otherwise.} \end{cases} \qquad \widehat{b}_3(i,j) = \begin{cases} 2\nu, & i = 1, \quad j = 3, \\ 0, & \text{otherwise.} \end{cases} \tag{6.125}$$

$$\widehat{b}_4(i,j) = \begin{cases} 3\nu, & i = 1, j = 4, \\ 0, & \text{otherwise.} \end{cases} \quad \cdots \quad \widehat{b}_K(i,j) = \begin{cases} (K-1)\nu, & i = 1, j = K, \\ 0, & \text{otherwise.} \end{cases} \quad \ldots, \tag{6.126}$$

$$\widehat{b}_{K+1}(i,j) = \begin{cases} K\nu, & i = j - K, j = K+1, K+2, \ldots, \\ 0, & \text{otherwise.} \end{cases} \tag{6.127}$$

Example 6.5.3 The system matrix (6.121) for the special case $K = 3$, $k = 1, 2, 3$, is as follows (note that if $k = 2$ or 3, or both, then the system needs to be adjusted accordingly):

(i)

$$\lambda \Psi_{0,0} = q_o \mu \Psi_{1,0}, m = 0; \ n = 0; \ k = 1, 2, 3,$$

(ii)

$$(\lambda + \upsilon)\Psi_{0,1} = q_o \mu \Psi_{1,1} + (p_f + p_s q_{so})\mu \Psi_{1,0}, \quad m = 0; \ n = 1; \ k = 1, 2, 3,$$

(iii)

$$(\lambda + \upsilon)\Psi_{0,n} = q_o \mu \Psi_{1,n} + (p_f + p_s q_{so})\mu \Psi_{1,n-1} + p_s p_{sf} \mu \Psi_{1,n-2}, m = 0; \ n \geq 2; \ k = 1, 2, 3,$$

(iv)

$$(\lambda + \mu)\Psi_{1,0} = \lambda \Psi_{0,0} + q_o \mu \Psi_{1,0} + \upsilon \Psi_{0,1}, m = 1; \ n = 0; \ k = 1, 2, 3,$$

(v)

$$(\lambda+\mu+\upsilon)\Psi_{1,n} = \lambda\Psi_{0,n} + q_o\mu\Psi_{2,n} + (p_f + p_s q_{so})\mu\Psi_{2,0}, m = 1;$$
$$n = 1;\ k = 1, 2, 3,$$

(vi)

$$(\lambda+\mu+\upsilon)\Psi_{1,n} = \lambda\Psi_{0,n} + q_o\mu\Psi_{2,n} + (p_f + p_s q_{so})\mu\Psi_{2,n-1} + p_s p_{sf}\mu\Psi_{2,n-2},$$
$$m = 1;\ n \geq 2;\ k = 1, 2, 3, \tag{6.128}$$

(vii)

$$(\lambda+\mu)\Psi_{2,0} = \lambda\Psi_{1,0} + q_o\mu\Psi_{3,0} + \upsilon\Psi_{1,1} + \upsilon\Psi_{0,2}, m = 2;\ n = 0;\ k = 1, 2, 3,$$

(viii)

$$(\lambda+\mu+\upsilon)\Psi_{2,1} = \lambda\Psi_{1,1} + q_0\mu\Psi_{3,1} + (p_f + p_s q_{so})\mu\Psi_{3,0}, m = 2;$$
$$n = 1;\ k = 1, 2, 3,$$

(ix)

$$(\lambda+\mu+\upsilon)\Psi_{2,n} = \lambda\Psi_{1,n} + q_o\mu\Psi_{3,n} + (p_f + p_s q_{so})\mu\Psi_{3,n-1} + p_s p_{sf}\mu\Psi_{3,n-2},$$
$$m = 2;\ n \geq 2;\ k = 1, 2, 3,$$

(x)

$$(\lambda+\mu)\Psi_{m,0} = \lambda\Psi_{m-1,0} + q_o\mu\Psi_{m+1,0} + \upsilon(\Psi_{m-1,1} + \Psi_{m-2,2} + \Psi_{m-3,3}),$$
$$m \geq 3;\ n = 0;\ k = 1, 2, 3,$$

(xi)

$$(\lambda+\mu+\upsilon)\Psi_{m,1} = \lambda\Psi_{m-1,1} + q_o\mu\Psi_{m+1,1} + (p_f + p_s q_{so})\mu\Psi_{m+1,0} + \upsilon\Psi_{m-3,4},$$
$$m \geq 3;\ n = 1;\ k = 1, 2, 3,$$

(xii)

$$(\lambda+\mu+\upsilon)\Psi_{m,2} = \lambda\Psi_{m-1,2} + q_o\mu\Psi_{m+1,2} + (p_f + p_s q_{so})\mu\Psi_{m+1,1}$$
$$+ p_s p_{sf}\mu\Psi_{m+1,0} + \upsilon\Psi_{m-3,5}, m \geq 3;\ n = 2;\ k = 1, 2, 3,$$

(xiii)

$$(\lambda+\mu+\upsilon)\Psi_{m,n} = \lambda\Psi_{m-1,n} + q_o\mu\Psi_{m+1,n} + (p_f + p_s q_{so})\mu\Psi_{m+1,n-1}$$
$$+ p_s p_{sf}\mu\Psi_{m+1,n-2} + \upsilon\Psi_{m-3,n+3},$$
$$m \geq 3;\ n \geq 2;\ k = 1, 2, 3,$$

(xiv)

$$\sum_{m=0}^{\infty}\sum_{n=0}^{\infty}\Psi_{m,n} = 1.$$

The system (6.128) may be summarized in matrix form as follows:

$$\widehat{\mathbf{Q}} = \begin{bmatrix} \widehat{\mathbf{B}}_1 & \widehat{\mathbf{A}}_0 & 0 & 0 & 0 & \cdots & 0 & 0 & 0 & 0 & \cdots \\ \widehat{\mathbf{B}}_2 & \widehat{\mathbf{A}}_1 & \widehat{\mathbf{A}}_0 & 0 & 0 & \cdots & 0 & 0 & 0 & 0 & \cdots \\ \widehat{\mathbf{B}}_3 & \widehat{\mathbf{B}}_2 & \widehat{\mathbf{A}}_1 & \widehat{\mathbf{A}}_0 & 0 & \cdots & 0 & 0 & 0 & 0 & \cdots \\ \widehat{\mathbf{B}}_4 & \widehat{\mathbf{B}}_3 & \widehat{\mathbf{B}}_2 & \widehat{\mathbf{A}}_1 & \widehat{\mathbf{A}}_0 & \cdots & 0 & 0 & 0 & 0 & \cdots \\ 0 & \widehat{\mathbf{B}}_4 & \widehat{\mathbf{B}}_3 & \widehat{\mathbf{B}}_2 & \widehat{\mathbf{A}}_1 & \widehat{\mathbf{A}}_0 & \ddots & \cdots & \cdots & \cdots & \cdots \\ 0 & 0 & \ddots & \ddots & \ddots & \ddots & \ddots & \cdots & \cdots & \cdots & \cdots \\ \vdots & \vdots & 0 & \ddots & \ddots & \ddots & \ddots & \ddots & \ddots & \vdots & \cdots \\ 0 & 0 & \cdots & \cdots & \widehat{\mathbf{B}}_4 & \widehat{\mathbf{B}}_3 & \widehat{\mathbf{B}}_2 & \widehat{\mathbf{A}}_1 & \widehat{\mathbf{A}}_0 & \vdots & \cdots \\ 0 & 0 & \cdots & 0 & 0 & \widehat{\mathbf{B}}_4 & \widehat{\mathbf{B}}_3 & \widehat{\mathbf{B}}_2 & \widehat{\mathbf{A}}_1 & \widehat{\mathbf{A}}_0 & \cdots \\ 0 & 0 & \cdots & 0 & 0 & 0 & \widehat{\mathbf{B}}_4 & \widehat{\mathbf{B}}_3 & \widehat{\mathbf{B}}_2 & \widehat{\mathbf{A}}_1 & \cdots \\ \vdots & \vdots & \vdots & \vdots & \vdots & \vdots & \vdots & \vdots & \vdots & \vdots & \cdots \end{bmatrix}, \tag{6.129}$$

where block matrices $\widehat{\mathbf{A}}_0, \widehat{\mathbf{A}}_1, \widehat{\mathbf{B}}_1, \widehat{\mathbf{B}}_2, \mathbf{B}_3$, and $\widehat{\mathbf{B}}_4$ are square infinite-size matrices and elements of each described, respectively, below:

$$\widehat{a}_0(i,j) = \begin{cases} q_o\mu, & i = j, \quad j = 1, 2, \dots, \\ (p_f + p_s q_{so})\mu, & i = j+1, \quad j = 1, 2, \dots, \\ p_s p_{sf}\mu, & i = j+2, \quad j = 1, 2, \dots, \\ 0, & \text{otherwise}. \end{cases} \tag{6.130}$$

$$\widehat{a}_1(i,j) = \begin{cases} -(\lambda + \mu), & i = j = 1, \\ -(\lambda + \mu + \nu), & i = j = 2, \\ -(\lambda + \mu + 2\nu), & i = j = 3, \\ -(\lambda + \mu + 3\nu), & i = j = 4, 5, \dots, \\ 0, & \text{otherwise}. \end{cases} \tag{6.131}$$

$$\widehat{b}_1(i,j) = \begin{cases} -\lambda & i = j = 1, \\ -(\lambda + \nu), & i = j = 2, \\ -(\lambda + 2\nu), & i = j = 3, \\ -(\lambda + 3\nu), & i = j = 4, 5, \dots, \\ 0, & \text{otherwise}. \end{cases} \tag{6.132}$$

$$\widehat{b}_2(i,j) = \begin{cases} \lambda, & i = j, i = 1, 2, 3, \dots, \\ \nu, & i = 1, j = 2, \\ 0, & \text{otherwise}. \end{cases} \tag{6.133}$$

$$\hat{b}_3(i,j) = \begin{cases} 2\nu, & i = 1, j = 3, \\ 0, & \text{otherwise.} \end{cases} \tag{6.134}$$

$$\hat{b}_4(i,j) = \begin{cases} 3\nu, & i = j - 3, j = 4, 5, \dots, \\ 0, & \text{otherwise.} \end{cases} \tag{6.135}$$

The normalizing equation for this system is

$$\sum_{m=0}^{\infty}\sum_{n=0}^{\infty} \Psi_{m,n} = 1. \tag{6.136}$$

The literature shows that an exact solution for the system represented by Equation (6.121) and in the special case given by Equation (6.128) is forbidden. Thus, a step-wise algorithmic solution would be the second best. The method we use may be found in Haghighi et al. (2011). What is extra in this case is that, because the system is an infinite one and the matrix and its blocks are all of infinite sizes, we have to truncate the process for computation purpose and manually create infinity. In order to ensure theoretical convergence, we use the ***augmentation method*** that already exists in the literature; see Gibson and Seneta (1987). To be able to use some of the existing properties of Markov processes, another new idea has been introduced in this study, that of, tridiagonalization of our matrix $\mathbf{Q}$ by combining blocks within the matrix.

We now start the algorithm. Let

$$\mathbf{X} = \langle \mathbf{X}_0, \mathbf{X}_1, \mathbf{X}_2, \dots, \mathbf{X}_n, \dots \rangle^T$$

be an infinite-size column vector, where each of $\mathbf{X}_0$, $\mathbf{X}_1$, $\mathbf{X}_2$, …, is a column-vector, in its own right, as follows:

$$\begin{aligned} \mathbf{X}_0 &= \langle x_{0,0}, x_{0,1}, x_{0,2}, \dots \rangle^T, \\ \mathbf{X}_1 &= \langle x_{1,0}, x_{1,1}, x_{1,2}, \dots \rangle^T, \\ \mathbf{X}_2 &= \langle x_{2,0}, x_{2,1}, x_{2,2}, \dots \rangle^T, \\ &\vdots \qquad\qquad\qquad . \end{aligned} \tag{6.137}$$

Further, let matrix $\mathbf{Q}^T$ be a $K+2$ diagonal block matrix as follows:

$$\mathbf{Q}^T = \begin{bmatrix} \mathbf{B}_1^T & \mathbf{B}_2^T & \mathbf{B}_3^T & \cdots & \mathbf{B}_K^T & \mathbf{B}_{K+1}^T & 0 & 0 & 0 & \cdots & \cdots \\ \mathbf{A}_0^T & \mathbf{A}_1^T & \mathbf{B}_2^T & \mathbf{B}_3^T & \cdots & \mathbf{B}_K^T & \mathbf{B}_{K+1}^T & 0 & 0 & \cdots & \ddots \\ 0 & \mathbf{A}_0^T & \mathbf{A}_1^T & \mathbf{B}_2^T & \mathbf{B}_3^T & \cdots & \mathbf{B}_K^T & \mathbf{B}_{K+1}^T & 0 & \cdots & \ddots \\ 0 & 0 & \mathbf{A}_0^T & \mathbf{A}_1^T & \mathbf{B}_2^T & \mathbf{B}_3^T & \cdots & \mathbf{B}_K^T & \mathbf{B}_{K+1}^T & 0 & \ddots \\ 0 & 0 & 0 & \mathbf{A}_0^T & \mathbf{A}_1^T & \mathbf{B}_2^T & \mathbf{B}_3^T & \cdots & \mathbf{B}_K^T & \mathbf{B}_{K+1}^T & \ddots \\ \ddots & \ddots & \ddots & \ddots & \ddots & \ddots & \ddots & \ddots & \ddots & \ddots & \ddots \\ \vdots & \vdots & \vdots & \vdots & \vdots & \vdots & \vdots & \vdots & \ddots & \ddots & \ddots \end{bmatrix}, \tag{6.138}$$

where block matrices $\mathbf{A}_0, \mathbf{A}_1, \mathbf{B}_1, \mathbf{B}_2, \dots, \mathbf{B}_K$, and $\mathbf{B}_{K+1}$ are square infinite-size matrices, whose elements are described by Eqs. (6.140)–(6.146), respectively, with its transpose as

$$\mathbf{Q} = \begin{bmatrix} \mathbf{B}_1 & \mathbf{A}_0 & 0 & 0 & 0 & \cdots & 0 & 0 & 0 & 0 & \cdots\cdots \\ \mathbf{B}_2 & \mathbf{A}_1 & \mathbf{A}_0 & 0 & 0 & \cdots & 0 & 0 & 0 & 0 & \cdots\cdots \\ \mathbf{B}_3 & \mathbf{B}_2 & \mathbf{A}_1 & \mathbf{A}_0 & 0 & \cdots & 0 & 0 & 0 & 0 & \cdots\cdots \\ \mathbf{B}_4 & \mathbf{B}_3 & \mathbf{B}_2 & \mathbf{A}_1 & \mathbf{A}_0 & \cdots & 0 & 0 & 0 & 0 & \cdots\cdots \\ \ddots & \ddots & \ddots & \ddots & \ddots & \ddots & \ddots & \cdots & \cdots & \cdots & \cdots\cdots \\ \vdots & \vdots & \ddots & \ddots & \ddots & \ddots & \ddots & \ddots & \ddots & \vdots & \cdots\cdots \\ \mathbf{B}_K & \mathbf{B}_{K-1} & \mathbf{B}_{K-2} & \cdots & 0 & \mathbf{B}_3 & \mathbf{B}_2 & \mathbf{A}_1 & \mathbf{A}_0 & 0 & 0 & \cdots \\ \mathbf{B}_{K+1} & \mathbf{B}_K & \mathbf{B}_{K-1} & \cdots & 0 & 0 & \mathbf{B}_3 & \mathbf{B}_2 & \mathbf{A}_1 & \mathbf{A}_0 & 0 & \cdots \\ 0 & \mathbf{B}_{K+1} & \mathbf{B}_{K-2} & \cdots & 0 & 0 & 0 & \mathbf{B}_3 & \mathbf{B}_2 & \mathbf{A}_1 & \mathbf{A}_0 & \cdots \\ \vdots & \vdots & \vdots & \vdots & \vdots & \vdots & \vdots & \vdots & \vdots & \vdots & \vdots & \vdots \end{bmatrix}, \tag{6.139}$$

where the block matrices $\mathbf{A}_0, \mathbf{A}_1, \mathbf{B}_1, \mathbf{B}_2, \dots, \mathbf{B}_K$, and $\mathbf{B}_{K+1}$ are square infinite-size matrices, where the elements of each is described, respectively, below:

$$a_0(i,j) = \begin{cases} -q_o\mu, & i = j, \quad j = 1, 2, 3, \dots, \\ -(p_f + p_s q_{so})\mu, & i = j+1, \quad j = 1, 2, 3, \dots, \\ -p_s p_{sf}\mu, & i = j+2, \quad j = 1, 2, 3, \dots, \\ 0, & \text{otherwise.} \end{cases} \tag{6.140}$$

$$a_1(i,j) = \begin{cases} 1 + (\lambda + \mu), & i = j = 1, \\ 1 + (\lambda + \mu + \nu), & i = j = 2, \\ 1 + (\lambda + \mu + 2\nu), & i = j = 3, \\ \vdots & \\ 1 + [\lambda + \mu + (K-1)\nu], & i = j = K, \\ 1 + [\lambda + \mu + K\nu], & i = j = K+1, K+2, K+3, \dots, \\ 0, & \text{otherwise.} \end{cases} \tag{6.141}$$

$$b_1(i,j) = \begin{cases} 1 + \lambda, & i = j = 1, \\ 1 + \lambda + \nu, & i = j = 2, \\ 1 + (\lambda + 2\nu), & i = j = 3, \\ \vdots & \\ 1 + [\lambda + (K-1)\nu], & i = j = K, \\ 1 + \lambda + K\nu, & i = j = K+1, K+2, K+3, \dots, \\ 0, & \text{otherwise.} \end{cases} \tag{6.142}$$

$$b_2(i,j) = \begin{cases} -\lambda, & i = j, i = 1, 2, \ldots, \\ -\nu, & i = 1, j = 2, \\ 0, & \text{otherwise.} \end{cases} \tag{6.143}$$

$$b_3(i,j) = \begin{cases} -2\nu, & i = 1, j = 3, \\ 0, & \text{otherwise.} \end{cases} \tag{6.144}$$

$$b_K(i,j) = \begin{cases} -(K-1)\nu, & i = 1, j = K, \\ 0, & \text{otherwise.} \end{cases} \tag{6.145}$$

$$b_{K+1}(i,j) = \begin{cases} -K\nu, & i = j - K, j = K+1, K+2, K+3, \ldots, \\ 0, & \text{otherwise.} \end{cases} \tag{6.146}$$

Thus, we may summarize the system (6.128) in a matrix equation form as follows:

$$(\mathbf{I} - \mathbf{Q})\mathbf{X} = \mathbf{0}, \sum_{i=0}^{\infty} X_i = 1, \tag{6.147}$$

where $\mathbf{Q}$ is given by Equation (6.139) and the vector $\mathbf{X}$ is defined by Equation (6.137).

At this point, we need to use the "truncation" method together with the combining block matrices and transposes to obtain the truncated block matrix (6.148) below. We start with matrix $\mathbf{Q}$ defined by Equation (6.139) that is an infinite block matrix with each block as an infinite matrix. We choose τ rows and τ columns from each block starting from the northwest corner of matrix $\mathbf{Q}$. Hence, each block matrix is now finite of size $\tau \times \tau$. This ends the truncation part and we have created a new block matrix (6.148) as below:

$$ {}_\tau\mathbf{Q}^T = \begin{bmatrix} {}_\tau\mathbf{B}_1^T & {}_\tau\mathbf{B}_2^T & {}_\tau\mathbf{B}_3^T & \cdots & {}_\tau\mathbf{B}_K^T & {}_\tau\mathbf{B}_{K+1}^T & 0 & 0 & 0 & \cdots & \cdots \\ {}_\tau\mathbf{A}_0^T & {}_\tau\mathbf{A}_1^T & {}_\tau\mathbf{B}_2^T & {}_\tau\mathbf{B}_3^T & \cdots & {}_\tau\mathbf{B}_K^T & {}_\tau\mathbf{B}_{K+1}^T & 0 & 0 & \cdots & \ddots \\ 0 & {}_\tau\mathbf{A}_0^T & {}_\tau\mathbf{A}_1^T & {}_\tau\mathbf{B}_2^T & {}_\tau\mathbf{B}_3^T & \cdots & {}_\tau\mathbf{B}_K^T & {}_\tau\mathbf{B}_{K+1}^T & 0 & \cdots & \ddots \\ 0 & 0 & {}_\tau\mathbf{A}_0^T & {}_\tau\mathbf{A}_1^T & {}_\tau\mathbf{B}_2^T & {}_\tau\mathbf{B}_3^T & \cdots & {}_\tau\mathbf{B}_K^T & {}_\tau\mathbf{B}_{K+1}^T & 0 & \ddots \\ 0 & 0 & 0 & {}_\tau\mathbf{A}_0^T & {}_\tau\mathbf{A}_1^T & {}_\tau\mathbf{B}_2^T & {}_\tau\mathbf{B}_3^T & \cdots & {}_\tau\mathbf{B}_K^T & {}_\tau\mathbf{B}_{K+1}^T & \ddots \\ \ddots & \ddots & \ddots & \ddots & \ddots & \ddots & \ddots & \ddots & \ddots & \ddots & \ddots \\ \vdots & \vdots & \vdots & \vdots & \vdots & \vdots & \vdots & \vdots & \ddots & \ddots & \ddots \end{bmatrix}, \tag{6.148}$$

where ${}_\tau\mathbf{A}_0$, ${}_\tau\mathbf{A}_1$, ${}_\tau\mathbf{B}_1$, ${}_\tau\mathbf{B}_2$, ${}_\tau\mathbf{B}_3, \ldots,$ ${}_\tau\mathbf{B}_K$, and ${}_\tau\mathbf{B}_{K+1}$ are $\tau \times \tau$ matrices located in the northwest corners of $\mathbf{A}_0, \mathbf{A}_1, \mathbf{B}_1, \mathbf{B}_2, \mathbf{B}_3, \ldots, \mathbf{B}_K$, and $\mathbf{B}_{K+1}$, respectively.

Truncation only does not complete the trick. Therefore, we combine blocks as appropriate to make the matrix $\mathbf{Q}$ a tridiagonal block matrix. This is a novel way of creating a tridiagonal matrix (6.149) with elements as in Eqs. (6.150)–(6.153) below, after truncation:

$$
{}_\tau\mathbf{Q} = \begin{bmatrix}
\mathbf{B}_{1c} & \mathbf{A}_{0c} & 0 & 0 & 0 & 0 & \cdots \\
\mathbf{B}_{2c} & \mathbf{A}_{1c} & \mathbf{A}_{0c} & 0 & 0 & 0 & \cdots \\
0 & \mathbf{B}_{2c} & \mathbf{A}_{1c} & \mathbf{A}_{0c} & 0 & 0 & \cdots \\
0 & 0 & \mathbf{B}_{2c} & \mathbf{A}_{1c} & \mathbf{A}_{0c} & 0 & \cdots \\
0 & 0 & 0 & \mathbf{B}_{2c} & \mathbf{A}_{1c} & \mathbf{A}_{0c} & \cdots \\
0 & 0 & \ddots & \ddots & \ddots & \ddots & \ddots \\
\vdots & \vdots & \vdots & \vdots & \vdots & \vdots & \cdots
\end{bmatrix}, \tag{6.149}
$$

where $\mathbf{B}_{1c}$, $\mathbf{B}_{2c}$, $\mathbf{A}_{0c}$, and $\mathbf{A}_{1c}$, are $K\tau \times K\tau$ matrices as follows:

$$
\mathbf{B}_{1c} = \begin{bmatrix}
{}_\tau\mathbf{B}_1 & {}_\tau\mathbf{A}_0 & 0 & 0 & 0 & 0 \\
{}_\tau\mathbf{B}_2 & {}_\tau\mathbf{A}_1 & {}_\tau\mathbf{A}_0 & 0 & 0 & 0 \\
{}_\tau\mathbf{B}_3 & {}_\tau\mathbf{B}_2 & {}_\tau\mathbf{A}_1 & {}_\tau\mathbf{A}_0 & 0 & 0 \\
\cdots & \cdots & \cdots & \cdots & \cdots & \vdots \\
{}_\tau\mathbf{B}_{K-1} & {}_\tau\mathbf{B}_{K-2} & {}_\tau\mathbf{B}_{K-3} & \cdots & {}_\tau\mathbf{A}_1 & {}_\tau\mathbf{A}_0 \\
{}_\tau\mathbf{B}_K & {}_\tau\mathbf{B}_{K-1} & {}_\tau\mathbf{B}_{K-2} & \cdots & {}_\tau\mathbf{A}_2 & {}_\tau\mathbf{A}_1
\end{bmatrix}, \tag{6.150}
$$

$$
\mathbf{B}_{2c} = \begin{bmatrix}
{}_\tau\mathbf{B}_{K+1} & {}_\tau\mathbf{B}_K & {}_\tau\mathbf{B}_{K-1} & \cdots & {}_\tau\mathbf{B}_3 & {}_\tau\mathbf{B}_2 \\
0 & {}_\tau\mathbf{B}_{K+1} & {}_\tau\mathbf{B}_K & \cdots & {}_\tau\mathbf{B}_4 & {}_\tau\mathbf{B}_3 \\
0 & 0 & {}_\tau\mathbf{B}_{K+1} & \cdots & {}_\tau\mathbf{B}_5 & {}_\tau\mathbf{B}_4 \\
\vdots & \vdots & \vdots & \ddots & \ddots & \ddots \\
\vdots & \vdots & \vdots & \ddots & {}_\tau\mathbf{B}_{K+1} & {}_\tau\mathbf{B}_K \\
0 & 0 & 0 & 0 & 0 & {}_\tau\mathbf{B}_{K+1}
\end{bmatrix}, \tag{6.151}
$$

$$
\mathbf{A}_{0c} = \begin{bmatrix}
0 & 0 & 0 & \cdots & 0 & 0 \\
0 & 0 & 0 & \cdots & 0 & 0 \\
0 & 0 & 0 & \cdots & 0 & 0 \\
\vdots & \vdots & \vdots & \vdots & 0 & \vdots \\
0 & 0 & \vdots & \ddots & 0 & 0 \\
{}_\tau\mathbf{A}_0 & 0 & 0 & \cdots & 0 & 0
\end{bmatrix}, \tag{6.152}
$$

$$
\mathbf{A}_{1c} = \begin{bmatrix}
{}_\tau\mathbf{A}_1 & {}_\tau\mathbf{A}_0 & 0 & 0 & 0 & 0 \\
{}_\tau\mathbf{B}_2 & {}_\tau\mathbf{A}_1 & {}_\tau\mathbf{A}_0 & 0 & 0 & 0 \\
{}_\tau\mathbf{B}_3 & {}_\tau\mathbf{B}_2 & {}_\tau\mathbf{A}_1 & {}_\tau\mathbf{A}_0 & 0 & 0 \\
\cdots & \cdots & \cdots & \cdots & \cdots & \vdots \\
{}_\tau\mathbf{B}_{K-1} & {}_\tau\mathbf{B}_{K-2} & {}_\tau\mathbf{B}_{K-3} & \cdots & {}_\tau\mathbf{A}_1 & {}_\tau\mathbf{A}_0 \\
{}_\tau\mathbf{B}_K & {}_\tau\mathbf{B}_{K-1} & {}_\tau\mathbf{B}_{K-2} & \cdots & {}_\tau\mathbf{B}_2 & {}_\tau\mathbf{A}_1
\end{bmatrix}. \tag{6.153}
$$

The transpose of Equation (6.149) is

$$
{}_{\tau}\mathbf{Q}^T = \begin{bmatrix}
\mathbf{B}_{1c}^T & \mathbf{B}_{2c}^T & 0 & 0 & 0 & 0 & \cdots \\
\mathbf{A}_{0c}^T & \mathbf{A}_{1c}^T & \mathbf{B}_{2c}^T & 0 & 0 & 0 & \cdots \\
0 & \mathbf{A}_{0c}^T & \mathbf{A}_{1c}^T & \mathbf{B}_{2c}^T & 0 & 0 & \cdots \\
0 & 0 & \ddots & \mathbf{A}_{1c}^T & \mathbf{B}_{2c}^T & 0 & \cdots \\
0 & 0 & \ddots & \ddots & \ddots & \ddots & \cdots \\
0 & 0 & \ddots & \ddots & \ddots & \ddots & \cdots \\
\vdots & \vdots & \vdots & \vdots & \vdots & \vdots & \cdots
\end{bmatrix}. \tag{6.154}
$$

It is important to note that the sum of rows of ${}_{\tau}\mathbf{Q}^T$ is not 1. For that, we use the first-column augmentation (Bean and Latouche, 2010). Thus, we modify the individual blocks of ${}_{\tau}\mathbf{Q}^T$, denoting it by ${}_{\tau}\widetilde{\mathbf{Q}}$, as follows: let

$$
\begin{aligned}
&\widetilde{\mathbf{A}}_0 = \mathbf{A}_{0c}^T, \widetilde{\mathbf{A}}_1 = \mathbf{A}_{1c}^T, \widetilde{\mathbf{B}}_1 = \mathbf{B}_{1c}^T + (\mathbf{I} - \mathbf{B}_{1c}^T - \mathbf{B}_{2c}^T) \cdot \mathbf{1} \cdot \mathbf{e}_1^T, \\
&\widetilde{\mathbf{B}}_2 = \mathbf{B}_{2c}^T, \mathbf{D} = (\mathrm{I} - \mathbf{A}_{0c}^T - \mathbf{A}_{1c}^T - \mathbf{B}_{2c}^T) \cdot \mathbf{1} \cdot \mathbf{e}_1^T, \widetilde{\mathbf{B}}_0 = \mathbf{A}_{0c}^T + \widetilde{\mathbf{D}},
\end{aligned} \tag{6.155}
$$

where $\mathbf{e}_1$ is a column vector with the first element 1 and all others as 0, and $\mathbf{1}$ is a column vector with all elements 1 each. Therefore,

$$
{}_{\tau}\widetilde{\mathbf{Q}} = \begin{bmatrix}
\widetilde{\mathbf{B}}_1 & \widetilde{\mathbf{B}}_2 & 0 & 0 & 0 & 0 & 0 \\
\widetilde{\mathbf{B}}_0 & \widetilde{\mathbf{A}}_1 & \widetilde{\mathbf{B}}_2 & 0 & 0 & 0 & 0 \\
\widetilde{\mathbf{D}} & \widetilde{\mathbf{A}}_0 & \widetilde{\mathbf{A}}_1 & \widetilde{\mathbf{B}}_2 & 0 & 0 & 0 \\
\widetilde{\mathbf{D}} & 0 & \widetilde{\mathbf{A}}_0 & \widetilde{\mathbf{A}}_1 & \widetilde{\mathbf{B}}_2 & 0 & 0 \\
\widetilde{\mathbf{D}} & 0 & 0 & \widetilde{\mathbf{A}}_0 & \widetilde{\mathbf{A}}_1 & \widetilde{\mathbf{B}}_2 & \vdots \\
\widetilde{\mathbf{D}} & 0 & \ddots & \ddots & \ddots & \ddots & \ddots \\
\vdots & \vdots & \vdots & \vdots & \vdots & \vdots & \vdots
\end{bmatrix}. \tag{6.156}
$$

We note that as τ approaches infinity, the stationary distributions ${}_{\tau}\mathbf{X}$ of ${}_{\tau}\widetilde{\mathbf{Q}}$ converges to that of $\mathbf{Q}$. Having constructed the truncated stochastic matrix ${}_{\tau}\widetilde{\mathbf{Q}}$ (in the sense that the sum of the rows equals 1), the next move is to determine the steps to solve the system (6.128).

Now we should find a matrix $\mathbf{G}$ from the minimal nonnegative solution of the nonlinear matrix equation $\mathbf{G} = \sum_{i=0}^{\infty} \mathbf{G}^i \mathbf{A}_i$ (Neuts, 1989; Ramaswami, 1988b). Then, using the matrix $\mathbf{G}$ along with Ramaswami formula (Ramaswami, 1988a), we recursively compute the components of the vector $\mathbf{X} = \langle \mathbf{X}_0, \mathbf{X}_1, \mathbf{X}_2, \ldots \rangle^T$. For that, let $\mathbf{R}_j$ be an approximation for $\mathbf{G}$ and apply the following algorithm:

Step 1. Write matrices $\widetilde{\mathbf{A}}_0$, $\widetilde{\mathbf{A}}_1$, $\widetilde{\mathbf{B}}_1$, and $\widetilde{\mathbf{B}}_2$, as defined in Equation (6.155).

Step 2. Let $\mathbf{C} = (\mathbf{I} - \widetilde{\mathbf{A}}_1)^{-1}$, where $\mathbf{I}$ is a $K\tau \times K\tau$ identity matrix.

Step 3. Choose $\mathbf{R_0} = \mathbf{0}$, where $\mathbf{0}$ is a $K\tau \times K\tau$ zero matrix. Assume that $\mathbf{R}_0^0 = 1$. Then, find $\mathbf{R}_{m+1}$ for each $m = 0, 1, 2, \ldots$, under the condition of Step 4 below, as follows:

$$\mathbf{R}_{m+1} = (\widetilde{\mathbf{B}}_2 + \mathbf{R}_m^2 \cdot \widetilde{\mathbf{A}}_0) \cdot \mathbf{C}, m = 0, 1, 2, \ldots . \tag{6.157}$$

Step 4. If Equation (6.158) is true, choose the last $\mathbf{R}_{m+1}$ and write $\mathbf{G} = \mathbf{R}_{m+1}$. Move on to the next step, if all eigenvalues of $\mathbf{G}$ are within the unit circle.

Step 4.1 Choose a desired ϵ.

Step 4.2 For successive values of m, as $0, 1, 2, \ldots$, find the absolute value of difference between the two matrices found for m and $m + 1$.

Step 4.3 Find the largest element of this difference matrix (with i rows and j columns). Check to see if this element is less than ϵ, that is,

$$\max_{i,j} \ |(\mathbf{R}_{m+1} - \mathbf{R}_m)| < \epsilon, m = 0, 1, 2, \ldots . \tag{6.158}$$

Step 5. Compute the following sum of the matrices, denoted by $\overline{\mathbf{B}}_1$:

$$\overline{\mathbf{B}}_1 = \widetilde{\mathbf{B}}_1 + \mathbf{G} \cdot \widetilde{\mathbf{B}}_0 + \mathbf{G}^2 \cdot (\mathbf{I} - \mathbf{G})^{-1} \cdot \widetilde{\mathbf{D}}.$$

Step 6. Solve equation $\mathbf{M}\overline{\mathbf{X}}_0 = 0$, where $\mathbf{M} = \overline{\mathrm{B}}_1 - \mathbf{I}$, and $\overline{\mathbf{X}}_0 = \langle 1, \widetilde{\mathbf{X}}_0 \rangle$, with $\widetilde{\mathbf{X}}_0 = \langle \mathbf{X}_1, \mathbf{X}_2, \ldots, \mathbf{X}_n, \ldots \rangle$, as follows:

Step 6.1 Delete the first row of $\mathbf{M}$ and denote the remaining matrix as $\mathbf{M}_1$.

Step 6.2 Delete the first column of $\mathbf{M_1}$ and denote the remaining matrix as $\mathbf{M_2}$.

Step 6.3 Choose the first column of $\mathbf{M_2}$, multiply it by -1, and denote this matrix as $\mathbf{M_3}$.

Step 6.4 Write $\widetilde{\mathbf{X}}_0 = \mathbf{M}_2^{-1} \cdot \mathbf{M}_3$.

Step 7. Write $\mathbf{X}_0 = \langle 1, \widetilde{\mathbf{X}}_0 \rangle$. Find $\mathbf{X}_n, n = 1, 2, 3, \ldots$, as a matrix products as follows:

$$\mathbf{X}_1 = \mathbf{X}_0 \cdot \mathbf{G}, \ldots, \mathbf{X}_n = \mathbf{X}_{n-1} \cdot \mathbf{G}, \ldots .$$

Step 8. Compute the sum, denoted by S, as $S = \mathbf{X}_0 \cdot (\mathbf{I} - \mathbf{G})^{-1} \cdot \mathbf{e}$, where $\mathbf{e}$ is a $K\tau \times 1$ column vector with all elements as 1, or $S = \sum_{i=0}^{\infty} \overline{\mathbf{X}}_i^T$.

Step 9. From Steps 7 and 8, write

$$\boldsymbol{\Psi}_i = \frac{\mathbf{X}_i}{S}, i = 0, 1, 2, \ldots ,$$

which is the vector of the unknown joint probabilities as follows:

$$\boldsymbol{\Psi}_i = \langle \boldsymbol{\Psi}^*_{iK}, \boldsymbol{\Psi}^*_{iK+1}, \boldsymbol{\Psi}^*_{iK+2}, \ldots, \boldsymbol{\Psi}^*_{iK+K-1} \rangle^T, i = 0, 1, 2, \ldots ,$$

where T denotes the transpose of the vector, and

$$\begin{aligned}
\boldsymbol{\Psi}^*_{iK} &= \langle \psi_{iK,0}, \psi_{iK,1}, \psi_{iK,2}, \cdots, \psi_{iK,\tau} \rangle^T, \\
\boldsymbol{\Psi}^*_{ik+1} &= \langle \psi_{iK+1,0}, \psi_{iK+1,1}, \psi_{iK+1,2}, \cdots, \psi_{iK+1,\tau} \rangle^T, \\
\boldsymbol{\Psi}^*_{ik+2} &= \langle \psi_{iK+2,0}, \psi_{iK+2,1}, \psi_{iK+2,2}, \cdots, \psi_{iK+2,\tau} \rangle^T, \\
&\vdots \\
\boldsymbol{\Psi}^*_{iK+K-i-1} &= \langle \psi_{iK+K-1,0}, \psi_{iK+K-1,1}, \psi_{iK+K-1,2}, \cdots, \psi_{iK+K-1,\tau} \rangle^T.
\end{aligned}$$

We have now algorithmically obtained the joint distribution of the queue length of the system as in Step 9.

Example 6.5.4 In order to show how the steps work, we offer a numerical example with the following data: $\lambda = 5$, $\mu = 25$ and $\nu = 7$, $K = 3$, that is, $k = 1$, 2, or 3. For distribution of the batch size, represented by a random variable X, we choose a discrete uniform distribution:

$$P(X = 1) = P(X = 2) = P(X = 3) = \frac{1}{3}.$$

Hence, $\overline{\alpha} = (1 + 2 + 3)/3 = 2$. In the computation, we denote the maximum value of x by X, and let it equal to 3. We also choose the following probabilities:

$$p_f = 0.4, p_s = 0.2, q_o = 1 - (p_f + p_s), p_{sf} = 0.3, \text{ and } q_{so} = 1 - p_{sf}.$$

With the chosen data, we have $\rho_1 = 0.7600$, $\rho_2 = 0.8957$, and $\rho_{\text{sys}} = 0.4872$. Hence, the stability conditions are met.

For Steps 3 and 4, as m has to be infinity, we choose 50 for infinity as computation shows it is sufficient. For the infinite sum in Step 8, we choose the infinity as 10τ. It should be noted that, in the programming of the numerical example, a computation of the infinite matrices is done through approximating finite square $\tau \times \tau$ matrices. However, we use the ***first-column augmentation method*** so that the convergence of the method for the infinite case will be checked through Step 4. We recall that in order to obtain the distribution of the queue length of the system to be very close to the real values, we have to choose τ large enough so that the error is less than a preassigned value, say $\varepsilon = 10^{-15}$. Hence, we assumed $\tau = 700$ and the error analysis shows that we have met this condition (Tables 5.2 and 6.3).

For tabulation presentation of probabilities, we choose only 18 rows and 12 columns of the distribution matrix to fit in a page. Results are recorded in Table 6.4. However, for a three-dimensional graphic presentation of the probabilities, as in Figure 6.9, we have chosen 60 rows of the probability distribution matrix.

With $\tau = 700$, for the queue length of the system, the mean, second moment, variance, and standard deviation are, respectively, 4.5691, 27.0521, 6.1752, and 2.485.

In order to validate that the probabilities found form a pdf, we compute and graph the cumulative probabilities as in Figures 6.9 and 6.10, and to present a

TABLE 6.3 Expected Value and Variance of Length of a Busy Period for Three Sets of Data Points

λ	μ	υ	ρ_1	ρ_2	H_n with $h=0.0002$	H_n with $h=0.0001$	Error	$E(B)$	$H_n * E(B)$ with $h=0.0002$	$H_n * E(B)$ with $h=0.0001$
15	180	30	0.6667	0.8686	123.3951	122.3515	1.0436	0.0167	1.7645	1.7496
10	60	10	0.7500	0.9771	32.9822	32.9716	0.0106	0.0637	0.3496	2.0508
3	30	3	0.9000	0.9771	7.1935	7.1934	0.0001	0.1818	0.0007	1.3079

two-dimensional graph of cdf (Fig. 6.9) to ensure we are adding all elements of the matrix in some manner, we choose the following method: take the first element of the matrix, that is, $\psi_{0,0}$, add all the remaining elements of the first squared box, and elements to be added to make the next squared box and so on to the end of the matrix. For instance, take $\psi_{0,0}$, and denote it by S_1, then add $\psi_{0,0}$, $\psi_{0,1}$, $\psi_{1,0}$ and $\psi_{1,1}$, to S_1 and denote it by S_2. Then, add $\psi_{0,2}$, $\psi_{2,0}$ and $\psi_{2,2}$ to S_2 and denote it by S_3. Continue this method to cover all elements of the matrix and graph the $S_1, S_2, \ldots$, to obtain the graph in Figure 6.10. However, for the three-dimensional cumulative probabilities, we take each row and add its elements one by one. Result is presented in Figure 6.11.

We take the matrix of block matrices of size 10 times of that of a block matrix. The difference between the elements of two consecutive matrices of probabilities (error) starts as a large number. However, as the length of the truncation line increases, the number reduces and soon reaches zero. Figure 6.11 shows a three-dimensional graph of the differences for a set of values of τ at 60, 100, 300, 500, 600, and 700. The absolute value of the differences of probability matrices for $\tau = 700$ and 600 are listed in Table 6.5. The absolute values of the maximum differences of all elements between every two of these values, in the order written, are listed in Table 6.6. The fluctuation of errors for large values of τ such as 600 and 700 is negligible as it occurs at the 16th decimal place. Perhaps running the program with a much larger value of τ, that is, severalfold, we may see a complete stability of the values (Fig. 6.12).

Remarks The model we have described may be interpreted as a two-station tandem queue with the possibility of immediate splitting after exiting the service station. Some of the particular features of the system that make the system unique are

1. two sources of arrivals: singly from outside and by varying batch sizes internally;
2. two types of feedbacks: one directly after the service and another from after going through splitting, both types with delay;
3. both buffers are of infinite sizes and thus, no blocking; and
4. consideration of the delay station as the second station to make the system a two-station queue in series.

TABLE 6.4 Distribution of the Number of Tasks in the System (Both Stations), the First 18 Rows and 12 Columns of the Probability Distribution Matrix

		$\psi_{i,0}$ through $\psi_{i,11}$, $i = 0, 1, 2, \ldots, 17$											
		0	1	2	3	4	5	6	7	8	9	10	11
ψ_{0j}	0	0.1772	0.1468	0.0595	0.0184	0.0064	0.0022	0.0008	0.0003	0.0001	0.0000	0.0000	0.0000
through	1	0.0886	0.0565	0.0235	0.0078	0.0027	0.0009	0.0003	0.0001	0.0000	0.0000	0.0000	0.0000
ψ_{17j},	2	0.0745	0.0351	0.0149	0.0050	0.0017	0.0006	0.0002	0.0001	0.0000	0.0000	0.0000	0.0000
$j = 0, 1, \ldots, 11$	3	0.0563	0.0258	0.0105	0.0035	0.0012	0.0004	0.0001	0.0000	0.0000	0.0000	0.0000	0.0000
	4	0.0355	0.0164	0.0067	0.0022	0.0008	0.0003	0.0001	0.0000	0.0000	0.0000	0.0000	0.0000
	5	0.0232	0.0106	0.0043	0.0014	0.0005	0.0002	0.0001	0.0000	0.0000	0.0000	0.0000	0.0000
	6	0.0152	0.0069	0.0028	0.0009	0.0003	0.0001	0.0000	0.0000	0.0000	0.0000	0.0000	0.0000
	7	0.0099	0.0045	0.0018	0.0006	0.0002	0.0001	0.0000	0.0000	0.0000	0.0000	0.0000	0.0000
	8	0.0064	0.0029	0.0012	0.0004	0.0001	0.0000	0.0000	0.0000	0.0000	0.0000	0.0000	0.0000
	9	0.0042	0.0019	0.0008	0.0003	0.0001	0.0000	0.0000	0.0000	0.0000	0.0000	0.0000	0.0000
	10	0.0027	0.0012	0.0005	0.0002	0.0001	0.0000	0.0000	0.0000	0.0000	0.0000	0.0000	0.0000
	11	0.0018	0.0008	0.0003	0.0001	0.0000	0.0000	0.0000	0.0000	0.0000	0.0000	0.0000	0.0000
	12	0.0011	0.0005	0.0002	0.0001	0.0000	0.0000	0.0000	0.0000	0.0000	0.0000	0.0000	0.0000
	13	0.0007	0.0003	0.0001	0.0000	0.0000	0.0000	0.0000	0.0000	0.0000	0.0000	0.0000	0.0000
	14	0.0005	0.0002	0.0001	0.0000	0.0000	0.0000	0.0000	0.0000	0.0000	0.0000	0.0000	0.0000
	15	0.0003	0.0001	0.0001	0.0000	0.0000	0.0000	0.0000	0.0000	0.0000	0.0000	0.0000	0.0000
	16	0.0001	0.0001	0.0000	0.0000	0.0000	0.0000	0.0000	0.0000	0.0000	0.0000	0.0000	0.0000
	17	0.0001	0.0000	0.0000	0.0000	0.0000	0.0000	0.0000	0.0000	0.0000	0.0000	0.0000	0.0000

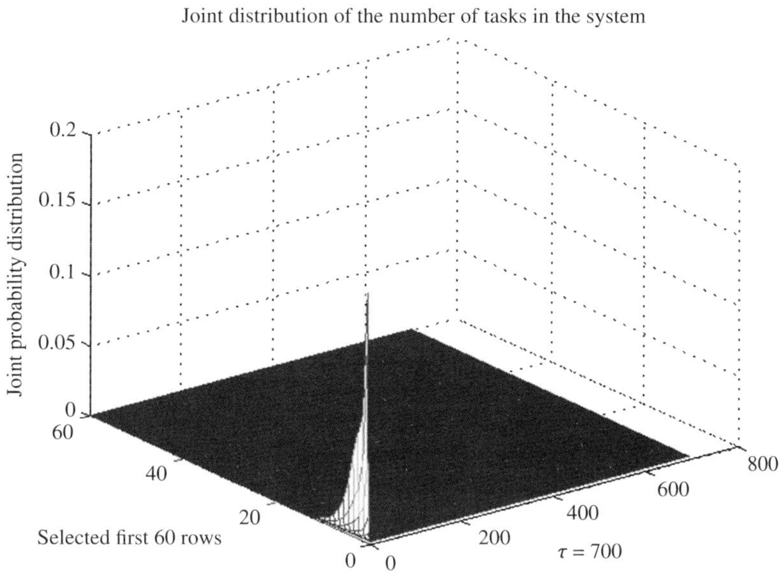

Figure 6.9 Graph of the joint probability distribution of the number of tasks in the system.

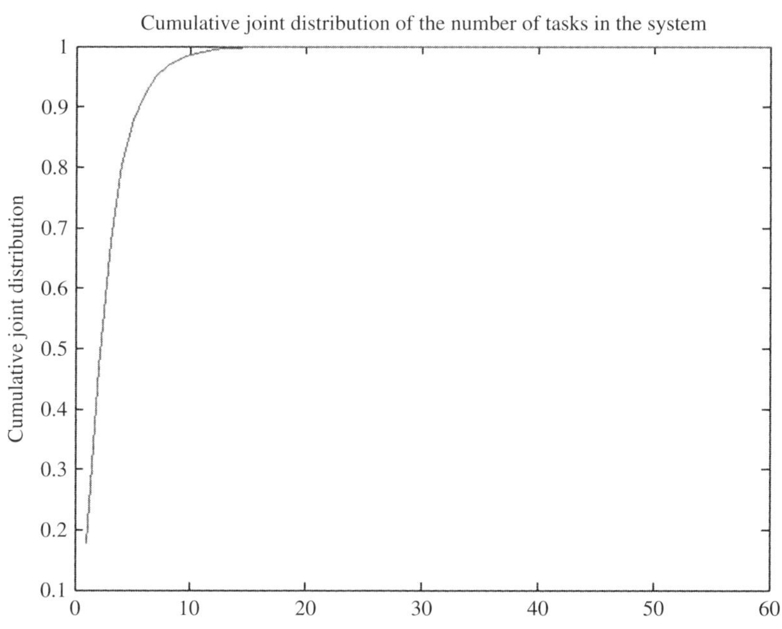

Figure 6.10 Graph of the cumulative joint probability distribution of the number of tasks in the system.

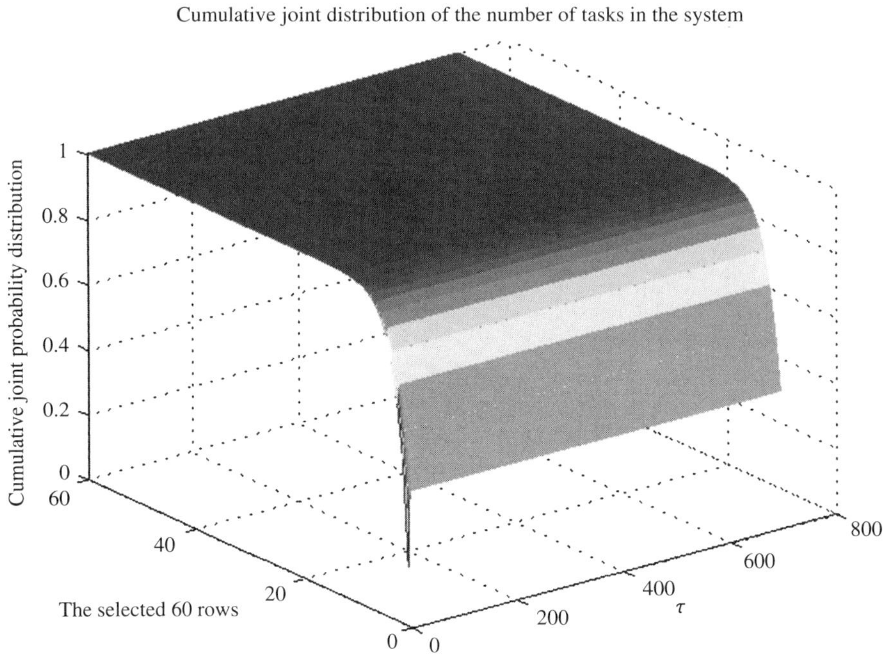

Figure 6.11 Three-dimensional graph of the cumulative joint probability distribution of the number of tasks in the system.

The unique fact about solution of this model is that we offered a stepwise explicit joint probability distribution of the system's queue length that we have accomplished through approximation of the infinite matrices with infinite matrix elements. For the busy-period distribution of the service station, in Haghighi and Mishev (2014b), we used two ideas:

1. Duality of two systems $G/M/1$ and $M/G/1$ and
2. Independence of the two sources of arrivals: external and internal while the external arrivals are generated by the departures from the service station.

Similar use enabled us to reduce our model to an $MAP/M/1$ and, thus, the use of some of the existing properties of $MAP/M/1$ to obtain the stepwise joint distribution of the queue length of the system. In order to facilitate the approximations involved, we had to choose the values of some of the parameters. We have performed an error analysis to show how the error approaches zero as the matrix sizes approach infinity and also offered for the first time, to our knowledge, an explicit joint distribution of a tandem queue.

TABLE 6.5 Differences between Two Consecutive pdfs of the Queue with $\tau = 700$ and $\tau = 600$

1	2	3	4	5
0.166533453693773	0.166533453693773	0.055511151231258	0.020816681711722	0.006071532165919
0.111022302462516	0.076327832942980	0.027755575615629	0.010408340855861	0.002168404344971
0.111022302462516	0.041633363423443	0.013877787807814	0.000867361737988	0.001734723475977
0.076327832942980	0.034694469519536	0.008673617379884	0.001734723475977	0.001084202172486
0.041633363423443	0.013877787807814	0.006938893903907	0.002168404344971	0.000758941520740
0.017347234759768	0.010408340855861	0.003469446951954	0.001084202172486	0.000542101086243
0.010408340855861	0.006071532165919	0.002602085213965	0.000867361737988	0.000325260651746
0.008673617379884	0.003469446951954	0.001734723475977	0.000542101086243	0.000108420217249
6	**7**	**8**	**9**	**10**
0.001301042606983	0.000650521303491	0.000216840434497	0.000081315162936	0.000027105054312
0.001192622389734	0.000271050543121	0.000162630325873	0.000033881317890	0.000016940658945
0.000758941520740	0.000243945488809	0.000081315162936	0.000027105054312	0.000006776263578
0.000542101086243	0.000135525271561	0.000067762635780	0.000020328790734	0.000005929230631
0.000216840434497	0.000081315162936	0.000027105054312	0.000010164395367	0.000004235164736
0.000135525271561	0.000054210108624	0.000016940658945	0.000005929230631	0.000002541098842
0.000108420217249	0.000033881317890	0.000010164395367	0.000004235164736	0.000001058791184
0.000067762635780	0.000023716922523	0.000008470329473	0.000002541098842	0.000001058791184

Numbers shown are to be multiplied by 10^{-15}.

TABLE 6.6 Maximum Differences between Two Consecutive Probabilities in the Distribution of Queue Size

$\tau = 60, 100, 300, 500, 600, 700$	Max Difference = Error
pdf with $\tau = 100$ − pdf with $\tau = 60$	2.775557561562891e−17
pdf with $\tau = 300$ − pdf with $\tau = 100$	1.249000902703301e−15
pdf with $\tau = 500$ − pdf with $\tau = 300$	1.110223024625157e−16
pdf with $\tau = 600$ − pdf with $\tau = 500$	3.330669073875470e−16
pdf with $\tau = 700$ − pdf with $\tau = 600$	1.665334536937735e−16

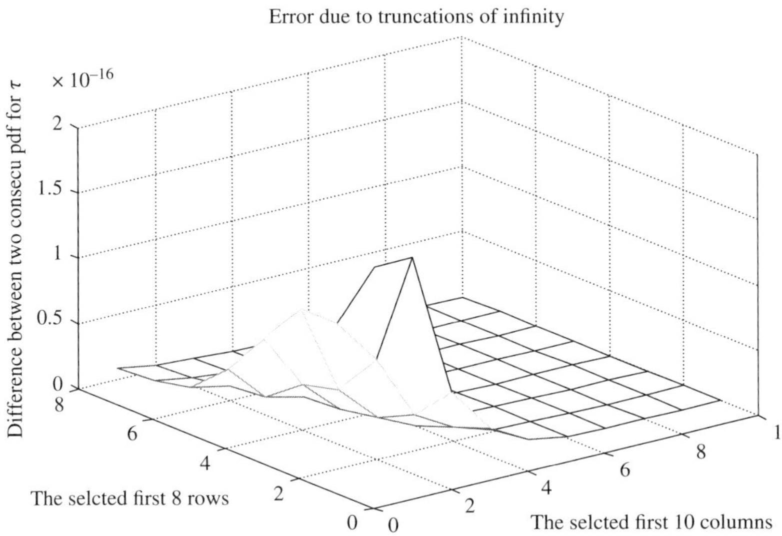

Figure 6.12 Three-dimensional graph of differences between two consecutive probabilities of the queue length with $\tau = 700$ and 600.

6.6 UNRELIABLE NETWORKS OF QUEUEING SYSTEM MODELS

Another important characteristic of real-life systems is nonreliability, which also has a negative influence on these measures, because most of the components of the systems are subject to random breakdowns and require repairs. Nonreliability has been extensively studied for traditional queues with waiting lines, but only for infinite-source queues with returning customers. For instance, it is known that wireless ad hoc networks are built of varying sets of mobile users with wireless communication capabilities without relying on a preexisting infrastructure. As described by Schopp and Viallon (2003), the emergence of wireless ad hoc networks introduces problems concerning the availability of transmission nodes next to a user. In such networks, overload arises from too many users entering a region and applying for transmission. Thus, in such networks, the required high quality of service for speech and data transmission has to be guaranteed based on not necessarily reliable network nodes.

A node is ***unreliable*** if it cannot be accessed. A ***nonaccessible station*** is where (1) the server is broken down and is under repair, (2) the station is taken off the network and will be replaced by another station in later time, or (3) the server is serving jobs from another network.

Jackson's product form theory for stochastic networks is based on stations (nodes) that have completely reliable servers. When a node breaks down, there is a need for some policy to move the tasks, keeping then put, of course is one policy that it causes delay in service. Hence, when unreliable stations are present in a system, the ***method of adjusted service rates*** is one policy to be used. This method involves:

(i) Using reliability theoretical methods, the reliability of the stations is computed that yields the percentage of time the stations are broken down.
(ii) Service rates of the stations are decreased by this factor.
(iii) Performance evaluation methods for product form networks are applied (Chakka and Mitrani, 1996).
However, a disadvantage of using this method is that the interaction of performance and reliability cannot be studied in a unified model.

Methods used to study the interaction of reliability and performance of complex systems in integrated models are simulation and numerical evaluation (Haverkort et al., 2001; Mylosz and Daduna, 2009).

6.6.1 Unreliable Network Model of Goodman and Massey

In an ***unreliable process***, the condition of ergodicity is violated and hence, the system is ***nonergodic*** and the network has ***unstable nodes***. Goodman and Massey (1984) proposed the following general traffic equation for the Jackson networks, which includes the nonreliable and, hence, nonergodic cases:

$$\eta_j = \lambda_j + \sum_{i=1}^{N} \min(\eta_i, \mu_i) p_{i,j}, \quad j = 1, 2, \ldots, N, \tag{6.159}$$

Jonathan B. Goodman

It can be shown that the general traffic equation (6.159) has a unique solution (Goodman and Massey, 1984). We denote this unique solution by

$$\eta = (\eta_j \, : j = 1, 2, \ldots, N). \tag{6.160}$$

The idea behind the general traffic equation (6.159) is based on Burke's theorem (Burke, 1956) for the $M/M/1$ queueing system. At a station, if $\lambda \geq \mu$, the output is yet Poisson for a long duration, not with rate λ as in the ergodic case, but with rate μ, instead. Thus, in an $M/M/1$ queueing system, in the general case, when time is long enough, the output is a Poisson process with rate $\min(\lambda, \mu)$, see Goodman and Massey (1984, p. 861).

Thus, a node j is called ***stable***, if η_j defined in Equation (6.159) is strictly less than μ_j, otherwise it is called ***unstable***. The network is, then, unstable if at least one node in it is unstable.

We should note that the condition $\eta_j \geq \mu_j$ is only a necessary condition of unstability. In fact, when there are more than one unstable node in the network, the stability condition may be violated because of interaction among the nodes and very high departure rates.

Because of the complication in determining stability of the network, Goodman and Massey (1984) proposed an algorithm. We bring this algorithm here from Mylosz (2013) as follows:

Algorigthm 6.6.1 For Network Stability [Goodman and Massey (1984)]

Step 1. Assume all nodes are unstable and, hence, let $(\eta_j(1), j = 1, 2, \ldots, N)$ be the first estimate for solution $(\eta_j, j = 1, 2, \ldots, N)$ of Equation (6.159). In other words, we have the following:

$$\eta_j(1) = \lambda_j + \sum_{i=1}^{N} \mu_i p_{i,j}, \quad j = 1, 2, \ldots, N. \tag{6.161}$$

This estimate exists and is unique since all factors on the right-hand side of Equation (6.161) are known parameters. A comparison of Eqs. (6.161) and (6.160) shows that

$$\min(\eta_j, \mu_j) \leq \mu_j \Rightarrow \eta_j(1) \geq \eta_j \text{ for all } j = 1, 2, \ldots, N. \tag{6.162}$$

Thus, there are two cases:

(i) If $\eta_j(1) \geq \eta_j$ is true for all $j = 1, 2, \ldots, N$, then all nodes are unstable and, hence, for the first estimate it is true that $\eta_j = \eta_j(1)$ for all $j = 1, 2, \ldots, N$. Therefore, stop here.

(ii) If $\eta_j(1) \geq \eta_j$ is not true for at least one node $j *= 1, 2, \ldots, N$, that is, $\eta_{j*}(1) < \mu_{j*}$ for some $j *, j *= 1, 2, \ldots, N$, because of $\eta_j(1) \geq \eta_j$ is true for all $j = 1, 2, \ldots, N$, it is true that $\mu_{j*} > \eta_{j*}(1)$. Thus, the node $j *$ is a stable node.

Assuming all nodes are unstable, let $\eta_j(1), j = 1, 2, \ldots, N$ be the first estimate for solution $\eta_j, j = 1, 2, \ldots, N$ of (6.6.1), that is,

$$\eta_j(1) = \lambda_j + \cdots \tag{6.163}$$

and proceed to the next step.

Step 2. All nodes j in $S(1)$ are eventually stable. Hence, assume that all other nodes i in $\widetilde{N} \setminus S(1)$ are unstable. Let $(\eta_j(2), j = 1, 2, \ldots, N)$ be the second estimate for solution $(\eta_j, j = 1, 2, \ldots, N)$ of Equation (6.159). Thus, it is the solution of the traffic equation

$$\eta_j(2) = \lambda_j + \sum_{i \in \widetilde{N} \setminus S(1)} \mu_i p_{i,j} + \sum_{i \in S(1)} \eta_i(2) p_{i,j}, \quad j = 1, 2, \ldots, N, \tag{6.164}$$

which uniquely exists. [For proof of this statement, see Goodman and Massey (1984) or Mylosz (2013, Lemma 1.19).] Again, a comparison of Eqs. (6.164) and (6.160) shows that

$$\min(\eta_j, \mu_j) \le \mu_j \Rightarrow \eta_j(2) \ge \eta_j \quad \text{for all } j = 1, 2, \ldots, N. \tag{6.165}$$

Thus, $\eta_j \le \eta_j(2) \le \eta_j(1)$ and $\eta_j(2) \le \mu_j$ are true for all $j \in S(1)$. Thus, there are two cases again:

(i) If $S(1) = S(2) = \{j : \eta_j(2) < \mu_j\}$, for all $j = 1, 2, \ldots, N$, then $\eta_j(2) = \eta_j$, for all $j = 1, 2, \ldots, N$. Therefore, stop here.

(ii) If $S(1) \neq S(2)$, then $S(1) \subset S(2)$. Hence, $\eta_{j*}(2) > \eta_j$ is true for at least one node $j*, j* = 1, 2, \ldots, N$. Therefore, proceed to the next step with $S(2)$.

It should be noted that for the analysis of a Jackson network, one should identify the stable and unstable nodes. This identification will help to choose the proper traffic equations. The following algorithm will be used to identify the set of stable nodes in a Jackson network:

Algorigthm 6.6.2 To Identify the Set of Stable Nodes of Networks' Network [Goodman and Massey (1984)]

Step 1. Solve the standard traffic equation (6.161). There are two possible cases:

(i) If $\eta_j < \mu_j$, for all $j = 1, 2, \ldots, N$, then all nodes are stable, the standard traffic equations are appropriate, and stop here.

(ii) If there is only one node, say i_*, for which $\eta_j \ge \mu_j$, then this node is the only unstable node and the appropriate traffic equations are

$$\eta_j = \lambda_j + \sum_{i \in \widetilde{N} \setminus \{i_*\}} \eta_i p_{i,j} + \mu_{i_*} p_{i_*,j}, j = 1, 2, \ldots, N. \tag{6.166}$$

Hence, stop here.

(iii) If there are two or more nodes for which $\eta_j \ge \mu_j$, proceed to the next step.

Step 2. Run Algorithm 6.6.1 to solve the general traffic equation (6.161). Let $S = \{i : \eta_i < \mu_i\}$ and $U = \widetilde{N} \setminus S$ denote the set of stable and unstable nodes, respectively. Then, the traffic equations are

$$\eta_j = \lambda_j + \sum_{i\in S} \eta_i p_{i,j} + \sum_{i\in U} \mu_i p_{i,j}, \quad j = 1, 2, \ldots, N. \tag{6.167}$$

It should be noted that if $\eta = (\eta_j \; : \; j = 1, 2, \ldots, N)$ is the unique solution of Equation (6.161), $S = \{i \; : \; \eta_i < \mu_i\}$ and $U = \widetilde{N} \backslash S$ denote the set of stable and unstable nodes, respectively. Then, summing Equation (6.161) over all values of $j = 1, 2, \ldots, N$, shows that the sum of departure rates of the system is lower than the rate of external arrival, λ, into the system if not all unstable nodes operate at full capacity, that is, $\eta_j = \mu_j$. However, at least one unstable node is overloaded, that is, $\eta_j > \mu_j$. This fact shows that the maximal departure rate at each node is limited by the maximal capacity of its server.

It should also be noted that, for a network system that the ergodicity condition is violated, that is, if there is at least one unstable node in the network, the Jackson's solution does not exist for the system, that is, there is no limiting solution for the queue length distribution for the process. However, Goodman and Massey (1984) prove that: if $X = \{X(t) \; : \; t \in \mathbb{R}_+\}$ denotes the queue length process of a Jackson network (that is, it is a Markov process with state space $\mathbb{Z}_+^N$), then $\eta = (\eta_j \; : \; j = 1, 2, \ldots, N)$ is the unique solution of the general traffic equation (6.161), $S = \{i \; : \; \eta_i < \mu_i\} \subseteq \widetilde{N}$ is the set of stable nodes of the system, and $U = \widetilde{N} \backslash S$ is the set of unstable nodes of the system, then ***the limiting marginal distribution for each queue*** in the network is:

$$\lim_{t\to\infty} P\{X_j(t)\} = n_j = \begin{cases} \prod_{j=1}^{N} \left(1 - \frac{\eta_j}{\mu_j}\right) \left(\frac{\eta_j}{\mu_j}\right)^{n_j}, & j \in S, \\ 0, & j \in U, \end{cases} \tag{6.168}$$

independent of the initial distribution for all $n_j \in \widetilde{\mathbb{N}}$.

6.6.2 Unreliable Network of Queues Model of Mylosz and Daduna

Despite, Sauer and Daduna (2003) who considered a network system in which all stations are unreliable, Jennifer Mylosz considered a system that some stations may break down at any time. In her doctoral dissertation, Mylosz (2013), that contains the contents of Mylosz and Daduna (2009) and Goodman and Massey (1984), under guidance of her adviser, Hans Daduna, at Universität Hamburg, studied the globally unstable networks with unreliable nodes, that is, with breakdowns of nodes. Mylosz analyzed nodes' long-time behavior toward a local stabilization. She also studied the analysis of the local stabilization of globally unstable networks, particularly, Jackson's network of queues. She started with results from Sauer and Daduna (2003) on Jackson networks with unreliable nodes, which have product form equilibrium distribution. However, she did not assume stability for the global network, which may be due to breakdowns or overload of nodes. Mylosz assumed that if a node breaks

Hans Daduna

down, the following policy would be in effect: (1) no service is provided, (2) tasks in the node stay put (delay in service), and (3) the station will not accept any new arrival. However, if some tasks insist to attend this broken node, they will be rerouted to an orbit until the station is repaired, and service is resumed.

Before we describe Mylosz's model, we define the following notations, some of which have been defined earlier:

Notation	Definition
$\mathbb{N}$:	$\{1, 2, \ldots\}$, the set of natural numbers
$\mathbb{Z}_+$:	$\{0, 1, 2, \ldots\}$, the set of nonnegative integers
$\mathbb{R}$:	$(-\infty, +\infty)$, the set of real numbers
$\mathbb{R}_+$:	$[0, \infty)$, the set of nonnegative real numbers
$\emptyset$:	$\{\quad\}$, the empty set
$\widetilde{N}$:	The set of all possible nodes in the system, $\{0, 1, 2, \ldots, N\}$, where 0 stands for the "outside"
D:	The set of unreliable nodes, $D \subseteq \widetilde{N}$, that is, all nodes in the system that may break down
$\mathbb{P}(D)$:	The power set of D, that is, the set of all subsets of D
I:	The set of down nodes, $I \subseteq D \subseteq \widetilde{N}$, that is, the set of broken down nodes, that is, the states in I are not available and are under repair
Set $B \setminus A$:	The set $B - A$, the difference between two sets B and A
$\mathrm{I}_A(x)$:	$= \begin{cases} 1, & \text{if } x \in A, \\ 0, & \text{if } x \notin A, \end{cases}$ the indicator function
$\mathbb{Z}_+^N$:	The state space for the joint queue lengths
X on $\mathbb{Z}_+^N$:	The queue lengths in the network
Y on $\mathbb{P}(D)$:	The availability status of the network
$\mathbb{P}(D) \times \mathbb{Z}_+^N$:	The state space for the joint available queue length process
$(I;\ n_1, n_2, \ldots, n_N)$: $\in \mathbb{P}(D) \times \mathbb{Z}_+^N$:	The states of the Markov process under consideration
(Y, X):	The joint availability queue length process

We now describe Mylosz's model, which is a Jackson network with some unreliable nodes, as follows: A Jackson network with N nodes is considered. For the node $n, n = 1, 2, \dots, N$, the following are assumed:

1. External arriving tasks arrive at the node n according to a Poisson distribution with rate λ_n, $n = 1, 2, \dots, N$, such that

$$\sum_{n=1,2,\dots,N} \lambda_n = \lambda > 0, \tag{6.169}$$

   where λ represents the expected total external arrivals at the system.
2. There is a single server in the node $n, n = 1, 2, \dots, N$, and service times at that node is exponential with parameter μ_n, $n = 1, 2, \dots, N$.
3. There is an infinite-sized buffer (waiting room) in front of each station.
4. Service discipline is FCFS.
5. Arriving tasks are indistinguishable.
6. All interarrival and service times are mutually independent random variables.
7. Routing is Markovian, that is, if i is the departure node, the next node to be visited is independent of the past transitions. In other words, a task departing node i will independently move to node j with transition probability $p_{i,j} \geq 0$, $i, j = 1, 2, \dots, N$.
8. Probability of a task departing the network is denoted by $p_{i,0}$, $i = 1, 2, \dots, N$, where node 0 stands for the outside.
9. $p_{0,0} = 0$.
10. $p_{0,j} = \lambda_j / \lambda$, $i = 1, 2, \dots, N$.
11. The transition matrix (or routing matrix) denoted by $\mathbf{R}$, where $\mathbf{R} = [p_{i,j}]$, $i, j = 0, 1, 2, \dots, N$ is a stochastic and nonreducible matrix.
12. Denoting the set of all nodes by $\widetilde{N}$, that is, $\widetilde{N} = \{1, 2, \dots, N\}$, there is a subset D of $\widetilde{N}(D \subseteq \widetilde{N})$, in which nodes are unreliable, that is, they may break down randomly. If a node breaks down, tasks are rerouted according to one of the three cases: (a) ***stalling***, (b) ***skipping***, and (c) ***blocking***. These cases are defined as follows:
    (a) ***Stalling*** is the case when a node breaks down and causes the system to ***freeze***. Hence, arrival processes are interrupted and the services across the system stops. The network system returns to ***working condition*** after the broken down node is repaired and has entered the system again. While the broken down node is being repaired, all other nodes are in standby mode and are subject to break down, as well, while in that mode. Thus, if $I \neq \emptyset$, then for all $i \in \widetilde{N}$, the following is true: $\lambda_i^I = \mu_i^I = 0$, that is, again, no arrival to and no service at that node will occur.

    Automotive production is an example of this case. The production system is set up so that if in a division, a part is identified as defective or below the standard quality, the whole system will stop.

(b) ***Skipping*** is the case when a node breaks down and no arrival is allowed to that node. However, an arrival will be routed to pass this broken down node and move on until it finds a working node. If no working node is available, the job will leave the network. In this case, although no delay will occur, not all servers will be used either. Nonetheless, transition matrix in this case would be assumed to be

$$\mathbf{R}^I = [p^I_{i,j}], i,j = 0, 1, 2, \ldots, \widetilde{N}\backslash I, \tag{6.170}$$

where $\widetilde{N} \setminus I$ denotes the set of working nodes (which is the set of all nodes minus the set of broken down nodes) and elements of the transition matrix $\mathbf{R}^I$ are as follows:

$$p^I_{i,j} = p_{i,j} + \sum_{k\in I} p_{i,k}p^I_{k,j}, i,j = 0, 1, 2, \ldots, \widetilde{N}\backslash I \tag{6.171}$$

with

$$p^I_{k,j} = p_{k,j} + \sum_{l\in I} p_{k,l}p^I_{l,j}, k \in I, j = 0, 1, 2, \ldots, \widetilde{N}\backslash I. \tag{6.172}$$

External arrival rates during a breakdown period are assumed as follows:

$$\begin{cases} \lambda^I_j = \lambda_j + \sum_{i\in I}\lambda_i, p^I_{i,j}, & j \in \widetilde{N} \setminus I, \\ \lambda^I_k = 0, & k \in I. \end{cases} \tag{6.173}$$

Service rates are also assumed as follows:

$$\mu^I_i = \begin{cases} \mu_i, & i \in \widetilde{N} \setminus I, \\ 0, & \text{otherwise}. \end{cases} \tag{6.174}$$

(c) ***Blocking*** is the case when a node breaks down and it becomes blocked. In this case, a task whose next destination is this down node stays put and receives repeated services (that is the queue becomes a ***retrial system***). However, this task after completion of its repeated services will choose its next destination according to the following transition matrix:

$$\mathbf{R}^I = [p^I_{i,j}], i,j = 0, 1, 2, \ldots, \widetilde{N}\backslash I, \tag{6.175}$$

where

$$p^I_{i,j} = \begin{cases} p_{i,j}, & i,j = 0, 1, 2, \ldots, \widetilde{N}\backslash I, \ i \neq j \\ p_{i,j} + \sum_{k\in I} p_{i,k}, & i,j = 0, 1, 2, \ldots, \widetilde{N}\backslash I, \ i = j. \end{cases} \tag{6.176}$$

For this case, the external arrival rates during a breakdown are

$$\begin{cases} \lambda_j^I = \lambda_j, & j \in \widetilde{N} \setminus I, \\ 0, & \text{otherwise}. \end{cases} \tag{6.177}$$

Service rates for this case are as follows:

$$\mu_i^I = \begin{cases} \mu_i, & i \in \widetilde{N} \setminus I, \\ 0, & \text{otherwise}. \end{cases} \tag{6.178}$$

Blocked cases appear in telecommunications systems. When the blocked case is used, it has been shown that to obtain the Jackson networks, the routing should be reversible (Sauer and Daduna, 2003; Mylosz and Daduna, 2009, p. 1254).

In addition to the block case just defined, other possibilities are studied in the literature, such as rejection blocking, retransmission, and repeat blocking.

Before we discuss the Mylosz (2013) model described by the aforementioned 12 items, we make a note on the open Jackson network. A Jackson network may have nodes with ***unlimited or infinite supply of work***. That is, ***a node i is with infinite supply of work, if when all the tasked queued at node i departed, it will process a task from this supply***. In other words, nodes such as i will be servicing tasks all the time, and hence they will be fully utilized and will experience a traffic intensity of 1.

The general traffic equations for Jackson networks with unlimited supply of work are

$$\eta_i = \lambda_i + \sum_{j \in W} \min(\eta_j, \mu_j) p_{j,i} + \sum_{j \in V} \mu_j p_{j,i}, j \in \widetilde{N}. \tag{6.179}$$

Now, let V be a subset of the set of nodes, $\widetilde{N}$. That is,

$$V = \{i \in \widetilde{N}, \text{ such that it has infinite supply of work}\}. \tag{6.180}$$

Let also W be the subset of $\widetilde{N}$ such that

$$W = \widetilde{N} \backslash V, \tag{6.181}$$

that is, W is the subset of nodes without an infinite supply of work. Then, the queue size process

$$X = [(X_1(t), \ldots, X_N(t)) : t \in \mathbb{R}_+] \tag{6.182}$$

is a Markov process on $\mathbb{Z}_+^N$ with transition probability matrix,

$$\mathbf{Q} = (q(r, s) : r, s \in \mathbb{Z}_+), \tag{6.183}$$

which is defined by the following:

For all $i, j \in \widetilde{N}$, $i \neq j$:

$$q(n_1, \ldots, n_i, \ldots, n_N;\ n_1, \ldots, n_i + 1, \ldots, n_N) = \lambda_i + \sum_{j\in V} \mu_j p_{j,i} I_{\{0\}}(n_j), \tag{6.184a}$$

$$q(n_1, \ldots, n_i, \ldots, n_N;\ n_1, \ldots, n_i - 1, \ldots, n_N) = \mu_j p_{i,0} I_{\mathbb{Z}_+}(n_i). \tag{6.184b}$$

$$q(n_1, \ldots, n_i, \ldots, n_j, \ldots, n_N;\ n_1, \ldots, n_i - 1, \ldots, n_j + 1, \ldots, n_N) = \mu_i p_{i,j} I_{\mathbb{Z}_+}(n_i), \tag{6.184c}$$

$$q(n_1, \ldots, n_N;\ n_1, \ldots, n_N) = -\sum_{i\in\widetilde{N}} \lambda_i - \sum_{i\in\widetilde{N}} \sum_{i\in\widetilde{V}} \mu_j p_{j,i} I_{\{0\}}(n_j) - \sum_{i\in\widetilde{N}} \mu_i (1 - p_{i,i}) I_{\mathbb{Z}_+}(n_i), \tag{6.184d}$$

and

$$q(r, s) = 0, \text{otherwise}. \tag{6.184e}$$

Weiss (2005) proved that a Jackson network in which nodes in V have unlimited supply of work, departure processes from nodes are independent Poisson process with rates μ_j and, hence, the departure from node $j \in V$ to node $i \in \widetilde{N}$ is Poisson with rare $\mu_j p_{j,i}$.

On the basis of the Mylosz (2013) model described by the aforementioned 12 items, the queue length distribution may be derived. It should be noted that to specify transition of the ***availability status*** of the system, $\boldsymbol{Y}$, from a ***repair status***, $\boldsymbol{I}$, to the ***next state***, say I', denoted by $I \to I'$, there are two possible cases: (1) the successor node is a down node (in this case, I will increase) and (2) the successor node is an up node (in this case, I will decrease). Thus, th ***joint available queues length process*** can be described by the Markov process:

$$(Y, X) = \{Y(t);\ X_1(t), X_2(t), \ldots, X_N(t), t \in \mathbb{R}_+\}, \tag{6.185}$$

on the state space $\mathbb{P}(D) \times \mathbb{Z}_+^N$ with transition probability matrix, $\mathbf{Q}$, as follows:

$$\mathbf{Q} = [q_{r,s}], r, s \in \mathbb{P}(D) \times \mathbb{Z}_+^N. \tag{6.186}$$

It is assumed that the homogenous Markov processes (6.184) is a regular ***jump process***, that is,

(i) the sum of each row of the $\mathbf{Q}-$ matrix is zero;
(ii) the diagonal entries of the $\mathbf{Q}-$ matrix are finite; and
(iii) the sequence of jump times diverges almost surely.

The elements of $\mathbf{Q}$ are defined below for three cases using the following notations and assumptions from Sauer and Daduna (2003): Assuming $\frac{0}{0} = 0$, A and B are defined as nonnegative functions as follows:

$$\left.\begin{aligned} A : \mathbb{P}(D) \to [0, \infty), \\ A(\emptyset) = 1, \ \ \forall K \subseteq D \subseteq \widetilde{N}, \\ \tfrac{A(K\cup G)}{A(K)} < \infty, \ \ \forall K \cap G = \emptyset, \end{aligned}\right\} \tag{6.187}$$

and

$$\left.\begin{aligned} B : \mathbb{P}(D) \to [0, \infty), \\ B(\emptyset) = 1, \ \ \forall K \subseteq D \subseteq \widetilde{N}, \\ \tfrac{B(K)}{B(K\setminus H)} < \infty, \ \ \forall H \subseteq K. \end{aligned}\right\}. \tag{6.188}$$

In addition, breakdown and repair rates (or death and birth rates, respectively, in B–D terminology) are defined as follows:

$$\alpha(K, K \cup G) = \frac{A(K \cup G)}{A(K)}, \quad K \cap G = \emptyset, \ \ \forall K \subseteq D, \tag{6.189}$$

and

$$\beta(K, K\setminus H) = \frac{B(K)}{B(K\setminus H)}, \quad \forall H \subseteq K \subseteq D. \tag{6.190}$$

Case 1 For all $i, j \in \widetilde{N}$, $i \neq j$:

$$q(\emptyset, n_1, \ldots, n_i, \ldots, n_N;\ \ \emptyset, n_1, \ldots, n_i + 1, \ldots, n_N) = \lambda_i, \tag{6.191a}$$

$$q(\emptyset, n_1, \ldots, n_i, \ldots, n_N;\ \ \emptyset, n_1, \ldots, n_i - 1, \ldots, n_N) = \mu_i p_{i,0} \mathrm{I}_{\mathbb{N}}(n_i), \tag{6.191b}$$

$$\begin{aligned} &q(\emptyset, n_1, \ldots, n_i, \ldots, n_j, \ldots, n_N;\ \ \emptyset, n_1, \ldots, n_i - 1, \ldots, n_j + 1, \ldots, n_N) \\ &\quad = \mu_i p_{i,0} \mathrm{I}_{\mathbb{N}}(n_i), \end{aligned} \tag{6.191c}$$

$$\begin{aligned} &q(\emptyset, n_1, \ldots, n_N;\ \ \emptyset, n_1, \ldots, n_N) = -\sum_{i\in\widetilde{N}} \lambda_i - \sum_{i\in\widetilde{N}} \mu_i (1 - p_{i,i}) \mathrm{I}_{\mathbb{N}}(n_i) \\ &\qquad - \sum_{I\subseteq D} \alpha(\emptyset, n_1, n_2, \ldots, n_N;\ \ I, n_1, n_2, \ldots, n_N) \mathrm{I}_{\mathbb{N}}(n_i), \end{aligned} \tag{6.191d}$$

$$q(\emptyset, n_1, \ldots, n_N;\ \ I, n_1, \ldots, n_N) = \alpha(\emptyset, n_1, \ldots, n_N;\ \ I, n_1, \ldots, n_N), \ \ I \subseteq D. \tag{6.191e}$$

Case 2 For all $\emptyset \neq I \subseteq D$, $i, j \in \widetilde{N}\setminus I$, $i \neq j$:

$$q(I, n_1, \ldots, n_i, \ldots, n_N;\ \ I, n_1, \ldots, n_i + 1, \ldots, n_N) = \lambda_i^I, \tag{6.192a}$$

$$q(I, n_1, \ldots, n_i, \ldots, n_N;\ \ I, n_1, \ldots, n_i - 1, \ldots, n_N) = \mu_i^I p_{i,0}^I \mathrm{I}_{\mathbb{N}}(n_i), \tag{6.192b}$$

$$q(I, n_1, \ldots, n_i, \ldots, n_j, \ldots, n_N;\ I, n_1, \ldots, n_i - 1, \ldots, n_j + 1, \ldots, n_N)$$

$$= \mu_i^I p_{i,0}^I \mathrm{I}_{\mathbb{N}}(n_i), \tag{6.192c}$$

$$q(I, n_1, \ldots, n_N;\ I, n_1, \ldots, n_N) = -\sum_{i \in \widetilde{N} \setminus I} \lambda_i^I - \sum_{i \in \widetilde{N} \setminus I} \mu_i^I (1 - p_{i,i}^I) \mathrm{I}_{\mathbb{N}}(n_i)$$

$$- \sum_{I \subset H \subseteq D} \alpha(I, n_1, n_2, \ldots, n_N;\ H, n_1, n_2, \ldots, n_N)$$

$$- \sum_{K \subset I \subseteq D} \beta(I, n_1, n_2, \ldots, n_N;\ K, n_1, n_2, \ldots, n_N), \tag{6.192d}$$

$$q(I, n_1, \ldots, n_N;\ H, n_1, \ldots, n_N) \tag{6.193a}$$

$$= \alpha(I, n_1, \ldots, n_N;\ H, n_1, \ldots, n_N),\ I \subset H \subseteq D, \tag{6.192e}$$

$$q(I, n_1, \ldots, n_N;\ K, n_1, \ldots, n_N) \tag{6.193b}$$

$$= \alpha(I, n_1, \ldots, n_N;\ K, n_1, \ldots, n_N),\ K \subset I \subseteq D. \tag{6.192f}$$

Case 3 Other than cases (1) and (2):

$$q(r, s) = 0. \tag{6.193}$$

Let $\lambda_j, j = 1, 2, \ldots, N$, denote the arrival rate at node j and $\eta_j, j = 1, 2, \ldots, N$, denote the total arrival to a Jackson network. We say a homogeneous Markov process on discrete states $X = \{X(t) : t \in \mathbb{R}_+\}$ is ergodic if and only if $\eta_j < \mu_j$, where $\mu_j, j = 1, 2, \ldots, N$, denotes the service rate at node j, which is true for all $j = 1, 2, \ldots, N$. Hence, as in Section 6.3, for a steady-state **ergodic** Jackson network, the arrival rate at a node j is given by

$$\eta_j = \lambda_j + \sum_{i=1}^{N} \eta_i p_{i,j}, j = 1, 2, \ldots, N, \tag{6.194}$$

where $\lambda_j = \lambda_0 p_{0,j}$ and λ_0 denotes the external arrival rate. Thus, for an open Jackson network, the mean arrival rate at a node j is the sum of external and internal arrivals, $\lambda = (\lambda_j, j = 1, 2, \ldots, N)$ is the solution of Equation (6.194). The joint probability distribution of the number of tasks in the system is given by Equation (6.7). This unique steady-state product solution for the homogeneous Markov process on discrete states $X = \{X(t) \ : \ t \in \mathbb{R}_+\}$, based on the notations of this section, denoted by π, on $\mathbb{Z}_+^N$, can be rewritten as

$$\pi_{n_1, n_2, \ldots, n_N} = \lim_{t \to \infty} P\{X_j(t) = n_j, j \in \widetilde{N}\} = \prod_{j=1}^{N} \left(1 - \frac{\eta_j}{\mu_j}\right) \left(\frac{\eta_j}{\mu_j}\right)^{n_j}. \tag{6.195}$$

We conclude this part by considering a Jackson network with unreliable nodes as in the model defined above by Mylosz. Sauer and Daduna (2003) proved the following theorem for these modes as in Mylosz (2013, Theorem 2.14):

Theorem 6.6.1 Consider a Jackson network with unreliable nodes as defined in Mylosz's model defined above, in which D, $D \subseteq \widetilde{\mathbb{N}}$, denotes the set of unreliable nodes. Let $\eta = (\eta_j : j = 1, 2, \ldots, N)$ be the unique solution of the general traffic equation $\eta_j = \lambda_j + \sum_{i=1}^{N} \eta_i p_{i,j}, j = 1, 2, \ldots, N$, (Equation (6.161)). It is assumed that when breakdown of a node occurs, tasks are rerouted according to stalling, skipping, or blocking, as defined above. In case of blocking, it is assumed that

$$\eta_i p_{i,j} = \eta_j p_{j,i}, \text{ for all } i, j \in \widetilde{\mathbb{N}}, \tag{6.196}$$

sometimes called ***reversibility constraint***. Then, if the availability of queue length process $(Y, X) = \{Y(t);\ X_1(t), X_2(t), \ldots, X_N(t), t \in \mathbb{R}_+\}$ (Equation (6.185)) is ergodic, then its steady-state distribution has a product form as follows:

$$\pi(I;\ n_1, \ldots, n_N) = C^{-1} \frac{A(I, n_i : i \in I)}{B(I, n_i : i \in I)} \prod_{j=1}^{N} \left(\frac{\eta_j}{\mu_j}\right)^{n_j}, \tag{6.197}$$

for all $(I;\ n_1, \ldots, n_N) \in \mathbb{P}(D) \times \mathbb{Z}_+^N$, where

$$C = \sum_{(n_1,\ldots,n_N) \in \mathbb{Z}_+^N} \left[\prod_{j=1}^{N} \left(\frac{\eta_j}{\mu_j}\right)^{n_j} \sum_{I \subseteq D} \frac{A(I, n_i : i \in I)}{B(I, n_i : i \in I)} \right] < \infty, \tag{6.198}$$

and A and B are nonnegative functions defined in Eqs. (6.187) and (6.188), respectively.

6.6.3 Unreliable Network of Queues Model of Gautam Choudhury, Jau-Chuan Ke, and Lotfi Tadj: A Queueing System with Two Network Phases of Services, Unreliable Server, Repair Time Delay under N-Policy

In this section, we discuss a model analyzed by Choudhury et al. (2009). They considered the following queueing system model:

Jau-Chuan Ke

Lotfi Tadj

1. There are two phases of heterogeneous services and unreliable server under N-policy.
2. Tasks arrive by batches of random sizes represented by iid random variables $X_1, X_2, \ldots$, according to a compound Poisson process with rate λ with pmf a_n, that is,
$$a_n = P\{X = n\}, n \geq 1. \tag{6.199}$$
3. The server turns the system off each time the system becomes empty and returns to work when the queue size builds up to threshold value N. This indicates that once the system is turned off, the server will be idle and remains idle until the number of arriving tasks accumulates to size N ($N \geq 1$,, the threshold). The server then begins to serve the first phase of regular service (FPS) to all arriving tasks. This event is denoted by B_1.
4. It is assumed that input process; lifetime, repair time, and delay time of the server; and service time random variables are independent of each other.
5. The service discipline is assumed as FCFS.
6. After completion of FPS, the task may leave the system with probability q or may be provided with a second phase of optional service (SPS) with probability p. This event is denoted by B_2, $0 \leq p \leq 1, p + q = 1$.
7. The system may break down at any time while working with any phase of service and the server will be idle for a short interval of time.
8. The breakdowns are generated by an exogenous Poisson process with rates α_1 for FPS and α_2 for SPS.
9. As soon as breakdown occurs, it is sent for repair.
10. During the repair time, the server stops providing service to the tasks waiting in the queue until the service station is repaired.

11. The task being served just before server breakdown waits for repair to start (due to nonavailability of the broken down server), which is called ***delayed-service period***.
12. The following notations with their meanings are defined:

$B_i(x)$, $i = 1, 2$: the cumulative distribution functions of the service times at phases 1 and 2, respectively;
$\beta_i^*(s)$, $i = 1, 2$: the Laplace–Stieltjes transform of the distribution functions of the service times at phases 1 and 2, respectively; and
$\beta_i^{(k)}$, $i = 1, 2$: the kth finite moment of the distribution functions of the service times at phases 1 and 2, respectively.

13. Define the generating function of sequence of batch sizes by

$$a(z) = E(z^X) = \sum_{n=0}^{\infty} a_n z^n. \tag{6.200}$$

14. Also denote by $a_{[1]}$ and $a_{[2]}$ the first two factorial moments (Chapter 2).
15. The system with Markovian batch arrivals, general service time distribution, and a single server is denoted by $M^x/G/1$.
16. The delay time of the server for ith phase of service is denoted by D_i, $i = 1, 2$.
17. We also let

$D_i(y)$, $i = 1, 2$, denote the cumulative distribution functions of the delay times at phases 1 and 2, respectively,
$\gamma_i^*(s)$, $i = 1, 2$, denote the Laplace–Stieltjes transform of the delay times at phases 1 and 2, respectively,
$\gamma_i^{(k)}$, $i = 1, 2$, denote the kth finite moment of distribution functions of the delay times at phases 1 and 2, respectively,

18. We further let the repair time of the service station be denoted by R_1 for FPS and R_2 for SPS, and

$G_i(y)$, $i = 1, 2$, denote the cumulative distribution functions of the repair times of the service station at phases 1 and 2, respectively,
$G_i^*(s)$, $i = 1, 2$, denote the Laplace–Stieltjes transform of the repair times at phases 1 and 2, respectively, and
$g_i^{(k)}$, $i = 1, 2$, denote the kth finite moment of distribution functions of the repair times at phases 1 and 2, respectively.

19. The server starts to work and serves the remaining service to tasks in both phases of service immediately after the service station is up and running. In

this case, the service times are cumulative, that is, preemptive resume for service time, which is referred to as ***generalized service times***. This type of model is known as ***queue with a second optional service with unreliable server***, during which a breakdown period and a delayed-service time will occur.

20. The generalized service time for the ith phase of service is denoted by H_i, $i = 1, 2$.
21. We finally let

$H_i(x)$, $i = 1, 2$,	denote the cumulative distribution functions of the total (or generalized) service times of the service station at phases 1 and 2, respectively,
$H_i^*(s)$, $i = 1, 2$,	denote the Laplace–Stieltjes transform of the total (or generalized) service times at phases 1 and 2, respectively,
$h_i^{(k)}$, $i = 1, 2$,	denote the kth finite moment of distribution functions of the total (or generalized) service times at phases 1 and 2, respectively.

Thus, we have

$$\begin{aligned} H_i^*(s) = E(e^{-sH_i(x)}) &= \sum_{n=0}^{\infty} \int_0^{\infty} e^{-sx}\ e^{-\alpha_i x} \left[\frac{(\alpha_i x)^n}{n!}\right] [\gamma_i^*(s) G_i^*(s)]^n dB_i(x) \\ &= B_i^*(s + \alpha_i(1 - \gamma_i^*(s)G_i^*(s))), i = 1, 2, \end{aligned} \tag{6.201}$$

and the first two moments as

$$h_i^{(1)} = -\left.\frac{dH_i^{(*)}(s)}{ds}\right|_{s=0} = \beta_i^{(1)}\{1 + \alpha_i(\gamma_i^{(1)} + g_i^{(1)})\} \tag{6.202}$$

and

$$\begin{aligned} h_i^{(2)} &= (-1)^2 \left.\frac{d^2 H_i^{(*)}(s)}{ds^2}\right|_{s=0} \\ &= \beta_i^{(2)}\{1 + \alpha_i(\gamma_i^{(1)} + g_i^{(1)})\}^2 + \alpha_i \beta_i^{(1)}\{\gamma_i^{(2)} + g_i^{(2)} + 2\gamma_i^{(1)} g_i^{(1)}\}. \end{aligned} \tag{6.203}$$

The method Choudhury et al. (2009) used to find the queue length is the standard method deriving the probability generating functions (pgfs) of the stationary queue size distribution. Of course, in this case, they have a system of integral equations, since the service distribution is general. In order to set up this system, the authors set up a system of state equations for the stationary queue size distribution through treating ***elapsed service time***, ***elapsed repair time***, and ***elapsed delay time*** of the server for both phases of service as supplementary variables. Therefore, let

$N(t) \equiv$	the number of tasks in the system, including one being served, if any, at time t.

We also introduce the following supplementary variables:

$B_i^0(t) \equiv$	the elapsed service time of the task for ith phase of service at time t for $i = 1, 2$, (that is, for FPS and SPS, respectively);
$R_i^0(t) \equiv$	the elapsed repair time of the server for ith phase of service during which breakdown occurs in the system at time t; and
$D_i^0(t) \equiv$	the elapsed delay time of the server for ith phase of service during which breakdown occurs in the system at time t.

$$Y(t) = \begin{cases} 0, & \text{if the server is idle at time } t, \\ 1, & \text{if the server is busy with FPS at time } t, \\ 2, & \text{if the server is busy with SPS at time } t, \\ 3, & \text{if the server is waiting for repair during FPS at time } t, \\ 4, & \text{if the server is waiting for repair during SPS at time } t, \\ 5, & \text{if the server is under repair during FPS at time } t, \\ 6, & \text{if the server is under repair during SPS at time } t. \end{cases}$$

The introduction to supplementary variables is to obtain a bivariate Markov process $\{N(t), X(t)\}$, where

$$X(t) = \begin{cases} 0, & \text{if } Y(t) = 0, \\ B_1^0(t), & \text{if } Y(t) = 1, \\ B_2^0(t), & \text{if } Y(t) = 2, \\ D_1^0(t), & \text{if } Y(t) = 3, \\ D_2^0(t), & \text{if } Y(t) = 4, \\ R_1^0(t), & \text{if } Y(t) = 5, \\ R_2^0(t), & \text{if } Y(t) = 6. \end{cases}$$

Now, we define the following limiting probabilities:

$$\Psi_n = \lim_{t\to\infty} P\{N(t) = n, X(t) = 0\}, \quad n = 0, 1, 2, \ldots, N-1. \tag{6.204}$$

$$P_{i,n}(x)dx = \lim_{t\to\infty} P\{N(t) = n, X(t) = B_i^0(t);\ \ x < B_i^0(t) \le x + dx\},$$

$$x > 0,\ i = 1, 2,\ n = 1, 2, \ldots . \tag{6.205}$$

$$Q_{i,n}(x, y)dy = \lim_{t\to\infty} P\{N(t) = n, X(t) = D_i^0(t);\ y < D_i^0(t) \le y + dy|B_i^0(t) = x\},$$

$$(x, y) > 0,\ i = 1, 2,\ n = 1, 2, \ldots . \tag{6.206}$$

$$R_{i,n}(x, y)dy = \lim_{t\to\infty} P\{N(t) = n, X(t) = R_i^0(t);\ \ y < R_i^0(t) \le y + dy|B_i^0(t) = R\},$$

$$(x, y) > 0,\ i = 1, 2,\ n = 1, 2, \ldots . \tag{6.207}$$

The following are more assumptions:

$$B_i(0) = 0, \quad B_i(\infty) = 1, \quad i = 1, 2,$$

$$D_i(0) = 0, \quad D(\infty) = 1, \quad i = 1, 2,$$

$$G_i(0) = 0, \quad G_i(\infty) = 1, \quad i = 1, 2,$$

$B_i(x)$ is continuous at $x = 0$, and $D_i(y)$ and $G_i(y)$ are continuous at $y = 0$, respectively, so that

$$\mu_i(x)dx = \frac{dB_i(x)}{1 - B_i(x)},\ \eta_i(y)dy = \frac{dD_i(y)}{1 - D_i(x)}, \xi_i(y)dy = \frac{dG_i(y)}{1 - G_i(x)}, \quad i = 1, 2, \tag{6.208}$$

are the first-order differential (hazard rate) functions.

On the basis of all the aforementioned assumptions and properties, the Kolmogorov forward equations governing the system under steady-state conditions (Cox, 1955) can be written as follows:

$$\frac{d}{dx}P_{i,n}(x) + [\lambda + \alpha_i + \mu_i(x)]P_{i,n}(x)$$

$$= \lambda \sum_{k=1}^{n} a_k P_{i,n-k}(x) + \int_0^\infty \xi_i(y) R_{i,n}(x, y)dy, n \ge 1, \tag{6.209}$$

$$\frac{d}{dx}Q_{i,n}(x, y) + [\lambda + \eta_i(y)]Q_{i,n}(x, y) = \lambda \sum_{k=1}^{n} a_k Q_{i,n-k}(x, y), n \ge 1, \tag{6.210}$$

$$\frac{d}{dx}R_{i,n}(x, y) + [\lambda + \xi_i(y)]R_{i,n}(x, y) = \lambda \sum_{k=1}^{n} a_k R_{i,n-k}(x, y), n \ge 1, \tag{6.211}$$

$$\lambda\Psi_n = \delta_{n,0} \int_0^\infty \mu_2(x) P_{2,n+1}(x)dx + \delta_{n,0} q \int_0^\infty \mu_1(x) P_{1,n+1}(x)dx$$

$$+ \lambda(1 - \delta_{n,0}) \sum_{k=1}^{n} a_k \Psi_{n-k}, \quad n = 0, 1, 2, \ldots, N - 1, \tag{6.212}$$

where $\delta_{m,n}$ denotes Kronecker's function and

$$P_{i,0}(x) = 0,\ Q_{i,0}(x,y) = 0, \text{ and } R_{i,0}(x,y) = 0, \text{ for } i = 1,2, \tag{6.213}$$

occurring in Eqs. (6.209)–(6.211).

The goal is to solve the system of equations (6.209)–(6.211) under the following initial conditions:

$$P_{1,n}(0) = \int_0^\infty \mu_2(x)P_{2,n+1}(x)dx + q\int_0^\infty \mu_1(x)P_{1,n+1}(x)dx,\ 1 \le n \le N-1, \tag{6.214}$$

$$P_{1,n}(0) = \int_0^\infty \mu_2(x)P_{2,n+1}(x)dx + q\int_0^\infty \mu_1(x)P_{1,n+1}(x)dx + \lambda\sum_{k=0}^{N-1} a_{n-k}\Psi_k,\ n \ge N, \tag{6.215}$$

$$P_{2,n}(0) = p\int_0^\infty \mu_1(x)P_{1,n}(x)dx,\ n \ge 1, \tag{6.216}$$

$$Q_{i,n}(x,0) = \alpha_i P_{i,n}(x),\ x > 0,\ n \ge 1,\ y = 0, \text{ for } i = 1,2, \text{ and fixed value of } x, \tag{6.217}$$

$$R_{i,n}(x,0) = \int_0^\infty \eta_i(y)Q_{i,n}(x,y)dy,\ x > 0, \tag{6.218}$$

$$n \ge 1,\ y = 0 \text{ for } i = 1,2, \text{ and fixed value of } x, \tag{6.219}$$

with normalization condition:

$$\sum_{n=0}^{N-1}\Psi_n + \left[\sum_{n=1}^{\infty}\sum_{i=1}^{2}\left\{\int_0^\infty P_{i,n}(x)dx + \int_0^\infty\int_0^\infty Q_{i,n}(x,y)dx\,dy + \int_0^\infty\int_0^\infty R_{i,n}(x,y)dx\,dy\right\}\right] = 1. \tag{6.220}$$

Thus, we need to solve the system of equations (6.209)–(6.220) for the unknown probabilities. For this purpose, we define the following generating functions:

$$P_i(x,y) = \sum_{n=1}^{N-1} z^n P_{i,n}(x),\ P_i(0,y) = \sum_{n=1}^{N-1} z^n P_{i,n}(0), \tag{6.221}$$

$$Q_i(x,y;z) = \sum_{n=1}^{\infty} z^n Q_{i,n}(x,y),\ Q_i(x,0;z) = \sum_{n=1}^{\infty} z^n Q_{i,n}(x,0), \tag{6.222}$$

$$R_i(x,y;z) = \sum_{n=1}^{\infty} z^n R_{i,n}(x,y),\ R_i(x,0;z) = \sum_{n=1}^{\infty} z^n R_{i,n}(x,0), \tag{6.223}$$

and

$$\Psi_N(z) = \sum_{n=0}^{N-1} z^n \Psi_n. \tag{6.224}$$

Let

$$b(z) = \lambda(1 - a(z)). \tag{6.225}$$

Then, applying generating functions defined in Eqs. (6.222)–(6.225) on the system (6.209)–(6.220), we leave it as an exercise for the reader to show that we obtain a set of differential equations of Lagrangian type. Solving the obtained set of equations, we have the following:

$$Q_i(x, y; z) = Q_i(x, 0; z)[1 - D_i(y)]e^{-b(z)y}, (x, y) > 0 \text{ for } i = 1, 2, \tag{6.226}$$

$$R_i(x, y; z) = R_i(x, 0; z)[1 - G_i(y)]e^{-b(z)y}, (x, y) > 0 \text{ for } i = 1, 2, \tag{6.227}$$

where $Q_i(x, 0; z)$ and $R_i(x, 0; z)$, for $i = 1, 2$, can be obtained from Eqs. (6.216) and (6.217).

We leave it as an exercise for the reader to show that Eqs. (6.216) and (6.217), after some simplification, lead to

$$Q_i(x, 0; z) = \alpha_i P_i(x; z) \tag{6.228}$$

and

$$R_i(x, 0; z) = Q_i(x; z)\gamma_i^*(b(z)). \tag{6.229}$$

Solving the differential equation (6.209), we have

$$P_i(x; z) = P_i(0, z)[1 - B_i(x)]e^{-\lambda_i(z)x}, x > 0, \text{ for } i = 1, 2, \tag{6.230}$$

where

$$\lambda_i(z) = b(z) + \alpha_i\{1 - G_i^*[b(z)]\gamma_i^*[b(z)]\}, i = 1, 2. \tag{6.231}$$

Substituting Eqs. (6.228)–(6.230) into Eqs. (6.226) and (6.227), we obtain

$$Q_i(x, y; z) = \alpha_i P_i(0; z)[1 - B_i(x)]e^{-\lambda_i(z)x}[1 - D_i(y)]e^{-b(z)y}, \text{ for } i = 1, 2, \tag{6.232}$$

$$R_i(x, y; z) = \alpha_i \gamma_i^*(b(z))P_i(0; z)[1 - B_i(x)]e^{-\lambda_i(z)x}[1 - G_i(y)]e^{-b(z)y}, i = 1, 2. \tag{6.233}$$

We note that

$$\sum_{n=N}^{\infty} z^n \sum_{k=0}^{N-1} (\Psi_k a_{n-k}) = \Psi_0 - \frac{\Psi_N(z)b(z)}{\lambda}. \tag{6.234}$$

Multiplying Eqs. (6.214) and (6.215) by z^n, summing over all possible values of n, and using Eqs. (6.212) and (6.233), after some simplification, we obtain

$$zP_1(0; z) = qP_1(0; z)\beta_1^*(\lambda_1(z)) + P_2(0; z)\beta_2^*(\lambda_2(z)) - z\Psi_N(z)b(z). \tag{6.235}$$

Similarly, with Equation (6.216), we obtain

$$P_2(0;z) = pP_1(0;z)\beta_1^*(\lambda_1(z)). \tag{6.236}$$

Substituting Equation (6.236) into Equation (6.235) and simplifying, we have

$$P_1(0;z) = \frac{z\Psi_N(z)b(z)}{[q + p\beta_2^*(\lambda_2(z))]\beta_1^*(\lambda_1(z)) - z}. \tag{6.237}$$

Now, letting z approach infinity and applying L'Hôpital's rule, we have

$$P_1(0;1) = \frac{\lambda a_{[1]}\left[\sum_{n=0}^{N-1}\Psi_n\right]}{1-\rho_0}, \tag{6.238}$$

where

$$\rho_0 = \rho_1[1 + \alpha_1(g_1^{(1)} + \gamma_1^{(1)})] + p\rho_2[1 + \alpha_2(g_2^{(1)} + \gamma_2^{(1)})] \tag{6.239}$$

is the utilizing factor of the system and

$$\rho_1 = \lambda a_{[1]}\beta_i^{(1)}, i = 1, 2. \tag{6.240}$$

Thus, we have the following:

$$P_i(x;1) = \frac{\lambda a_{[1]}(1-\delta_{i,2}q)\left[\sum_{n=0}^{N-1}\Psi_n\right][1 - B_i(x)]}{1-\rho_0}, \quad i = 1, 2, \tag{6.241}$$

$$Q_i(x,y;1) = \frac{\alpha_i\lambda a_{[1]}(1-\delta_{i,2}q)\left[\sum_{n=0}^{N-1}\Psi_n\right][1 - B_1(x)][1 - D_i(y)]}{1-\rho_0}, i = 1, 2, \tag{6.242}$$

and

$$R_i(x,y;1) = \frac{\alpha_i\lambda a_{[1]}(1-\delta_{i,2}q)\left[\sum_{n=0}^{N-1}\Psi_n\right][1 - B_1(x)][1 - G_i(y)]}{1-\rho_0}, i = 1, 2. \tag{6.243}$$

Now we define

$$\Psi_n = K_0\phi_n, n = 0, 1, 2, \ldots, N-1, \tag{6.244}$$

where

$$\phi_n = P\{\text{A batch of } n \text{ tasks arrives in the system during an idle period}\} \tag{6.245}$$

and K_0 is a normalizing constant that can be found using Equation (6.189) and some simplification can be found as

$$K_0 = \frac{1-\rho_0}{\left[\sum_{n=0}^{N-1}\phi_n\right]}. \tag{6.246}$$

From Equation (6.246), we know that

$$\rho_0 < 1, \tag{6.247}$$

which is the stability condition under which a steady-state solution exists.

Thus, we have proved the following theorem:

Theorem 6.6.2 Choudhury et al. (2009) Under the stability condition $\rho_0 < 1$, in Equation (6.247), the joint distribution of the number of tasks in the system and the server's state has the following partial pgfs:

$$\Psi_N(z) = \frac{(1-\rho_0)\left[\sum_{n=0}^{N-1} z^n \phi_n\right]}{\left[\sum_{n=0}^{N-1} \phi_n\right]}, \tag{6.248}$$

$$P_1(x;z) = \frac{z\ b(z)\Psi_N(z)[1-B_1(x)]e^{-\lambda_1(z)x}}{[q+p\beta_2^*(\lambda_2(z))]\beta_1^*(\lambda_1(z))-z}, \tag{6.249}$$

$$P_2(x;z) = \frac{pz\ b(z)\beta_1^*(\lambda_1(z))\Psi_N(z)[1-B_2(x)]e^{-\lambda_2(z)x}}{[q+p\beta_2^*(\lambda_2(z))]\beta_1^*(\lambda_1(z))-z}, \tag{6.250}$$

$$Q_1(x,y;z) = \frac{\alpha_1\ zb(z)\Psi_N(z)[1-B_1(x)]e^{-\lambda_1(z)x}[1-D_1(y)]e^{-b(z)\ y}}{[q+p\ \beta_2^*(\lambda_2(z))]\beta_1^*(\lambda_1(z))-z}, \tag{6.251}$$

$$Q_2(x,y;z) = \frac{p\alpha_2\ z\ b(z)\ \beta_1^*(\lambda_1(z))\ \Psi_N[1-B_2(x)]e^{-\lambda_2(z)x}[1-D_2(y)]e^{-b(z)y}}{[q+p\ \beta_2^*(\lambda_2(z))]\beta_1^*(\lambda_1(z))-z}, \tag{6.252}$$

$$R_1(x,y;z) = \frac{\alpha_1\ z\ \gamma_1^*(b(z))b(z)\beta_1^*(\lambda_1(z))\ \Psi_N[1-B_1(x)]e^{-\lambda_1(z)x}[1-G_1(y)]e^{-b(z)y}}{[q+p\ \beta_2^*(\lambda_2(z))]\beta_1^*(\lambda_1(z))-z}, \tag{6.253}$$

$$R_2(x,y;z) = \frac{p\alpha_2\ z\ \gamma_2^*(b(z))b(z)\beta_1^*(\lambda_1(z))\Psi_N(z)[1-B_2(x)]e^{-\lambda_2(z)x}[1-G_2(y)]e^{-b(z)y}}{[q+p\ \beta_2^*(\lambda_2(z))]\beta_1^*(\lambda_1(z))-z}, \tag{6.254}$$

where $a(z)$ and $b(z)$ are defined in Eqs. (6.200) and (6.225), respectively, and

$$\lambda_i(z) = b(z) + \alpha_i[1 - G_i^*(b(z))\gamma_i^*(b(z))],\ i = 1, 2. \tag{6.255}$$

The next theorem provides the distribution of the queue size.

Theorem 6.6.3 Choudhury et al. (2009) Under the stability condition $\rho_0 < 1$, in Equation (6.247), the marginal pgfs of the distribution of the number of tasks in the system are given as follows:

$$P_1(z) = \frac{(1-\rho_0)\left[\sum_{n=0}^{N-1} z^{n+1}\phi_n\right] b(z)[1-\beta_1^*(\lambda_1(z))]}{\lambda_1(z)[\{q+p\beta_2^*(\lambda_2(z))\}\beta_1^*(\lambda_1(z))-z]}, \tag{6.256}$$

$$P_2(z) = \frac{p(1-\rho_0)\left[\sum_{n=0}^{N-1} z^{n+1}\phi_n\right] b(z)\beta_1^*(\lambda_1(z))[1-\beta_2^*(\lambda_2(z))]}{\lambda_2(z)[\{q+p\beta_2^*(\lambda_2(z))\}\beta_1^*(\lambda_1(z))-z]}, \tag{6.257}$$

$$Q_1(z) = \frac{\alpha_1(1-\rho_0)(1-\gamma_1^*(b(z)))\left[\sum_{n=0}^{N-1} z^{n+1}\phi_n\right][1-\beta_2^*(\lambda_2(z))]}{\lambda_1(z)[\{q+p\beta_2^*(\lambda_2(z))\}\beta_1^*(\lambda_1(z))-z]}, \tag{6.258}$$

$$Q_2(z) = \frac{p\ \alpha_2(1-\rho_0)(1-\gamma_2^*(b(z)))\left[\sum_{n=0}^{N-1} z^{n+1}\phi_n\right]\beta_1^*(\lambda_1(z))[1-\beta_2^*(\lambda_2(z))]}{\lambda_2(z)[\{q+p\beta_2^*(\lambda_2(z))\}\beta_1^*(\lambda_1(z))-z]}, \tag{6.259}$$

$$R_1(z) = \frac{\alpha_1(1-\rho_0)\gamma_1^*(b(z))(1-G_1^*(b(z)))\left[\sum_{n=0}^{N-1} z^{n+1}\phi_n\right][1-\beta_1^*(\lambda_1(z))]}{\lambda_1(z)[\{q+p\beta_2^*(\lambda_2(z))\}\beta_1^*(\lambda_1(z))-z]}, \tag{6.260}$$

$$R_2(z) = \frac{p\ \alpha_2(1-\rho_0)\gamma_2^*(b(z))(1-G_2^*(b(z)))\left[\sum_{n=0}^{N-1} z^{n+1}\phi_n\right]\beta_1^*(b(z))[1-\beta_2^*(\lambda_2(z))]}{\lambda_2(z)[\{q+p\beta_2^*(\lambda_2(z))\}\beta_1^*(\lambda_1(z))-z]}, \tag{6.261}$$

Proof: The Laplace transform of $[1 - B_i(x)]$ is as follows:

$$\mathcal{L}[1 - B_i(x)] = \frac{1-\beta_i^*(s)}{s}, i = 1, 2. \tag{6.262}$$

Thus, integrating Eqs. (6.248) and (6.249) with respect to x and using Equation (6.262), we obtain Eqs. (6.256) and (6.257). Similarly, integrating Eqs. (6.261) and (6.262) with respect to y, we have

$$Q_i(x;z) = \int_0^\infty Q_i(x,y;z)dy = \frac{\alpha_i}{b(z)}[1-\gamma_i^*(b(z))]P_i(0;z)[1-B_i(x)]e^{-\lambda_i(z)x}, i = 1, 2 \tag{6.263}$$

and

$$R_i(x;z) = \int_0^\infty R_i(x,y;z)dy \tag{6.264}$$

$$= \frac{\alpha_i}{b(z)}\gamma_i^*(b(z))[1-G_i^*(b(z))]P_i(0,z)[1-B_i(x)]e^{-\lambda_i(z)x}, \quad i = 1, 2. \tag{6.265}$$

Then, integrating Eqs. (6.263) and (6.265) with respect to x yields Eqs. (6.258)–(6.261), and this completes the proof.

Now, denote the probability that the server is waiting by P_{W_i}, $i = 1, 2$. Then, noting that

$$P_{B_i} = \lim_{z\to 1} P_i(z), \quad P_{W_i} = \lim_{z\to 1} Q_i(z), \quad P_{R_i} = \lim_{z\to 1} R_i(z), \quad i = 1, 2,$$

and

$$P\{\text{server is idle}\} = 1 - \sum_{i=1}^{2} [P_{B_i} + P_{D_i} + P_{R_i}].$$

The authors show that the following is true for the steady-state case of the system:

(i)

$$P\{\text{server is idle}\} \\ = 1 - \rho_1\{1 + \alpha_1(\gamma_1^{(1)} + g_1^{(1)})\} - p\ \rho_2\{1 + \alpha_2(\gamma_2^{(1)} + g_2^{(1)})\}, \tag{6.265}$$

(ii) the probability that the server is busy with FPS is

$$P_{B_1} = \lambda\ a_{[1]}\beta_1^{(1)}, \tag{6.265}$$

(iii) the probability that the server is busy with SPS is

$$P_{B_2} = p\lambda\ a_{[1]}\beta_2^{(1)}, \tag{6.266}$$

(iv) the probability that the server is waiting for repair during FPS is

$$P_{W_1} = \lambda\ a_{[1]}\beta_1^{(1)}\alpha_1\gamma_1^{(1)}, \tag{6.267}$$

(v) the probability that the server is waiting for repair during SPS is

$$P_{W_2} = p\lambda\ a_{[1]}\beta_2^{(1)}\alpha_2\gamma_2^{(1)}, \tag{6.268}$$

(vi) the probability that the server is under repair during FPS is

$$P_{R_1} = \lambda\ a_{[1]}\beta_1^{(1)}\alpha_1 g_1^{(1)}, \tag{6.269}$$

(vii) the probability that the server is under repair during SPS is

$$P_{R_2} = p\lambda\ a_{[1]}\beta_2^{(1)}\alpha_2 g_2^{(1)}. \tag{6.270}$$

Finally, in order to address the distribution of the number of tasks in the system, let the steady-state distribution of the queue length be denoted by P_j. Furthermore, let the pgf of P_j be denoted by P_Q, that is,

$$P_Q(z) = \sum_{j=0}^{\infty} P_j z^j. \tag{6.271}$$

Then, from Theorem 6.6.2, with the help of pgfs Ψ_n, $P_i(z)$, $Q_i(z)$, and $R_i(z)$, for $i = 1, 2$, the distribution of the number of tasks in the system has pgf as

$$P_Q(z) = \Psi_N(z) + \sum_{i=1}^{2} [P_i(z) + Q_i(z) + R_i(z)]. \tag{6.272}$$

From Equation (6.272) and direct calculation, we obtain the pgf of the number of tasks in the waiting line at a random time is (as in Theorem 3.3 of Choudhury et al. (2009)):

$$P_Q(z) = \frac{(1-\rho_0)(1-z)\left[\sum_{n=0}^{N-1} z^n \phi_n\right][q + p\beta_2^*(\lambda_2(z))]\beta_1^*(\lambda_1(z))}{\left[\sum_{n=0}^{N-1} z^n \phi_n\right][\{q + p\beta_2^*(\lambda_2(z))\}\beta_1^*(\lambda_1(z)) - z]}. \tag{6.273}$$

Note:

1. If the batch size is 1, that is, if $P\{X = 1\} = 1$, in other words, arrivals are singly, Equation (6.273) reduces to the result obtained in Choudhury and Tadj (2009).
2. If $p = 0$, that is, there is no SPS in the system, then Equation (6.273) reduces to the pgf of steady-state distribution at a random epoch of queue size of an $M^x/G/1$ queueing system with unreliable server under N-policy.
3. If there is no threshold in the system, that is, $N = 1$, then Equation (6.273) reduces to the pgf of the queue length distribution at a random epoch for an $M^x/G/1$ queue with the second optional service and unreliable server.
4. For the value of $z = \beta_2^*(\lambda_2(z))$, Equation (6.273) reduces to the pgf of the steady-state distribution at the service completion epoch of an $M^x/G/1$ queueing system with unreliable server and Bernoulli feedback under N-policy.
5. The pgf of the queue size distribution at random epoch for $M^x/G/1$ queueing system with a second optional service and random breakdown without delayed repair can be obtained from Equation (6.273) using the following facts:

$$\rho_0 = \rho_1(1 + \alpha_1 g_1^{(1)}) + p\rho_2(1 + \alpha_2 g_2^{(1)}) \tag{6.274}$$

and

$$\gamma_i^*(b(z)) = 1, \quad i = 1, 2. \tag{6.275}$$

6. Let us denote by ψ_n, $n = 0, 1, 2, \ldots, N-1$, the probability that a batch of tasks finds n tasks in the system during an idle period such that

$$\psi_n = \sum_{k=1}^{n} a_k \psi_{n-k}, \quad n = 1, 2, \ldots, N-1, \ \psi_0 = 1, \tag{6.276}$$

where a_n is defined in Equation (6.199). Then, Choudhury et al. (2009) showed that the stochastic decomposition property for their model can be demonstrated by showing the following:

$$P_Q(z) = \frac{(1-\rho_0)(1-z)[q+p\ \beta_2^*(\lambda_2(z))]\beta_1^*(\lambda_1(z))}{[q+p\ \beta_2^*(\lambda_2(z))]\beta_1^*(\lambda_1(z))-z}\left[\frac{\sum_{n=0}^{N-1} z^n\phi_n}{\sum_{n=0}^{N-1}\phi_n}\right]$$
$$= \zeta(z)\nu_N(z), \tag{6.277}$$

where ν_n, $n = 0, 1, 2, \ldots, N-1$, is the probability that a batch of size n is in the system during an idle period given that the server is idle and $\zeta(z)$ is defined as

$$\zeta(z) = \frac{(1-\rho_0)(1-z)[q+p\beta_2^*(\lambda_2(z))]\beta_1^*(\lambda_1(z))}{[q+p\beta_2^*(\lambda_2(z))]\beta_1^*(\lambda_1(z))-z} \tag{6.278}$$

is the pgf of the steady-state queue length distribution at a random epoch of an $M^x/G/1$ queueing system with a second optional service and an unreliable server.

7. Interestingly, from Equation (6.277), it can be observed that the steady-state queue length distribution at a random epoch of an $M^x/G/1$ queueing system with two phases of service and an unreliable server under N-policy can be decomposed into distributions of two independent random variables through the following:
   (i) the queue length distribution of an $M^x/G/1$ queueing system with a second optional service and an unreliable server (represented by the first term of the right-hand side of Equation (6.277)), and
   (ii) the additional queue length distribution due to N-policy (represented by the second term of the right-hand side of Equation (6.277)).
8. The pgf of the steady-state queue length distribution at a departure epoch of an $M^x/G/1$ queueing system with two phases of service and an unreliable server under N-policy, denoted by $P(z)$, is given by (Choudhury et al., 2009):

$$P(z) = \frac{(1-\rho_0)\left[\sum_{n=0}^{N-1} z^n\phi_n\right][1-a(z)][q+p\beta_2^*(\lambda_2(z))]\beta_1^*(\lambda_1(z))}{a_{[1]}\left[\sum_{n=0}^{N-1}\phi_n\right][\{q+p\beta_2^*(\lambda_2(z))\beta_1^*(\lambda_1(z))-z]}. \tag{6.279}$$

9. From Equation (6.279), when $z = 0$, that is, the probability that no task is waiting in the system at the departure epoch can be obtained as

$$P(0) = \frac{(1-\rho_0)\phi_0}{a_{[1]}\left[\sum_{n=0}^{N-1}\phi_n\right]} = P_0. \tag{6.280}$$

10. From Equation (6.280), it can be observed that

$$P_0 a_{[1]} = \psi_0. \tag{6.281}$$

In other words, a random observer is more likely to find the system empty than a departing task departing the system.

11. Let us denote by T_0 and T_b the length of the idle period and the busy period, respectively. We know that the mean arriving batch sizes during an idle period with arrival rate of λ is $\sum_{j=0}^{N-1} \phi_j$. Hence, the mean idle period is

$$E(T_0) = \frac{\sum_{j=0}^{N-1} \psi_j}{\lambda}. \tag{6.282}$$

As the mean number of arrivals during an idle period is $E(T_0)\lambda a_{[1]}$, the mean busy period is

$$E(T_b) = \frac{\rho_0 E(T_0)}{1 - \rho_0}. \tag{6.283}$$

Therefore, denoting the busy cycle, that is, a busy period is followed by an idle period, by T_c, we have

$$E(T_c) = E(T_0) + E(T_b) = \frac{\sum_{j=0}^{N-1} \phi_j}{\lambda(1 - \rho)}. \tag{6.284}$$

Example 6.6.1 Choudhury et al. (2009) If we assume the arrival rate $\lambda = 0.4$ and batch size X to follow a geometric distribution with parameter 0.25, that is

$$a_k = P\{X = k\} = (0.75)(0.25^{k-1}), \quad k = 1, 2, \dots,$$

then, the first moment is $a_{[1]} = 4$. Furthermore, let us assume that the times of FPS and SPS follow exponential distributions with means

$$\beta_1^{(1)} = 0.1 \text{ and } \beta_2^{(1)} = 0.2,$$

respectively.

We further assume that the breakdown rates of the server in FPS and SPS are

$$\alpha_1 = \alpha_2 = 0.01.$$

Furthermore, if the delay times of the broken down server for FPS and SPS follow a Erlang distribution of order 2 or with shape parameter 2 (E_2, see Chapter 2) with

$$dD_1(y) = 4^2 y\ e^{-4y} \text{ and } dD_2(y) = 2^2 y\ e^{-2y},$$

then

$$\gamma_1^{(1)} = 0.5 \text{ and } \gamma_2^{(1)} = 1.$$

Finally, if the repair times of the broken down server for FPS and SPS follow a hyperexponential distribution (Chapter 2) with

$$dG_1(y) = (0.7)(2)e^{-2y} + (0.3)(4)e^{-4y} \text{ and } dG_2(y) = (0.8)(5)e^{-5y} + (0.2)(10)e^{-10y},$$

then $g_1^{(1)} = 0.425$ and $g_2^{(1)} = 0.18$.

6.7 ASSESSMENT OF RELIABILITY OF A NETWORK OF QUEUES

Reliability of networks is extremely important these days more than ever. Queueing theory being the central axis of networks, it is natural to be concerned about the modern terroristic action of disabling networks, hacking, and obtaining personal information of individuals unwantedly. Hence, increasing incidents of network terrorism and natural disasters worldwide have led to heightened interest in identifying and reducing the vulnerability of infrastructure networks (Auerswald et al., 2005; Matisziw and Murray 2009). Reliability of a network is also important from viewpoint that breakdown of a node in the network may cause a delay in service performance in the network.

It is interesting to refer to the discussion of the reliability of networks by Günneç and Salman (2011). They considered that the postdisaster functionality of infrastructure networks, such as a telecommunication, power, water, and transportation network, is critical for effective disaster response. In response to a disaster, in addition to the time-critical operations carried out by the agencies involved, some residents will be on the roads trying to evacuate the affected areas while others will try to reach the area to provide humanitarian aid and help others. Thus, a proper functionality of the transportation network is essential for the success of the rescue and relief operations. Therefore, what all these have to do with reliability of networks? In order to answer this question, we should note that a disaster may disrupt some of the links of the transportation network and make them nonfunctional, leading to the blockage of some routes and/or disconnectedness of some areas in need of aid. Therefore, it is important to assess the postdisaster performance of the network under possible disaster scenarios for the purpose of both strengthening the components of the network and planning the activities of the postdisaster logistics.

Sibel Salman

The interesting part of the work of Günneç and Salman (2011) is that they measure the reliability and the expected postdisaster performance of a network under disaster risk. In particular, measuring is the uncertainty in the predisaster stage that is due to both the magnitude and location of the disaster to occur, and furthermore, how the infrastructure network would be affected from the disaster. They look at scenarios characterized in terms of their intensity and geographic area of influence.

In addition, they evaluate the vulnerability of the civil components of the network, mostly by structural analysis and statistical predictive methods. For that purpose, they assume that each link of the network will be randomly in one of the following two states after the disaster: (1) operational (survival) and (2) nonoperational (failure), as is customary in the network reliability literature.

In general, a partially damaged link will not be available for immediate use (appearance of delay factor) because of the associated risk and inconvenience. However, an important concern for analysis is whether the links would fail independently or not. In postdisaster response, several nodes in the network act as supply and demand points, for instance, in terms of relief aid distribution and casualty transportation, thereby creating pairs of origin–destination (O–D) nodes, whose connectedness carries high priority. The reliability and expected performance measures that they propose in this study vary in terms of the number of disaster scenarios and the number of O–D pairs under consideration. These measures generalize the expected network performance defined in Peeta et al. (2010) for disaster response by extending it to multiple disaster scenarios.

When statistical dependency exists among link failures, a joint probability distribution is needed. However, as major disasters are rare events, in most cases, a sufficient amount of data does not exist to fit a joint probability distribution. Yet, a practical method is needed for quantitative analysis. In this study, these authors define a novel link dependency structure that allows the existence of a polynomial-time algorithm when the number of O–D paths in the network is fixed. They propose a structure that serves the practicality requisite while providing a reasonable approach to cope with insufficiency of data. The proposed method is as follows: the network links are partitioned into sets by geographic proximity and the degree of disaster risk they are exposed to. Links in different sets are assumed to fail independently. Then, within each set, the vulnerability of the link components is taken into consideration to define a dependency structure that they call ***vulnerability-based dependency***. They characterize the joint probability distribution of the network realizations under the proposed dependency structure and present efficient algorithms to compute the proposed measures.

The methodology they used is as follows: When a natural disaster affects an area enclosing the network, the causes of link failure are both internal and external to the network. Vulnerability of the components in a link, due to factors such as the strength of a bridge and the ground soil type, is internal. On the contrary, the magnitude of force, for example, ground acceleration or wind at the link location, as well as the consequences of the disaster, such as the collapse of buildings, an explosion or fire, are external factors, which are typically effective throughout an area. For example, in an earthquake scenario, a city may be divided into areas of varying risk. One can expect links in the same area to show a similar behavior, creating dependency. On the contrary, the internal factors will create differences among the links exposed to the same level of risk. They propose a dependency model that aims to combine these two aspects.

In summary, the authors study a network for its reliability and performance after a disaster. They focus on eight different measures in the context of postdisaster logistics, particularly considering the condition of a transportation network after an earthquake, in which the links are likely to collapse. They designed a framework for dependent link failures that can be adapted in different contexts, where dependent sets can be determined by the decision maker with respect to the network and how the environment affects the dependency relationship.

In the proposed framework, in order to identify the dependencies within a set, they defined a novel vulnerability-based link dependency structure that ranks the links according to their probabilities of survival. They assumed that the failure of a stronger link, that is, the link with higher probability of survival, implies the failure of weaker links (links with smaller probability of survival) with certainty. This assumption seems to be reasonable in the earthquake context where links within the same area of risk with respect to an expected earthquake scenario show a weakness/strength ranking according to the vulnerability of the components in each link, such as the structural strength of bridges. Furthermore, this assumption allows the existence of a polynomial-time algorithm when the number of paths connecting an O–D pair is fixed. They illustrated the applicability of the proposed analysis method by means of a case study of the Istanbul (Turkey) highway system under earthquake risk.

For the dependent link failure case, they applied our exact polynomial-time algorithm for Istanbul's sparse main highway network in the region most likely to be affected by an expected earthquake, by selecting a reasonable number of O–D pairs according to the expected earthquake scenarios. In addition, they used a Monte Carlo sampling algorithm to estimate the measures of interest for the computationally difficult case of independent link failures for comparison. These results also support that reliability and performance of a network of realistic size can be estimated with high accuracy in moderate computation time with the proposed Monte Carlo simulation method. As a result, they obtained very promising results in terms of proving practicality of the proposed approach.

6.8 EFFECT OF NETWORK SERVICE BREAKDOWN

We conclude this chapter by presenting a model that investigates how the network connection service problems can affect the performance of a complex cognitive info-communication system, that is, the effect of the communication network breakdowns to the system performance. The model is proposed by Kuki et al. (2014).

As an example of a cognitive info-communication system, consider two doctors ("telesurgeons"), A and B, located on different (and probably remote) places performing a medical operation by using a telesurgical robot. In the background, different pieces of information are transmitted: The camera view of the robot's environment

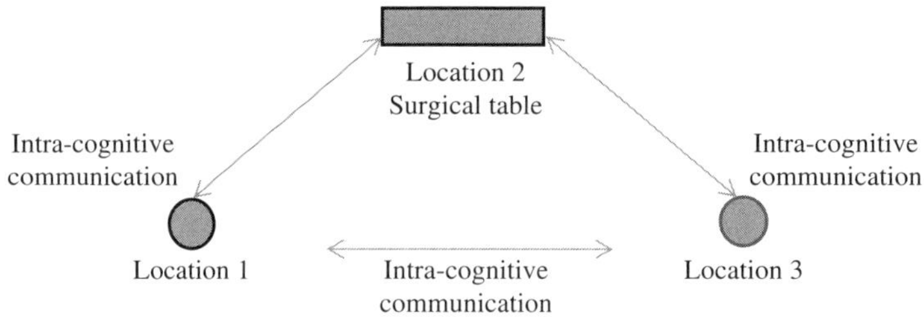

Figure 6.13 A CoginfoCom scenario.

and data of the temperature, blood pressure, and heartbeat rate of the patient are transmitted to the telesurgeons (intercognitive communication). Similarly, the speaking between the doctors is also transmitted using the network services (intracognitive communication).

The sensibility of the system to the communication failures can be highly different: For instance, if doctor "A" tells something, but the voice is not understandable by doctor "B," then doctor "A" will repeat the text and the operation may continue. On the contrary, suppose that the camera view of the robot's environment is lost, then the operation must be stopped, or even it may produce unpredictable results.

The model is a Markovian one, constructed focusing on the question that how the breakdowns of the communication network ("Network Cloud") influence the system's performance. The service can be performed in three different locations, fully operational, limited operation, broken offering different communication capabilities to the "sources" (Fig. 6.13). The server is referred to "service element." The purpose of the model is to investigate the main steady-state performance measures of the system. In order to achieve this goal, the authors use the ***MOSEL*** (MOdeling, Specification, and Evaluation Language) tool to formulate and solve the problem. An illustrative numerical example will be given to show the effect of the server's failure rate, such as on the mean number of tasks in the system, mean orbit size, mean response, and waiting time of the requests.

6.8.1 The Model (CoginfoCom System)

A single-server (communication network service) queueing system with two finite-sources is being considered. A common network transmission finite buffer with capacity of B is placed in front of the service station. The sources generate requests (jobs) toward the service station. The server (communication network service) receives tasks from these sources. The sources are classified into two classes: (1) "Intelligent" class, containing human being or equipment that can respond to the changes of the communication environment. This group may have information on the network's state, and so they can retry the transmission in some special cases. (For instance, if the information was lost due to the full state of the network transmission buffer.) (2) "Normal" class, containing bare equipment (such as simple sensors and some input devices), which transmit information to the network without knowing the state of the communication service. For this class, if the network transmission buffer is full, then the sensor cannot send the sensed or measured data to the server. Hence, it is rejected and sent back to the source.

The life cycles of the requests arriving from the "Normal" and "Intelligent" classes are different. The total number of entities of "Intelligent" and "Normal" classes in the system at any time are denoted by N_1 and N_2, respectively.

For the entities in "Intelligent" and "Normal" classes, the request generation times (source times) are assumed to be exponentially distributed with parameters λ_1 and λ_2, respectively. Let $k_1(t)$ and $k_2(t)$ denote the number of working entities in the "Intelligent" and "Normal" classes at time t, respectively. Thus, the total generation rate will be $\lambda_1 k_1(t) + \lambda_2 k_2(t)$ with a maximum of $\lambda_1 N_1 + \lambda_2 N_2$.

Tasks (entities) arriving from the "Normal" class can enter the buffer only if the buffer is not full. In case of a full buffer (a queue with a maximum number of B tasks), the generated request will be rejected and returned to the source where a new request generation starts.

Let $q_1(t)$ and $q_2(t)$ represent the number of requests (waiting) in the queue for requests from the "Intelligent" and "Normal" classes at time t, respectively. Thus, total requests in queue (waiting line) will be $q(t) = q_1(t) + q_2(t)$ with maximum of B. Thus, the total number of tasks in the system at time t is $q(t) + 1$ with the maximum of $B + 1$.

The entities of the "Intelligent" class can react to the changes in the servicing environment. If the buffer is full, their requests will be retransmitted. An orbit is used for this feature. When a request originated from the "Intelligent" class finds the buffer is full, it will be sent to the orbit. After an exponentially distributed time period with parameter ν (retrial rate), the request retries joining the queue. If it finds the buffer is full, it will return to the orbit and starts generating a new retrial. Let $o(t)$ denote the number of waiting jobs in the orbit at time t. Then, the orbit capacity will be N_1.

It is assumed that the service deterioration condition can be described by three levels. The best condition is referred to as the Level 1. In this case, the service rate is the highest. A less efficient service is referred to as the Level 2. When the system is broken down, the service is referred to as the Level 3. In this case, the service stops. It is further assumed that the network communication service times are exponentially distributed with parameters μ_1 and μ_2 with $\mu_1 > \mu_2$. The operating condition of the service station may switch from a specific level to any higher level, or in case of a repair, to the Level 1. The operation times at Levels 1 and 2 are assumed to be identical and exponentially distributed with parameter δ. The transitions probabilities among different levels are as follows: In case of a failure, the service state changes from Level 1 to Level 2 or Level 3 with probability p and $1 - p$, respectively. It is clear that a state can transit from Level 2 to Level 3 and from Level 3 to Level 1. After completing a service, a request is selected from either the Intelligent or Normal source, if there is any task in the queue, with probability 0.5. That is, tasks in either source have an equal chance to be called for service. If the server breaks down in a busy state, the interrupted request will stay at the service station and its service resumes when the server is repaired. The repair times are assumed to be exponentially distributed with parameters β_i, $i = 2, 3$. When the server is broken, only service is interrupted; however, all the other operations are continued (new and repeated requests can be generated).

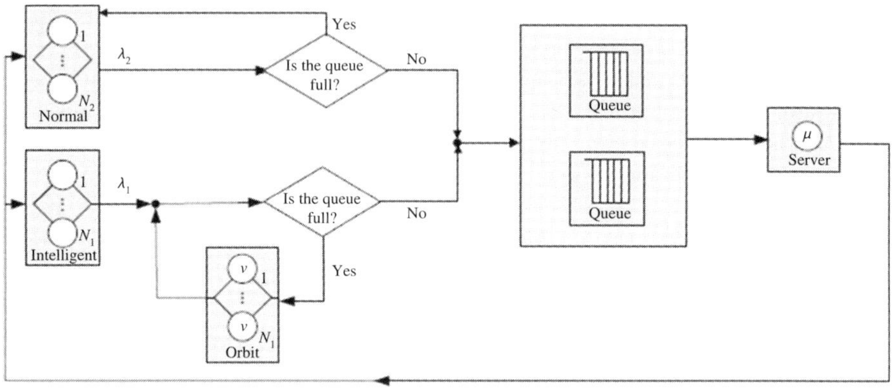

Figure 6.14 A retrial queue with components.

Thus, the service station can be in three states as follows:

(i) Fully operational, that is, the service is at Level 1. In this case, the service of a request is highly efficient.
(ii) Limited operation, that is, the service is at Level 2. The service of a request is not highly efficient.
(iii) Totally failed operational, that is, the service station is broken and it cannot start serving any arriving requests until it is repaired.

Figure 6.14 illustrates the functionality of this communication network.

Letting the operation level of the service station at time t by $y(t)$, it will be clear that $y(t) = 1$, if the service station is in fully operational mode, $y(t) = 2$, if the service station is in limited operation mode, and $y(t) = 3$, if the service station is broken at time t. Also, letting the state of the server as being busy or idle at time t be denoted by $c(t)$, we can see that $c(t) = 0$, if the server is idle, and $c(t) = 1$, if the server is busy with a request coming from the "Intelligent" class. However, $c(t) = 2$, when the server is busy with a request coming from the "Normal" class.

6.8.2 Analysis

We leave it as an exercise for the reader to show that

$$k_1(t) = \begin{cases} N_1 - q_1(t) - o(t) - 1, & c(t) = 1, \\ N_1 - q_1(t) - o(t), & c(t) = 0, 2 \end{cases} \tag{6.285}$$

and

$$k_2(t) = \begin{cases} N_2 - q_2(t) - o(t) - 1, & c(t) = 2, \\ N_2 - q_2(t) - o(t), & c(t) = 0, 1. \end{cases} \tag{6.286}$$

The objective is to make the model a Markovian one, the distributions of the interevent times (i.e., request generation times, operation times, service times, retrial times, and repair times) presented in the network are assumed to be exponentially distributed and independent of each other. The state of the communication network at a time t can be described with a five-dimensional continuous-time Markov chain (CTMC) with finite (and very large) state space, denoted by $X(t)$, that is,

$$X(t) = \langle y(t), c(t), q(t), q_2(t), o(t)\rangle. \tag{6.287}$$

Denote the steady-state distribution by

$$P\{y, c, q_1, q_2, 0\} = \lim_{t\to\infty} P\{y(t) = y; c(t) = c; q_1(t) = q_1; q_2(t) = q_2; o(t) = 0\}. \tag{6.288}$$

Since the state space is very large, calculation of the system measures, using solving the system of difference equations is not with much difficulty, if not formidable. Thus, instead, the software MOSEL is used (Begain et al., 2001; Bolch et al., 2015; Wuechner et al., 2011). As soon as the distribution in Equation (6.288) has been calculated, the most important steady-state system measures: ***utilization of the server***, U_S, ***availability of the server***, A_S, ***average number of request in the orbit***, $\overline{O} = E[o(t)]$, ***average number of request in the queue***, $\overline{Q} = E[q_1(t) + q_2(t)]$, ***average number of request in the network***, $\overline{M}$, ***average number of active Intelligent entities***, $\overline{\Lambda}_1$, ***average number of Intelligent entities in the network***, $\overline{\Lambda}_2$, ***average number of Normal entities in the network***, $\overline{N}_2$, ***average generation rate of Intelligent entities***, $\overline{\lambda}_1$, and ***mean response time for Intelligent entities***, $\overline{T}_1$. These measures are found using the following relations:

$$U_S = E[o(t)] = \sum_{y=1}^{2}\sum_{c=1}^{2}\sum_{q_1=0}^{B}\sum_{q_2=0}^{B-q_1}\sum_{o=0}^{N_1-q_1} P(y, c, q_1, q_2, o). \tag{6.289}$$

$$A_S = E[o(t)] = \sum_{y=1}^{2}\sum_{c=0}^{2}\sum_{q_1=0}^{B}\sum_{q_2=0}^{B-q_1}\sum_{o=0}^{N_1-q_1} P(y, c, q_1, q_2, o). \tag{6.290}$$

$$\overline{O} = E[o(t)] = \sum_{y=1}^{3}\sum_{c=0}^{2}\sum_{q_1=0}^{B}\sum_{q_2=0}^{B-q_1}\sum_{o=0}^{N_1-q_1} oP(y, c, q_1, q_2, o). \tag{6.291}$$

$$\overline{Q} = E[q_1(t) + q_2(t)] = \sum_{y=1}^{3}\sum_{c=0}^{2}\sum_{q_1=0}^{B}\sum_{q_2=0}^{B-q_1}\sum_{o=0}^{N_1-q_1} q_1P(y, c, q_1, q_2, o)$$
$$+ \sum_{y=1}^{3}\sum_{c=0}^{2}\sum_{q_1=0}^{B}\sum_{q_2=0}^{B-q_1}\sum_{o=0}^{N_1-q_1} q_2P(y, c, q_1, q_2, o). \tag{6.292}$$

$$\overline{M} = \overline{O} + \overline{Q} + \sum_{y=1}^{2}\sum_{c=1}^{2}\sum_{q_1=0}^{B}\sum_{q_2=0}^{B-q_1}\sum_{o=0}^{N_1-q_1} P(y, c, q_1, q_2, o). \tag{6.293}$$

$$\overline{\Lambda}_1 = N_1 - E[q_1(t)] - \sum_{y=1}^{2}\sum_{q_1=0}^{B}\sum_{q_2=0}^{B-q_1}\sum_{o=0}^{N_1-q_1} P(y,1,q_1,q_2,o)$$

$$= N_1 - \sum_{y=1}^{3}\sum_{c=0}^{2}\sum_{q_1=0}^{B}\sum_{q_2=0}^{B-q_1}\sum_{o=0}^{N_1-q_1} q_1 P(y,c,q_1,q_2,o)$$

$$- \sum_{y=1}^{2}\sum_{q_1=0}^{B}\sum_{q_2=0}^{B-q_1}\sum_{o=0}^{N_1-q_1} P(y,1,q_1,q_2,o). \tag{6.294}$$

$$\overline{N}_1 = N_1 - \overline{\Lambda}_1. \tag{6.295}$$

$$\overline{\Lambda}_2 = N_2 - E[q_2(t)] - \sum_{y=1}^{2}\sum_{q_1=0}^{B}\sum_{q_2=0}^{B-q_1}\sum_{o=0}^{N_1-q_1} P(y,2,q_1,q_2,o)$$

$$= N_2 - \sum_{y=1}^{3}\sum_{c=0}^{2}\sum_{q_1=0}^{B}\sum_{q_2=0}^{B-q_1}\sum_{o=0}^{N_1-q_1} q_2 P(y,c,q_1,q_2,o)$$

$$- \sum_{y=1}^{2}\sum_{q_1=0}^{B}\sum_{q_2=0}^{B-q_1}\sum_{o=0}^{N_1-q_1} P(y,2,q_1,q_2,o). \tag{6.296}$$

$$\overline{N}_2 = N_2 - \overline{\Lambda}_2. \tag{6.297}$$

$$\overline{\lambda}_1 = \lambda_1 \overline{\Lambda}_1. \tag{6.298}$$

$$\overline{T}_1 = \frac{N_1}{\overline{\lambda}_1}. \tag{6.299}$$

It is left as an exercise for the reader to verify relations(6.289)–(6.299).

6.8.3 Numerical Example

Using the data listed in Table 6.7, Figures 6.15–6.19 show the effect of the service station's breakdown on a cognitive information system. Verification of these graphs is left as an exercise for the reader.

Figure 6.15 shows the mean orbit size versus service's failure rate. The effect of increasing value of failure rate on the size of the orbit is shown. It can be observed, how the service's breakdown fills up the orbit.

Figure 6.16 shows the probability that the server is in limited state versus service's failure rate. The graph should display the limited operation mode of the service, that is, the service is in state 2. The probability of this state should be computed as a function of the failure rate.

Figure 6.17 shows the probability that the server fails versus service's failure rate. The graph deals with the broken state of the service. In this state, the service is interrupted. However, all other operations are still being performed. As for the previous graph, the probability of this state should be computed.

TABLE 6.7 Data Used for Numerical Example

Measure/Parameter	Notation	Value
Buffer size	B	3
Overall generation rate	λ	1.5
Intelligent generation rate	$\lambda_1 = \frac{2}{10}\lambda$	0.3
Normal generation rate	$\lambda_2 = \frac{8}{10}\lambda$	1.2
Number of Intelligent entities	N_1	3
Number of Normal entities	N_2	50
Retrial rate	ν	4
Service rate	μ_1	20
Service rate in limited state	μ_2	5
Server's failure rate	δ	$[0.01, \ldots, 0.1]$
Server's repair rate	β_2, β_3	0.1
Probability that service state changes from Level 1 to Level 2	p	0.5

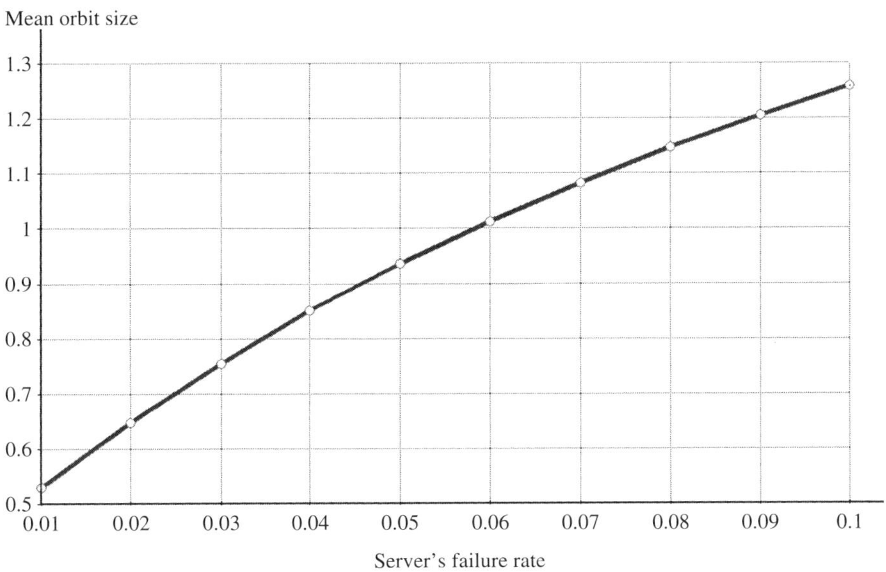

Figure 6.15 Mean orbit size versus service's failure rate.

Figure 6.18 shows the mean response time of Intelligent entities versus service's failure rate, where the effect of service's breakdown with mean response time of Intelligent entities is displayed.

Figure 6.19 shows the wasted time of Intelligent entities versus service's failure rate. The graph in this case shows the wasted time of the Intelligent entities in function of service's failure rate. The entities can waste time in the case, when the service

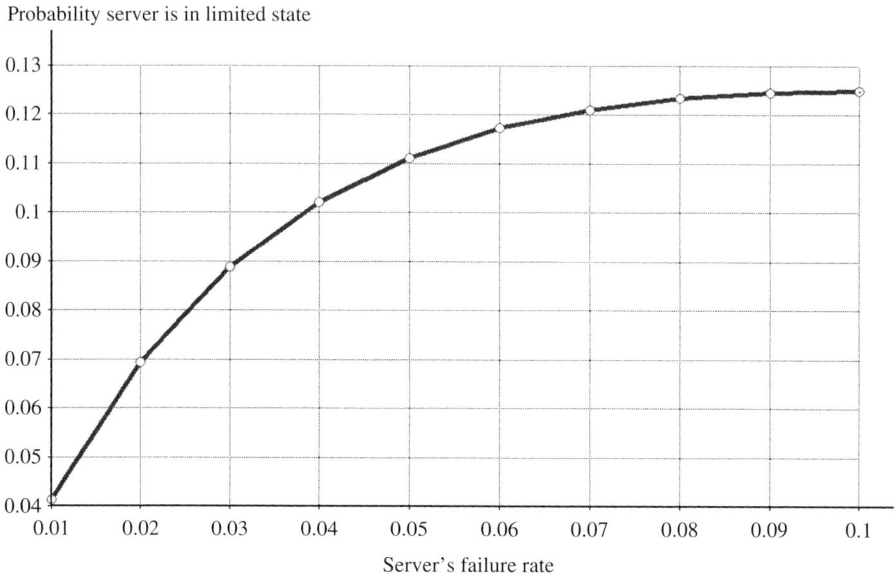

Figure 6.16 Probability that the server is in limited state versus service's failure rate.

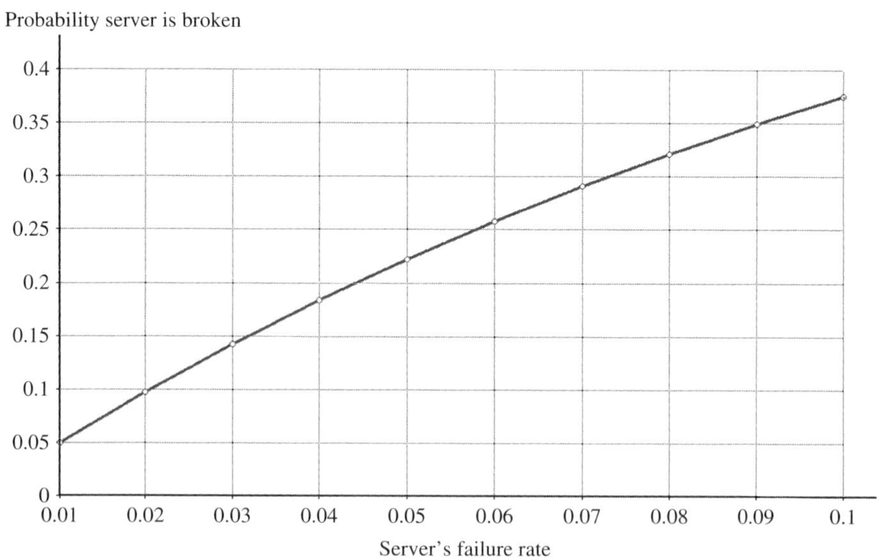

Figure 6.17 Probability that the server fails versus service's failure rate.

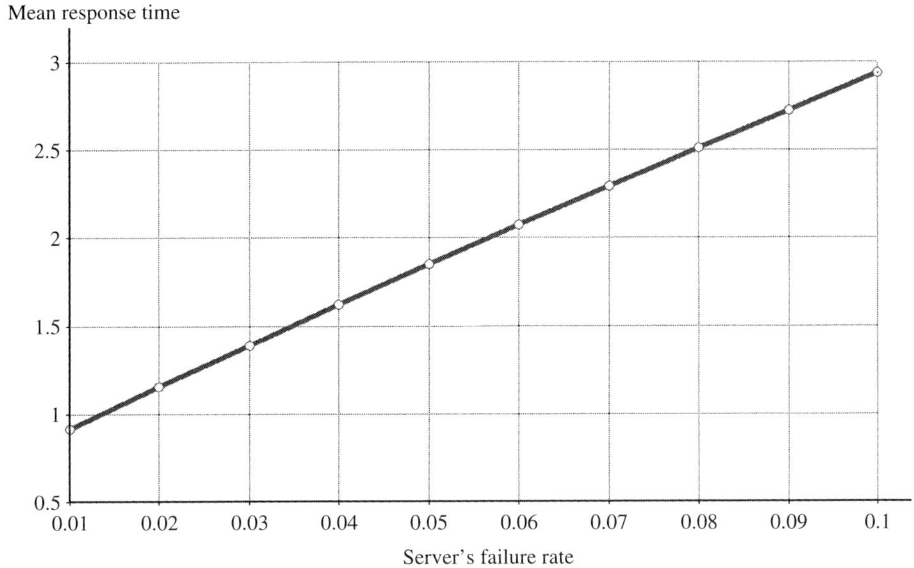

Figure 6.18 Mean response time of Intelligent entities versus service's failure rate.

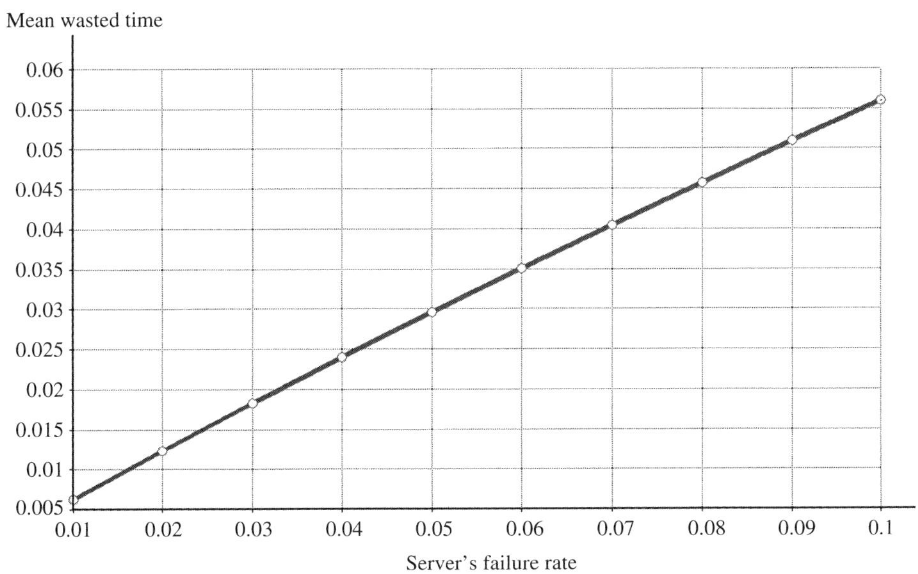

Figure 6.19 Wasted time of Intelligent entities versus service's failure rate.

becomes inoperable while the server is busy. The wasted time should be calculated as $\overline{T}_1 - \overline{W}_1 - (1/\mu_1)$.

Thus, a cognitive communication system was considered. A model was built with two classes of entities. The "Intelligent" tasks could be aware of any changes of the network environment, while the "Normal" ones could not. The service is subject to breakdowns. Multiple deterioration levels of service was introduced and investigated.

EXERCISES

6.1 Use recurrent relations (6.3.28)–(6.3.30) to find $N_i(s)$, $i = 1, 2, 3, 4, 5$, and Equation (6.3.25) to find expressions for expected incomes of the network's systems.

6.2 For Example 6.3.3, show that the busy period process is completely described by Equation (6.3.35), which results in Equation (6.3.36).

6.3 Applying generating functions defined in Eqs. (6.222)–(6.225) on Eqs. (6.209)–(6.220), obtain a set of differential equations of Lagrangian type.

6.4 Using Eqs. (6.216) and (6.217) obtain Eqs. (6.228) and (6.229).

6.5 The acronym NACK (Negative ACKnowledgment) let the sending system know that there was an error while processing the message. Consider a switching facility that transmits messages to a required destination. When a packet is not properly received, a NACK is sent by the destination. In this case, the packet in error is retransmitted as soon as the NACK is received. Assume that the time to send a message and the time to receive a NACK are both exponentially distributed with parameter μ. Also, assume that packets arrive at the switch according to a Poisson process with rate λ. Suppose that the probability that a message is received correctly is p, $0 < p \leq 1$. Show that the mean number, N, of packet and mean response time, T, are, respectively, as

$$N = \frac{\lambda}{p\mu - \lambda} \text{ and } T = \frac{1}{p\mu - \lambda}, \quad \frac{\lambda}{p\mu} < 1.$$

6.6 Refer to Exercise 6.5. Suppose now the switching facility is composed of K nodes in tandem, each model as $M/M/1$ queue with switching rate μ. Show that the mean number of packet in queue i and the mean response time are, respectively,

$$E(N_i) = \frac{\lambda}{p\mu - \lambda}, \ i = 1, 2, \ldots, K, \text{ and } E(T) = \frac{K}{p\mu - \lambda}, \ \frac{\lambda}{p\mu} < 1.$$

6.7 Consider a queueing network, Figure 6.20, consisting of $K = 4$ single-server station with FCFS service discipline. The interarrival times are assumed to be exponentially distributed random variables with rate of $\lambda = 4$ jobs per second. The service time at each node is exponentially distributed with means $1/\mu_1 = 0.04$ s, $1/\mu_2 = 0.03$ s, $1/\mu_3 = 0.06$ s, and $1/\mu_1 = 0.05$ s. The transition (routing) probabilities are $p_{12} = p_{13} = 0.5$, $p_{41} = p_{21} = 1$, $p_{31} = 0.6$, and $p_{30} = 0.4$.

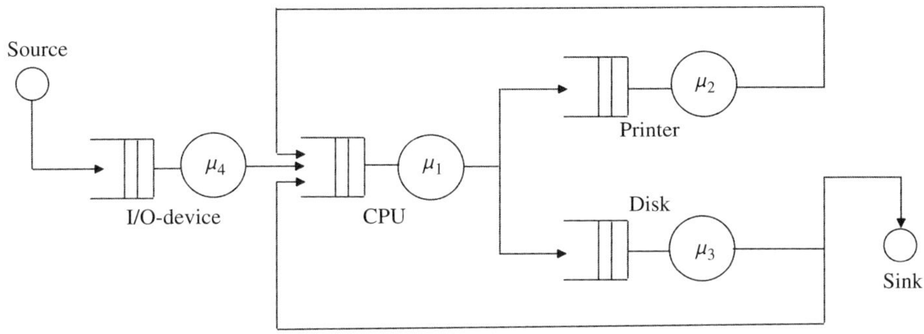

Figure 6.20 Queueing network of Exercise 6.7.

a. Compute the arrival rates for each node.
b. Compute the state probabilities for each node.
c. What is the steady-state probability of state $(k_1, k_2, k_3, k_4) = (3, 2, 4, 1)$?
d. Compute the mean number of jobs in each queue.
e. Compute the mean response time in each queue.
f. Compute the mean overall response time, using Little's formula.

6.8 Verify relations (6.285) and (6.286).

6.9 Verify relations (6.289)–(6.299).

6.10 Verify the graphs in Figures 6.15–6.19.

REFERENCES

Abate J, Whitt W. Transient behavior of the M/M/1 queue; Starting at the Origin, AT&T Bell Laboratories USA. Queueing Syst 1986;2:41–65.

Abate J, Whitt W. Approximations for the *M/M/*1 busy-period distribution. In: Boxma OJ, Syski R, editors. *Queueing Theory and Its Applications*. North-Holland, Amsterdam: Liber Amicorum for J.W. Cohen; 1988. p 149–191.

Abolnikov LM, Dshalalow JH. On a multilevel controlled bulk queueing system $M^x/G^{r,R}/1$. JAMSA 1992;5(3):237–260.

Abramowitz M, Stegun IA, editors. Modified Bessel Functions I and K. In: *§9.6 in Handbook of Mathematical Functions with Formulas, Graphs, and Mathematical Tables*. 9th printing ed. New York: Dover; 1972. p 374–377.

Addario-Berry L, Reed BA. Ballot theorems, old and new. In: *Horizons of Combinatorics*. Bolyai Society Mathematical Studies. Vol. 17. Berlin: Springer; 2008. p 9–35.

Aeppli A. A propos de l'interprétation géometrique du problèm du scruin. L'Enseignement Mathématique. Zurich: Gebr. Leemann & Co. A.-G. 1923;23:328–329.

Aeppli A. Zur Theorie verketteter Wahrscheinlichkeiten: Markoff-Ketten höherer Ordnung. Zurich: Gebr. Leemann & Co. A.-G. 1924.

Alfa AS, Srinivasa Rao TSS. Supplementary variable technique in stochastic models. Probab Eng Inform Sci 2000;14(2):203–218, Cambridge University Press New York, NY, USA.

Allen LJS. *An Introduction to Stochastic Processes with Biology Applications*. Upper Saddle River, NJ: Prentice Hall; 2003.

Ammar SI, Helan MM, Al Amri FT. The busy period of an *M/M/*1 queue with balking and reneging. Appl Math Modell 2013;37:9223–9229.

Ancker CJ Jr, Gafarian AV. Some queueing problems with balking and reneging I. Oper Res 1963a;11(1):88–100.

Delayed and Network Queues, First Edition. Aliakbar Montazer Haghighi and Dimitar P. Mishev.

Ancker CJ Jr, Gafarian AV. Some queueing problems with balking and reneging II. Oper Res 1963b;11(6):928–937.

Anderson WJ. *Continuous-Time Markov Chains, An Applications-Oriented Approach*. New York: Springer-Verlag; 1991.

André D. Solution directe du problèm résolu par M. Bertrand. C R Acad Sci Paris 1887; 105:436–437.

Andrews LC. *Special Functions of Mathematics for Engineers*. 2nd ed. New York: McGraw-Hill, Inc.; 1992.

Apianus P. *Ein newe und wolgegründete underweisung aller Kauffmanns Rechnung in dreyen Büchern*. Ingolstadt: Hans Holbein the Younger; 1527.

Arivudainambi D, Godhandaraman P. A batch arrival retrial queue with two phases of service, feedback and K optional vacations. Appl Math Sci 2012;6(22):1071–1087.

Ash R. *Basic Probability Theory*. John Wiley & Sons; 1970.

Auerswald P, Branscomb LM, La Porte TM, Michel-Kerjan E. The challenge of protecting critical infrastructure. Issues Sci Technol 2005;22:77–80.

Avi-ltzhak B, Naor P. Some queueing problems with the service station subject to breakdowns. Oper Res 1963;11:303–320.

Ayyapan G, Muthu Ganapathi Subramanian A, Sekar G. *M*/*M*/1 retrial queueing system with loss and feedback under non-pre-emptive priority service by matrix geometric method. Appl Math Sci 2010;4(48):2379–2389.

Baeka JW, Moona SK, Lee HW. A time-dependent busy period queue length formula for the M/Ek/1 queue. Stat Probaby Lett 2014;87:98–104.

Bailey NTJ. On queueing processes with bulk service. J Roy Stat Soc Ser B 1954;16(1): 80–97.

Baker KR. A note on operating policies for the queue *M*/*M*/*1 with* exponential startups. Inf Syst Oper Res 1973;11:71–72.

Balanov AG, Janson NB, Scholl E. *Control of noise-induced oscillations by delayed feedback*. Leicestershire, UK: Loughborough University; 2004.

Bandi C, Bertsimas D, Youssef N. Robust queueing theory. Oper Res Informs 2015; 63(3):676–700.

Barbier É. Généralization du problèm résolu par M. Bertrand. C R Acad Sci Paris 1887;105: 407.

Baruah M, Madan KC, Eldabi T. A batch arrival single server queue with server providing general service in two fluctuating modes and reneging during vacation and breakdowns. J Probab Stat 2014;2014 Article ID 319318:1–12, Hindawi Publishing Corporation, 10.1155/2014/319318.

Bean N, Latouche G. Approximations to quasi-birth-and-death processes with infinite blocks. Adv Appl Prob 2010;42:1102–1125.

Begain K, Bolch G, Herold H. *Practical performance modeling, application of the MOSEL language*. Boston: Kluwer Academic Publisher; 2001.

Bellman RE. *Dynamic Programming*. Princeton: Princeton University Press; 1957.

Ben-Tal A, El-Ghaoui L, Nemirovski A. *Robust Optimization*. Princeton, New Jersey: Princeton University Press; 2009.

Bertrand J. *Calcul des probabilités*. Paris: Gauthier-Villars; 1889. p 5–6.

Bertsimas D, Brown D, Caramanis C. Theory and application of robust optimization. SIAM Rev 2011;53:464–501.

Bhattacharya RN, Waymire EC. *Stochastic Processes with Applications*. New York: John Wiley & Sons; 1990.

Birge JR, Louveaux FV. *Introduction to Stochastic Programming*. Springer; 2011.

Böhm W. Lattice path counting and the theory of queues, Department of Statistics and Mathematics, Wirtschaftsuniversität Wien; 2008. Research Report Series Report 74.

Böhm W, Mohanty SG. On the transient solution of N-policy queues. McMaster University; 1990, Statistics Research Report, No. 11.

Böhm W, Mohanty SG. The transient solution of *M*/*M*/1 queues under (*M*, *N*)-policy, A combinatorial approach. J Stat Plann Inference 1993;34:23–33.

Bolch G, Greiner S, de Meer H, Trivedi K. *Queueing Networks and Markov Chains*. 2nd ed. New York: John Wiley & Sons; 2015.

Borel É. Sur l'emploi du théorème de Bernoulli pour faciliter le calcul d'une infinité de coefficients. Application au problème de l'attente à un guichet. C R Acad Sci Paris 1942;214:452–456.

Bouchentouf AA, Sakhi H. A note on an *M*/*M*/s queueing system with two reconnect and two redial orbits. AAM: Intern J 2015;10(1):1–12.

Bouchentouf AA, Kadi M, Rabhi A. Analysis of two heterogeneous server queueing model with balking, reneging and feedback. Math Sci Appl E-Notes 2014;2(2):10–21.

Bouchentouf AA, Rabhi A, Yahiaoui L. Stability condition of a retrial queueing system with abandoned and feedback customers. AAM: Intern J 2015;10(2):634–647.

Boxma OJ, Daduna H. The cyclic queue and the tandem queue. Queueing Syst 2014; 77(3):275–295.

Brémaud P. Streams of a M/M/1 feedback queue in statistical equilibrium. Z Wahrsch Verw Gebiete 1978;45(1):21–33.

Brockmeyer E, Halstrom HL, Jensen A. *The Life and Works of A.K. Erlang*. Transactions of the Danish Academy of Technical Science. Vol. 2. Akademiet for de Tekniske Videnskaber; 1948.

de Bruin AM, van Rossum AC, Visser MC, Koole GM. Modeling the emergency cardiac inpatient flow, an application of queuing theory. Health Care Manag Sci 2007;10(2):125–137.

Bunday BD, EL-Badri WK. The busy period for the M/M/1 machine interference model. Stoch Anal Appl 1985;3:1–13.

Burke PJ. The output of a queueing system. Oper Res 1956;4:699–704.

Cao J, Cheng K. Analysis of M/G/1 queueing system with repairable service station. Acta Math Appl Sinica 1982;5:113–127.

Cardano H. *Opus Novum de Proportionibus Numerorum*. Basel: Artis magnae, sive, de regulis algebraicis, De aliza regula liber; 1570.

Chakka R, Mitrani I. Approximate solutions for open networks with breakdowns and repairs. In: Kelly FP, Zachary S, Ziedins I, editors. *Stochastic Networks, Theory and Applications*. volume 4 of Royal Statistical Society Lecture Notes Series chapter 16. Oxford: Clarendon Press; 1996. p 267–280.

Chakravarthy S. Analysis of a finite *MAP*/*G*/1 queue with group services. Queueing Syst 1993; 13:385–407.

Chakravarthy SR, Alfa AS. A finite capacity queue with Markovian arrivals and two servers with group services. J Appl Math Stoch Anal 1994;7(2):161–178.

Champernowne DG. An elementary method of solution of the queueing problem with a single server and constant parameters. J Roy Stat Soc 1956;B18:125–128.

Chan WC, Lin YB. Waiting time distribution for the M/M/m queue. Communications, IEE Proceedings 2003;150(3):159–162.

Chen H, Yao DD. *Fundamentals of Queueing Networks: Performance, Asymptotics, and Optimization*. Springer; 2001.

Chen X, Li L, Zhang Y. A Markov model for headway/spacing distribution of road traffic. IEEE Trans Intell Transp Syst 2010;11(4):773–785.

Choi BD, Kim YC. The M/M/c retrial queue with geometric loss and feedback. Comput Math Appl 1998;36(6):41–52.

Choi BD, Hwang GU, Han DH. Supplementary variable method applied to the *MAP*/*G*/1 queueing system. J Austral Math Soc 1998;B40:86–96.

Choudhury G, Tadj L. An *M*/*G*/1 queue with two phases of service subject to the server breakdown and delayed repair. Appl Math Modell 2009;33:2699–2709.

Choudhury G, Ke J-C, Tadj L. The *N*-policy for an unreliable server with delaying repair and two phases of service. J Comput Appl Math 2009;231(1):349–364.

Cohen JW. *The Single-Server Queue*. 2nd ed. Amsterdam: North-Holland; 1982.

Cohen JW, Boxma OJ. *Boundary Value Problems in Queueing System Analysis*Comm. 28, No. 2. Elsevier Science Ltd; 1983. p 165–172.

Conolly BW. The busy period for the infinite capacity service system $M/G/\infty$. In: *In studii diprobabilitâ statistica e Ricerca opertiva in onore di G. Pompilj. Instituto di calcolo delle probabilitâ Universita di Roma, Oderisi Gubbio, Roma*; 1971. p 128–130.

Conolly BW. The generalized state dependent Erlangian queue: the busy period. J Appl Prob 1974;11:618–623.

Cox DR. The analysis of non-Markovian stochastic processes by the inclusion of supplementary variables. Proc Camb Phil Soc (Math Phys Sci) 1955;51:433–441.

Cox DR. The relation between theory and application in statistics. Test 1995;4(2):207–261.

Cox DR, Miller HD. *The Theory of Stochastic Processes*. London: Chapman and Hall; 1984.

Dai JG, Harrison JM. Reflected Brownian motion in an orthant, Numerical methods for steady-state analysis. Ann Appl Probab 1992;2:66–86.

Dantzig GB. Programming of independent activities –II: Mathematical model. Econometrica 1949;17:200–211.

Disney RL. Some multichannel queueing problems with order entry. J Ind Eng 1962;13:46–48.

Dixon WJ. A method for obtaining and analyzing sensitivity data. J Am Stat Assoc 1948; 43:109–126.

Dshalalow JH, editor. Special Jubilee issue in honor of Lajos Takács. J Appl Math Stoch Anal 1994;7:3.

Dshalalow JH. In: Jewgeni H, editor. *Frontiers in Queueing Models and Applications in Science and Engineering*. Dshalalow: CRC Press, Inc.; 1997.

Durham SD, Flournoy N. Convergence results for an adaptive ordinal urn design. J Theory Probab Its Appl 1993;37:14–17.

Durham SD, Flournoy N. Random walks for quantile estimation. In: *Statistical Decision Theory and Related Topics V (West Lafayette, IN, 1992)*. New York: Springer; 1994. p 467–476.

Durham SD, Flournoy N, Montazer-Haghighi AA. Up-and-down designs. Computing and Statistics Interface 1993;25:375–384.

Durham SD, Flournoy N, Rosenberger WF. A random walk rule for phase I clinical trials. Biometrics 1997;53:745–760.

Ebrahimi N. System reliability based on system wear. Stochastic Models 2006;22(1):21–36.

Edwards AWF. *Pascal's Arithmetical Triangle: The Story of a Mathematical Idea*. John Hopkins University Press; 2002.

Erlander S. The remaining busy period for a single server queue with Poisson input. Oper Res 1965;14:444–459.

Erlang AK. On the rational determination of the number of circuits. In: Brockmeyer E, Halstrom HL, Jensen A, editors. *The Life and Works of A.K. Erlang*. Copenhagen, Denmark: The Copenhagen Telephone Company; 1948.

Falin GI. A survey of retrial queues. Queueing Syst Theory Appl 1990;7(2):127–167.

Falin GI, Artalejo GR. A finite source retrial queue. Eur J Oper Res 1998;108(2):409–424.

Falin GI, Templeton JGC. *Retrial queues*. London: Chapman and Hall; 1997.

Fayolle G. Méthodes Analytiques pour les filess d'attente couplées [Theses], Université Paris VI; 1979.

Feller W. Zur Theorie der stochastischen Prozesse. Math Ann 1936;113:116–160.

Feller W. *An Introduction to Probability Theory and its Applications*. 3rd ed. Vol. I. Wiley & Sons; 1968.

Flajolet P, Guillemin F. The Formal Theory of Birth-and-Death Processes, Lattice Path Combinatorics, and Continued Fractions. Rapport de recherche n-3667, Avril, Institut National de Recherche en Informatique et en Automatique (INRIA); 1999.

Flournoy N. *Workshop on the Design and Analysis of Clinical Trials*. National University of Singapore; 2001. , 24–28 October 2011.

Foley RD, Disney RL. Queues with delayed feedback. Adv Appl Prob 1983;15:162–182.

Gakhov FD. *Boundary Value Problems*. Oxford, UK: Pergamon Press; 1966.

Galambos J, Gani J, editors. Studies in applied probability, (Papers in Honor of L. Takács). J Appl Prob 1994;31A:xi–xiii.

Gao S, Wang J. Performance and reliability analysis of an $M/G/1$-G retrial queue with orbital search and non-persistent customers. Eur J Oper Res 2014;236(2):561–572.

Gaver DP. The influence of service times in queueing processes'. Oper Res 1954;2(2):139–149.

Gavish B, Schweitzer PJ. The Markovian queue with bounded waiting time. Manag Sci 1977;2(12):1349–1357.

Gharbi N, Ioualalen M. GSPN analysis of retrial systems with server breakdowns and repairs. Appl Math Comput 2006;174(2):1151–1168.

Gibson D, Seneta E. Augmented truncations of infinite stochastic matrices. J Appl Prob 1987;24:600–608.

Goodman JB, Massey WA. The nonergodic Jackson network. J App Probab 1984;21:860–869.

Goodman E, Narayana TV. Lattice paths with diagonal steps. Canad. Math. Bull. 1969;12:847–855.

Gordon WJ, Newell GF. Cyclic queueing networks with exponential servers. Oper Res 1967;15(2):254–265.

Grassmann WK. Transient solutions in Markovian queues; an algorithm for finding them and determining their waiting-time distributions. Eur J Oper Res 1977a;1:396–402.

Grassmann WK. Transient solutions in Markovian queueing systems. Computers Oper Res 1977b;4:47–53.

Grassmann WK. Computational methods in probability theoryChapter 5. In: *Handbooks in OR and MS*. Vol. 2. North-Holland: Elsevier; 1990. p 200.

Greenberg I. Some duality results in the theory of queues. J Appl Prob 1969;6:99–121.

Grimmet GR, Stirzaker DR. *Probability and Random Processes*. 3rd ed. New York: Oxford University Press Inc.; 2004(Reprint with correction of 2001).

Gross D, Harris CM. *Fundamentals of Queueing Theory*. 2nd ed. New York, NY, USA: John Wiley & Sons, Inc.; 1985.

Gross D, Harris CM. *Fundamentals of Queueing Theory*. 3rd ed. John Wiley & Sons; 1998.

Günneç D, Salman FS. Assessing the reliability and the expected performance of a network under disaster risk. OR Spectrum 2011;33:499–523, Published online: 10 May, Springer-Verlag.

Gupta SM. Interrelationship between controlling arrival and service in queueing systems. Comput Opera Res 1995;22:1005–1014.

Gupta UC, Banerjee A. New results on bulk service queue with finite-buffer: $M/G^{(a,b)}/1/N$. OPSEARCH 2011;48(3):279–296, 10.1007/s12597-011-0051-4.

Hága P, Diriczi K, Vattay G, Csabai I. Understanding packet pair separation beyond the fluid model: The key role of traffic granularity. Proceedings of the INFOCOM 2006, 25th IEEE International Conference on Computer Communications; 2006.

Haghighi AM, Mishev DP. A parallel priority queueing system with finite buffers. J Parallel Distrib Comput 2006;66:379–392.

Haghighi AM, Mishev DP. Analysis of a two-station task-splitting feedback tandem queue with infinite buffers by functional equation. Int J Math Oper Res 2009;1(1/2): 246–277.

Haghighi AM, Mishev DP. *Difference and Differential Equations with Applications in Queueing Theory*. Hoboken, New Jersey, United State of America and simultaneously in Canada: John Wiley & Sons, Inc.; 2013.

Haghighi AM, Mishev DP. *Queueing Models in Industry and Business*. 2nd ed. New York: Nova Science Publishers, Inc.; 2014.

Haghighi AM, Mishev DP. Busy period of a single-server Poisson queueing system with splitting and batch delayed-feedback. Int J Math Oper Res 2016;8(2):239–256.

Haghighi AM, Mishev DP. Stepwise explicit solution for the joint distribution of queue length of a MAP single-server service queueing system with splitting and varying batch size delayed-feedback. Int J Math Oper Res 2016b;9(1):39–64.

Haghighi AM, Chukova S, Mishev DP. A two station tandem queueing system with delayed-service. Int J Oper Res 2008;3(4):363–383.

Haghighi AM, Chukova S, Mishev DP. Single-server Poisson queueing system with delayed-feedback: Part 1. Int J Math Oper Res (IJMOR) 2011;3(1):1–21.

Haight FA. Queueing with balking. Biometrika 1957;44:360–369.

Halfin S, Whitt W. Heavy-traffic limits for queues with many exponential servers. Oper Res 1981;29:567–587.

Hannibalsson I, Disney RL. An $M/M/1$ queues with delayed feedback. Naval Res Logist Quart 1977;24:281–291.

Haridass M, Arumuganathan R. Analysis of a batch arrival general bulk service queueing system with variant threshold policy for secondary jobs. Int J Math Oper Res 2011;3(1):56–77.

Harris CM. Queues with state-dependent stochastic service rates [PhD dissertation]. Polytechnic Institute of Brooklyn; 1966.

Haverkort BR, Marie R, Rubino G, Trivedi K. *Performability Modeling, Technique and Tools*. New York: Wiley; 2001.

Hoel PG, Port SC, Stone CJ. *Introduction to Stochastic Processes*. Boston: Houghton Mifflin Company; 1972.

Hogg RV, Tanis EA. *Probability and Statistical Inference*. 4th ed. NY: Macmillan Publishing Company; 1993.

Howard RA. *Dynamic Programming and Markov Processes*. Cambridge: MIT Press; 1960.

Howard RA. Comments on the origin and application of Markov decision processes. Oper Res Informs 2002;50(1):100–102.

Howard RA. Comments on the origin and application of Markov decision processes. Oper Res 2002;50(1):100–102.

Hunt GC. Sequential arrays of waiting lines. Oper Res 1957;4:674–683.

Hunter JJ. *Mathematical Techniques of Applied Probability*. Vol. 2. New York: Academic Press; 1983.

IMSL, Problem-Solving Software System. *User's Manual, Math/Library, FORTRAN Subroutines for Mathematical Applications*. Vol. 1 Chapter 2, version 1.0. IMSL; 1987. p 345.

Jackson JR. Networks of waiting lines. Oper Res 1957;5:518–521.

Jackson JR. Job-shop like queueing systems. Manage Sci 1963;10(1):131–142.

Jackson JR. How networks of queues came about. Oper Res Informs 2002;50(1):112–113.

Jackson RRP, Henderson JC. The time-dependent solution to the many-server Poisson queue. Oper Res. 1966;14(4):720–722.

Jain N, Schawrtz M, Bashkow T. Transport protocol processing at GBPS rates. In: *ACM SIGCOMM Computer Communications*. Philadelphia, PN: ACM; 1990. p 188–199.

Jain JL, Mohanty SG, Böhm W. *A course on queueing models*. Chapman & hall/CRC, Taylor & Francis Group; 2007.

Jain M, Upadhyaya Shweta. Optimal repairable $M^X/G/1$ queue with Bernoulli feedback and setup. Int J Math Oper Res 2012;4(6):679–702.

Jennings OB, de Véricourt F. Fluid and diffusion approximations of large-scale membership services, Received to Operations Research: 2007.

Jensen GL, Paulson AS, Sullo P. Explicit stationary state solutions for a particular $M^{(x)}/M/1$ queueing system. Naval Res Logist Online 2006;24:651–659, 10.1002/nav.30024012.

Kaczynski WH, Leemis LM, Drew JH. Transient queueing analysis. INFORMS J Comput 2012;24(1):10–28.

Karlin S, McGregor J. Many server queueing processes with Poisson input and exponential service times. Pac J Math 1958;8:87–118.

Karlin S, Taylor HM. *A First Course in Stochastic Processes*. 2nd ed. New York: Academic Press; 1975.

Ke JC. Operating characteristic analysis on the $M^{[X]}/G/1$ system with a variant vacation policy and balking. Appl Math Model 2007;31(7):1321–1337.

Ke JC, Chang FM. Modified vacation policy for M/G/1 retrial queue with balking and feedback. Comput Ind Eng 2009;57(1):433–443.

Ke JC, Wang KH. A recursive method for N-policy $G/M/1$ queueing system with finite capacity. Eu J Oper Res 2002;142:577–594.

Keilson J. The role of Green's functions in congestion theory. In: *Proceedings of Symposium on Congestion Theory*. University of North Carolina Press; 1965. p 43–71.

Kendall DG. Some problems in the theory of queues. J Roy Stat Soc Ser B 1951;13:151–185.

Kendall DG. Some problems in the theory of queues. J Roy Stat Soc Ser 1953;B13:151–185.

Khalaf RF, Madan KC, Lukas CA. An M[X]/G/1 queue with Bernoulli schedule, general vacation times, random breakdowns, general delay times and general repair times. Appl Math Sci 2011;5(1):35–51.

Khintchine AY. Mathematical theory of a stationary queue. Mat Sb 1932;39(4):73–84, Retrieved 2011 July 14.

Kijima M. *Markov Processes For Stochastic Modeling*. London: Chapman and Hall; 1997.

Kim S-H, Whitt W. Statistical analysis with Little's law. Oper Res 2013;61(4):1030–1045.

Kingman JKC. The single server queue in heavy traffic. Math Proc Cambridge Phil Soc 1961;57(04):902–904, published online: 24 October 2008.

Kingman JKC. The first erlang century – and the next. Queueing Syst 2009;63(1–4):3–12.

Kleinrock L. Information flow in large communication nets [PhD thesis proposal]. Cambridge, MA: Massachusetts Institute of Technology; 1961a.

Kleinrock L. Information flow in large communication nets. RLE Quart. Progress Report. Cambridge, MA: Massachusetts Institute of Technology; 1961b.

Kleinrock L. *Communication Nets: Stochastic Message Flow and Delay*. New York: McGraw-Hill; 1964. (reprinted by Dover Publications, 1972).

Kleinrock L. *Queueing Systems, Volume 1: Theory*. NY: John Wiley & Sons; 1975.

Kleinrock L, Gail R. *Queueing Systems, Problems and Solutions*. John Wiley & Sons; 1996.

Kobayashi H. Stochastic modeling, queueing models. In: Louchard G, Latouche G, editors. *Probability Theory and Computer Science*. New York: Academic Press; 1983.

Kolmogorov A. Uber die analytischen methoden in der Wahrscheinlichkeitsrechnung. Math Ann 1931;104:415–458.

Koole GM, Mandelbaum A. Queueing models of call centers: an introduction. Ann Oper 2002;113:41–59.

Krinik A, Mortensen C, Rubino G. Connections between birth-death processes. In: *Stochastic Processes and Functional Analysis*. Lecture Notes in Pure and Applied Mathematics. Vol. 238. New York: Dekker; 2004. p 219–240.

Krinik A, Rubino G, Marcus D, Swift RJ, Kasfy H, Lam H. Dual processes to solve single server systems. J Stat Plann Inference 2005;135(1):121–147.

Krishna Kumar B. The busy period of an *M/M/*1 queue with balking. J Appl Stat Sci 1996;3:209–218.

Krishna Kumar B, Arivudainambi D. The M/G/1 retrial queue with Bernoulli schedules and general retrial times. Comput Math Appl 2002;43(1–2):15–30.

Krishnakumar B, Pavai Madheswari S, Anantha Lakshmi SR. An *M/G/*1 Bernoulli feedback retrial queueing system with negative customers. Oper Res Int J 2013;13:187–210.

Kuehn PJ. On the calculation of waiting times in switching and computer system. 15th Report on Studies in Congestion Theory, University of Stuttgart; 1972.

Kuehn PJ. On a combined delay and loss system with different queue disciplines. In: *Transaction of the Sixth Prague Conf. on Information Theory, Statistical Decision Function and Random Processes*. Prague: Academia Publishing House of the Czech Academy of Sciences; 1973. p 501–528.

Kuehn PJ. Analysis of busy period and response time distributions in queuing networks. Informatik-Fachberichte 1983;61:135–149, Springer Berlin Heidelberg.

Kuki A Bérczes T, Almási B, Sztrik J. A queueing model to study the effect of network service breakdown in a CogInfoCom system, 5th IEEE Conference on Cognitive Infocommunications (CogInfoCom); 2014, pp. 237-241, INSPEC Accession Number: 14881516, IEEE.

Kumar R, Sharma SK. *M/M/*1*/N* queuing system with retention of reneged customers. Pak J Stat Oper Res 2012;8(4):859–866.

Kuo C-C, Wang K-H, Pearn WL. The interrelationship between *N*-policy *M*/*G*/1/*K* and *F*-policy *G*/*M*/1/*K* queues with startup time. Qual Tech Quant Manag 2011;8(3):237–251.

Laxmi PV, Yesuf OM. Renewal input infinite buffer batch service queue with single exponential working vacation and accessibility to batches. Int J Math Oper Res 2011;3(2): 219–243.

Lederman W, Reuter GEH. Spectral theory for the differential equations of simple birth-and-death processes. Phil Trans Roy Soc London 1954;A246:321–369.

Leguesdron P, Pellaumail J, Rubino G, Sericola B. Transient solution of the *M*/*M*/1 queue. Adv Appl Probab 1993;25:702–713.

Leon-Garsia A. *Probability and Random Processes for Electrical Engineering*. 3rd ed. Pearson Prentice Hall; 2008.

Lerch M. Sur un point de la théorie des fonctions génératrices d'Abel. Acta Mathematica 1903;27:339.

Li W, Shi D, Chao X. Reliability analysis of *M*/*G*/1 queueing system with server breakdowns and vacations. J Appl Probab 1997;34:546–555.

Liau P-Y. The queuing model with balking index and reneging rate. Int J Serv Oper Manag 2011;10(1):1–12.

Libman L, Orda A. Optimal retrial and timeout s trategies for accessing network resources. IEEE/ACM Trans Netw 2002;10(4):551–564.

Lindley DV. The theory of queues with single-server. Proc Cambridge Phil Soc 1952;48:227–289.

Little JDC. A proof of a queueing formula: $L = \lambda W$. Oper Res 1961;9:383–387.

Madan KC. An *M/G/*1 queue with second optional service. Queueing Syst 2000;34:37–46.

Madan KC. An *M*/*G*/1 queue with time homogeneous breakdown and deterministic repair times. Soochow J Math 2003;29:103–110.

Mandelbaum A, Massey W, Reiman M, Stolyar A, Rider B. Queue lengths and waiting times for multiserver queues with abandonment and retrials. Telecommun Syst 2002;21(2–4):149–171.

Maragatha SS, Srinivasan S. Analysis of M/M/I queueing model for ATM facility. Global J Theor Appl Math Sci 2012;2(1):41–46.

Matalytski MA. On some results on the analysis and optimization of Markov networks with incomes, and their applications (Russian). Avtomat i Telemekh 2009;10:97–113; translation in Autom Remote Control; 2009; 70(10), 1683–1697.

Matalytski MA. Analysis and forecasting of expected incomes in Markov networks with bounded waiting time for the claims. Autom Remote Control 2015;76(6):1005–1017. Pleiades Publishing, Ltd; Original Russian Text, Matalytski MA, published in Avtomatika i Telemekhanika, 2015: 6, 75–90.

Matalytski MA, Pankov AV. Application of operation calculus for the investigation of banking networks models. Proc. 17th Int. Conf. "Modern Mathematical Methods of Analysis and Optimization of Telecommunication Networks;" Queues: Flows, Systems, Networks, Minsk: BSU; 2003a, vol. 17, pp. 172–177.

Matalytski MA, Pankov AV. Incomes probabilistic model of the banking network. Sci Res Inst Math Comput Sci Czestochowa Univ Technol 2003b;1(2):99–104.

Matalytski MA, Pankov AV. Analysis of the stochastic model of the changing of incomes in the open banking network. Comput Sci 2003c;3(5):19–29.

Matalytski MA, Pankov AV. *Probabilistic Income Analysis in Banking Networks*. Vestnik, BGUno. 2; 2004a. p 86–91.

Matalytski MA, Pankov AV. Numerical analysis of probabilistic model of incomes changing in banking networks. Comput Sci 2004b;4(6):7–11.

Matalytski MA, Pankov AV. Research of Markov queueing network with central system and incomes. Comput Sci 2004c;4(7):23–32.

Matalytski M, Pankov A. Finding of expected incomes in closed stochastic network with central system. Informatyka Teoretyczna i Stosowana 2007;7(11):21–31.

Matalytski M, Pankov A, Statkiewicz S. Investigation of HM-network with limited waiting time in queueing systems. Comput Sci 2008;8(13):29–40.

Matisziw TC, Murray AT. Modeling *s-t* path availability to support disaster vulnerability assessment of network infrastructure. Comput Oper Res 2009;36:16–26; Simul Comput 20(1): 129–148.

Medhi J. *Stochastic Models in Queueing Theory*. 2nd ed. Amsterdam: Academic Press; 2003.

Medhi J. *Stochastic Processes*. 3rd Revised ed. New Age Science Publisher; 2009.

Mehmet Ali MK, Gu H. Performance analysis of a wireless sensor Network. In: *Wireless Communications and Networking Conference, 2006, WCNC 2006*. Vol. 2. IEEE; 2006. p 1166–1171.

Melamed B. Characterizations of Poisson traffic streams in Jackson queueing networks. Adv Appl Prob 1979;11:422–438.

Mickens RE. *Difference Equations: Theory and Applications*. 2nd ed. Chapman and Hall/CRC; 1990.

Mode CJ. *Theory and Applications of Multitype Branching Processes (Modern Analytic and Computational Methods in Science and Mathematics)*. New York: American Elsevier; 1971.

Mohanty SG. *Lattice Path Counting and Applications*. New York: Academic Press; 1979.

Mohanty SG, Panny W. A discrete time analog of the *M*/*M*/1 queue and the transient solution: a geometric approach. Sankyiā Ser 1990;A52:364–370.

Mohanty SG, Montazer-Haghighi A, Trrueblood R. On the transient behavior of a finite birth-death process with an application. Comput Ops Res 1993;20(3):239–248.

Montazer Haghighi A. A many-server queueing system with feedback. Bull Iranian Math Soc 1981;9(I):65–74 (serial No. 16).

Montazer-Haghighi A. *Many-Server Queueing System with Feedback, doctorate dissertation*. Case Western Reserve University; 1976.

Montazer-Haghighi A, Medhi J, Mohanty SG. On a multi-server Markovian queueing system with balking and reneging. Comput Oper Res 1986;13(4):421–425.

Muh DCR. A bulk queueing system under N-policy with bilevel service delay discipline and start-up time. J Appl Math Stoch Anal 1993;4(Winter):359–384.

Murphy JA, O'Donohoe MR. Properties of continued fractions with applications in Markov processes. J Inst Math Appl. 1975;16:57–71.

Murugan S, Pazhani Bala, Santhi K. An *M*/*G*/1 queue with server breakdown and multiple working vacation. Appl Appl Math 2015;10(2):678–693.

Muskhelishvili, N.I. (1992). Singular integral equations: boundary problems of function theory and their application to mathematical physics/Translation from the Russian edited by J. R. M. Radok, Groningen, English translation of originally published by P. Noordhoff in 1953 published by Dover first in 2008 and reprint in 1992.

Mylosz J. Local stabilization of non-Ergodic Jackson networks with unreliable nodes [PhD Dissertation]. Germany: Fakultät für Mathematik, Informatik und Naturwissenschaften, der Universität Hamburg; 2013.

Mylosz J, Daduna H. On the behavior of stable subnetworks in nonergodic networks with unreliable nodes. Comput Netw 2009;53(8):1249–1263.

Nakamura G. A feedback queueing model for an interactive computer system. AFIPS Proceedings of the Fall Joint Conference; 1971.

Nance RE, Bhat VN, Claybrook BG. Busy period analysis of a time-sharing system: transform inversion. J Assoc Comput Mach 1972;19:453–462.

Neuts MF. The distribution of the maximum length of a Poisson queue during a busy period. Oper Res 1964;18:388–395.

Neuts MF. A general class of bulk queues with Poisson input. Ann Math Stat 1967;38(3):759–770.

Neuts MMF. *Structured Stochastic Matrices of M/G/1 Type and Their Application*. New York: Marcel Dekker, Inc.; 1989.

Neuts MF, Ramalhoto MF. A service model in which the server is required to search for customers. J Appl Probab 1984;21:157–166.

Niu S-C, Cooper RB. Duality and other results for *M*/*G*/1 and *GI*/*M*/1 queues, via a New Ballot theorem. Math Oper Res 1989;14:281–293.

Plam C. Intensity fluctuations in telephone traffic. Ericsson Technol 1943;1:1–18.

Parathasarthy PR, Sharafali M. Transient solution to the many sever Poisson queue. J Appl Prob 1989;26:584–594.

Parlett BN. *The* Symmetric Eigenvalue Problem. New York: Prentice-Hall; 1980.

Pascal B. *Traié du Ttiangle Arithmétique, avec quelques autres petits traitez sur la mesme matiere*. Paris: G. Desprez; 1908a. p 433–593.

Pascal B. In: Brunschwicg L, Boutroux P, Gazier F, editors. *Oeuvres, Tomes. I-XIV*. Paris: Hachette; 1908b.

Peeta S, Sibel Salman F, Gunnec D, Viswanath K. Pre-disaster investment decisions for strengthening a highway network. Comput Oper Res 2010;37(10):1708–1719.

Plam C. The distribution of repairmen in serving automatic machines. Industritidningen Nordern 1947;75:75–80 (in Swedish).

Poincaré H. *Calculus des Probabilities*. Paris: Gauthier-Villars; 1896. (1912, second edition, pp. 21-26).

Pollaczek F. Über eine Aufgabe der Wahrscheinlichkeitstheorie. Math Z 1930;32:64–100.

Prabhu NU. *Queues and Inventories*. New York: Wiely; 1965.

Prabhu N. *Stochastic Processes: Basic Theory and its Applications*. World Scientific Publishing Co Ltd; 2007.

Prékopa A. *Stochastic Programming*. Kluwer Academic Publishers; 1995.

Probhn NU. *Queues and Inventories, A Study of Their Basic Stochastic Process*. New York: Wiley; 1965.

Rajadurai P, Saravanarajan MC, Chandrasekaran VM. Analysis of an $M^{[X]}/(G_1,G_2)/1$ retrial queueing system with balking, optional re-service under modified vacation policy and service interruption. Ain Shams Eng J 2014;5(3):935–950.

Rajadurai R, Saravanarajan MC, Chandrasekaran VM. Analysis of repairable $M^{[X]}/(G_1,G_2)/1$-feedback retrial *G*-queue with balking and starting failures under at most *J* vacations. Appl Appl Math 2015;10(2):694–717.

Rajadurai R, Chandrasekaran VM, Saravanarajan MC. Analysis of an $M^{[X]}/G/1$ unreliable retrial G-queue with orbital search and feedback under Bernoulli vacation schedule. Oper Res Soc India 2016;53(1):197–223, Springer.

Ramaswami V. A stable recursion for the steady state vector in Markov chains of *M/G/*1 type. Commun Statist Stoch Models 1988a;4(1):183–188.

Ramaswami V. Nonlinear matrix equations in applied probability – solution techniques and open problems. SIAM Rev 1988b;30(2):256–263.

Reisman A, Trivedi K. Numerical transient analysis of Markov models. Comput Oper Res 1988;15(1):19–36.

Rice SO. Single server systems–II. Busy period. Bell Syst Technol J 1962;41:279–310.

Saaty TL. Time dependent solution of the many server Poisson queue. Oper Res 1960;8:755–771.

Sauer C, Daduna H. Availability formulas and performance measures for separable degradable networks. Economic Quality Control 2003;18:165–194.

Schinazi RB. *Classical and Spatial Stochastic Processes: With Applications to Biology*. Springer; 1999.

Schopp M, Viallon J. Service availability in distributed network architectures: evaluation and planning issues. In: Charzinski J, Lehnert R, Tran-Gia P, editors. *Providing Quality of Service in Heterogeneous Environments*. Volume 5a of Teletrac Science and Engineering. Amsterdam: Elsevier; 2003. p 251–260; Proceedings of the 18th International Teletrac Congress.

Shanbhag DN. On infinite server queues with batch arrivals. J Appl Prob 1966;3:274–279.

Sharma O. *Markovian Queues*. NY: Ellis Harwood; 1990.

Sharma OP, Dass J. Multi-server Markovian queue with finite waiting space. Sankhya Ser 1988;B50:328–331.

Sheng Y and Sheng Z. Items selection with biased-coin up-and-down designs. In: Weiss DJ, editor. Proceedings of the 2009 GMAC Conference on Computerized Adaptive Testing; 2009.

Shi DH. A new method for calculation of the mean failure numbers of a repairable system during (0, t]. Aeta Math Appl Sinica 1985;8:101–110.

Singh CJ, Jain M, Kumar B. Queueing model with state-dependent bulk arrival and second optional service. Int J Math Oper Res 2011;3(3):322–340.

Singh CJ, Jain M, Kumar B. $M^X/G/1$ queuing model with state dependent arrival and Second Optional Vacation. Int J Math Oper Res 2012;4(1):78–96.

Smirnov VI. *A Course of Higher Mathematics*. Vol. 3 Part 2. Oxford: Pergamon Press; 1964.

Sobolov B, Levy A, Kuramoto L. Access to surgery and medical consequences of delays. In: Hall RW, editor. *Patient Flow: Reducing Delay in Healthcare Deliverychapter 3. Springer*; 2006.

Soyster AL. Convex programming with set-inclusive constraints and applications to inexact linear programming. Oper Res 1973;25(5):1154–1157.

Stadje W. The busy period of some queueing systems. Stoch Process Appl 1995;55:159–167.

Statkevich SE, Matalytski MA. Optimization of closed networks with bounded waiting time for queues in transition mode. Vestn GrGU 2012;2(1):134–142.

Syski R. *Introduction to Congestion Theory in Telephone Systems*. Oliver and Boyd; 1960.

Syski R. Markovian queues. In: Smith WL, Wilkinson WE, editors. *Symposium on Congestion Theory*. Chapel Hill: The University of North Carina; 1964. p 170–227.

Täcklind S. Elementare Behandlung vom Erneuerungsproblem. Skand Aktuarietidsk 1944;27:1–15.

Takács L. On a probability problem in the theory of counters. Ann Math Statist 1958;29:1257–1263.

Takács L. The probability law of the busy period for two types of queuing processes. Oper Res 1961;9(3):402–407.

Takács L. *Introduction to the Theory of Queues*. Oxford University Press; 1962.

Takács L. A single server queue with feedback. Bell System Tech J 1963;42:505–519.

Takács L. *Combinational Methods in the Theory of Stochastic Processes*. NY: John Wiley & Sons; 1967.

Takács L. A single-server queue with limited virtual waiting time. J Appl Prob 1974; 11:612–617.

Takács L. Ballot problems. In: Balakrishnan N, editor. *Advances in Combinatorial Methods and Applications to Probability and Statistics*. Boston: Birkhäuser; 1997. p 97–114.

Takács L. The distribution of the sojourn time of the Brownian excursion. Methodol Comput Appl Probab 1999;1(1):7–28.

Tarabia AMK. A new formula for the busy period of anon-empty multiserver queueing system. Appl Math Comput 2003;143:401–408.

Tijms HC. *A First Course in Stochastic Models*. West Sussex, England: John Wiley & Sons Ltd; 2003.

Vaulot E. Extension des formules d'Erlang aù cas ou les durées des conversations suivent une loie quelconque. Révue Générale de l'Electricité 1927;22:1164–1171.

de Véricourt F, Jennings OB. Dimensioning large-scale membership services. Oper Res Inform 2008;56(1):173–187.

Vijaya Laxmi P, Yesuf OM. Renewal input infinite buffer batch service queue with single exponential working vacation and accessibility to batches. Int J Math Oper Res 2011;3(2):219–243.

Walrand J. *Communication Networks: A First Course*. The Aksen Associates Series in Electrical and Computer Engineering. Homewood, IL and Boston, MA: Richard D. Irwin, Inc., and Aksen Associates, Inc.; 1991.

Wang J. An M/G/1 queue with optional service and server breakdown. Comput Math Appl 2004;47:1713–1723.

Wang KH, Ke JC. A recursive method to the optimal control of an *M/G/*1 queueing system with finite capacity and infinite capacity. Appl Math Model 2000;24:899–914.

Wang J, Li J. A single server retrial queue with general retrial times and two phase of service. J Syst Sci Complexity 2009;22(2):291–302.

Wang K, Li N and Jiang Z. Queuing system with impatient customers: a review. 2010 IEEE International Conference on Service Operations and Logistics and Informatics, 2010 July 15–17, Shandong; 2010. p 82-87.

Weiss G. Jackson-networks with unlimited supply of work. J Appl Probab 2005;42:879–882.

Whitt W. The queueing network analyzer. Bell Syst Tech J 1983;62(9):2779–2813.

Whitt W. Open and Closed Models for Networks of Queues. AT&T Bell Lab Tech J 1984;63(9):1911–1979.

Whitt W. Fluid models for multiserver queues with abandonments. Oper Res 2006; 54(1):37–54.

Wilkinson JH. *The* Algebraic Eigenvalue Problem. Oxford: Clarendon Press; 1965.

Wuechner P, Sztrik J, de Meer H. Modeling wireless sensor networks using finite-source retrial queues with unreliable orbit. Springer Lect Notes Comput Sci 2011;6821:275–285.

Xie M. On the solution of renewal-type integral equations. Commun Stat 1989;B18:281–293.

Yadin M, Naor P. Queueing systems with a removable service station. Oper Res Q 1963; 14:393–405.

Yang T, Templeton JGC. A survey on retrial queue. Queueing Syst 1987;2:201–233.

Yom-Tov G. Queues in hospitals: semi-open queueing networks in the quality- and efficiency-driven regime [doctoral dissertation] (Avishai Mandelbaum, Advisor). Faculty of Industrial Engineering and Management, Technion-Israel Institute of Technology; 2008.

INDEX

Delayed and Network Queues, First Edition. Aliakbar Montazer Haghighi and Dimitar P. Mishev.